MACHINE LEARNING TECHNIQUES FOR SPACE WEATHER

MACHINE LEARNING TECHNIQUES FOR SPACE WEATHER

Edited by

Enrico Camporeale

Simon Wing

Jay R. Johnson

Elsevier
Radarweg 29, PO Box 211, 1000 AE Amsterdam, Netherlands
The Boulevard, Langford Lane, Kidlington, Oxford OX5 1GB, United Kingdom
50 Hampshire Street, 5th Floor, Cambridge, MA 02139, United States

Library of Congress Cataloging-in-Publication Data
A catalog record for this book is available from the Library of Congress

British Library Cataloguing-in-Publication Data
A catalogue record for this book is available from the British Library

ISBN: 978-0-12-811788-0

For information on all Elsevier publications visit our website at https://www.elsevier.com/books-and-journals

Publisher: Candice Janco
Acquisition Editor: Marisa LaFleur
Editorial Project Manager: Katerina Zaliva
Production Project Manager: Nilesh Kumar Shah
Cover Designer: Matthew Limbert

Typeset by SPi Global, India

Contents

4. Regression

ALGO CARÈ, ENRICO CAMPOREALE

5. Supervised Classification: Quite a Brief Overview

MARCO LOOG

III APPLICATIONS

6. Untangling the Solar Wind Drivers of the Radiation Belt: An Information Theoretical Approach

SIMON WING, JAY R. JOHNSON, ENRICO CAMPOREALE, GEOFFREY D. REEVES

7. Emergence of Dynamical Complexity in the Earth's Magnetosphere

GIUSEPPE CONSOLINI

8. Applications of NARMAX in Space Weather

RICHARD BOYNTON, MICHAEL BALIKHIN, HUA-LIANG WEI, ZI-QIANG LANG

9. Probabilistic Forecasting of Geomagnetic Indices Using Gaussian Process Models

MANDAR CHANDORKAR, ENRICO CAMPOREALE

10. Prediction of MeV Electron Fluxes and Forecast Verification

ADAM KELLERMAN

11. Artificial Neural Networks for Determining Magnetospheric Conditions

JACOB BORTNIK, XIANGNING CHU, QIANLI MA, WEN LI, XIAOJIA ZHANG, RICHARD M. THORNE, VASSILIS ANGELOPOULOS, RICHARD E. DENTON, CRAIG A. KLETZING, GEORGE B. HOSPODARSKY, HARLAN E. SPENCE, GEOFFREY D. REEVES, SHRIKANTH G. KANEKAL, DANIEL N. BAKER

12. Reconstruction of Plasma Electron Density From Satellite Measurements Via Artificial Neural Networks

IRINA S. ZHELAVSKAYA, YURI Y. SHPRITS, MARIA SPASOJEVIC

13. Classification of Magnetospheric Particle Distributions Via Neural Networks

VITOR M. SOUZA, CLAUDIA MEDEIROS, DAIKI KOGA, LIVIA R. ALVES, LUIS E. A. VIEIRA, ALISSON DAL LAGO, LIGIA A. DA SILVA, PAULO R. JAUER, DANIEL N. BAKER

14. Machine Learning for Flare Forecasting

ANNA M. MASSONE, MICHELE PIANA, FLARECAST CONSORTIUM

15. Coronal Holes Detection Using Supervised Classification

VÉRONIQUE DELOUILLE, STEFAN J. HOFMEISTER, MARTIN A. REISS, BENJAMIN MAMPAEY, MANUELA TEMMER, ASTRID VERONIG

16. Solar Wind Classification Via *k*-Means Clustering Algorithm

VERENA HEIDRICH-MEISNER, ROBERT F. WIMMER-SCHWEINGRUBER

Contributors

Livia R. Alves National Institute for Space Research—INPE, São José dos Campos, SP, Brazil

Vassilis Angelopoulos Institute of Geophysics and Planetary Physics/Earth, Los Angeles, CA, United States

Daniel N. Baker University of Colorado Boulder, Boulder, CO, United States

Ramkumar Bala Rice University, Houston, TX, United States

Michael Balikhin Department of Automatic Control and Systems Engineering, University of Sheffield, Sheffield, UK

Jacob Bortnik University of California, Los Angeles, Los Angeles, CA, United States

Richard Boynton Department of Automatic Control and Systems Engineering, University of Sheffield, Sheffield, UK

Enrico Camporeale Centrum Wiskunde & Informatica, Amsterdam, The Netherlands

Algo Carè University of Brescia, Brescia, Italy

Mandar Chandorkar Centrum Wiskunde & Informatica, Amsterdam, The Netherlands

Xiangning Chu University of California, Los Angeles, Los Angeles, CA, United States

Giuseppe Consolini National Institute for Astrophysics, Institute for Space Astrophysics and Planetology, Rome, Italy

FLARECAST Consortium Academy of Athens, Trinity College Dublin, Università di Genova, Consiglio Nazionale delle Ricerche, Centre National de la Recherche Scientifique, Université Paris-Sud, Fachhochschule Nordwestschweiz, Met Office, Northumbria University

Véronique Delouille Royal Observatory of Belgium, Brussels, Belgium

Richard E. Denton Dartmouth College, Hanover, NH, United States

Mike Hapgood Lancaster University, Lancaster; RAL Space, Harwell, Didcot, United Kingdom

Verena Heidrich-Meisner Institute of Experimental and Applied Physics, Kiel, Germany

Stefan J. Hofmeister University of Graz, Graz, Austria

George B. Hospodarsky University of Iowa, Iowa City, IA, United States

Paulo R. Jauer National Institute for Space Research—INPE, São José dos Campos, SP, Brazil

Jay R. Johnson Andrews University, Berrien Springs, MI, United States

Shrikanth G. Kanekal NASA Goddard Space Flight Center, Greenbelt, MD, United States

Adam Kellerman UCLA, Los Angeles, CA, United States

Craig A. Kletzing University of Iowa, Iowa City, IA, United States

Daiki Koga National Institute for Space Research—INPE, São José dos Campos, SP, Brazil

Alisson Dal Lago National Institute for Space Research—INPE, São José dos Campos, SP, Brazil

Zi-Qiang Lang Department of Automatic Control and Systems Engineering, University of Sheffield, Sheffield, UK

Wen Li Boston University, Boston, MA, United States

Marco Loog Delft University of Technology, Delft, The Netherlands; University of Copenhagen, Copenhagen, Denmark

Qianli Ma Boston University, Boston, MA, United States; University of California, Los Angeles, Los Angeles, CA, United States

Benjamin Mampaey Royal Observatory of Belgium, Brussels, Belgium

Anna M. Massone CNR—SPIN, Genova, Italy

Claudia Medeiros National Institute for Space Research—INPE, São José dos Campos, SP, Brazil

Michele Piana CNR—SPIN; Università di Genova, Genova, Italy

Geoffrey D. Reeves Los Alamos National Laboratory; Space Science and Applications Group, Los Alamos, NM, United States

Patricia Reiff Rice University, Houston, TX, United States

Martin A. Reiss Space Research Institute, Graz, Austria

Yuri Y. Shprits Helmholtz Centre Potsdam, GFZ German Research Centre for Geosciences; University of Potsdam, Potsdam, Germany; University of California Los Angeles, Los Angeles, CA, United States

Ligia A. Da Silva National Institute for Space Research—INPE, São José dos Campos, SP, Brazil

Vitor M. Souza National Institute for Space Research—INPE, São José dos Campos, SP, Brazil

Harlan E. Spence University of New Hampshire, Durham, NH, United States

Maria Spasojevic Stanford University, Stanford, CA, United States

Manuela Temmer University of Graz, Graz, Austria

Richard M. Thorne University of California, Los Angeles, Los Angeles, CA, United States

Astrid Veronig University of Graz, Graz, Austria

Luis E. A. Vieira National Institute for Space Research—INPE, São José dos Campos, SP, Brazil

Hua-Liang Wei Department of Automatic Control and Systems Engineering, University of Sheffield, Sheffield, UK

Robert F. Wimmer-Schweingruber Institute of Experimental and Applied Physics, Kiel, Germany

Simon Wing Johns Hopkins University, Laurel, MD, United States

Xiaojia Zhang University of California, Los Angeles, Los Angeles, CA, United States

Irina S. Zhelavskaya Helmholtz Centre Potsdam, GFZ German Research Centre for Geosciences; University of Potsdam, Potsdam, Germany

Introduction

Enrico Camporeale[*], *Simon Wing*[†], *Jay R. Johnson*[‡]

[*]Centrum Wiskunde & Informatica, Amsterdam, The Netherlands [†]Johns Hopkins University, Laurel, MD, United States [‡]Andrews University, Berrien Springs, MI, United States

A common goal of scientific disciplines is to understand the relationships between observable quantities and to construct models that encode such relationships. Eventually any model, and its supporting hypothesis, needs to be tested against observations—the celebrated Popper's falsifiability criterion (Popper, 1959). Hence, experiments, measurements, and observations—in one word *data*—have always played a pivotal role in science, at least since the time of Galileo's experiment dropping objects from the leaning tower of Pisa.

Yet, it is only in the last decade that libraries' bookshelves have started to pile up with books about the data revolution, big data, data science, and various modifications of these terms. While there is certainly a tendency both in science and publishing to re-brand old ideas and to inflate buzzwords, one cannot deny that the unprecedented large amount of collected data of any sort—be it customer buying preferences, health and genetic records, high energy particle collisions, supercomputer simulation results, or of course, space weather data—makes the time we are living in unique in history. The discipline that benefits the most from the explosion of the *data revolution* is certainly machine learning. This field is traditionally seen as a subset of artificial intelligence, although its boundaries and definition are somehow blurry. For the purposes of this book, we broadly refer to machine learning as the set of methods and algorithms that can be used for the following problems: (1) make predictions in time or space of a continuous quantity (regression); (2) assign a datum to a class within a prespecified set (classification); (3) assign a datum to a class within a set that is determined by the algorithm itself (clustering); (4) reduce the dimensionality of a dataset, by exposing relationships among variables; and (5) establish linear and nonlinear relationships and causalities among variables.

Machine learning is in its golden age today for the simple reason that methods, algorithms, and tools, studied and designed during the last two decades (and sometimes forgotten), have started to produce unexpectedly good results in the last 5 years, exploiting the historically unique combination of big data availability and cheap computing power.

The single methodology that has been popularized the most by nonspecialist media as the archetype of machine learning's groundbreaking promise is probably the massive multilayer neural network, which is often referred to as *deep learning* (LeCun et al., 2015). For instance, deep learning is the technology behind the recent successes in image and speech recognition (with the former recently achieving better-than-human accuracy; He et al., 2015) and the first computer ever defeating a world champion in the game of Go (Silver, 2016).

The popular media often focus on the technological applications of machine learning, which has propelled recent advances in many areas, such as self-driving cars, online fraud detection, personalized advertisement and recommendation, real-time translation, and many others (Bennett and Lanning, 2007; Sommer and Paxson, 2010; Guizzo, 2011). However, we believe that it makes sense to ask whether machine learning could even change the process of scientific discovery.

Looking specifically at physics, the process of developing a model often relies on some form of the well-known Occam's razor: the simplest model that can explain the data is preferred. As a consequence, an important characteristic of most physics models is that every step of the process that led to their development is completely intelligible by the human mind. Such models are referred to as *white-box* models, suggesting that each component (including the set of assumptions) is transparent. Despite its marvelous achievements, the human brain has a very limited ability to process data, especially in high dimensions. This might be trivially related to the fact that the basic way of understanding data is graphical, and it is hard to visualize more than three variables in a single plot. Hence, the relationships between observable quantities that are encoded in white-box physics models usually do not explore high dimensional spaces. This human limitation does not mean that such models are "simple"; on the contrary they can be quite complicated, sometimes requiring formidable numerical methods to produce results that can be compared against observations. Essentially, all first-principles physics models are white-box models.

Contrary to the modus operandi of the white boxes (one could perhaps say of the human mind), machine learning algorithms focus essentially on two characteristics: being accurate and being robust against new data (i.e., being able to generalize). Indeed, the guiding principle concerns the trade-off between complexity and accuracy to avoid overfitting (see Chapter 4).

Hence, in contrast to white-box models, machine learning methods are often referred to as *black-box*, signifying that the mathematical structure and the relationships between variables are so complicated that it is often not useful to try to understand them, as long as they deliver the expected results. For example, and referring again to deep learning, one can certainly unroll a neural network to the point of deriving a single closed formula that relates inputs and outputs. However, such a formula would generally be incomprehensible and completely useless from a science-based perspective, although some features may be related to physical processes.

We need to mention a third, in-between paradigm, obviously called *gray-box modeling* that has recently emerged. Whereas white-box models are accurate but computationally slow (often much slower than real time when it comes to forecasting), and black-box models are fast but very sensitive to noise and outliers, the idea of gray box is to employ reduced physics models, and to calibrate the assumptions or the free parameters of the models via machine learning techniques. Gray box is often used in engineering modeling, and it is gradually making its way into more fundamental physics. In particular, we believe that the skepticism that surrounds machine learning in certain physics communities will be eventually overcome by embracing gray-box models, which allow the use of prior physical information in a more transparent way.

MACHINE LEARNING AND SPACE WEATHER

Space weather is the study of the effect of the Sun's variability on Earth, on the complex electromagnetic system surrounding it, on our technological assets, and eventually on human life. It will be more clearly introduced in Chapter 1, along with its societal and economic importance.

This book presents state-of-the-art applications of machine learning to the space weather problem. Artificial intelligence has been applied to space weather at least since the 1990s. In particular, several attempts have been made to use neural networks and linear filters for predicting geomagnetic indices and radiation belt electrons (Baker, 1990; Valdivia et al., 1996; Sutcliffe, 1997; Lundstedt, 1997, 2005; Boberg et al., 2000; Vassiliadis, 2000; Gleisner and Lundstedt, 2001; Li, 2001; Vandegriff, 2005; Wing et al., 2005). Neural networks have also been used to classify space boundaries and ionospheric high frequency radar returns (Newell et al., 1991; Wing et al., 2003), and total electron content (Tulunay et al., 2006; Habarulema et al., 2007). A feature that makes space weather very remarkable and perfectly posed for machine learning research is that the huge amount of data is usually collected with taxpayer money and is therefore publicly available. Moreover, the released datasets are often of very high quality and require only a small amount of preprocessing. Even data that have not been conceived for operational space weather forecasting offer an enormous amount of information to understand processes and develop models. Chapter 2 will dwell considerably on the nature and type of available data.

In parallel to the above-mentioned machine learning renaissance, a new wave of methods and results have been produced in the last few years, which is the rationale for collecting some of the most promising works in this volume.

The machine learning applications to space weather and space physics can generally be divided into the following categories:

- *Automatic event identification*: Space weather data is typically imbalanced, with many hours of observations covering uninteresting/quiet times, and only a small percentage of data of useful events. The identification of events is still often carried out manually, following time-consuming and nonreproducible criteria. As an example, techniques such as convolutional neural networks can help in automatically identifying interesting regions like solar active regions, coronal holes, coronal mass ejections, and magnetic reconnection events, as well as to select features.
- *Knowledge discovery*: Methods used to study causality and relationships within highly dimensional data, and to cluster similar events, with the aim of deepening our physical understanding. Information theory and unsupervised classification algorithms fall into this category.
- *Forecasting*: Machine learning techniques capable of dealing with large class imbalances and/or significant data gaps to forecast important space weather events from a combination of solar images, solar wind, and geospace in situ data.
- *Modeling*: This is somewhat different from forecasting and involves a higher level approach where the focus is on discovering the underlying physical and long-term behavior of the system. Historically, this approach tends to develop from reduced descriptions based on first principles, but the methods of machine learning can in theory also be used to discover the nonlinear map that describes the system evolution.

We will certainly see increasing applications of machine learning in space physics and space weather, falling in one of these categories. Yet, we also believe it is still an open question whether the amount and the kind of data at our disposal today is sufficient to train accurate models.

SCOPE AND STRUCTURE OF THE BOOK

The aim of this book is to bridge the existing gap between space physicists and machine learning practitioners. On one hand, standard machine learning techniques and off-the-shelf available software are not immediately useful to a large part of the space physics community that is not familiar with the jargon and the potential use of such methods; on the other hand, the data science community is eager to apply new techniques to challenging and unsolved problems with a clear technological impact, such as space weather.

The first part of the book is intended to provide some context to the latter community which might not be familiar with space weather forecasting. Chapter 1 summarizes the *Societal and Economic Importance of Space Weather*, while Chapter 2 describes the *Data Availability and Forecast Products for Space Weather.*

The second part offers a short, high-level overview of the three main topics that will be discussed throughout the book: *Information Theory* (Chapter 3), *Regression* (Chapter 4), and *Classification* (Chapter 5). Obviously, we refer the reader to more specific textbooks for in-depth explanation of these concepts.

The last part is devoted to applications covering a broad range of subdomains.

Chapter 6, *Untangling the Solar Wind Drivers of Radiation Belt: An Information Theoretical Approach*, is concerned with an application of information theory to study the classical problem of discerning different solar wind input parameters and quantifying their different roles in driving the radiation belt electrons.

Chapter 7, *Emergence of Dynamical Complexity in the Earth's Magnetosphere*, tackles the Earth's magnetosphere complexity from the standpoint of system science, studying classical concepts such as scale-invariance, self-similarity, and multifractality in the context of the analysis of time series of geomagnetic data.

Chapter 8, *Application of NARMAX to Space Weather*, reviews the several uses of the methodology based on Nonlinear AutoRegressive Moving Average with eXogenous inputs models to space weather, focusing on geomagnetic indices and radiation belt electrons.

Chapter 9, *Probabilistic Forecasting of Geomagnetic Indices Using Gaussian Process Models*, presents an application of Gaussian process (GP) regression with a particular emphasis on model selection and design choice. GP can be understood in the context of Bayesian inference, and it is a particularly promising tool for space weather prediction, for its natural ability to provide probabilistic forecasts.

Chapter 10, *Prediction of MeV Electron Fluxes With Autoregressive Models*, focuses on relativistic electrons in the radiation belts and on relevant forecasting verification techniques for autoregressive models. The approach employed in this chapter represents a nice example of a gray-box modeling discussed earlier.

Chapter 11, *Artificial Neural Network for Magnetospheric Conditions*, discusses an application of feed-forward neural networks to the problems of electron density estimation in the radiation belt and the specification of waves and flux properties.

Chapter 12, *Reconstruction of Plasma Electron Density From Satellite Measurement Via Artificial Neural Networks*, is also concerned with the study of radiation belt electron density via neural networks, although using a completely different approach to derive input features, and emphasizing model selection and verification.

Chapter 13, *Classification of Magnetospheric Particle Distribution Using NN*, tackles an unsupervised multicategory classification problem: clustering particle distribution in pitch-angle from Van Allen Probes data. The machine learning method chosen for this task is a class of neural networks called self-organizing map.

Chapter 14, *Machine Learning for Flare Forecasting*, discusses the recent progresses in solar flare forecasting, comparing several types of machine learning algorithms, and some relevant computing aspects.

Chapter 15, *Coronal Holes Detection Using Supervised Classification*, presents results on the problem of coronal holes detection, comparing different techniques including support vector machine and decision trees. The chapter has a useful hands-on approach, with a direct link to MATLAB software available on the author's website.

Finally, Chapter 16, *Solar Wind Classification Via the K-Means Clustering*, presents an unsupervised clustering technique to divide the solar wind in different types, based on their characteristics measured by instruments on the Advanced Composition Explorer.

In conclusion, we believe that this book provides an up-to-date portrait of some state-of-the-art applications of machine learning to space weather. However, some important works have inevitably been left out. In particular, we would like to mention the recent progress in the prediction of solar flares and coronal mass ejections using Solar Dynamic Observatory data via support vector machine and automatic feature extraction (Bobra and Couvidat, 2015; Muranushi, 2015; Bobra and Ilonidis, 2016; Jonas et al., 2017); the use of data assimilation (Koller et al., 2007; Shprits et al., 2007; Arge et al., 2010; Innocenti et al., 2011; Godinez et al., 2016; Lang et al., 2017); and uncertainty quantification and ensemble techniques (Schunk, 2014; Guerra et al., 2015; Knipp, 2016; Camporeale, 2016).

ACKNOWLEDGMENTS

The authors would like to thank several colleagues that have helped in reviewing the chapters: George Balasis, Shaun Bloomfield, Monica Bobra, Joe Borovsky, Jacob Bortnik, Algo Carè, Michele Cash, Veronika Cheplygina, Xiangning Chu, Gregory Cunningham, Rob Decker, Veronique Delouille, Mariusz Flasinski, Simone Garatti, Manolis Georgoulis, Larisza Krista, Naoto Nishizuka, Juan Valdivia, Shawn Young, and Chao Yue.

The authors would like to thank Mathworks for providing complimentary MATLAB licenses to some of the authors.

This work was partially supported by the NWO-VIDI grant 639.072.716 and NASA grants (NNX15AJ01G, NNX16AR10G, and NNX16AQ87G).

References

Arge, C.N., et al., 2010. Air force data assimilative photospheric flux transport (ADAPT) model. In: AIP Conference Proceedings, vol. 1216, 1.

Baker, D.N., 1990. Linear prediction filter analysis of relativistic electron properties at 6.6 RE. J. Geophys. Res. Space Phys. 95 (A9), 15133–15140.

Bennett, J., Lanning, S., 2007. The netflix prize. In: Proceedings of KDD Cup and Workshop.

Boberg, F., Wintoft, P., Lundstedt, H., 2000. Real time Kp predictions from solar wind data using neural networks. Phys. Chem. Earth Part C 25 (4), 275–280.

Bobra, M.G., Couvidat, S., 2015. Solar flare prediction using SDO/HMI vector magnetic field data with a machine-learning algorithm. Astrophys. J. 798 (2), 135.

Bobra, M.G., Ilonidis, S., 2016. Predicting coronal mass ejections using machine learning methods. Astrophys. J. 821 (2), 127.

Camporeale, E., 2016. On the propagation of uncertainties in radiation belt simulations. Space Weather 14 (11), 982–992.

Gleisner, H., Lundstedt, H., 2001. A neural network-based local model for prediction of geomagnetic disturbances. J. Geophys. Res. Space Phys. 106 (A5), 8425–8433.

Godinez, H.C., et al., 2016. Ring current pressure estimation with RAM-SCB using data assimilation and Van Allen Probe flux data. Geophys. Res. Lett. 43 (23), 11948–11956.

Guerra, J.A., Pulkkinen, A., Uritsky, V.M., 2015. Ensemble forecasting of major solar flares: first results. Space Weather 13 (10), 626–642.

Guizzo, E., 2011. How Google's self-driving car works. In: IEEE Spectrum, vol. 18.

Habarulema, J.B., McKinnell, L.A., Cilliers, P.J., 2007. Prediction of global positioning system total electron content using neural networks over South Africa. J. Atmos. Sol. Terr. Phys. 69 (15), 1842–1850.

He, K., et al., 2015. Delving deep into rectifiers: surpassing human-level performance on imagenet classification. In: Proceedings of the IEEE International Conference on Computer Vision.

Innocenti, M.E., et al., 2011. Improved forecasts of solar wind parameters using the Kalman filter. Space Weather 9, S10005.

Jonas, E., et al., 2017. Flare prediction using photospheric and coronal image data. ArXiv preprint arXiv:1708.01323.

Knipp, D.J., 2016. Advances in space weather ensemble forecasting. Space Weather 14 (2), 52–53.

Koller, J., et al., 2007. Identifying the radiation belt source region by data assimilation. J. Geophys. Res. Space Phys. 112, A06244.

Lang, M., et al., 2017. Data assimilation in the solar wind: challenges and first results. Space Weather 15, 1490–1510.

LeCun, Y., Bengio, Y., Hinton, G., 2015. Deep learning. Nature 521 (7553), 436–444.

Li, X., 2001. Quantitative prediction of radiation belt electrons at geostationary orbit based on solar wind measurements. Geophys. Res. Lett. 28 (9), 1887–1890.

Lundstedt, H., 1997. AI techniques in geomagnetic storm forecasting. In: Tsurutani, B.T., Gonzalez, W.D., Kamide, Y., Arballo, J.K. (Eds.), Magnetic Storms. American Geophysical Union, Washington, D.C.

Lundstedt, H., 2005. Progress in space weather predictions and applications. Adv. Space Res. 36 (12), 2516–2523.

Muranushi, T., 2015. UFCORIN: a fully automated predictor of solar flares in GOES X-ray flux. Space Weather 13 (11), 778–796.

Newell, P.T., Wing, S., Meng, C.I., Sigillito, V., 1991. The auroral oval position, structure and intensity of precipitation from 1981 onwards: an automated on-line data base. J. Geophys. Res. 96, 5877–5882.

Popper, K., 1959. The Logic of Scientific Discovery. Routledge, London.

Schunk, R.W., 2014. Ensemble modeling with data assimilation models: a new strategy for space weather specifications, forecasts, and science. Space Weather 12 (3), 123–126.

Shprits, Y., et al., 2007. Reanalysis of relativistic radiation belt electron fluxes using CRRES satellite data, a radial diffusion model, and a Kalman filter. J. Geophys. Res. Space Phys. 112, A12.

Silver, D., 2016. Mastering the game of go with deep neural networks and tree search. Nature 529 (7587), 484–489.

Sommer, R., Paxson, V., 2010. Outside the closed world: on using machine learning for network intrusion detection. In: 2010 IEEE Symposium on Security and Privacy (SP).

Sutcliffe, P.R., 1997. Substorm onset identification using neural networks and Pi2 pulsations. Ann. Geophys. 15 (10), 1257–1264.

Tulunay, E., et al., 2006. Forecasting total electron content maps by neural network technique. Radio Sci. 41 (4), RS4016.

Valdivia, J.A., Sharma, A.S., Papadopoulos, K., 1996. Prediction of magnetic storms by nonlinear models. Geophys. Res. Lett. 23 (21), 2899–2902.

Vandegriff, J., 2005. Forecasting space weather: predicting interplanetary shocks using neural networks. Adv. Space Res. 36 (12), 2323–2327.

Vassiliadis, D., 2000. System identification, modeling, and prediction for space weather environments. IEEE Trans. Plasma Sci. 28 (6), 1944–1955.

Wing, S., et al., 2003. Neural networks for automated classification of ionospheric irregularities from HF radar backscattered signals. Radio Sci. 38 (4), 1–8.

Wing, S., et al., 2005. Kp forecast models. J. Geophys. Res. Space Phys. 110, A4.

PART I

SPACE WEATHER

CHAPTER

1

Societal and Economic Importance of Space Weather

Mike Hapgood

RAL Space, Harwell, Didcot, United Kingdom; Lancaster University, Lancaster, United Kingdom

CHAPTER OUTLINE

1 WHAT IS SPACE WEATHER?

The past few decades have seen a growing awareness that phenomena in space can affect human activities on Earth. For example, this includes the impact of near-Earth objects, such as the small asteroid that exploded above Chelyabinsk in Russia that occurred on February 15, 2013. However, it also includes the range of phenomena that we call space weather, which are variations in a number of natural environments. In this chapter we emphasize three environments that can directly disrupt the operation of many technologies important for the smooth running of modern societies and their economies:

Machine Learning Techniques for Space Weather
https://doi.org/10.1016/B978-0-12-811788-0.00001-9

- the electromagnetic fields that exist within the solid body of the Earth;
- the radiation environments in Earth's atmosphere and near-Earth space; and
- the density, composition, and dynamics of the upper atmosphere; both its neutral and ionized components (thermosphere and ionosphere).

The headline example of these disruptions is the susceptibility of electric power grids to variations in natural electromagnetic fields, but other important examples include the impact of natural radiation environments on the digital systems that now control so many technological systems, and the potential disruption of satellite navigation systems by changes in the upper atmosphere. We will discuss these and other examples in detail herein, presenting first some of the physics through which these technologies are affected by space weather and then showing why this has societal and economic importance.

But first we should outline why these environments are influenced by phenomena in space. Most space weather has its origin on the Sun—in the magnetic fields produced in the outer "convection" layer of the Sun as hot ionized gas (plasma) rises toward the Sun's surface, transporting heat to that surface. These magnetic fields are distorted by the differential rotation of the Sun, and emerge into the solar atmosphere (the "corona") to form the complex magnetic fields that can be easily seen in modern extreme ultraviolet (EUV) images of the Sun, such as Fig. 1. It is the energy in these fields that drives space weather, in particular as the result of a plasma process called magnetic reconnection. This reconfigures the magnetic field topology in the solar corona to a simpler state, thus releasing energy from the magnetic field, and increasing the kinetic energy of the electrons and ions that form the plasma of the corona. This reconfiguration of the field can lead to parts of the corona becoming magnetically disconnected from the Sun and ejected into interplanetary space, forming a coronal mass ejection (CME).

CMEs are the most important of the several phenomena by which solar activity can drive space weather effects toward Earth. A CME arriving at Earth will first interact with the magnetosphere, the region of space dominated by Earth's own magnetic field. This region typically extends some 60,000 km Sunward of the Earth, where it is confined by the tenuous plasma continuously flowing from the Sun (the solar wind). On the anti-Sunward side, the

FIG. 1 Extreme ultraviolet image of a tangle of arched magnetic field lines in the Sun's corona, taken in January 2016 by NASA Solar Dynamics Observatory. *Credit: Solar Dynamics Observatory, NASA.*

magnetosphere is stretched out into a long tail (the magnetotail) that extends to perhaps a million kilometers or more. The arrival of a strong CME (i.e., speed and density higher than the background solar wind) will compress the magnetosphere, producing a sharp increase (sudden impulse) in the magnetic field at Earth's surface (see left-hand side of Fig. 2). If the CME contains regions where its own magnetic field points southward (opposite to the northward pointing field of the Earth), magnetic reconnection will allow the CME field to interlink with the Earth's magnetic field so that CME energy can enter the magnetosphere. That inflowing energy can then drive a cycle of energy storage in the magnetotail, followed by explosive release toward Earth, where it produces aurora, heating of the upper atmosphere, and strong electric currents in the ionosphere. This "substorm cycle" drives many space weather impacts, as we discuss below. Substorm cycles are a fundamental dynamical cycle of planetary magnetospheres and, in the case of the Earth, have a typical period of 1–3 h. But a large CME can take 12–24 h to pass the Earth. Thus a large CME can drive a series of substorms, producing an ensemble of space weather effects that we term a geomagnetic storm. It is important to appreciate this temporal relationship between substorms and storms

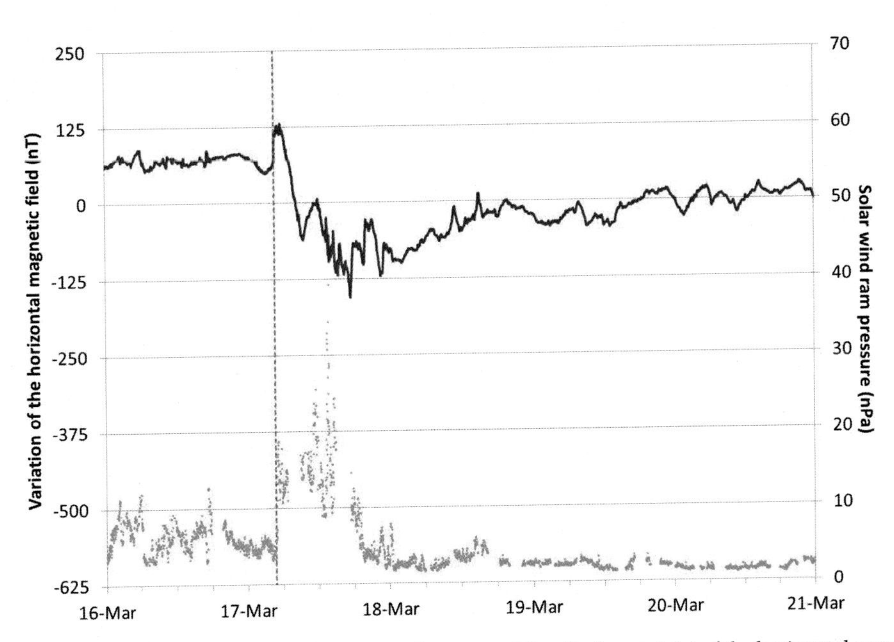

FIG. 2 The *red trace* shows the variation (relative to the mean of the displayed data) of the horizontal component of the geomagnetic field as measured at Kanoya observatory (31.4 degrees N, 130.9 degrees E) during the St. Patrick's Day storm of 2015. The sharp rise in the field at 04:45 UTC on March 17 (marked by the *vertical-dashed line*) is a sudden impulse as the magnetosphere was compressed by the arrival of a CME (as shown by the simultaneous sharp rise in solar wind ram pressure marked by the *blue trace*). The sudden impulse is followed by a large, more gradual decrease as the ring current grows in intensity during the St. Patrick's Day storm, and then a very slow return to normal conditions after the storm. Kanoya was well placed to show these effects well as a low-latitude observatory with local time just past midday at the time of the sudden impulse. Because the sudden impulse was followed by a magnetic storm, it is also termed a sudden storm commencement. Kanoya magnetometer data courtesy of Japan Meteorological Agency, solar wind data courtesy of NASA Space Physics Data Facility, and sudden impulse timing courtesy of the International Service on Rapid Magnetic Variations hosted by Observatori de l'Ebre.

as it has profound implications for many space weather effects—with the detail for each effect depending on the physical timescales associated with each effect.

Another striking consequence of magnetic reconnection in the solar corona is the occurrence of solar flares. In this case reconnection generates high fluxes of very energetic electrons, some of which propagate down toward the solar surface where they collide with dense plasma, producing the intense burst of X-ray and EUV radiation, often extending to optical emissions, which characterizes a solar flare. The X-ray and EUV emissions from strong flares can cause a sharp increase in the density of the ionosphere on the dayside of the Earth: the X-rays causing increases at low levels around 80–90 km altitude, while the EUV causes increases at a higher level (150–400 km). These changes in density have impacts on a number of radio technologies, as we discuss below.

Both CMEs and solar flares are associated with solar radiation storms—bursts of particle radiation (protons, alphas, and heavier ions), which can cause a marked rise in the radiation environments in near-Earth space, and sometimes deep inside Earth's atmosphere, even down to sea-level. In the case of CMEs the particles are energized at the shocks that form in front of fast CMEs, while in the case of flares the particle energization is thought to arise from the reconnection event that also produces the flare (Drake, 2009). The radiation storms produced by flares tend to be short-lived, lasting less than a day (Reames, 1999), and occur only if the solar magnetic fields over the flare site allow the particles to escape into interplanetary space. There are examples of strong flares not producing radiation storms when the overlying magnetic field did not allow particle escape. A notable recent example was the huge sunspot seen in October 2014; this produced several strong flares, but no radiation storms (and no significant CMEs). The radiation storms produced by CMEs often last several days, with particles being energized throughout the journey of the CME, and its shock, to and beyond Earth (Reames, 1999). Radiation storms have impacts on a range of electrical and electronic systems, and can also pose a minor radiation hazard to humans, in space and in aircraft.

There are also a number of other solar phenomena that can cause space weather effects at Earth (e.g., high-speed solar wind streams can also generate substorms and geomagnetic storms while solar radio bursts can interfere with radio technologies on Earth). In addition, some space weather effects have their origin on Earth. The dynamics of the upper atmosphere are also modulated by energy and momentum that propagates upward from the lower atmosphere, for example, in the form of atmospheric gravity waves generated by convective activity in the troposphere. In addition, it is increasingly recognized that the strong electric fields in thunderstorms can produce energetic events that propagate toward the upper atmosphere and even into space (sprites, gamma rays, etc.). Thus space weather includes a wide range of different physical phenomena and makes this a fascinating subject for study and a challenge for scientists developing methods that help us understand and forecast when adverse conditions will occur. Thus it is a fitting topic for machine learning as discussed in this book.

2 WHY NOW?

The phenomena that cause space weather have existed since the Sun and the Earth were formed 5 billion years ago. So why have they come to public importance in the past couple of decades? There are a number of reasons, not least that our scientific understanding has advanced greatly in those decades; also there has been a much greater political recognition of the need for societal resilience against natural hazards. But the outstanding reason is the way that modern societies have become much more dependent on advanced technologies over the past 50 years; and many of those technologies are vulnerable to space weather, as discussed as follows.

This technological vulnerability has grown over the past 170 years. The earliest records of space weather impacts date back as far as the 1840s with the deployment of the first electric telegraph network based on metal wires. This technology was vulnerable to geomagnetically induced currents (GIC), just as are modern power grids, and space weather disruption was reported as early as 1847 (Barlow, 1849). Early telephone systems also proved vulnerable to these currents (Preece, 1894). Both systems continued to be vulnerable until the late 20th century when telegraph networks were phased out and telephone networks switched from metal wires to optical fiber (which is not vulnerable to GIC). The advent of long-distance radio communications in the early 20th century was another valuable technology that is vulnerable to space weather, in this case disturbances of the upper atmosphere that disrupt the propagation of radio signals at the megahertz frequencies suited to long-distance communications. Thus users of this technology have long been aware of space weather—and continue to need that awareness today. The development of power grids, networks for long-distance transmission of electric power (rather than every town having a small power station), in the middle of the 20th century began what is still our biggest space weather concern. The first space weather impacts on this technology were reported as early as 1940 (McNish, 1940), but gained limited attention until the spectacular failure of the Hydro-Québec power grid on March 13, 1989. An intense substorm over eastern Canada set off a cascade of failures such that the whole grid went from nominal operation to fully switched-off in 92 s (Bolduc, 2002). This left six million people without power on a cold day and thus was a major news story. Combined with other problems reported in the United Kingdom (problems with two transformers during another intense substorm 18 h later—see Erinmez et al., 2002) and the United States (destruction of a transformer at Salem in New Jersey—see Wrubel, 1992; Boteler, 2001), the events of March 1989 acted as a wake-up call for the power industry and its regulators. It has led to many efforts to improve engineering resilience and operator awareness around the world, and also to much of the governmental interest in space weather (because electric power is the fundamental infrastructure of almost every country). The space weather impact on the power grid continues to be a major concern to industry and governments, and thus stimulates their engagement with the space weather expert community.

Since the advent of the space age, the range of technologies vulnerable to space weather has continued to grow. Satellites are highly exposed to space weather, particularly to radiation effects that can damage and disrupt many key components, and, for this reason, are designed with high levels of resilience. Thus direct impacts of space weather on satellite services are unusual, but not unknown. But indirect impacts on satellite services are more common. This arises because radio signals between satellites and the ground must pass through the ionosphere, and thus can be disrupted by space weather disturbances in the ionosphere. In severe space weather conditions, this can have major impacts on key satellite services such as mobile communications (e.g., to ships and aircraft), satellite navigation, and maritime surveillance.

In summary, our modern vulnerability to space weather is a consequence of the technological revolutions of the late 20th century. These revolutions have helped to raise the standard of living for billions of people around the world. But they have also created a vulnerability in which space weather can disrupt people's lives. Thus it is vital to mitigate that vulnerability through better understanding of the science of space weather—and then to encourage better engineering to reduce the vulnerability of the technologies at risk, and better forecasting so that people can manage those vulnerabilities that cannot be engineered out.

3 IMPACTS

We now discuss a number of space weather impacts in detail—focusing on some of the most critical impacts, including power grids, satellite navigation systems, digital systems, other radio technologies, and finally on drag effects that have major impacts on the operation of satellites in low Earth orbit. These are not a complete set of impacts—but they will give good insight into the biggest risks that arise from space weather.

3.1 Geomagnetically Induced Currents

One of the most important impacts of space weather is to generate significant electric fields in the solid body of the Earth. These geoelectric fields drive electric currents through both the solid body of the Earth and through any human-built structures that are electrically conducting and electrically connected to the Earth. These electric currents are the geomagnetically induced currents, or GICs, that can affect a range of infrastructures, including power grids, railway circuits, pipelines, and even the power systems of communications cables that run under the oceans.

Space weather generates geoelectric fields through the process of magnetic induction. The substorm cycle outlined above will drive large variations in the electric currents that flow through the ionosphere, resulting in large magnetic field variations that can be observed on the surface of the Earth at high to mid-latitudes (see Fig. 3). These changes can penetrate deep into the solid body of the Earth as they have low frequencies (tens of millihertz), and the electrically conducting material that forms the Earth has a skin depth of hundreds of kilometers at these frequencies. Thus intense substorms can induce significant geoelectric fields ($\sim$1–10 V km^{-1}) in high to mid-latitude regions.

The substorm cycle can also enhance the ring current, a torus of electric current that flows in Earth's magnetosphere, some 10,000–20,000 km above equatorial regions, resulting in a changing magnetic field that can be observed on the surface of the Earth in low-latitude regions. Typically each substorm will add more strength to the ring current, where it will persist for several days as shown in Fig. 2. Thus the ring current integrates the inputs from substorms, providing a natural overview of a geomagnetic storm as an ensemble of substorms. But it is important to understand that this integration does not affect the geoelectric field; that arises only from the changes in the magnetic field produced by the ring current. As with substorms, it is the low-frequency component of those changes that penetrates deep into the Earth to induce significant electric fields, in this case in lower-latitude regions.

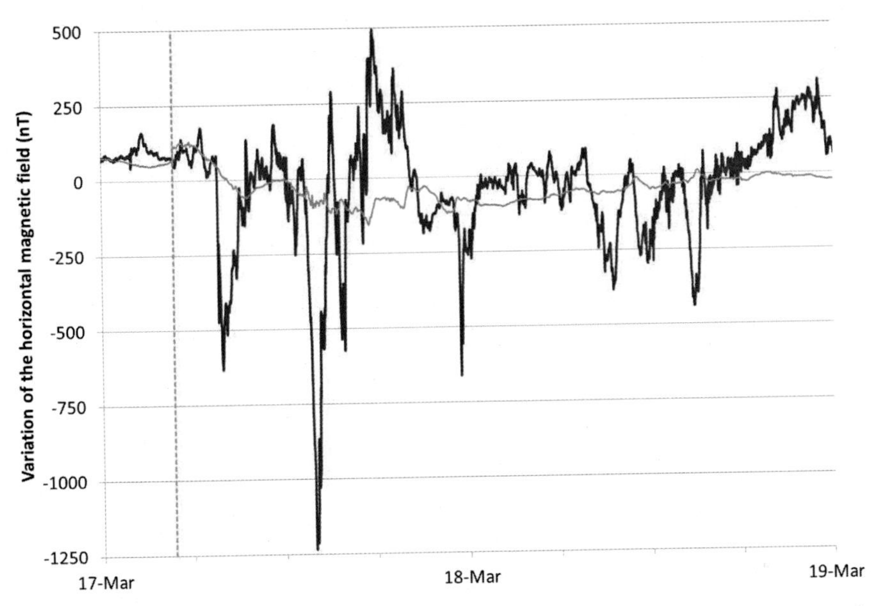

FIG. 3 The *red trace* shows the variation (relative to the mean of the displayed data) of the horizontal component of the geomagnetic field as measured at Fort Churchill observatory (58.8 degrees N, 94.1 degrees W) during the St. Patrick's Day storm of 2015. The sudden impulse seen on the dayside at 04:45 UTC on March 17 (marked by the *vertical-dashed line*) is not seen at Churchill as the observatory was then near midnight local time. Instead a substorm (the first deep dip in the *red trace*) commences around 07:20 UTC, indicating the explosive release of magnetic energy in the tail of the magnetosphere, energy that had built up over 2.5 h following the arrival of the CME that caused the sudden impulse. A second, even stronger substorm commences around 13:20 UTC, indicating a second cycle of energy storage and release in the magnetosphere as a result of the CME flowing past the Earth. Later in the day (16:00–21:00 UTC) Churchill is on the dayside of the Earth, so this observatory sees a positive variation due to strong eastward currents in the dayside ionosphere. But observatories on the nightside see a continuing series of smaller substorms; a series that reappears at Churchill on March 18 as the observatory returns to the nightside. For comparison the simultaneous Kanoya magnetic field data from Fig. 2 are displayed as a *gray trace*; this shows how substorm-driven variations in the magnetic field are usually much larger than those driven by sudden impulses or the ring current. Churchill data courtesy of Geological Survey of Canada.

Although there are other current systems induced by space weather, we have focused here on substorm currents and the ring current, as these are two current systems with proven impacts. It was substorm currents that were responsible for the failure of the Hydro-Québec power grid in 1989, as well as the transformer problems reported in the United Kingdom and in New Jersey. By contrast, the ring current was responsible for multiple transformer failures that severely reduced the capability of the South African power grid during another very severe event in October 2003 (Gaunt and Coetzee, 2007). This latter event provided another wake-up call: showing space weather impacts on power grids were not limited to high to mid-latitudes.

Why do GICs disrupt power grids? The key factor is to recognize that these are relatively low-frequency currents, much lower than the 50/60 Hz frequency of the currents that deliver power across these grids. Thus, when these currents enter transformers, they will behave as quasi-DC currents and have the potential to drive transformers into half-phase saturation. This then leads a number of adverse effects—heating and vibration that has the potential to damage transformers, but also generation of harmonics (of the basic 50/60 Hz frequency) that can disrupt other grid devices. Strong GIC, therefore, has the potential to destabilize grid operation, triggering safety systems to switch off parts of the grid, quickly leading to a cascade effect in which the grid shuts down—as happened in Quebec in 1989. If safety systems switch fast enough, the amount of actual damage will be limited, and the grid can be restarted (a procedure much practiced by grid operators). This recovery will take from hours to many days, depending on the types of power generation (e.g., many renewables such as hydro can restart in hours but some nonrenewables such as nuclear can take many days to restart). But where there is significant damage, it may take many weeks to fully restore power (e.g., it can take a month or two to replace a damaged transformer, even if a spare is available). Thus grid operators today make extensive efforts to avoid disruption and damage by space weather. This is done in part by better engineering, for example, gradually replacing vulnerable transformers with designs that have high resilience to GIC, and in part by operational procedures that will temporarily increase grid resilience when adverse space weather is expected, for example, by all-on procedures that spread GIC as widely as possible, thus reducing the risk of large values at any point in the network. These procedures, often taken in discussion with government risk managers, reinforce the need for better scientific understanding of the conditions that lead to adverse space weather and to develop models that can help space weather forecasters provide advice to grid operators and government.

As noted previously, space weather can also drive GICs into rail systems: the track circuits that detect the locations of trains, as well as the circuits that control color light signals. There are several cases in which lights have incorrectly switched from green to red during major space weather events in 1982, 1989, and 2003 (Wik et al., 2009; Eroshenko et al., 2010). These "right-side failures" appear to arise from GIC interfering with operation of relays and due to good design have failed to the safe condition of red. The complementary case of failure from red to green, termed "wrong-side failure" because it puts lives at risk, is thought to be possible, but has thankfully not yet been observed (Krausmann et al., 2015). The impact of space weather on rail systems remains a major area for study, especially with the growth of high-speed rail systems. A recent paper (Liu et al., 2016) reported the first actual measurement of GIC in a rail system.

GICs can also pass through pipelines (where they can interfere with the operation of electric systems that act to reduce corrosion = cathodic protection), and the power lines on modern optical fiber-based transoceanic cables (where they can act to increase the voltages in the supply of power to the repeaters boost the optical signals). Both effects have been reported in the scientific literature (Gummow and Eng, 2002; Medford et al., 1989) but are still poorly studied, possibly reflecting low awareness of the issue. Thus this is an area in which future studies could be productive.

3.2 Global Navigation Satellite Systems

More popularly known as satnav or satellite navigation, the provision of precise location and timing services via satellites has become a mainstay of modern societies over the past 20 years. Its importance today is shown by the provision of at least four different constellations of Global Navigation Satellite Systems (GNSS) satellites to deliver these services: the original Global Positioning System (GPS) developed by the United States, the European Galileo system, plus Beidou from China and Glonass from Russia.

A satnav receiver determines its location by detecting signals from a number of satellites, analyzing each signal to determine the signal travel time between the satellite and the receiver, and then using this set of travel times to deduce its own position relative the known positions of the satellites. The satellite signals include a pulse code that enables the receiver to determine travel time, as well as information on satellite positions and current time. A minimum of four satellite signals are needed to determine position and time, but accurate solutions typically use eight or more satellites as shown in Fig. 4.

However, it is not quite so simple. There are a number of other issues that have to be resolved to make GNSS services work. Two important but straightforward issues are to correct for the slow running of the clocks on the GNSS satellites due to the lower gravity that they experience, and to correct positions for the variable rotation speed of the Earth. In contrast,

FIG. 4 *Screenshot* from a GNSS app running on an Android tablet. The *large circle* is a sky view of the satellites visible to the GNSS receiver, zenith at the center, and horizon at the circumference. Eleven satellites are visible and eight (marked *green*) were contributing to the position fix, which is shown at the right. Note that the receiver failed to get data from three satellites at low elevation. Image by the author using an Ulysee Gizmos app running on a Nexus 7 tablet with Android 6.0.1.

space weather poses a more complex set of challenges to making GNSS work well—through its impact on the propagation of signals from the satellites down to the Earth. The preceding simple overview assumes that these signals propagate at the speed of light in a vacuum. This is not quite true, as they have to pass through the plasma that forms Earth's ionosphere, and the group refractive index of this plasma at GNSS signal frequencies (1–2 GHz) is slightly greater than unity. The difference is only a few parts in 100,000, but this is sufficient to delay signal arrival by tens of nanoseconds, leading to position errors of several meters. This would not be a problem if the signal group delay was known accurately (as with the preceding relativistic clock correction), but it is not. It depends on the state of the ionosphere (specifically the column density of plasma along the signal path), and varies markedly with space weather conditions. The importance of this ionospheric correction has led governments around the world to invest billions of dollars in error correction systems to aid accurate use of GNSS by aviation and shipping. Examples of these systems include satellite-based augmentation systems such WAAS in the United States (Loh et al., 1995), EGNOS in Europe (Gauthier, 2001), and MTSAT in Japan, and differential GPS systems such as that used by ships approaching ports in the United Kingdom and Ireland (Trinity House, 2016).

But this ionospheric correction is not the only problem caused by the ionosphere. In some conditions, the plasma in the ionosphere can become unstable, for example, through Rayleigh-Taylor breakdown of plasma upflows (as commonly happens around dusk in equatorial regions) or in the presence of sharp horizontal gradients in plasma density (as often happens at high latitude due to the interplay of plasma production, decay, and transport). The presence of these instabilities causes the ionosphere to act as a diffraction screen for signals at GNSS frequencies, so that their signals scintillate when observed on the ground. Amplitude scintillation can pose problems when signals fade, but phase scintillation can challenge the ability of receivers to maintain a phase lock on the signal. In extreme conditions that lock will be lost, thus denying access to the GNSS service until it can be reacquired. Thus GNSS services should always be considered at risk of interruption if scintillation is expected. It is important to design GNSS applications so they can cope with at least short breaks in the signals from satellites.

In extreme space weather conditions, all these impacts on GNSS services will be exacerbated for many hours to a few days. Error correction systems will struggle to keep up with rapidly changing conditions in the ionosphere and are likely declare a loss of integrity, a design feature that allows them to advise users to avoid using their services until conditions improve. In addition, there is likely to be extensive loss of signal due to scintillation. Some signals will likely be available on an intermittent basis, but that is a poor basis for many important activities. In such conditions, it is vital to have alternative systems for navigation and timing. If not, it is best to stop activities that use GNSS services, and wait for conditions to improve.

Although GNSS services are best known for their navigational uses, the most economically important applications of GNSS services are probably in providing accurate timing. Timing services are vital to many high-profile economic activities. Most prominent is the financial services sector, where activities such as high-speed trading now require that transactions are time-stamped to microsecond accuracy, and rely on GNSS as the primary source for those times. Fortunately, this sector has increasing awareness of space weather and is implementing solutions in the form of holdover clocks—high-quality clocks that can maintain microsecond accuracy for several days between contacts with GNSS timing services. Accurate timing is also vital for synchronization of many kinds of networks, most obviously in mobile/cellular phone networks, where accurate timing is vital to smooth handover of calls from one cell to another. The resilience of these networks to loss of GNSS depend on details of local implementation, for example, a UK study (Cannon et al., 2013) showed that UK phone networks were resilient against loss of GNSS, but raised concerns that US networks might assume continuous access to GNSS timing.

As we discussed, there has been, and continues to be, extensive investment in systems to ensure the accuracy of GNSS services for navigation of aircraft and shipping. For example, systems for the correction of ionospheric error have been implemented in India (Rao, 2007) and are being developed for use in Africa (Avanti, 2017). The aviation sector has increasing awareness of space weather and is factoring this awareness into its strong safety culture. The latter includes the recognition that aircraft should always carry at least two different technologies for vital functions such as communications and navigation. So aviation is well positioned to exploit the benefits of GNSS, while maintaining good resilience against space weather and other factors that can disrupt GNSS. In contrast shipping seems to be heading toward sole reliance of GNSS as its main system for navigation (other than traditional methods of navigation via sextant sightings of the Sun and bright stars). Many experts have advised that the VLF radio systems previously used for maritime navigation should be modernized to provide a backup to GNSS (Johnson et al., 2007). This would have ensured the resilience of maritime navigation against GNSS loss to space weather or human interference, but in practice these systems have now largely ceased operation (Maritime Executive, 2016). In the absence of a technological backup, good seamanship will be as important as it ever was.

GNSS is also now playing an important role in road transport and logistics—especially where drivers have to travel to unfamiliar locations, for example, as in the provision of emergency services and in the delivery of internet shopping (a mundane but vital part of modern economies). Widespread loss of GNSS due to space weather would have a profound impact on such services. The emergency services would work harder to ensure the safety of life services that they provide. But, if the space weather event continued for many hours, this could lead to escalating problems as emergency teams became too tired to work safely. This impact on emergency services was highlighted by the Organisation for Economic Co-operation and Development (OECD) in their 2011 study on the impact of geomagnetic storms (OECD (2011), see Fig. 5). In contrast, the disruption of road delivery services is not generally a threat to life, but could cause enormous societal disruption. It would have to be managed carefully by both industry and government.

The disruption of GNSS by space weather is an issue of increasing societal importance, perhaps second only to the disruption of power grids, and likely to escalate as GNSS becomes ever more embedded in the technologies that support everyday life. Thus it is important to develop tools that can raise awareness of space weather and improve forecasting of adverse conditions. These will be essential in helping governments prepare for disruption, and to provide useful advice to the public ahead of and during a severe event. This advice could just be "please don't panic, this is a natural event and we need to let it pass"—but that simple message needs good scientific support.

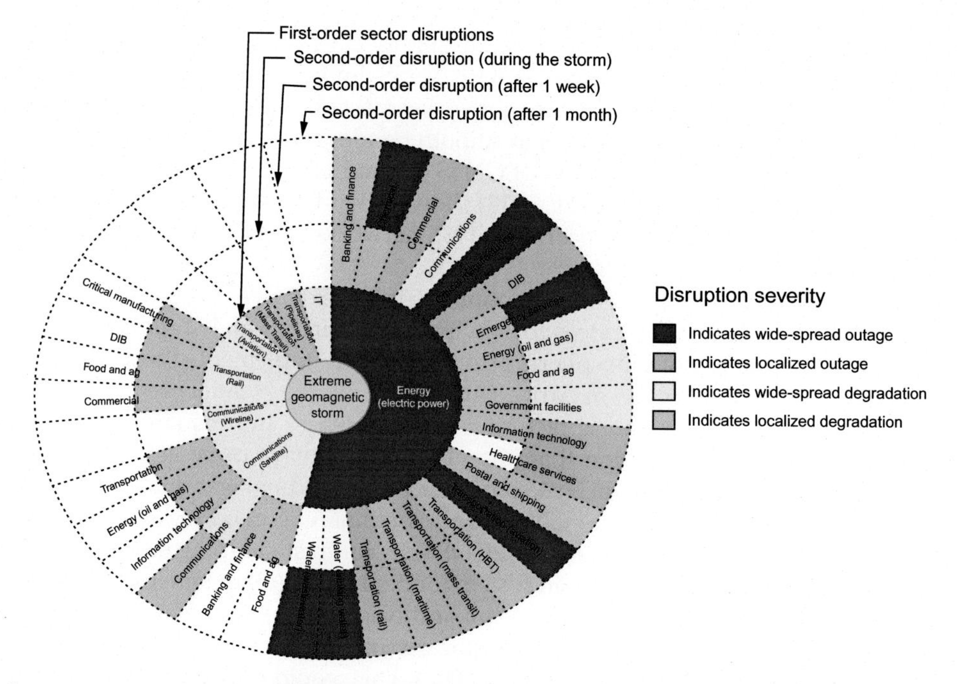

FIG. 5 First- and second-order critical infrastructure disruptions as assessed in a 2011 study on the impacts of geomagnetic storms, as part of OECD's Future Global Shocks project: a set of studies on events that have the potential to cause major disruption to societies and their economies around the world (OECD, 2011). There are many interesting features in this plot, but here we want to highlight the impact on emergency services, the segment at around the 2 o'clock position in the *pie chart*. This segment highlights how disruption of emergency services will start as localized degradation, but would escalate to wide-spread outage if a severe space weather event were to continue several days (OECD, 2011).

3.3 Single-Event Effects

An important but subtle aspect of space weather is that it can significantly increase the levels of particle radiation in space, in the atmosphere, and on the surface of the Earth. These increases can disrupt the operation of the digital devices that are now embedded in a wide range of technological systems. This disruption is a consequence of single-event effects (SEE)—that a single high-energy ion passing through a computer chip or other digital device can deposit electrical charge in sensitive regions of the device, sufficient to change its operation in a variety of ways. The best known of these ways is the single-event upset, where the deposited charge is sufficient to change a bit of data stored in the device (e.g., from 0 to 1 or the reverse). As digital devices have become more sophisticated, a wide range of other SEE have become common, including latch-up, where the device gets locked into a particular state, as well as effects in which parts of the device are damaged (a concise overview of these effects is given on page 37 of Cannon et al., 2013). The key driver for the growth in SEE has been the reduction of the feature sizes: the sizes of the active elements on the silicon chip that is the heart of the device. This reduction means that smaller amounts of charge, maybe only a few hundred electrons, are needed to disrupt a feature on the chip.

SEE will therefore occur whenever digital devices are exposed to particle radiation. This is obviously most common on satellites that are highly exposed to radiation, in particular during solar radiation storms, but also with a low background due to cosmic rays. As a result satellites are designed to have a high resilience to SEEs, and satellite operators are adept at resolving problems caused by SEEs. But the radiation that causes SEEs is not confined to space; it can penetrate into Earth's atmosphere right down to the surface. It is attenuated both by the Earth's magnetic field and by the atmosphere. The magnetic field provides a latitude-dependent radiation shielding such that most particle radiation can reach the top of the polar atmosphere, but only higher energy particles can reach the atmosphere at lower latitudes. The atmosphere then acts a physical barrier in much the same way as the shielding around radiation sources here on Earth, just operating on a much larger scale. The atmosphere provides roughly the equivalent of 4 m of concrete as radiation shielding for the Earth's surface, falling to about 80 cm of shielding for aircraft at a cruising altitude of 12 km. The markedly lower shielding at aircraft altitudes means that SEEs are a more serious issue for aircraft, especially as digital devices now play an important role in systems used to control aircraft, as well as for navigation and communications (see Fig. 6). Thus it is important that aircraft systems are designed to be resilient against SEEs and that this resilience is tested by controlled exposure to radiation (Hambling, 2014; STFC, 2017).

The greater radiation shielding at Earth's surface means that SEEs are less frequent here. There is good evidence that they do occur, but are perhaps poorly reported due to a lack of awareness of the issue. One of the best known examples of an SEE occurring at ground level came from an election at Schaerbeek in Belgium where an electronic voting system recorded an extra 4096 votes for one candidate, more than the votes that could have been cast

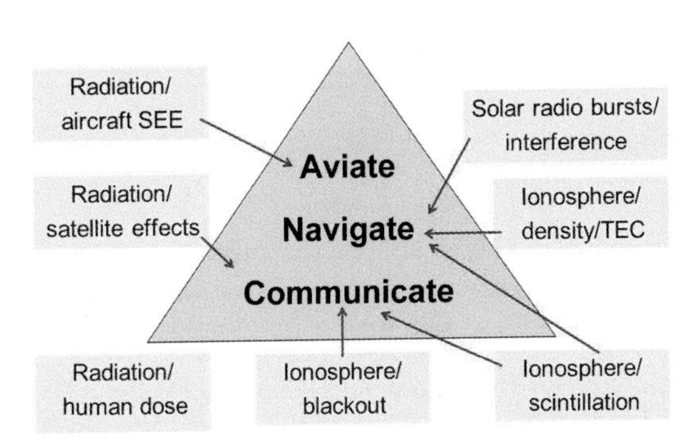

FIG. 6 How space weather impacts a pilot's control of aircraft. The pilot's mantra of "aviate, navigate, and communicate" summarizes key priorities for safe flying. Space weather impacts all three priorities. SEE in avionics could disrupt the pilot's top priority of maintaining stable flight, so the mitigation of these effects is also a top priority. GNSS is increasingly used in aircraft navigation, the pilot's second priority of knowing their location and course, and is vulnerable to many space weather effects, so mitigation of those effects through error correction services and use of complementary technologies is vital. Once these priorities are satisfied, pilots must communicate with ground control and other pilots to ensure safety through coordination of aircraft movements. Aircraft communications are vulnerable to space weather disruption when flying over remote areas, such as the oceans and the Arctic, areas that support busy air routes despite their remoteness.

(PourEVA, 2003) . This stands out as an obvious bit-flip in a very public forum. The risk of SEEs is well recognized by the manufacturers of digital devices; they generally make it clear that radiation is a potential source of error in their devices and that mitigating this error is the responsibility of the system designers who utilize their devices.

How does one make digital devices resilient to SEEs? There are a range of solutions including:

- error correction codes that can detect and fix simple bit flips;
- use of parallel redundancy so that key functions are replicated in triplicate, or more, and hence, an anomalous function is outvoted by two or more correct functions; and
- additional circuitry that allows operators to unlock devices that have latched-up.

These engineering solutions greatly reduce the impact of SEEs and are widely used in critical systems, especially on satellites and aircraft. But they are not perfect, there will always be some SEEs that slip through resilience, especially when radiation fluxes are very high (such as during the most intense radiation storms). In high flux conditions, the SEEs rates may overwhelm engineering resilience, for example, creating more errors than can be handled by correction codes. In these conditions, it will fall to people operating technological systems to deal with the errors, for example, through a system reset, the equivalent of rebooting a computer to clear a problem (indeed some everyday computer problems may be due to SEEs). Thus it is important that operators are prepared to recognize when a reset is needed and to do it efficiently. As with other space weather impacts, this demonstrates the need for tools that can raise awareness of the risks from SEEs and to warn operators of conditions when SEEs may be common, for example, the International Civil Aviation Organization is developing standards for briefing pilots on current space weather conditions as part of the briefing that they receive prior to every flight (ICAO, 2012). This is a good example of how space weather information needs to be well focused on user needs—pilots typically have only 20 min to read and review their preflight briefing. Thus good tools are essential.

3.4 Other Radio Systems

There is a wide range of other radio systems (i.e., besides GNSS) that are vulnerable to disruption by space weather. The disruption primarily arises from problems with radio propagation through the ionosphere, which, as we have seen with GNSS, is heavily influenced by space weather. The key systems at risk are

- Radio communications at low frequencies, 30–100 kHz. These frequencies can propagate as interface waves following the curvature of the Earth, but can support only a very low bandwidth of data. Thus they are very attractive for long-distance communication of small amounts of data. In the 20th century, they were widely used for transmission of navigation and timing signals, but now are largely superseded by use of GNSS. They remain important for communications with submarines as the low frequency allows them to penetrate deep into sea water. Space weather, in the form of energetic particles and X-rays, can disrupt these systems by producing a dense ionospheric layer at altitudes around 70–90 km. This layer can reflect low-frequency radio signals back down to the ground where they can interfere with the signals propagating as an interface wave.

This interference between sky and ground waves was a significant source of disruption for navigation systems in the late 20th century, for example, leading to disruption and danger for professional fishing activities (News Fishing, 1983). It is unclear whether this is a serious problem for current uses of low-frequency radio, but should not be ignored as new applications may move into this part of the radio spectrum.

- Radio communications at high frequencies (HF), 3–30 MHz. These frequencies have been used for long-distance radio communications since the beginning of radio transmissions in the early 20th century. Signals at these frequencies can reflect off the F-region of the ionosphere at altitudes of 250–350 km, which allows propagation over much greater distances, and that usually is available both day and night (the loss rates for ionization at these heights are usually very slow); 3000 km is a typical distance for a single hop from Earth's surface to the ionosphere, and back to the surface. Multihop propagation (where the signal reflects off the surface, then the ionosphere and back to the surface again) allows propagation over even greater distances. Although satellite communications have now taken over much of the work of long-distance radio communications, HF is still widely used by civil aviation over the oceans and remote land areas; indeed it remains mandatory for most aircraft to be equipped with a HF system. HF is also widely used by the military who value its stealthiness and resilience to hostile interference, when compared with satcom. Space weather plays an important role in the use of HF radio through its impact on the global morphology of the ionosphere. This morphology determines the optimal frequencies for HF radio links in any particular location. In quiet space weather conditions, it leads to regular diurnal and seasonal patterns, but a geomagnetic storm will disrupt those patterns in ways that depend on the size, start time, and phase of the storm. Thus users of HF need forecasts of how they should adapt their operational frequencies in response to a geomagnetic storm. In the worst cases, adverse space weather can deny the use of HF communications—via a number of different mechanisms. For example, a geomagnetic storm can enhance the loss rates for ionization leading to the disappearance of the ionosphere and loss of communications during the night. Another example is that enhanced occurrence of ionospheric instabilities (as under GNSS) that can distort HF radio signals, making effective communication impossible. A third example is that low altitude ionization (produced by energetic particles and X-rays) will absorb, rather than reflect HF radio signals, leading to a blackout of radio communications (Hargreaves, 1992). The users of HF need awareness that space weather can deny the use of HF (not least to distinguish this from equipment failure) and forecasts of when this may happen.
- Satellite communications (satcom) are extensively used for transmission of voice and data communications over long distances. The use of satcom for data is increasingly important with the growth in satellite applications used by industry and government. Satcom uses a wide range of frequencies, with L-band (1–2 GHz) heavily used for what we might call traditional satcom applications, including mobile voice communications (satellite phones) and datalink services to aircraft. This is the same radio band as used by GNSS—unsurprisingly, as GNSS developed alongside these other applications. Thus, like GNSS, L-band satcom services are vulnerable to ionospheric scintillation. Thus, even in normal conditions, they are vulnerable to interruption at high and equatorial latitudes, and during severe space weather conditions, this will extend to mid-latitudes. Thus users of applications based on satcom need to be aware that space weather can disrupt their use

of those applications. If reliable operation of those applications is critical to the success of their activities, they need to obtain forecasts of adverse space weather, and to factor those forecasts into their planning. L-band is not the only frequency used for satellite communications. Some satellite applications use lower frequencies in the VHF band (e.g., telemetry from ships, giving ship's identity, position, course, and speed) and in the UHF band (voice communications); these are at even higher risk from scintillation as the effect is inversely related to frequency. But there is also a growing use of high frequencies greater than 4 GHz or even 20 GHz, to exploit the higher bandwidth available—a key example being internet links to aircraft. A valuable side effect of using these frequencies is to eliminate the impact of ionospheric scintillation (though it will also increase risk of disruption by heavy rain). Thus we may expect more use of these higher frequencies in the future as operators try to balance the impacts of space weather and normal weather.
- Radar is often used in ways that send signals through the ionosphere, for example, radar on the ground is the primary means of tracking satellites in low Earth orbit (below 2000 km altitude), while radar on spacecraft are extensively used for observation of the Earth. In these cases, it is essential to measure the time that the radio signal takes to travel from the radar to the target and back again. Thus, as with GNSS, the ionospheric group delay is an important factor. Even in quiet space weather conditions, it is essential to apply ionospheric corrections to raw measurements of the distance and direction of a satellite from a radar (Hapgood, 2010). These corrections can be substantial, up to a kilometer for VHF satellite radar, such as GRAVES in France (Michal et al., 2005). They require tools based on current space weather conditions, especially when the ionosphere is disturbed by geomagnetic storms. In addition to suffering group delay, radar signals passing through the ionosphere can be distorted by plasma instabilities, again as with GNSS. This can make it more difficult to track satellites from the ground and can distort the images produced from satellite radar observations of the Earth's surface. Here the need is, as ever, for awareness that these problems can occur, and for forecasts that allow users to make plans to work around periods when radar data will be poor, for example, rescheduling to a quiet space weather period.

3.5 Satellite Drag

Space weather is an important source of uncertainty in our awareness of the locations of satellites and debris in low Earth orbits. This is an increasingly important topic, a major part of what is now called space situational awareness. Obviously the operators of satellites in these orbits need forecasts of the location of their satellites, so that they can plan operations (e.g., to observe specific regions on the Earth's surface) and schedule ground station contacts (to upload commands for future operations and downlink data from previous operations). But there are also important risk management activities:

- to avoid collisions between objects in orbit. Any collisions involving a working satellite are likely to cause damage, leading to a partial or total loss of the functionality it provides. Furthermore collisions will produce more space debris, gradually leading to increased collision risks for other satellites, and eventually to a cascade in which random collisions between objects determine the collision risk from debris (Kessler and Cour-Palais, 1978). The risk of that cascade is today considered a substantial threat to the sustainability of satellite operations in low Earth orbit (Kessler et al., 2010). Thus space agencies are working with satellite operators to avoid collisions that could create more debris and, where possible, to remove satellites from orbit within a few years after they cease operations.

- To assess when and where satellites will reenter the Earth's atmosphere, especially if those satellites include large elements, such as fuel tanks that could survive reentry. The fall to Earth of such objects poses a risk for aviation, and for people and physical assets on the surface of the Earth. Thus space agencies and satellite operators need reentry forecasts so they can provide appropriate warnings to people at risk and can try to manage the risk (e.g., where possible, force reentry over remote areas).

Space weather is important for forecasting satellite locations and reentries. It can have a profound effect on the density of the thermosphere, which is the region of the neutral atmosphere between 90 and 800 km altitudes. Large changes can occur during geomagnetic storms and substorms, when the aurora causes strong heating in the lower thermosphere over polar regions. This drives both an upwelling of denser material into the thermosphere, and winds that blow toward the equator (i.e., away from the heating). Thus denser material can be transported across the global thermosphere, leading to increases in satellite drag. Auroral heating can also create large-scale atmosphere gravity waves that then propagate toward the equator—waves that can create measurable changes in the drag of a satellite as it travels through the peaks and troughs of density in those waves (Guo et al., 2015).

Thus space agencies and satellite operators need tools that can provide information and advice on thermospheric densities, so that they can properly include satellite drag in their models of satellite orbits. In quiet conditions, climatological models will probably be adequate for this task. The challenge is for models to be able to reflect changes in density driven by space weather, in particular, the profound effects of geomagnetic storms. These models also need to take into account the less-marked, but still important, changes in thermospheric density driven by changes in EUV irradiance from the Sun. This is the main heat source for the thermosphere in the absence of geomagnetic storms, and will vary significantly with the solar cycle, the morphology of solar active regions, and the intensity of solar flares. Strong EUV will heat the thermosphere, causing it to expand, so the role of EUV needs to be factored into thermospheric models.

4 LOOKING TO THE FUTURE

The history of space weather, as discussed earlier in this chapter, shows how the risk posed by space weather has grown as a side effect of our increasing dependence on advanced technologies. This dependence has grown enormously over the past 50–60 years; for example, as electric power and electronic communications have become fundamental infrastructures of society. If you doubt this, just look back 60 years and consider the role of coal as a local energy source (e.g., for heating and steam power), and of paper mail for long-distance communications.

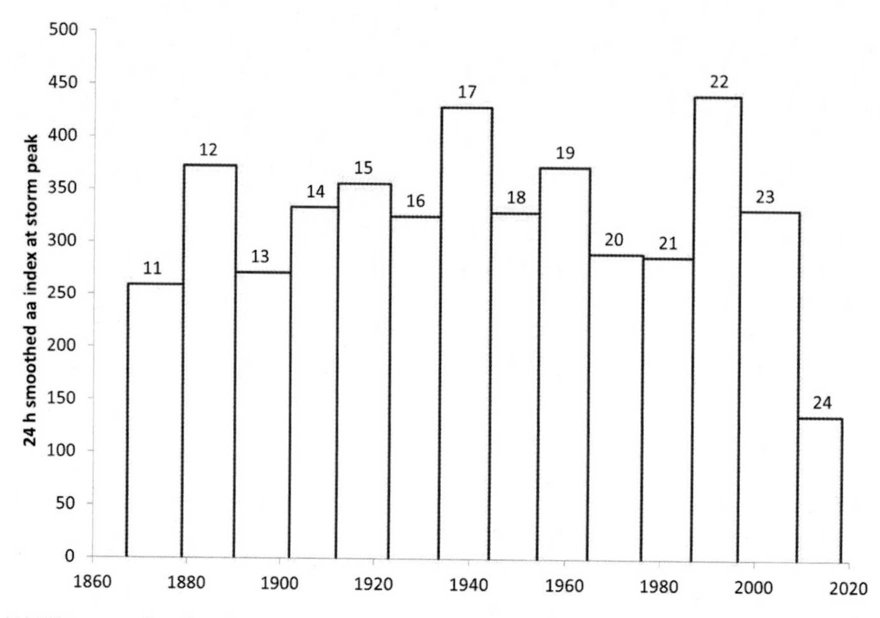

FIG. 7 (A) Histogram showing the maximum geomagnetic storm intensity for each of the 14 solar cycles since 1868, using a 24-h running mean of the aa geomagnetic index as an indicator of storm intensity, as discussed in Appendix A of Hapgood (2011). The solar cycle number is shown above each block in the histogram. You can immediately see that the largest storm of current solar cycle 24 (the St. Patrick's Day storm of 2015) was much weaker than the largest storm in any of the previous cycles since 1868 (indeed it was weaker than even the fifth strongest storm in any of those previous cycles). For completeness of reporting, the dates and intensities of the largest storm in each solar cycle are given in Table 1.

Our dependence on advanced technologies continues to grow today as digital systems, and GNSS and other satellite applications, become ever more deeply embedded in the economy and in society as a whole. It is therefore a cause for concern that space weather conditions near Earth have been so quiet for more than a decade—with the largest event of that decade, the St. Patrick's Day storm of 2015, being a very modest event in the historical record (see Fig. 7). Thus the huge growth of technology in that decade has been only weakly challenged by space weather. In particular, the growth of GNSS has not yet been challenged by a disruptive space weather event in the way that electric power was challenged by the geomagnetic storm on March 13–14, 1989. This is, in part, a matter of luck: on July 23, 2012, the Sun produced a CME that could have generated a highly disruptive event, but fortunately the CME travelled away from the Earth. It did, though, pass over NASA's STEREO A satellite, giving the scientific community the opportunity to study such a dangerous event without Earth being at risk (Russell, 2013). In scientific terms, it was a wake-up call, strong evidence for the risk from space weather (Baker et al., 2013); but much work remains to convert that evidence into concrete action.

Where are the areas to watch for future growth of space weather impacts? GNSS is a particular concern for the future because there appear to be, as yet, no published studies

TABLE 1 The Date and Intensity of the Strongest Geomagnetic Storm in Each of the Solar Cycles 11–24 as Determined From the 24-h Running Mean of the aa Index

Solar Cycle	Maximum Storm Date	24-h aa Index at Maximum
11	February 4, 1872	259
12	November 17, 1882	372
13	February 13, 1892	271
14	September 25, 1909	333
15	May 14, 1921	356
16	July 7, 1928	325
17	September 18, 1941	429
18	March 28, 1946	329
19	November 12, 1960	372
20	August 4, 1972	290
21	February 8, 1986	287
22	March 13, 1989	441
23	October 29, 2003	332
24	March 17, 2015	137

on the economic impact that would follow an event that severely disrupted this technology, irrespective of the cause of the disruption. (The author was involved in 2016 in a socioeconomic study of space weather that searched for such studies.) Severe disruption of GNSS is a serious concern, whether from space weather or from human action such as jamming (Curry, 2014). However, it will be difficult to make progress in preparing for such disruption until we better understand the potential economic impact. That understanding will inevitably guide the level of investment in how GNSS disruption is mitigated and how mitigation is built into future use of GNSS, for example, better GNSS receivers, backup technologies, and warnings of potential disruption. Thus the scope to apply science to develop better space weather tools for GNSS mitigation will depend on a better understanding of economic impact, for example, through socioeconomic studies that are now being encouraged.

Another important area for the future is space weather impact on rail systems. This has been the subject of a number of studies (Atkins, 2014; Krausmann et al., 2015) that have explored how space weather can disrupt rail systems. Most obvious is that GIC can disrupt key electrical systems used for motive power, track circuits, and signals, and even just the general power supply at train stations. The latter is primarily a safety concern, particularly for night-time operations. The former are complex because the motive power and track circuits of modern electric railways (not least modern high-speed rail systems) have shared Earth connections, and hence a common route for GICs to intrude into these systems. We have limited understanding of how vulnerable these systems are to disruption by GIC. There is significant circumstantial evidence that disruptions have occurred during severe geomagnetic storms, as discussed herein. But there is considerable scope for future studies, including: (a) analysis of anomaly reports from rail systems (if these can be made available for study); (b) theoretical analyses of how rail electrical systems may respond to geoelectric fields induced by space weather; and (c) more measurements of GIC in rail systems. Another potential source of disruption from space weather is the push for more use of GNSS applications on rail systems (European Global Navigation Satellite System Agency, 2016). Examples of such applications include reporting of train location, and ensuring that passenger train doors open only when the train is at a station. These developments are in their infancy, so there is a need to ensure that space weather is included in risk scenarios and that appropriate measures are taken. A simple example is that train managers must be able to override GNSS advice when they it see is unhelpful, for example, GNSS claiming that a train is not at a station, perhaps because of adverse space weather, when the manager can see it is at the station. A more serious example is that train-based GNSS applications must be able to cope with interruptions of the GNSS signals, whether from space weather, going through tunnels, or even signal blocking by dense trackside vegetation. Thus advances in rail technology (not least those supporting high-speed rail) may create new vulnerabilities to space weather, and hence a need for engagement with the space weather community. In particular, there is likely to be a need for tools to support the inclusion of space weather in the risk scenarios that drive the design of safe and reliable rail systems.

Our third example of a technology trend where space weather matters is the development of driverless cars. This trend has gained much momentum in recent years, though a number of accidents have highlighted the challenge of making this technology safe (Stilgoe, 2017). Driverless cars face a number of challenges from space weather, as shown in Fig. 8. These cars incorporate at least two advanced technologies that are vulnerable to space weather:

- Use of GNSS for route determination. This is vulnerable to position errors and loss of signal due to space weather. Thus the driverless cars will need high-quality GNSS receivers to minimize these effects, and control systems that can mitigate these effects when they occur. In particular, the control systems need to be able to respond safely to loss of signal—ideally continuing the journey using alternative navigational data, at least until reaching a safe and convenient place to stop. What it absolutely must not do is stop providing route control.
- Extensive use of digital systems. Digital systems in cars, both human-driven and driverless, are vulnerable to SEE from atmospheric radiation. The adverse impact of radiation on these systems was noted more than a decade ago as a critical issue that must be considered by the designers of electronic control systems for cars (Wood and Caustin, 2006). SEEs in car control systems can occur at any time, but are likely to be more common during a major ground level enhancement, when severe space weather strongly enhances the atmospheric radiation environment for a few hours. The control systems must therefore be able to cope with the frequent SEEs that could arise during a severe space weather event. This is a concern because there is evidence from court cases that parts of the automotive industry have been slow to understand the importance of safety critical software (Safety Research & Strategies, 2013).

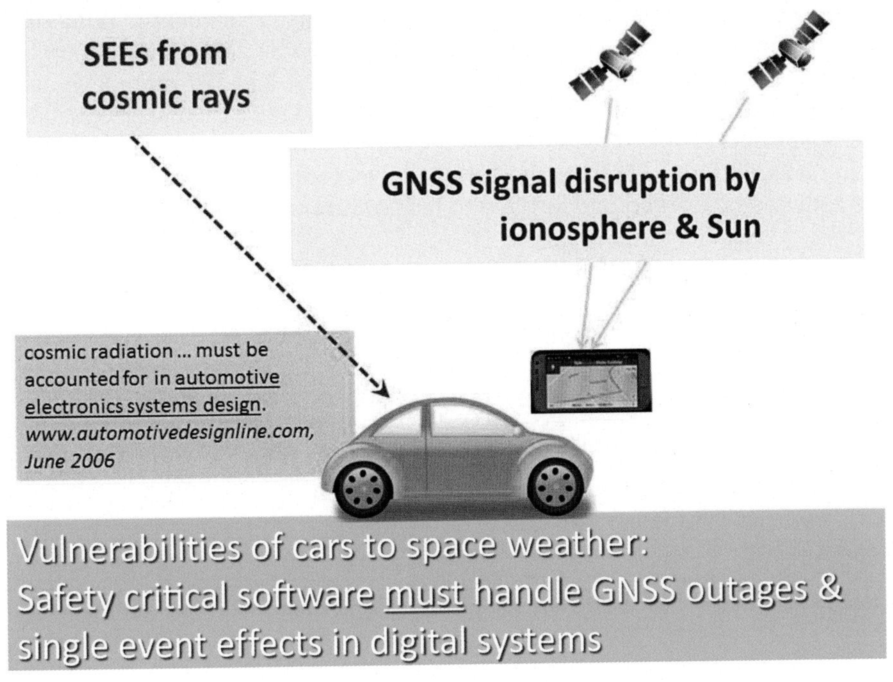

FIG. 8 Summary of space weather impacts on modern automotive technologies. Modern cars make extensive use of electronic systems that are vulnerable to SEE from cosmic rays, a design constraint that has been stressed for more than a decade. Such effects will likely be exacerbated when a severe space weather event leads to a large ground-level enhancement, a many-fold increase in the natural radiation levels on Earth's surface. GNSS is now also widely used for road navigation and is vulnerable to space weather disruption. Both technologies are important in the development of driverless cars, and so it is important that space weather be included in the risk scenarios for this important new technology.

Thus there is a need for tools that can raise awareness of the risks that space weather poses to driverless car technology, and to support the inclusion of space weather in the risk scenarios for this technology. The mitigation of those risks will likely focus on engineering solutions, given the need to manage the risks in the context of automated systems.

5 SUMMARY AND CONCLUSIONS

Space weather is now widely recognized as a significant natural hazard, one that poses a risk to everyday life via its impact on technologies critical to the smooth functioning of modern societies and their economies. This chapter has shown how space weather arises on the Sun, but leads to a wide range of technological impacts on Earth. The diversity of space weather phenomena and impacts is an important challenge to all the different disciplines working to mitigate the level of risk that it poses. We need good tools to raise awareness that this risk is not something from science fiction (the vital first step), and to produce information needed by the operators, engineers, and policy makers who must deal with the societal and economic consequences of adverse space weather. This information includes the assessment of reasonable worst case conditions, say, at the 1-in-100 years level used in other risk management scenarios (important: not the absolute worst case that scientists can invent). It also includes forecasts and nowcasts needed to support operational actions to manage space weather impacts, and tools for after-the-event analysis of problems caused by space weather, so that lessons can be learned and applied to reduce future risks.

ACKNOWLEDGMENTS

The discussion presented in this paper relies on the data collected at the Kanoya and Fort Churchill observatories. The author thanks Japan Meteorological Agency and Geological Survey of Canada for supporting their operation and INTERMAGNET for promoting high standards of magnetic observatory practice (www.intermagnet.org).

Solar wind data courtesy of the Omniweb service (https://omniweb.gsfc.nasa.gov/) hosted by NASA's Space Physics Data Facility. Sudden impulse timing data courtesy of the International Service on Rapid Magnetic Variations hosted by Observatori de l'Ebre (http://www.obsebre.es/en/rapid).

The original dataset of geomagnetic storm dates and intensities derived from the aa index was obtained from the US National Centers for Environmental Information (https://www.ngdc.noaa.gov/stp/geomag/aastar.html) and updated to 2016 using aa data from the UK Solar System Data Centre (https://www.ukssdc.ac.uk/).

References

Atkins, 2014. Rail Resilience to Space Weather, Final Phase 1 Report, Report commissioned by the UK Department for Transport. Available from: https://www.sparkrail.org/Lists/Records/DispForm.aspx?ID=21810.

Avanti, 2017. SBAS-AFRICA. Available from: http://sbas-africa.avantiplc.com/.

Baker, D.N., Li, X., Pulkkinen, A., Ngwira, C.M., Mays, M.L., Galvin, A.B., Simunac, K.D.C., 2013. A major solar eruptive event in July 2012: defining extreme space weather scenarios. Space Weather 11, 585–591.

Barlow, W.H., 1849. On the spontaneous electrical currents observed in wires of the electric telegraph. Phil. Trans. R. Soc. Lond. 139, 61–72.

Bolduc, L., 2002. GIC observations and studies in the Hydro-Québec power system. J. Atmos. Sol. Terr. Phys. 64, 1793–1802.

Boteler, D.H., 2001. Space weather effects on power systems. In: Song, P., Singer, H.J., Siscoe, G.L. (Eds.), Space Weather. American Geophysical Union, Washington, D.C.

Cannon, P., Angling, M., Barclay, L., Curry, C., Dyer, C., Edwards, R., Greene, G., Hapgood, M., Horne, R.B., Jackson, D., Mitchell, C.N., 2013. Extreme space weather: impacts on engineered systems and infrastructure. Royal Academy of Engineering. Available from: https://www.raeng.org.uk/publications/reports/space-weather-full-report.

Curry, C., 2014. SENTINEL Project: Report on GNSS Vulnerabilities. Chronos Technology Ltd. Available from: http://www.chronos.co.uk/files/pdfs/gps/SENTINEL_Project_Report.pdf.

Drake, J.F., 2009. A magnetic reconnection mechanism for ion acceleration and abundance enhancements in impulsive flares. Astrophys. J. Lett. 700, L16.

Erinmez, I.A., Kappenman, J.G., Radasky, W.A., 2002. Management of the geomagnetically induced current risks on the national grid company's electric power transmission system. J. Atmos. Sol. Terr. Phys. 64, 743–756.

Eroshenko, E.A., Belov, A.V., Boteler, D., Gaidash, S.P., Lobkov, S.L., Pirjola, R., Trichtchenko, L., 2010. Effects of strong geomagnetic storms on Northern railways in Russia. Adv. Space Res. 46, 1102–1110.

European Global Navigation Satellite System Agency, 2016. GNSS Introduction in the Rail Sector. Available from: https://www.gsa.europa.eu/gnss-introduction-rail-sector.

Gaunt, C.T., Coetzee, G., 2007. Transformer failures in regions incorrectly considered to have low GIC-risk. In: Proceedings of the IEEE Powertech Conference, July 2007, Lausanne, Switzerland.

Gauthier, L., 2001. Egnos: the first step in Europe's contribution to the global navigation satellite system. ESA Bull. 105, 35–42. Available from: http://esamultimedia.esa.int/multimedia/publications/ESA-Bulletin-105/.
Gummow, R.A., Eng, P., 2002. GIC effects on pipeline corrosion and corrosion control systems. J. Atmos. Sol. Terr. Phys. 64, 1755–1764.
Guo, J., Forbes, J.M., Wei, F., Feng, X., Liu, H., Wan, W., Yang, Z., Liu, C., Emery, B.A., Deng, Y., 2015. Observations of a large-scale gravity wave propagating over an extremely large horizontal distance in the thermosphere. Geophys. Res. Lett. 42, 6560–6565.
Hambling, D., 2014. Burnout. New Sci. 223, 42–45.
Hapgood, M., 2010. Ionospheric correction of space radar data. Acta Geotech. 58, 453–467.
Hapgood, M.A., 2011. Towards a scientific understanding of the risk from extreme space weather. Adv. Space Res. 47, 2059–2072.
Hargreaves, J.K., 1992. The Solar-Terrestrial Environment: An Introduction to Geospace—The Science of the Terrestrial Upper Atmosphere, Ionosphere, and Magnetosphere. Cambridge University Press, Cambridge.
ICAO, 2012. Concept of operations for the provision of space weather information in support of International Air Navigation. International Civil Aviation Organization. Available from: http://www.icao.int/safety/meteorology/iavwopsg/Space%20Weather/Concept%20of%20Operations.V2.2%20(IAVWOPSG.7.WP.019).pdf (Accessed 28 April 2017).
Johnson, G.W., Swaszek, P.F., Hartnett, R.J., Shalaev, R., Wiggins, M., 2007. An evaluation of eLoran as a backup to GPS. In: 2007 IEEE Conference on Technologies for Homeland Security, Woburn, MA, pp. 95–100.
Kessler, D.J., Cour-Palais, B.G., 1978. Collision frequency of artificial satellites: the creation of a debris belt. J. Geophys. Res. 83 (A6), 2637–2646.
Kessler, D.J., et al., 2010. The Kessler syndrome: implications to future space operations. Adv. Astronaut. Sci. 137, 47–61.
Krausmann, E., Andersson, E., Russel, T., William, M., 2015. Space weather and rail: findings and outlook. EU Joint Research Centre Report 98155.
Liu, L., Ge, X., Zong, W., Zhou, Y., Liu, M., 2016. Analysis of the monitoring data of geomagnetic storm interference in the electrification system of a high-speed railway. Space Weather 14, 754–763.
Loh, R., Wullschleger, V., Elrod, B., Lage, M., Haas, F., 1995. The U.S. Wide-Area Augmentation System (WAAS). Navigation 42, 435–465.
Maritime Executive, 2016. Europe gives up on eLoran. Available from: http://maritime-executive.com/article/europe-gives-up-on-eloran.
McNish, A.G., 1940. The magnetic storm of March 24, 1940. Terr. Magn. Atmos. Electr. 45 (3), 359–364.
Medford, L.V., Lanzerotti, L.J., Kraus, J.S., Maclennan, C.G., 1989. Transatlantic Earth potential variations during the March 1989 magnetic storms. Geophys. Res. Lett. 16 (10), 1145–1148.
Michal, T., Eglizeaud, J.P., Bouchard, J., 2005. GRAVES: the new French system for space surveillance. In: Danesy, D. (Ed.), Proc. 4th European Conference on Space Debris (ESA SP-587), April 18–20, 2005, ESA/ESOC, Darmstadt, Germany, p. 61.
News Fishing, 1983. DECCA: more gear lost. Article published in 16 December 1983 issue of Fishing News.
OECD, 2011. Future Global Shocks: Geomagnetic Storms. Organisation for Economic Cooperation and Development. IFP/WKP/FGS(2011)4. Available from: http://www.oecd.org/gov/risk/46891645.pdf.
PourEVA, 2003. Le Ministre DEWAEL reconnait la faillibilité du vote électronique grâce à un rayon cosmique complice! Available from: http://www.poureva.be/article.php3?id_article=36.
Preece, W.H., 1894. Earth currents. Nature 49, 554.
Rao, K.N., 2007. GAGAN—the Indian satellite based augmentation system. Indian J. Radio Space Phys. 36, 293–302. Available from: http://nopr.niscair.res.in/handle/123456789/4707.
Reames, D.V., 1999. Particle acceleration at the Sun and in the heliosphere. Space Sci. Rev. 90, 413.
Russell, C.T., 2013. The very unusual interplanetary coronal mass ejection of 2012 July 23: a blast wave mediated by solar energetic particles. Astrophys. J. 770, 38.
Safety Research & Strategies, 2013. Toyota unintended acceleration and the big bowl of "Spaghetti" code. Available from: http://www.safetyresearch.net/blog/articles/toyota-unintended-acceleration-and-big-bowl-%E2%80%9C spaghetti%E2%80%9D-code.
STFC, 2017. ChipIR: instrument for rapid testing of effects of high energy neutrons. Available from: http://www.isis.stfc.ac.uk/instruments/chipir/chipir8471.html.

Stilgoe, J., 2017. Tesla crash report blames human error—this is a missed opportunity. The Guardian. Available from: https://www.theguardian.com/science/political-science/2017/jan/21/tesla-crash-report-blames-human-error-this-is-a-missed-opportunity (Accessed 28 April 2017).

Trinity House, 2016. Satellite navigation ground based augmentations. Available from: https://www.trinityhouse.co.uk/dgps.

Wik, M., Pirjola, R., Lundstedt, H., Viljanen, A., Wintoft, P., Pulkkinen, A., 2009. Space weather events in July 1982 and October 2003 and the effects of geomagnetically induced currents on Swedish technical systems. Ann. Geophys. 27, 1775–1787.

Wood, J., Caustin, E., 2006. Timely testing avoids cosmic ray damage to critical auto electronics. Available from: http://www.eetimes.com/document.asp?doc_id=1272752.

Wrubel, J.N., 1992. Monitoring program protects transformers from geomagnetic effects. IEEE Comput. Appl. Power 5, 10–14.

CHAPTER

2

Data Availability and Forecast Products for Space Weather

Ramkumar Bala, Patricia Reiff

Rice University, Houston, TX, United States

CHAPTER OUTLINE

1 INTRODUCTION

Space data is a large compendium of multimission and multidisciplinary data covering the vast span of the heliosphere. The region of the near-Earth space, the Sun, and the data that are of interest to a space weather forecaster is a subset of that. In the United States, space weather forecasting is done under the auspices of NOAA (National Oceanic and Atmospheric Administration). The Space Weather Prediction Center (SWPC) (http://www.swpc.noaa.gov) is the main center in the United States and the official government agency that monitors and forecasts solar storms and many other associated events. The SWPC is part of the National Weather Service dedicated to providing space weather products and services. They are the authorized agency to provide "official" actionable alerts and forecasts. They work closely

Machine Learning Techniques for Space Weather
https://doi.org/10.1016/B978-0-12-811788-0.00002-0

with other agencies, including the National Aeronautics and Space Administration (NASA), FAA, US Geological Survey (USGS), DOD, and the Department of Energy. The SWPC also has international partners and works with other agencies outside the United States (e.g., Australia Bureau of Meteorology [BoM], Korean Meteorological Administration [KMA], UK Meteorology Office [UKMO]) to share information and knowledge. The emerging market for the commercial aerospace sector is also seeing a rise in the need for space weather products and activities.

A major milestone in today's space weather forecasting is the technological capability to have data acquisition systems at strategic points in near-Earth surroundings that provide data in real time for an instant evaluation of the geospace environment. For example, the Synchronous Meteorological Satellites (SMS) and the Geostationary Operational Environmental Satellites (GOES-1, GOES-2, etc.) that superseded SMS, all carry on board the Space Environment Monitor (SEM) instrument subsystem. Although the SEM satellites provide magnetometer, soft X-ray, and energetic particle data; GOES, in addition to the data that SEM provided, also has X-ray imager and extreme ultraviolet (EUV) sensors. Unfortunately, such technological capabilities are only as good as their tolerance to potential radiation hazards in space; technical glitches are also not uncommon, if not widespread, causing operational delays. While accurate long-term end-to-end forecast models of the solar-terrestrial system are being developed by major research facilities such as the Center for Space Environment Modeling (CSEM), Laboratory for Atmospheric and Space Physics (LASP), it becomes more of a necessity than a matter of interest within the forecasting framework to have short-term predictions of specific parameters to satisfy the needs of various end users (e.g., satellite, electric power grid operators, and manned space flight missions).

Forecast data products and services rely on measurements received from ground- and/or space-based instruments, which by themselves are good metrics to represent the state of the system specified. Data, when combined with sophisticated space weather models to take inputs from such measurements, can produce actionable outputs or information that can be further processed. The models include, but certainly are not limited to, GPS predictions to aid with accuracy, communications predictions to help HF communication and radar navigation for aircraft pilots, geomagnetic predictions, and auroral forecasts.

2 DATA AND MODELS BASED ON MACHINE LEARNING APPROACHES

In space research and solar research, as well as in the space weather community, machine learning (ML), and in particular, artificial neural networks (ANNs), have proven effective for temporal prediction problems from the very early days; for example, Koons and Gorney (1991) computed relativistic electron flux at geosynchronous orbit, and Wu and Lundstedt (1997) predicted the Dst index. The fundamental difference between basic research and space weather research is the predictive nature of the latter, while also being applied. ANN methods have also been suitable for modeling and predicting solar cycle activity using high-time resolution data from SOHO/MDI (Lundstedt, 2001). Valach et al. (2009) have used an ANN-based model to predict geomagnetic activity using solar energetic particle flux measurements. Furthermore, it has also been applied for classification problems, for example, Newell et al. (1991) classified geospace physical boundaries in the plasma data. ML-based models and capabilities are seen throughout multiple subdomains across the field, for example, the ANN technique was applied to a series of magnetospheric field models for calculating L^* (Yu et al., 2012), and for prediction of the thermosphere density by forecasting exospheric temperature (Choury et al., 2013), and so on.

In the past, several studies have demonstrated a good correlation between various geomagnetic indices and the interplanetary magnetic field (IMF), and with other parameters of the solar wind, and it is now a fairly well-accepted fact that the magnetosphere responds to variations in the solar wind parameters (e.g., Papitashvili et al., 2000). Several researchers have applied these empirical findings and have taken advantage of advanced nonlinear techniques such as machine learning, to predict the magnetospheric activities by using upstream solar wind inputs from satellites such as the WIND and Advanced Composition Explorer (ACE), which are both located at the L1 point. Launched in 1997, ACE's primary mission objective was to provide continuous measurements of the solar wind, low-energy solar, and interplanetary particles, and cosmic rays, requiring an orbit outside the Earth's magnetosphere. The modified halo orbit about the Sun-Earth system's libration point, L1, meets this requirement (Stone et al., 1998); assuming nominal solar wind speed of $400\,\mathrm{km\,s^{-1}}$, any advance warning systems typically get a lead time of approximately 45 min before the leading edge of the solar wind hits the magnetopause. Most importantly, ANN-based models have been successful in making predictive estimates of Kp, Dst, and AE ahead of time using real-time solar wind data inputs (e.g., Bala et al., 2009; Boberg et al., 2000; Costello, 1997; Wing et al., 2005; Lazzus et al., 2017; Wu and Lundstedt, 1997).

Other ML-based models include prediction for the geosynchronous electron flux in a wide energy range (40 keV to >2 MeV) and at a high time resolution, by used flux measurements from GOES-13 and 15 (Shin et al., 2016). Another recent model applied ANNs to identify particle pitch angle distribution using the Van Allen Probes data (Souza et al., 2016). Very recently, Chandorkar et al. (2017) have used an autoregressive Gaussian process approach to come up with probabilistic forecasting of the Dst index. As the field of machine learning and artificial intelligence is making long strides, so is the area of applied research and the advanced prediction models that are produced as a result of the developments.

3 SPACE WEATHER AGENCIES

Forecasters rely on a variety of forecasting tools, ranging from simple nonlinear regression techniques to empirical, semiempirical, and physics-based models, which are primarily physical approximations of the system at the electrodynamic, MHD, or kinetic level. They come from various government, academic, or private sectors. Given the copious data available for forecasting and nowcasting and the advancements made in data mining and machine learning research, subsequent scientific analysis and forecast models have improved tremendously over the past two decades. The following sections describe the role of various government, nongovernmental, academic, and commercial entities delivering forecast data and products from around the world. Although the list may not be comprehensive, it does convey the importance of space weather agencies and the need for their products globally.

3.1 Government Agencies

3.1.1 NOAA's Data and Products

As one of the official agencies in the United States, NOAA's prime focus and its forecast products are targeted toward US commercial customers. SWPC's data products include forecasts, reports, observations, models, summaries, and alerts and warnings. It provides access to its data service through (http://www.swpc.noaa.gov/content/data-access). One of their services includes subscription service via emails (https://www.swpc.noaa.gov/content/subscription-services). This includes X-ray flare alerts; radio burst summaries; geomagnetic warnings, alerts, and watches; electron flux alerts; proton flux warnings; and other forecasts and summaries. NOAA has also introduced color-coded space weather scales, going from minor to extreme, as an easy way to communicate and alert the general public of an impending threat to the people and infrastructure that could be affected. The scales provide warnings for geomagnetic storms, solar radiation storms, and radio blackouts. SWPC's subscription services have grown over the past two solar cycles. A service that started in 2005 has a steady subscriber growth that has reached more than 52,000 customers to date. Table 1 provides a list of data products available in real time from SWPC that is specific to the needs of various domains within the space weather community.

SWPC uses a wide suite of both space- and ground-based instruments to provide useful and actionable forecast products. These instruments come from various internal and external sources, and many of their datasets are available from SWPC in near-real-time. Some operational datasets include ground magnetometer at Boulder, GOES electron flux, magnetometer, proton flux, solar X-ray imager, X-ray flux, LASCO coronagraph from the SOHO instrument,

TABLE 1 Some Domain-Specific Space Weather Forecast Products From NOAA SWPC

Domain	Data and Products
Solar flares and X-rays	GOES X-ray imager, X-ray flux; solar region summary; National Solar Obsv. GONG and SOLIS
Solar CME and solar proton events	LASCO coronagraph; GOES proton flux
Solar wind particles and fields	DSCOVR real-time solar wind
Magnetospheric particle and fields	Geomagnetic activity Ap, Kp, Dst; USAF magnetometers
Upper atmospheric and ionospheric disturbances	Storm time empirical; ionospheric correction; D region absorption predictions; total election content; ionosonde stations

the planetary K-index, real-time solar wind from DSCOVR (preceded by ACE), Synoptic maps featuring solar active regions, coronal holes, prominences, and so on.

In areas where forecast models are unavailable, there are "nowcasted" data and products to fill up the immediate need. As an example, Kp proxies have become common, and a particularly useful one is the estimated 3 h Kp provided by SWPC. The term "nowcasting" generally refers to specifying the state of a certain physical parameter in near-real time by means of a ground- or space-based instrument. Nowcasted Kp is available through http://www.swpc.noaa.gov/rt_plots/kp_3d.html where the estimated Kp index is derived at the US Air Force 55th Space Weather Squadron using data obtained through telelinks from ground-based magnetometers across the United States and Canada (Meanook, Canada; Sitka, Alaska; Glenlea, Canada; Saint Johns, Canada; Ottawa, Canada; Newport, Washington; Fredericksburg, Virginia; Boulder, Colorado; and Fresno, California) along with one European station (Hartland, UK). SWPC makes them available through the cooperation of its partners in the Geological Survey of Canada and the USGS.

SWPC forecasters use a whole range of models to understand the current state (nowcast) of the environment and also to get an understanding of the future state (forecast). These models have been developed either at SWPC or at an academic or private enterprise. The NOAA Space Weather Scales were introduced as a way to communicate to the general public the current and future space weather conditions and their possible effects on people and systems (http://www.swpc.noaa.gov/NOAAscales/). They describe the space environment for three different event types: geomagnetic storms, solar radiation storms, and radio blackouts. The duration of the event influences the severity of the storms, and the space weather advisories are issued using scales ranging from G1 (Minor; Kp = 5) to G5 (Extreme; Kp = 9). Therefore, forecasting geomagnetic indices, given the upstream solar wind conditions, has crucial importance from a space weather standpoint.

A notable example of a research model that has transitioned into the operational model at the SWPC is the Wing models (Wing et al., 2005) of Kp. They developed three real-time Kp models based on ANNs called the APL models: (1) a model that takes solar wind inputs and nowcast Kp and predicts Kp 1 h ahead; (2) a model with the same inputs as model 1 and predicts Kp 4 h ahead; and (3) a model that predicts Kp 1 h ahead using inputs from solar wind only. Fundamentally, their models include IMF |B|, B_z, V_x, and the dynamic pressure term, n in their inputs. These are presently running at SWPC in Boulder, Colorado since March 2011 (http://www.swpc.noaa.gov/wingkp), which superseded the Costello ANN Kp models. The magnetometer and the solar wind data are now available from the DSCOVR Space Weather Portal (https://www.ngdc.noaa.gov/dscovr/portal/index.html#/), which superseded the ACE real-time data since 2016. DSCOVR is currently NOAA's primary space weather monitor, and its data is available through its National Centers for Environmental Information.

Among long-term models is the WSA-Enlil model. It is the integration of the WSA (Wang-Sheeley-Arge) model that predicts the ambient solar wind parameters and the IMF and the Enlil code model, a time-dependent 3D model of the heliosphere. WSA-Enlil is a large-scale magnetohydrodynamic model of the heliosphere. It was adopted by SWPC as real-time forecast model in October 2012 (Pizzo et al., 2011) to provide long-term warning of the solar wind and the coronal mass ejections (CMEs) that could potentially be Earth-directed. The model relies on various data sources that are automated and redundant. The data includes synoptic magnetic maps for the ambient flow (via the US National Science Foundation supported Global Oscillations Network Group (GONG) and Synoptic Optical Long-term Investigations of the Sun (SOLIS) instruments) and white-light CME images (via the Large Angle and Spectrometric Coronagraph [LASCO] and Solar Terrestrial Relations Observatory [STEREO] spacecraft, which are supported by NASA and the European Space Agency [ESA]). This model (http://www.swpc.noaa.gov/products/wsa-enlil-solar-wind-prediction) has the ability to provide warning up to 4 days in advance.

Additional research models that are now part of the day-to-day activity at NOAA are the Aurora—30 min forecast (http://www.swpc.noaa.gov/products/aurora-30-minute-forecast) model (Newell et al., 2009), transitioned to forecast operations in 2014, showing the intensity and location of the aurora, and the probability of the forecast is based on prevailing solar wind conditions measured at L1. Other specification models include the North American Total Electron Content (http://www.swpc.noaa.gov/products/north-american-total-electron-content-us-region), a forecast product for GPS applications and satellite communications, D Region Absorption Predictions (D-RAP) (http://www.swpc.noaa.gov/products/d-region-absorption-predictions-d-rap) model, and the STORM Time Empirical Ionospheric Correction (http://www.swpc.noaa.gov/products/storm-time-empirical-ionospheric-correction) addresses the operational impact of the solar X-ray flux and SEP events on HF radio (3–30 MHz) communication, Relativistic Electron Forecast Model that predicts the >2 MeV 24-h electron fluence at geo-synchronous orbit and is based on a linear prediction filter (Baker, 1990) by using average solar wind speed as its input.

3.1.2 NASA

The NASA, since its inception in 1958, has been a leading force in scientific research and in space exploration, as well as science and technology in general. One of the areas of interest for NASA's science mission directorate (https://science.nasa.gov/heliophysics/) is Heliophysics, which encompasses the Sun, space weather, and the magnetosphere, among things that are of interest to a space weather forecaster. It functions as the nations's research arm for the space weather effort by closely engaging with NOAA, the USGS, and the US ARFL. In addition, it supports space weather models that are used by the SWPC by offering validation and improvements.

NASA's heliophysics missions have contributed tremendously to better understanding of the physical processes that are driving the space environment around Earth and throughout the solar system. In particular, missions such as the ACE and NOAA's Deep Space Climate Observatory (DSCOVR), which observe the solar wind, the Solar Dynamics Observatory, the Solar and Terrestrial Relations Observatory, the joint ESA/NASA Solar, and Heliospheric Observatory, which can all observe solar eruptions on the Sun; and the Van Allen Probes, which observes the radiation belts around Earth, helped improve our understanding of space weather. Although the missions such as ACE and DSCOVR were primarily aimed at providing useful forecasts by monitoring the upstream solar wind conditions from the L1 point, they have also helped with science in various other ways.

Research on improvements in space weather forecasting is performed at places such as the Goddard Space Flight Center (GSFC). Particularly, the Heliophysics Science Division at the GSFC conducts research of the Sun, the extended solar environment, and its interaction with the planetary objects, including the geospace environment. The Space Physics Data

Facility project (https://spdf.gsfc.nasa.gov) under the division serves as a large repository of high-resolution data products and provides a web-based service to allow researchers across all disciplines to access the facility. On the other hand, the Space Weather Research Center (http://swrc.gsfc.nasa.gov) under the GSFC provides space weather data and products to the larger community, and their main charter is to tend to the needs of NASA's robotic missions through experimental research forecasts and notifications. Also, the Community Coordinated Modeling Center (CCMC) situated at the GSFC is a multiagency partnership and research center (https://ccmc.gsfc.nasa.gov/). The CCMC provides access to modern space science simulations. It also facilitates the transition to space weather operations of modern space research models (R2O).

The Coordinated Data Analysis Web (CDAWeb) (https://cdaweb.sci.gsfc.nasa.gov/index.html/) portal is one of the largest repositories containing of all kinds of data coming from numerous space satellites and probes in addition to ground-based magnetometers, imagers, HF-radars, and so forth. One of the biggest benefits of accessing data from the CDAWeb is its strict guidelines and policies set for the dataset where, by definition, the data and descriptions form a logically complete and self-sufficient whole. For example, data from instruments onboard WIND and ACE are time-propagated to the magnetopause so the user is not required to make estimates of the time corrections or uncertainties associated with it. The data is easily accessible by the science community or any general public nonscience individual with its searchable database and the data and display retrieval interface. Fig. 1 shows the main components of the user interface: the home page, data selector, and data explorer.

3.1.3 *European Space Agency*

The ESA is Europe's premier space agency, whose primary mission is to shape the development of Europe's space capability. It is an international agency with 22 members. ESA's programs are designed to study the Earth, the near-Earth environment, the heliosphere, and the universe. It is also designed to develop satellite-based technologies and services in addition to supporting the European industries widely. Authorized in 2008, one of the programs of the

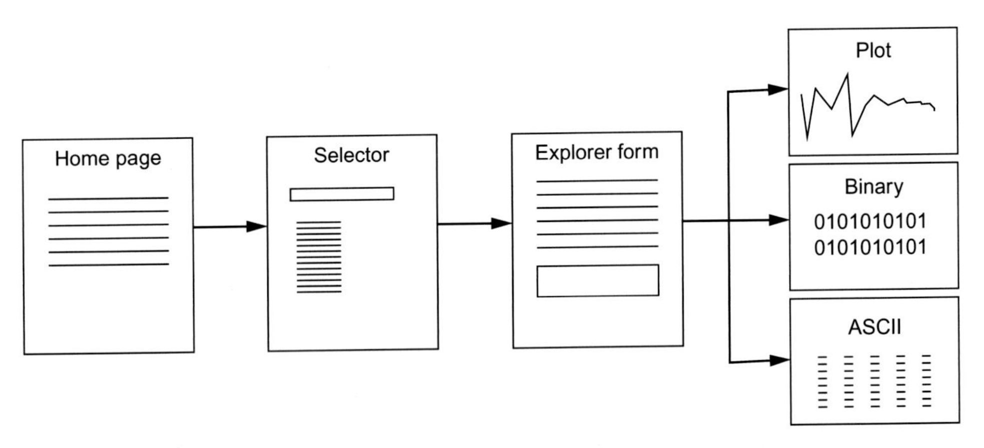

FIG. 1 CDAWeb system layout. *(Courtesy: https://cdaweb.sci.gsfc.nasa.gov/help.html.)*

TABLE 2 Some Domain-Specific Space Weather Forecast Products From ESA

Domain	Data and Products
Solar Weather	CME alerts; SW near-Earth forecasts; solar wind high-speed streams
Space Radiation	Ground-level enhancement (GLE) alert; ground-based neutron monitor data
Ionospheric Weather	TEC monitoring (IONMON); ionospheric scintillation monitoring; PROBA-V/EPT electron and proton flux data
Geomagnetic Conditions	Ground Magnetogrammes, K-index; provisional AA index; auroral alerts; modeled GIC; nowcast Kp
Heliospheric Weather	Solar radio burst detections; SDO/AIA Euv images

ESA is the Space Situational Awareness (SSA) that is within the scope and interests of this chapter. The objective of the SSA program is to provide timely and accurate information and data regarding the space environment, and particularly regarding hazards to infrastructure in orbit and on the ground.

The Space Weather (SWE) segment is an activity within SSA that deals with detecting and forecasting space weather events and their effects on the space- and ground-based infrastructure (http://swe.ssa.esa.int). The Space Weather Coordination Center, their main data processing center, is located in Space Pole, Brussels, Belgium and is responsible for coordinating space weather services at the SWE Data centers or at the Expert Service Centers. The applications supported by the SWE data centers are the European Debris Impact Database, Ionosphere Monitoring Facility, Space Environment Data System, Space Environment Information System for Operations, Space Weather Data Browsing and Analysis (SWE Data), Ionospheric Scintillation Monitoring, and the expert service areas include solar weather, space radiation, inospheric weather, geomagnetic conditions, and heliophysics.

Applications supported by the SWE data centers cater to both customer needs and to the general public for their use. Expert group services are those that currently provide nowcast and forecast products, which is done through the SSA SWE portal link provided herein. Table 2 gives a quick summary of the data and products provided through the SSA SWE program. A thorough documentation is also provided at http://swe.ssa.esa.int/DOCS/SSA-DC/SSA-SWE-SSCC-TN-0011_Product_Catalogue_summary_i4r2.pdf; Part 2 of the documentation provides readers with an exhaustive list of the forecast product providers (e.g., Athens Neutron Monitor Station [ANeMoS], NKU Athens, BIRA-IASB Space Weather Services, Helmholtz-centre Potsdam [GFZ], Institute of 4DTechnologies [FHNW], Solar Influences Data analysis Center [SIDC], UK Met Office [MET]).

3.1.4 *The US Air Force Weather Wing*

The US Air Force's 557th Weather Wing (WW) (http://www.557weatherwing.af.mil/About-Us/Space-Weather/) is another authorized agency other than NOAA SWPC that provides official alerts and warnings in addition to providing forecasts for mission planning and situation awareness. Although NOAA's focus is to primarily support the needs of the commercial customers within the United States and is geographically bounded, the AF WW's interests are far more comprehensive and global. Redesignated in 2015 from the AF Weather Agency, it is the only weather wing in the US Air Force and is comprised of more than 1450 active duty, reserve, civilian, and contract personnel. Their products are used by national agencies, Department of Defense Operators, warfighters, and decision makers.

3.2 Academic Institutions

The level of information and the plethora of spacecraft data available for the forecasters are immense, and so are the large challenges faced to predict space weather events. Arguably, at least in the United States, one of the challenges faced by major federal organizations such as NOAA's SWPC or NASA is that although they make the relevant observations and conduct some of the research, they alone cannot assimilate all the available data and deliver them in a format that is easily accessible for space weather forecasts. Therefore, there are great benefits to having active academic and community participation, from which some of the best ideas are generated. Thus the academic community has a broader impact because of its ability to bridge the gap by providing comprehensive physics-based models and deliver forecasts that not only complement, but also challenge a few models currently being run at the NOAA's Space Weather Prediction Center.

3.2.1 *Kyoto University, Japan*

Kyoto University's Graduate School of Science operates the Data Analysis Center for Geomagnetism and Space Magnetism (DACGSM). Established in 1957, in addition to being a facility of the university, it is also a part of an international organization, the World Data Center. DACGSM has a web archival of some of the historical geomagnetic data that also makes the AE index, Dst index, and ASY index available in real-time "quick-look" (unsuitable for scientific analysis) format through their worldwide web portal (http://wdc.kugi.kyotou.ac.jp/aedir/index.html; http://wdc.kugi.kyotou.ac.jp/dstdir/index.html; http://wdc.kugi.kyoto-u.ac.jp/aeasy/index.html). They also provide real-time observations of geomagnetic field data from observatories and also provide geomagnetic field models.

Another example of a nowcasted data is the AE index that is available in the form of a quick-look or near-real-time index, while the provisional and the final forms are made available at a later time; they are made available at the World Data Center C2 for Geomagnetism, Kyoto, Japan webpage (http://wdc.kugi.kyoto-u.ac.jp/aedir/index.html). The quick-look AE index is made available in quasireal-time, mostly for noncommercial purposes of monitoring and forecasting, that is, it is derived based on the number of stations that are currently reporting at a given point in time, and as more information becomes available for use, they are updated. At present, the maximum number of AE observatories reporting in real time is eight.

3.2.2 *Rice University, USA*

Rice's space physics department is one of the oldest in the nation and has pioneered in empirical space weather research, for example, the US Air Force and NOAA have used Rice models, developed in the 1990s, operationally for more than a decade. This is one of their long-term research foci and it culminated in the form of our own operational space weather webpage being established in October 2003, instituted for the purposes of education and awareness and for making the research models available more easily to the community. Rice's "spacalrt" system alerts subscribers to severe space weather events and has more than 1500 subscribers to its list. This later led to the implementation of some research models in late 2007. Rice has also created a space weather operational framework that is capable of delivering actual "forecasts." The Rice ANN models are empirical space weather prediction models that are capable of giving Kp, Dst, and AE forecasts up to 3 h ahead in near-real time (Bala et al., 2009; Bala and Reiff, 2012); the models are also capable of providing forecasts up to 6 h ahead, but with slightly less certainty. The Rice models are available in real time at http://mms.rice.edu/realtime/forecast.html.

A unique three-way effort between a government agency, academic institution, and a private industry has led to a forecast product (ENLIL-Dst) that enables long-term forecasting of a global geomagnetic index (the Dst) with a lead time of roughly 7 days. This end-to-end, data-to-model pipeline combines data from WSA-Enlil (http://www.swpc.noaa.gov/wsa-enlil/) obtained from the NOAA/SWPC E-SWDS data server that is fetched by the Space Environment Technologies (www.spacewx.com), at a 1–3 min cadence, which then reports the operational data to Rice University, which runs the model at its end. For model results and redundancy, the data and the forecast results are available at both SET (primary: http://sol.spacenvironment.net/~sam_ops/index.html?) and Rice (as backup).

3.2.3 *Laboratory for Atmospheric and Space Physics, USA*

The Laboratory for Atmospheric and Space Physics (lasp.colorado.edu) is an advanced space research lab at the University of Colorado Boulder. Their areas of research include basic science, engineering, mission operations, and scientific data analysis. Their involvement in space missions dates back to the 1960s and continues to play a large part even today. Their missions include planetary research (e.g., MAVEN, Cassini), heliophysics (e.g., Parker Solar Probe, New Horizons), solar (e.g., SDO, SORCE), and space weather. Because LASP has been involved in various missions and scientific analysis, it maintains data products that are publicly available for anyone to use. Some key LASP datasets from missions such as the Magnetospheric Multiscale (MMS) mission, Solar Dynamics Observatory (SDO), the EUV Variability Experiment (EVE) are listed in http://lasp.colorado.edu/home/mission-ops-data/data-systems/data-products/.

3.3 Commercial Providers

Government agencies such as NOAA, NASA, ESA, and the USGS are constantly looking for ways to improve their forecasting capabilities. The ENLIL-Dst model discussed herein was a fine example of a three-way collaboration between commercial service providers, academia, and the government as soon as the NOAA SWPC ESWDS becomes operational

TABLE 3 List of Some Commercial Space Weather Vendors and Their Areas of Risk Mitigation

Name of Vendor	Risk Mitigation Area	Website
Carmel Research Center	Satellite operations; radiation hazards; solar region summary	http://www.carmelresearchcenter.com/
Northwest Research Associates	Radio communications (HF); satellite communications	https://www.nwra.com/
ARINC		http://www.rockwellcollins.com
Solar Terrestrial Dispatch	Power grids	http://www.spacew.com/index.php
Space Environment Technologies	Space operations; aviation radiation; communication outages	http://www.spacew.com/index.php
Atmospheric and Environmental Research	Communication outages; spacecraft operations	http://www.aer.com/science-research/space/space-weather
Astra	GPS navigation, power grids	www.astraspace.net
GeoOptics	GPS, satellite systems	www.geooptics.com

for commercial Space Weather service providers. The American Commercial Space Weather Association (ACSWA) was formed in 2010 by commercial organizations (www.acswa.us) to represent private-sector commercial interests in space weather. There are now more than 15 companies that are part of the association, and provide a whole host of products ranging over a variety of space weather phenomenon. The main objective of the ACSWA is to strengthen the nation's R2O (Research-to-operations) capability. Table 3 shows a list of commercial vendors according to the products and services they offer.

In the past several years we have seen an expansion in commercial space weather ties to other industries. Some notable developments include ASTRA's DICES satellites to measure Space Weather SEDs (GPS), GeoOptics' CICERO satellite builds (RO/Weather and Space Weather Data), PlanetIQ's contracts with India (Weather and Space Weather services), and NextGen's SPRINTS SEP effects forecasts (satellite and ground operations).

3.4 Other Nonprofit, Corporate Research Agencies

3.4.1 USGS

The USGS is a science agency of the federal government for the Department of the Interior. Its data and products include geomagnetism plots, and real-time Dst index plots (https://www.usgs.gov/products/data-and-tools/real-time-data/geomagnetism). Through its cooperative effort with the international community by INTERMAGNET, a consortium dedicated to promoting the operation of modern geomagnetic observatories, the USGS provides real-time magnetometer data to NOAA's Space Weather Prediction Center, the US Air Force 557th Weather Wing, and NASA.

3.4.2 *JHU Applied Physics Lab*

The SuperMAG network (http://supermag.jhuapl.edu/) is a collaborative effort funded by the NASA, the National Science Foundation (NSF), and the ESA to enable scientists and the general public easy access to data of the Earth's magnetic field. It provides ground magnetic field perturbations in the same coordinate system, and identical time resolution, and with a common baseline removal approach collected across more than 300 ground-based magnetometers. They provide a new and unofficial version of the global magnetic indices such as the AU, AL, and SYMH, called the SMU, SML, and SMR indices.

The AMPERE (Active Magnetosphere and Planetary Electrodynamics Response Experiment) is a real-time tracking system from the JHU APL. It is a program funded by the NSF to the APL. The laboratory, working with Boeing, partnered with Iridium Communications to introduce this new capability by using Iridium's commercial satellite constellation. Its real-time capability started in 2010. A wealth of atmospheric data coming from the constellation of satellites operated by Iridim is collected and transmitted by Boeing to the APL, which becomes a real-time magnetic field mapping of the earth. AMPERE magnetic perturbation data and data products derived from the Iridium constellation are provided via the AMPERE Science Data Center to the scientific community for basic research in space weather and magnetosphere-ionosphere physics (ampere.jhuapl.edu).

3.4.3 *US Naval Research Lab*

The US Naval Research Lab is the corporate research laboratory for the Navy. It provides advanced scientific capabilities with a robust research portfolio. Of the various branches of research being conducted at the lab, the Solar and Heliophysics Branch (https://www.nrl.navy.mil/ssd/branches/) covers topics related to solar wind modeling, solar, space-based sensors such as coronagraphs, heliospheric imagers, and solar spectrometers. Their research has resulted in measurements and models in use by forecasting agencies for operational purposes.

3.4.4 *Other International Service Providers*

Space weather forecasting is a difficult and evolving science, and certainly it cannot be done in isolation. It requires international partnership to develop long-term sustainable solutions that will help increase the forecast horizons beyond simple global phenomenon. Fidelity of any regional forecast can be improved through well-coordinated international collaboration. Table 4 lists some major international thrusts that currently provide space weather forecast products.

The National Radio Research Agency (RRA) is the official source to provide space weather products and services in Korea. RRA is a government agency and a state-run research institute founded in 1966. The Korean Space Weather Center (KSWC) comes under the RRA and is the action agency providing space weather alerts and warnings (http://spaceweather.rra.go.kr). It is also a Regional Warning Center of the International Space Environment Service

TABLE 4 International Space Weather Providers

Australia Bureau of Meteorology (BoM)
Brazil National Institute for Space Research (INPE)
British Geological Survey (BGS)
China Meteorological Administration (CMA)
European Commission Joint Research Centre (JRC)
European Space Agency (ESA)
Finnish Meteorological Institute
German Aerospace Center (DLR)
GFZ Helmholtz-Zentrum POTSDAM
Japan National Institute of Information and Communications Technology (NICT)
Korea Meteorological Administration (KMA)
Korea National Radio Research Agency (RRA)
Natural Resources Canada (NRCan)
Royal Observatory of Belgium (SIDC)
Russian Federation Federal Service for Hydrometeorology and Environmental Monitoring (Roshydromet)
South African National Space Agency (SANSA)
United Kingdom Meteorology Office (UKMO)

(ISES). KSWC makes real-time observations using ionosondes, magnetometers, solar radio spectrographers, and TECs from instruments located in three different places across the country. KWSC is also responsible for tracking ACE, DSCOVR, and STEREO, working in conjunction with the US agencies in charge of those space instruments. It also provides alerts and warnings to its domestic customers ranging from aviation, satellite companies, and power to defense agencies.

In the United Kingdom, the Met Office Space Weather Operations Center (MOSWOC), is the official entity for space weather operations. Instituted in 2011, MOSWOC's responsibility is to provide UK operational space weather capabilities to mitigate space weather risks. It works in close collaboration with NOAA to build its knowledge and scope. UK's interest in space weather started recently following the addition of solar storms to the National Risk Register (NRR). The UK's current focus lies in placing a space instrument at L5. The motivation was sparked by the lack of forecast skill after STEREO B goes out of commission, and with more support coming from the government, hence the Carrington L5 mission is scheduled to be launched in 2020.

4 SUMMARY

The predictive nature of the space weather research of a complex system makes it an extremely difficult task. Space weather forecasting has made big progress in the past couple of decades in prediction capabilities and in promulgating increased awareness of its potential impact on space and ground infrastructures. However, there are still gaps and forecast challenges in several different areas, for example, uncertainty in IMF B_z, regional specification of geomagnetic forecasting, flare forecasts and prediction, and solar proton event warning of onset and occurrence. Also, not to mention the science challenges and the fundamental understanding of many issues, for example, predicting a flare, and forecasting a sunspot and its evolution is difficult today. Moreover, space weather risk is persistent and does not necessarily wane during solar minimums; extreme events can happen anytime. Mitigating the risks and improving space weather capability means that "R2O2R" is important for the space weather enterprise. Major thrusts such as the ESA SSA's roadmap are geared toward the Lagrange Mission L1/L5. High priorities have been set for IMF properties and dynamics measurements, plasma analyzers for solar wind, density and temperature, X-ray flux monitoring, an EUV imager for mapping the low corona, magnetograph measurements for mapping the magnetic field of the photosphere, and so forth. Additionally, enhancing particle flux measurements, and radiation monitoring with high-energy ions are also among its top priorities.

As a quick reminder, the high-speed CME, ejected on July 23, 2012 and observed by STEREO A, was unique both in respect to the observed plasma and magnetic structure and the large SEP flux that dynamically regulated the shock front. Because of its great intensity, many hailed it as "Carrington 2" for coming in the ranks of the famous 1859 Carrington storm. The plasma properties, magnetic structure, and the strong SEP flux observed in this CME makes it one of the strongest ever known to the space weather community. From a space weather perspective, had this been Earth-directed, the resulting impact would have been very devastating.

References

Baker, D.M., 1990. Wave-particle interactions for relativistic electrons in a recirculation acceleration model. In: Physics of Space Plasmas (1989) (A91-33008 13-46). Scientific Publishers, Inc., Cambridge, MA, pp. 215–227.

Bala, R., Reiff, P., 2012. Improvements in short-term forecasting of geomagnetic activity. Space Weather 10, S06001.

Bala, R., Reiff, P.H., Landivar, J.E., 2009. Real-time prediction of magnetospheric activity using the Boyle Index. Space Weather 7, S04003.

Boberg, F., Wintoft, P., Lundstedt, H., 2000. Real time Kp prediction from solar wind data using neural networks. Phys. Chem. Earth 25, 275–280.

Chandorkar, M., Camporeale, E., Wing, S., 2017. Probabilistic forecasting of the disturbance storm time index: an autoregressive Gaussian process approach. Space Weather 15, 1004–1019. https://doi.org/10.1002/2017SW001627.

Choury, A., Bruinsma, S., Schaeffer, P., 2013. Neural networks to predict exosphere temperature corrections. Space Weather 11, 592–602. https://doi.org/10.1002/2013SW000969.

Costello, K.A., 1997. Moving the Rice MSFM into a real-time forecast mode using solar wind driven forecast models. Ph.D. dissertation, Rice University, Houston, TZ.

Koons, H.C., Gorney, D.J., 1991. A neural network model of the relativistic electron flux at geosynchronous orbit. J. Geophys. Res. 96 (A1), 5549 5556. https://doi.org/10.1029/90JA02380.

Lazzús, J.A., Vega, P., Rojas, P., Salfate, I., 2017. Forecasting the Dst index using a swarm-optimized neural network. Space Weather 15, 1068–1089. https://doi.org/10.1002/2017SW001608.

Lundstedt, H., 2001. Solar activity predicted with artificial intelligence. Space Weather Geophys. Monogr. 125.
Newell, P.T., Wing, S., Meng, C.-I., Sigillito, V., 1991. The auroral oval position, structure, and intensity of precipitation from 1984 onward: an automated on-line data base. J. Geophys. Res. 96 (A4), 5877–5882. https://doi.org/10.1029/90JA02450.
Newell, P.T., Sotirelis, T., Wing, S., 2009. Diffuse, monoenergetic, and broadband aurora: the global precipitation budget. J. Geophys. Res. 114, A09207. https://doi.org/10.1029/2009JA014326.
Papitashvili, V.O., Papitashvili, N.E., King, J.H., 2000. Solar cycle effects in planetary geomagnetic activity: analysis of 36-year long OMNI dataset. Geophys. Res. Lett. 27, 2797–2800. https://doi.org/10.1029/2000GL000064.
Pizzo, V.J., Biesecker, D.A., Millward, G.H., Odstrcil, D., 2011. Heliospheric remote imaging and its relation to CME input to solar wind propagation models. Am. Geophys. Union, Fall Meeting. Abstract id: SH24A-04.
Shin, S., Lee, J., Moon, Y., Chu, H., Park, J., 2016. SoPh 291, 897.
Souza, V.M., Vieira, L.E.A., Medeiros, C., Da Silva, L.A., Alves, L.R., Koga, D., Sibeck, D.G., Walsh, B.M., Kanekal, S.G., Jauer, P.R., et al., 2016. A neural network approach for identifying particle pitch angle distributions in Van Allen Probes data. Space Weather 14, 275–284. https://doi.org/10.1002/2015SW001349.
Stone, E., Frandsen, A., Mewaldt, R., et al., 1998. The advanced composition explorer. Space Sci. Rev. 86, 1. https://doi.org/10.1023/A:1005082526237.
Valach, F., Revallo, M., Bochníček, J., Hejda, P., 2009. Solar energetic particle flux enhancement as a predictor of geomagnetic activity in a neural network-based model. Space Weather 7, S04004. https://doi.org/10.1029/2008SW000421.
Wing, S., Johnson, J.R., Jen, J., Meng, C.I., Sibeck, D.G., Bechtold, K., Freeman, J., Costello, K., Balikhin, M., Takahashi, K., 2005. Kp forecast models. J. Geophys. Res. 110, A04203.
Wu, J.-G., Lundstedt, H., 1997. Neural network modeling of solar wind-magnetosphere interaction. J. Geophys. Res. 102 (A7), 14457–14466. https://doi.org/10.1029/97JA01081.
Yu, Y., Koller, J., Zaharia, S., Jordanova, V., 2012. L* neural networks from different magnetic field models and their applicability. Space Weather 10, S02014. https://doi.org/10.1029/2011SW000743.

PART II

MACHINE LEARNING

CHAPTER

3

An Information-Theoretical Approach to Space Weather

Jay R. Johnson[*], *Simon Wing*[†]

[*]Andrews University, Berrien Springs, MI, United States [†]Johns Hopkins University, Laurel, MD, United States

CHAPTER OUTLINE

Machine Learning Techniques for Space Weather
https://doi.org/10.1016/B978-0-12-811788-0.00003-2

1 INTRODUCTION

System science is an emerging interdisciplinary field that studies the nature of systems from an overarching perspective that is concerned with the overall evolution of a system and how it interacts with its environment. Techniques developed in system science are particularly useful for complex systems in which the governing physical processes are not fully understood or in which modeling from first principles is not computationally feasible. The field aims to develop interdisciplinary foundations that are broadly applicable to physics, biology, medicine, and social science.

Systems thinking is somewhat different from the traditional reductionist approach utilized in science. Most scientists are trained to approach complex problems by reducing the system to components and trying to understand the dynamics of each component. With this approach, scientists specialize in having a deep understanding of each component of the system and then collaborating with experts on other components to develop a comprehensive view of the system as a whole. This approach can be seen in the space physics community, where models that were originally designed to understand the solar wind, global magnetosphere, and inner magnetosphere as reduced systems have been coupled together in a more comprehensive model (e.g., SWMF; Tóth et al., 2005). However, this approach is philosophically different than the holistic system approach that considers that the "whole is something other than the parts" (Aristotle, Book III, Ch. IX, pp. 168–169). A systems approach is based on considering the system as a dynamic and complex whole interacting as a structured functional unit.

One of the key objectives of system science is to identify variables that control system dynamics. For example, in space physics, it is of great interest to understand how the spatiotemporal variations of the solar wind control the dynamical response of the magnetosphere during storms and substorms and how the wave environment controls radiation belt energization and loss processes. Similar challenges also face fusion science, in which it is vitally important to understand processes that control disruptions and a variety of instabilities (Boozer, 2012; Zakharov et al., 2012). In medical science it is important to understand which factors causally control recurrence of debilitating diseases, such as rheumatoid arthritis (Salvarani et al., 2004; Doran et al., 2002). In neurological systems it is interesting to learn how the auditory cortex represents and processes information to enable selective attention in complex acoustic environments (Theunissen et al., 2000). Financial markets also undergo dynamical changes in response to socioeconomic factors, and instabilities can be disruptive, so it is particularly useful to understand causal factors that could bring greater stability to the world economy (Johnson et al., 2003). Political systems often exhibit repeatable patterns of instability. For example, the equilibrium properties of the two-party system of the United States involve a balance between representation and restraint (Midlarsky, 1984), which has led to reconfiguration at least six times since its inception (Aldrich, 1999). Understanding the dynamics of political systems could be useful for maintaining peace and stability.

2 COMPLEX SYSTEMS FRAMEWORK

The general objectives of system science encompass three fundamental tasks: characterization, forecasting, and modeling (Gershenfeld, 1998). *Characterization* involves determining what kind of system produced the signal. Characterization involves such tasks as identifying the degrees of freedom, patterns, dependency, randomness, and degree of nonlinearity. *Modeling* involves determining a set of governing equations that describe the evolution of the system. *Forecasting* involves predicting what the system will do next given its current state or history. This task also involves identifying causal variables, which are responsible for state transitions. Probabilistic forecasting often precedes modeling in systems that are highly complex, but determination of a suitable set of governing equations allows for greater predictive capability.

For complex systems, modeling can be physically or computationally difficult. For some systems such as the brain, physical equations describing neural interactions are not well specified. For other systems, such as the magnetosphere, the underlying physical equations may be known at the most fundamental level (particle simulation), but global computations are beyond present and/or future computational capabilities without appropriate approximations. Empirical models that employ intuition assume a priori a dynamical framework that may or may not apply to the system. Moreover, care must be taken because it may be possible to fit the data by choosing enough free parameters at the expense of losing physical understanding or overfitting.

Characterizing, modeling, and forecasting space weather is difficult because the coupled solar wind-magnetosphere-ionosphere system is nonlinear and complex, and it is not clear how much predictive information is actually available for accurate predictions of future states of the system. Within space weather systems there have been significant advances in physics-based modeling of solar, magnetospheric, and ionospheric processes, but the complexity of the coupled systems makes it difficult to pursue a first-principles approach to modeling, resulting in parameterization of coupling processes. The magnetospheric response during storms and substorms is not clearly understood, so it seems appropriate to apply statistical techniques that are unbiased a priori. The nonparametric, statistics-based methods described in this chapter provide a complimentary approach to prediction through the construction of a nonlinear map from causal driver variables to variables of interest to space weather users. The approach is not constrained by the assumption of underlying dynamics—rather, the underlying (physics-based) dynamics are discovered by the approach, and then ultimately utilized to improve predictions. These outcomes should be considered complimentary to other physics-based and empirical modeling efforts within the space weather community.

3 STATE VARIABLES

One of the most fundamental questions about a system that must first be addressed is how to describe it. Typically, a scientist observing a system will eventually identify "state variables" that describe the state of the system. Typically these variables are based on some characteristic of the system that an observer thinks is interesting. The choice of these observables are enabled by technology and informed by experience, while they are limited by spatial and temporal resolution as well as understanding. Some examples of "state variables" that have been used for heliospheric systems are flare waiting times, geomagnetic indices, radiation belt fluxes, ionospheric precipitation, and so forth.

One important question is: what is the smallest possible fundamental set of variables that are required to understand the system? We refer to this fundamental set of variables as the "state vector" of the system. The length of the "state vector" is known as the dimension of the system. Taken's theorem provides the remarkable concept that the dynamics of a system can be completely captured by looking at a single "state variable" by considering a vector that describes how that variable changes in time. In general, if the system can be characterized by a state vector of dimension d, the dynamics of the system can be reconstructed by monitoring the time dependence of a time-lagged state variable y such that $\mathbf{y} = [y(t), y(t-\Delta), \ldots, y(t-n\Delta)]$ where $n \leq 2d + 1$, and often much smaller. This means that for the purposes of characterization, it may be possible to only monitor a single state variable and the external inputs to a system.

4 DEPENDENCY, CORRELATIONS, AND INFORMATION

The key to characterizing a system is to understand the dependencies in a system. These dependencies can be spatial or temporal in nature. Temporal dependence is particularly relevant to determining causality and evaluating physical processes that cause system responses (as characterized by the dynamics of the state vector).

The standard approach from the theory of linear systems for evaluating the dependencies of the output, **y**, on the input, **x**, is to consider the covariance matrix for the variable $\mathbf{z} = (\mathbf{x}, \mathbf{y})$, where the covariance matrix is defined as $\mathbf{C}(\mathbf{z}) = \langle(\mathbf{z} - \langle\mathbf{z}\rangle) \cdot (\mathbf{z} - \langle\mathbf{z}\rangle)^T\rangle$. From the covariance matrix, we can define a measure of the dependency of the output variables on the input variables

$$\lambda(\mathbf{x}, \mathbf{y}) = \sqrt{1 - \frac{\det(\mathbf{C}(\mathbf{z}))}{\det(\mathbf{C}(\mathbf{x}))\det(\mathbf{C}(\mathbf{y}))}} \tag{1}$$

(the linear predictability), which is a generalization of the well-known correlation coefficient for one input and one output variable (Tsonis, 2001).

In a nonlinear system, correlation functions are not very useful because nonlinear systems tend to have broadband power spectra and hence featureless correlation structure. To illustrate this point we can consider a system in which we monitor two variables, which we can call the input and the output. Fig. 1 shows three sets of (input, output) pairs, which could be the result of plotting $(x(t), y(t))$ for all points in a time series. Panels (A) and (B) are constructed to have a high and low dependence, respectively, and yield correlation coefficients 1 and 0 accordingly. Panel (C) shows the probability distribution when x is determined by the Mackey-Glass equation (Mackey et al., 1977). This distribution obviously has some sort of order and information, but the correlation coefficient for this distribution is zero. In nonlinear systems, these types of higher-order, nonlinear correlations cannot be detected by correlation analysis.

The correlation coefficient is an example of a linear statistic that is limited to identifying linear relationships. On the other hand, probability provides a more powerful path to identification of dependence, even in nonlinear systems. A more general measure of dependency between an input and output is obtained by considering

$$P(\mathbf{x}, \mathbf{y}) \stackrel{?}{=} P(\mathbf{x})P(\mathbf{y}) \tag{2}$$

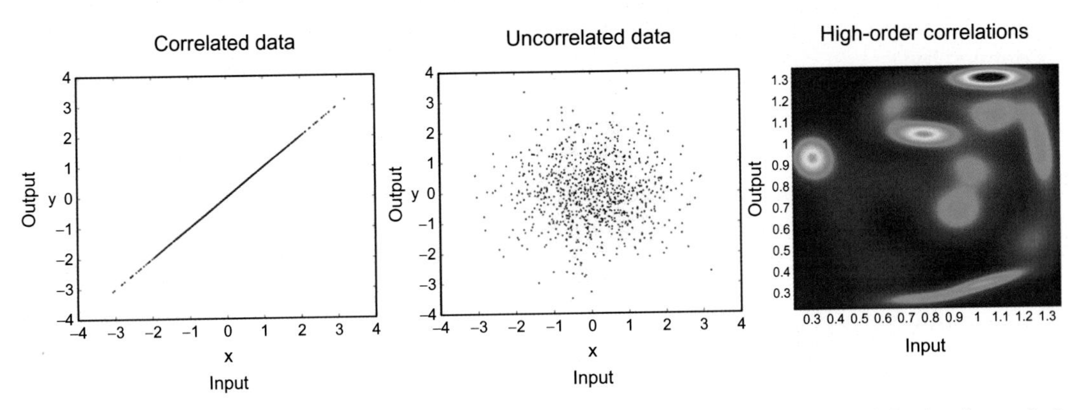

FIG. 1 Input-output pairs for correlated data (*left panel*), uncorrelated data (*center panel*), and the time lag analysis of the Mackey-Glass equation (*right panel*). The correlation coefficient of the correlated data is 1, but it vanished for both the uncorrelated data and the Mackey-Glass system. These results illustrated a weakness of correlation analysis and motivate the benefits of using an information theory approach.

where $P(\mathbf{x}, \mathbf{y})$ is the joint probability of $\mathbf{x}$ and $\mathbf{y}$ while $P(\mathbf{x})$ and $P(\mathbf{y})$ are the probabilities of $\mathbf{x}$ and $\mathbf{y}$, respectively. If the equality in Eq. (2) holds, then the variables $\mathbf{x}$ and $\mathbf{y}$ are independent. For all other cases, there is some measure of dependency. The advantage of considering Eq. (2) is that it is possible to detect the presence of higher order nonlinear dependencies between the input and output, even in the absence of linear dependencies (Gershenfeld, 1998).

4.1 Mutual Information as a Measure of Nonlinear Dependence

The mutual information, $\mathcal{M}(\mathbf{x}, \mathbf{y})$, between the input $\mathbf{x}$ and output $\mathbf{y}$ compares the uncertainty of measuring a particular input and its output together with the uncertainty of measuring the input and output independently. Computation of mutual information involves estimating the probability distribution function using clustering, kernel density estimation, or quantization methods (Tsonis, 2001; Li, 1990; Darbellay and Vajda, 1999).

Mutual information may be computed from measurements of input and output variables by binning the datasets—the number of bins may be different for each of the input and output variables. After quantization, we have the set of input ($\mathbf{x}$) and output ($\mathbf{y}$) variables that take on discrete values, $\hat{\mathbf{x}}$ and $\hat{\mathbf{y}}$, where

$$\hat{\mathbf{x}} \in \aleph_1; \quad \hat{\mathbf{y}} \in \aleph_2 \tag{3}$$

and $\aleph_{1,2}$ are vectors of alphabets of variable length determined by the number of bins for each input or output variable. The entropy associated with the input and output are, respectively,

$$H(\mathbf{x}) = -\sum_{\aleph_1} p(\hat{\mathbf{x}}) \log p(\hat{\mathbf{x}}); \quad H(\mathbf{y}) = -\sum_{\aleph_2} p(\hat{\mathbf{y}}) \log p(\hat{\mathbf{y}}) \tag{4}$$

where $p(\hat{\mathbf{x}})$ is the probability of finding the word $\hat{\mathbf{x}}$ in the set of $\mathbf{x}$-data and $p(\hat{\mathbf{y}})$ is the probability of finding word $\hat{\mathbf{y}}$ in the set of $\mathbf{y}$-data. To examine the relationship between the input and output, we extract the word combinations $(\hat{\mathbf{x}}, \hat{\mathbf{y}})$ from the dataset. The joint entropy is defined by

$$H(\mathbf{x}, \mathbf{y}) = -\sum_{\aleph_1, \aleph_2} p(\hat{\mathbf{x}}, \hat{\mathbf{y}}) \log p(\hat{\mathbf{x}}, \hat{\mathbf{y}}) \tag{5}$$

where $p(\hat{\mathbf{x}}, \hat{\mathbf{y}})$ is the probability of finding the word combination $(\hat{\mathbf{x}}, \hat{\mathbf{y}})$ in the set of $(\mathbf{x}, \mathbf{y})$-data. The mutual information is then defined as

$$\mathcal{M}(\mathbf{x}, \mathbf{y}) = H(\mathbf{x}) + H(\mathbf{y}) - H(\mathbf{x}, \mathbf{y}) \tag{6}$$

For a continuous probability distribution, the mutual information can be generalized to

$$\mathcal{M}(\mathbf{x}, \mathbf{y}) = \int p(\mathbf{x}', \mathbf{y}') \log \frac{p(\mathbf{x}', \mathbf{y}')}{p(\mathbf{x}')p(\mathbf{y}')} d\mathbf{x}' d\mathbf{y}' \tag{7}$$

and if the joint probability distribution is Gaussian, the mutual information collapses to

$$\mathcal{M}(\mathbf{x}, \mathbf{y}) = \frac{1}{2} \log \left(\frac{\det(\mathbf{C}(\mathbf{z}))}{\det(\mathbf{C}(\mathbf{x})) \det(\mathbf{C}(\mathbf{y}))} \right) \tag{8}$$

In this limit, the mutual information can be directly related to the linear correlation function of Eq. (1). The extension of this relationship,

$$\Lambda(\mathbf{x}, \mathbf{y}) = \sqrt{1 - e^{-2\mathcal{M}(\mathbf{x}, \mathbf{y})}} \tag{9}$$

introduces a generalized correlation function that includes linear, as well as nonlinear correlations (Li, 1990; Darbellay and Vajda, 1999). The mutual information may vary from 0 to ∞ so that Λ varies from 0 for independence to 1 for dependence. The difference between λ and Λ indicates the inadequacy of a linear model and the importance of nonlinear relationships. As such, the discriminating statistic, $D_{\mathcal{M}} = \Lambda - \lambda$ is an indicator of the presence of underlying nonlinear dynamics (Tsonis, 2001). The existence of nonlinear dependence in the data is established through comparison of the discriminating statistic with that obtained using realizations of surrogate data (Theiler et al., 1992; Prichard and Price, 1992; Prichard and Theiler, 1994) that share the same linear properties as the original data (Johnson and Wing, 2005). Mutual information has been used to study the dynamics of the solar cycle (Johnson and Wing, 2005) and timescale of the magnetospheric response to the solar wind (March et al., 2005).

4.2 Cumulant-Based Cost as a Measure of Nonlinear Dependence

Alternatively, we can investigate the nonlinear dependency in the system using higher order moments (cumulants) of the joint probability distribution (Deco and Schürmann, 2000; Johnson and Wing, 2005).

$$C_{i,\ldots,j} = \int d\mathbf{z} P(\mathbf{z}) z_i, \ldots, z_j \equiv \langle z_i, \ldots, z_j \rangle \tag{10}$$

where $i, \ldots, j \in 1, \ldots, m$. In particular, the cumulants, $K_{1 i_2 \ldots i_n}$, of the distribution defined by

$$
\begin{aligned}
K_i &= C_i = \langle z_i \rangle \\
K_{ij} &= C_{ij} - C_i C_j = \langle z_i z_j \rangle - \langle z_i \rangle \langle z_j \rangle \\
K_{ijk} &= C_{ijk} - C_{ij} C_k - C_{jk} C_i - C_{ik} C_j + 2 C_i C_j C_k \\
K_{ijkl} &= C_{ijkl} - C_{ijk} C_l - C_{ijl} C_k - C_{ilk} C_j - C_{ljk} C_i \\
&\quad - C_{ij} C_{kl} - C_{il} C_{kj} - C_{ik} C_{jl} + 2(C_{ij} C_k C_l \\
&\quad + C_{ik} C_j C_l + C_{il} C_j C_k + C_{jk} C_i C_l + C_{jl} C_i C_k \\
&\quad + C_{kl} C_i C_j) - 6 C_i C_j C_k C_l
\end{aligned}
\tag{11}
$$

should vanish unless $i_2 = \cdots = i_n = 1$ where n is the order of the cumulant. Therefore, a useful measure of the statistical independence of the components of z_1 on $(z_2, \ldots, z_n)$ is the cumulant-based cost function defined as:

$$
D_C = \sum_{n=1}^{\infty} \sum_{i_2,\ldots,i_n=1}^{m} (1 - \delta_{1 i_2 \ldots i_n}) \{K_{1 i_2 \ldots i_n}\}^2 \tag{12}
$$

where $\delta_{ij,\ldots,n}$ is the Kronecker delta, which eliminates the diagonal elements. In the absence of correlations, the cost function should vanish.

4.3 Causal Dependence

Up to this point, we have been concerned with dependence in a system. However, from a practical standpoint, it is particularly useful to understand whether the dependence is causal. Understanding causal dependencies also helps to simplify descriptions of highly complex physical processes because it identifies the most important coupling between dynamical variables, which can lead to improved understanding of the underlying physical processes that are most important for driving the system.

The most common method to establish causal dependencies between input and output data utilizes the time-shifted cross-correlation function, $\lambda(\mathbf{x}(t), \mathbf{y}(t+\tau))$, which has been used, for example, to compare solar wind and plasma sheet variables (Borovsky et al., 1998). A time-delay of the peak of the cross-correlation between the input and output variables suggests a causal response. This type of analysis has been used to relate solar wind structures with response in the radiation belt environment and is the basis for linear prediction filters (Baker et al., 1990; Vassiliadis et al., 2002; Rigler et al., 2004).

However, the procedure of detecting causal relationships based on linear cross-correlation suffers from a number of shortcomings. First, linear time series analysis ignores nonlinear correlations, which may be particularly important for describing energy transfer in the magnetospheric system associated with storm and substorm dynamics. Second, the cross-correlation may not be a particularly clear measure when there are multiple peaks in the forward and backward direction or if there is little asymmetry in the forward and backward directions. Finally, the cross-correlation does not provide any way to clearly distinguish between two variables that are passively correlated because of a common driver, rather than being causally related. As an example, if there are significant repetitive structures in the solar wind, the magnetospheric response can actually be well-correlated with solar wind variables at some later time, but these correlations are obviously not causal.

Alternatively, time-shifted entropy-based measures such as mutual information, $\mathcal{M}(\mathbf{x}(t), \mathbf{y}(t+\tau))$ (Prichard and Theiler, 1995) provide an ad hoc method for detecting linear, as well as nonlinear correlations in a directional sense, but suffer from the same problems as time-shifted cross-correlation when there are multiple peaks or long-range self-correlations in time series.

4.4 Transfer Entropy and Redundancy as Measures of Causal Relations

In this section and the following section, we discuss two methods to detect causal dependencies—transfer entropy and conditional redundancy. These methods utilize an entropy-based approach and, as such, include detection of both linear and nonlinear relationships. The transfer entropy (Schreiber, 2000; De Michelis et al., 2011; Materassi et al., 2014; Wing et al., 2016, 2018) can be considered a conditional mutual information that detects how much average information is contained in a source about the next state of a system that is not contained in the history of the system (Prokopenko et al., 2013). The conditional mutual information measures the dependence of two variables, x and y, given a conditioner variable, z,

$$\mathcal{M}_{\mathrm{C}}(x, y|z) \triangleq \sum_{x \in \aleph_1} \sum_{y \in \aleph_2} \sum_{z \in \aleph_3} p(x, y, z) \log \left(\frac{p(x, y, z) p(z)}{p(x, z) p(y, z)} \right) \tag{13}$$

If either x or y are dependent on z, the mutual information between x and y is reduced, and this reduction of information provides a method to eliminate coincidental dependence, or conversely to identify causal dependence.

Schreiber (2000) introduced transfer entropy to examine causality in information flow. The transfer entropy can be considered a specialized case of conditional mutual information.

$$\mathcal{T}_{\mathbf{x} \to \mathbf{y}}(\tau) \equiv \mathcal{M}_{\mathrm{C}}(y(t+\tau), x(t) | y_{\mathrm{p}}(t)) \tag{14}$$

where $y_{\mathrm{p}}(t) = [y(t), y(t-\Delta), \ldots, y(t-k\Delta)]$. The transfer entropy can be considered a conditional mutual information that detects how much average information is contained in an input, x, about the next state of a system, y, that is not contained in the history, y_{p}, of the system (Prokopenko et al., 2013). In the absence of information flow from x to y, $\mathcal{T}_{\mathbf{x} \to \mathbf{y}}$ vanishes. Also, unlike correlational analysis and mutual information, transfer entropy is directional, $\mathcal{T}_{\mathbf{x} \to \mathbf{y}} \neq \mathcal{T}_{\mathbf{y} \to \mathbf{x}}$. The transfer entropy accounts for static internal correlations, which can be used to determine whether x and y are driven by a common driver or whether x drives y or y drives x.

4.5 Conditional Redundancy

Fraser (1989) and Prichard and Theiler (1995) pioneered the use of redundancy as a generalization of mutual information to examine multidimensional systems. To examine dependency between a set of variables that are measured, it is useful to consider whether

$$P(x_1, x_2, \ldots, x_n) \stackrel{?}{=} P(x_1) P(x_2) \ldots P(x_n) \tag{15}$$

with $P(x_1, x_2, \ldots, x_n)$ the joint probability of measuring a combination of variables and $P(x_1), \ldots, P(x_n)$ the probabilities of measuring each of the variables separately. This relationship is preferable to examining cross-correlations because it allows more generally for nonlinear dependencies. The question posed in Eq. (15) can be quantified using the following definition of redundancy as a discriminating statistic (Prichard and Theiler, 1995)

$$R(x_1,\ldots,x_m) = \sum_i H_1(x_i) - H(x_1,\ldots,x_m) \tag{16}$$

which basically measures how much additional information is known about the relationship of set of variables $(x_1,\ldots,x_m)$ when they are measured simultaneously rather than independently.

In the case that none of the variables is related to that others, there is no redundancy and $R = 0$. Here, we are more interested in looking at conditional dependencies that are better described by marginal redundancy, which provides a measure of how much a variable, x_m depends on a set of other variables, $(x_1,\ldots,x_{m-1})$

$$R_M(x_1,\ldots,x_{m-1};x_m) = R(x_1,\ldots,x_m) - R(x_1,\ldots,x_{m-1}) \tag{17}$$

We would like to determine how an output, x_m, depends on another variable, x_1, given a vector of other inputs, $(x_2,\ldots,x_{m-1})$. The conditional redundancy

$$R_C(x_1|x_2,\ldots,x_{m-1};x_m) = R_M(x_1,\ldots,x_{m-1};x_m) - R_M(x_2,\ldots,x_{m-1};x_m) \tag{18}$$

provides such a measure and allows us to determine if a given variable provides additional information to what we know from another set of inputs, or whether that variable contains redundant information. Conditional redundancy would allow us to systematically assess the primary control of a magnetospheric state variable (x_m) given a set of solar wind variables $(n, V, B, \ldots) = (x_1,\ldots,x_{m-1})$ by considering the relative strength of the conditional redundancy for each potential driver.

4.6 Significance of Discriminating Statistics

The significance, S, of the discriminating statistics for a system is generally evaluated by constructing surrogate data consistent with a null hypothesis and evaluating the significance relative to a null hypothesis. For example, to determine whether there is nonlinear dependence, we assume that the system can be modeled with a linear model and construct surrogate data consistent with that hypothesis. We then compute the discriminating statistic for the data as well as many alternative realizations of the system that share the same linear statistical dependence (i.e., they have the same autocorrelation and/or cross-correlations). We then compare the discriminating statistic (also known as the cost function) of the actual data, D_0, with that of the surrogates, $S = \{s_i\}$, to obtain the significance

$$S = \frac{|D_0 - \mu_S|}{\sigma_S} \tag{19}$$

where μ_S and σ_S are the mean and variance of the costs computed with the surrogate datasets (subscript s refers to surrogate),

$$\mu_S = \frac{1}{N}\sum_{i=1}^{N} D_{s_i} \tag{20}$$

$$\sigma_S = \frac{1}{N-1}\sum_{i=1}^{N}(D_{S_i} - \mu_S)^2 \tag{21}$$

where N is the number of surrogate data sets.

4.7 Mutual Information and Information Flow

To illustrate the power of information-based measures of dependency, consider a simple nonlinear system known as the tent map. The tent map can be used to construct a sequence of points where

$$x_{j+1} = 1 - 2|x_j - 1/2| \tag{22}$$

The functional form of this map is shown in Fig. 2. It should be noted that the tent map is a function that maps $x_j \in [0,1]$ onto $x_{j+1} \in [0,1]$ with a uniform density, so the map can be sequentially applied to the same space. This relationship is clearly nonlinear. Moreover, the correlation coefficient between successive values of the map is obtained from

$$\langle x_{j+1}x_j\rangle - \langle x_{j+1}\rangle\langle x_j\rangle = \int_0^1 x(1-2|x-1/2|)dx - 1/4 = 0 \tag{23}$$

where $\langle x_{j+1}\rangle = \langle x_j\rangle = 1/2$. The mutual information in this case may be obtained by binning the input and output variables. If we partition the input and output space into N equally spaced bins (shown in Fig. 3), we can label the bins as $b_{n,m}$ where $n, m \in \{1, \ldots, N\}$, where the labels m and n refer to the binning of the input and output variable, respectively. There are a total of N^2 bins. The probability of a point starting in bin n is $1/N$, and the probability of ending in m is $1/N$. This result is a property of the map being one to one with a uniform density. For the case of a single iteration, it can be seen that the total number of jointly occupied bins is $2N$ leading to a probability of $1/2N$ for those bins that intersect the map and 0 for the other bins. The entropy is obtained from these probabilities

$$H_j = -\sum_{m=1}^{N} p_m \log p_m = \sum_{m=1}^{N}\left(\frac{1}{N}\right)\log N = \log N \tag{24}$$

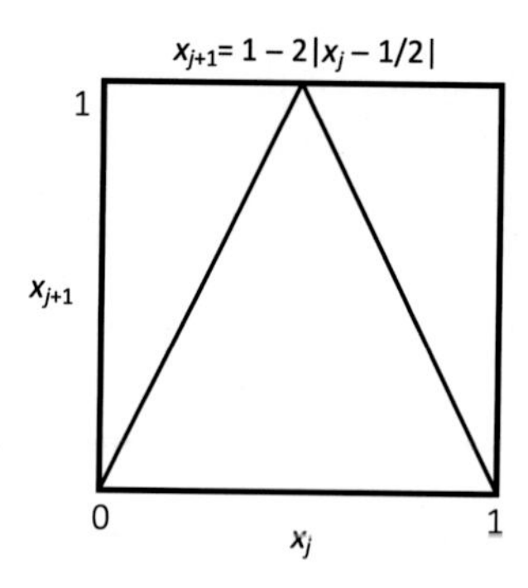

FIG. 2 The tent map provides a method to iterate a nonlinear system where $x_{j+1} = 1 - 2|x_j - 1/2|$.

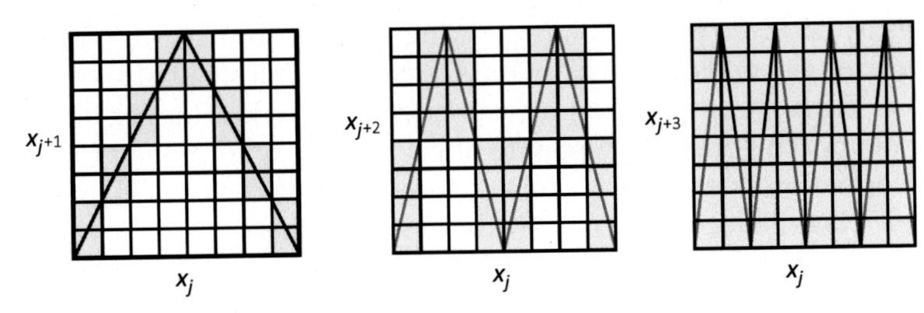

FIG. 3 Iterations 1, 2, and 3 of the nonlinear tent map. The input and output variables are partitioned into $n = 8$ bins for a total of 64 bins. For a single iteration, bin 1 of the input variable maps to bins 1 and 2 of the output variable, bin 2 of the input variable maps to bins 3 and 4 of the output variable, and so forth. While there is an equal probability (1/8) of x_j and x_{j+1} being in any of the bins, the joint probability is only nonzero in 16 of the *yellow bins* and each bin has probability 1/16. For two iterations, the map has two peaks and the joint probability is now nonzero in the 32 *yellow bins* with a value of 1/32 in each bin. With three iterations, it is impossible to tell anything about the output just from the bin number of the input variable and the mutual information vanishes.

$$H_{j+1} = -\sum_{n=1}^{N} p_n \log p_n = \sum_{n=1}^{N} \left(\frac{1}{N}\right) \log N = \log N \tag{25}$$

$$H_{j+1,j} = -\sum_{n,m} p_{m,n} \log p_{m,n} = \sum_{p_{m,n}\neq 0} \left(\frac{1}{2N}\right) \log 2N = \log 2N \tag{26}$$

where the summation should be over all combinations of m, n that have nonzero probability ($2N$ such bins) because $\lim_{p\to 0} p \log p \to 0$. It is therefore found that

$$\mathcal{M}_{j+1,j} = H_{j+1} + H_j - H_{j+1,j} = 2\log N - \log 2N = \log N - \log 2 \tag{27}$$

We may similarly consider what happens when we iterate the map two steps ahead. Successive iterations of the map are shown in Fig. 3. It is apparent that $p_{n,m}$ increases as $1/2^q N$ for q iterations as the uncertainty of the location of the point is doubled with each iteration. Then

$$H_{j+q,j} = -\sum_{n,m} p_{m,n} \log p_{m,n} = \sum_{p_{m,n}\neq 0} \left(\frac{1}{2^q N}\right) \log 2^q N = \log 2^q N \tag{28}$$

$$\mathcal{M}_{j+q,j} = H_{j+q} + H_j - H_{j+q,j} = 2\log N - \log 2^q N = \log N - q \log 2 \tag{29}$$

Choosing the log function base 2 shows that 1 bit of information is lost with each iteration. If we were to plot the information as a function of iteration, we would find that information is lost at a rate of 1 bit per iteration.

Three important points are illustrated by this example. First, the mutual information has provided a measure of dependence for this nonlinear relationship, even in the absence of any linear correlations. Second, there is a change in information with iterations. The projection of information into future iterations is referred to as *information flow*. It can be seen that at some future interaction, the information vanishes. If the information vanishes, there is no way to predict the future value of the tent-map dynamics beyond that interaction, which we call the *information horizon* or *prediction horizon* (e.g., Wing et al., 2016). Third, it can be seen that the information horizon depends not only on the system dynamics, but also on the resolution i.e., number of bins. For higher resolution measurements, there is a longer information horizon. Therefore, two important effects are seen to affect the ability to predict future dynamics of a system: the information flow (which is characteristic of the dynamical system) and the data resolution (which is a characteristic of how the system is measured). Obviously any attempts to build a predictive model beyond the prediction horizon are a waste of effort.

In the preceding case, we are able to compute the mutual information precisely because the system is described by a known analytical function. On the other hand, for a real system, data may be limited in quantity and resolution. In this case, it would be important to determine a baseline measure of significance for comparison. The information horizon would then be based on a comparison of the discriminating statistic with that of surrogates constructed based on the assumption that there is no relationship between the two variables (which can be accomplished by scrambling the data).

5 EXAMPLES FROM MAGNETOSPHERIC DYNAMICS

One of the problems of greatest practical importance in the area of space physics is that of understanding magnetospheric response to solar wind input. This response is expected to be highly nonlinear because magnetic energy is stored in the magnetotail and then suddenly released during violent events termed "substorms." During these violent releases of energy, energetic MeV electrons, which can damage satellite instrumentation, are injected into the ring current region. Power service and communications on the ground can also be interrupted due to induced currents generated during these massive events. It is therefore extremely important to be able to predict the magnetospheric response to solar wind input in order to be able to make provisions for the protection of scientific, communication, and defense satellite instrumentation, as well as ground-based power grids.

The dynamics of the solar-wind-driven magnetosphere-ionosphere system involve the complex coupling of the solar wind to the magnetosphere, convection, plasma entry, ionospheric outflow, mass transport, flux transport, the generation of plasma waves and the interaction of waves with particle populations, plasma heating, particle energization, production of aurora, dayside exhaust, downtail exhaust, charge exchange, charged-particle precipitation, atmospheric ionization, and energy dissipation into the thermosphere (e.g., Borovsky, 2018; Wing et al., 2005, 2014; Johnson and Wing, 2005; Johnson et al., 2014). These various processes are interconnected as the solar wind can drive the magnetosphere and the ionosphere; the ionosphere can drive magnetospheric convection and conversely, the magnetosphere can drive ionospheric convection. System-wide events such as geomagnetic storms and substorms can lead to substantially different outcomes, such as rapid changes in existing populations or the creation of new particle populations (Reeves, 2007). Particle energization during storms and substorms involves a complex interplay between particle injections from the magnetotail into the inner magnetosphere, excitation of waves, nonlinear growth and saturation of instabilities, and wave-particle interactions. Understanding these system-wide responses to solar interplanetary structures is particularly challenging because they are invariably associated with complex nonlinear physical processes that occur during storms and substorms (e.g., Wing et al., 2016).

One of the fundamental science questions addressed in system science is how energy, material, and information flow among the different elements that compose a system and from and to the surrounding environment via boundaries. Determining the information flow into and within the magnetospheric system is essential for describing the complex interactions between different subsystems of the magnetosphere, and is particularly useful for determining which interactions are causal. In the rest of this chapter, we illustrate how information-theoretical tools such as mutual information, transfer entropy, and conditional redundancy can be used to examine the response of the geospace state variables to solar wind driving to extract the response time, information flow, and prediction horizon. Linear and nonlinear dependencies can be detected in multivariate time series using information theoretical-based tools such as mutual information (Johnson and Wing, 2005; Wing et al., 2016; Materassi et al., 2011). Because we are particularly interested in the identification of causal dependencies, we will extend these measures by considering transfer entropy (Schreiber, 2000) and/or conditional redundancy (Prichard and Theiler, 1995), which can be used to distinguish between causal and coincidental dependencies. The redundancy provides a systematic method to quantify the dependence of a variable on various potential drivers, and the technique of maximizing information content provides a method to systematically construct coupling functions. Finally, information flow quantifies the underlying dynamics of a system and provides an estimate of the information horizon (Deco and Schürmann, 2000; Johnson and Wing, 2005; Wing et al., 2016), which ultimately defines the limits of predictive capability. The information flow detected in the data can also be compared with physical models to constrain free parameters and assess how well the models capture the system dynamics.

As an example of how these information-dynamical techniques can be used to identify changes in the dynamics of a system over a solar cycle, we show in Fig. 4 an analysis of the nonlinearity of the planetary activity level index, K_p (related to strength of geomagnetic storms) (Bartels, 1949), as a function of a solar cycle. Analysis of the discriminating statistic, $D_{\mathcal{M}}$, indicates that the nonlinearity becomes much more important in the declining phase of the solar cycle, as can be seen in the high significance of differences between the statistics of K_p and surrogate data sets that are constructed to share the same linear properties as the original dataset.

To illustrate the power of transfer entropy for identifying causal dependence, consider the relationship between sunspot number (SSN) and aa index, which gives a measure of geomagnetic activity (Mayaud, 1972). Both exhibit solar cycle variations. It is well known that SSN can cause variations in geomagnetic activity. For example, coronal mass ejections, which occur more frequently near solar maximum or solar wind high-speed streams, which occurs more frequently in the declining phase of the solar maximum, can increase geomagnetic activity or aa index. However, Ohl (1966) and Hathaway et al. (1999) showed that the aa index at the solar minimum is a good predictor of the maximum SSN of the next cycle. It was suggested that the low-level geomagnetic activity is correlated with the strength of the interplanetary magnetic field (IMF), which is related to the solar polar field (Schatten and Sofia, 1987; Layden et al., 1991; Schatten and Pesnell, 1993). The aa index here may act like a proxy for the polar field, which is causally related to the SSN (e.g., Babcock, 1961; Leighton, 1964; Dikpati and Gilman, 2006).

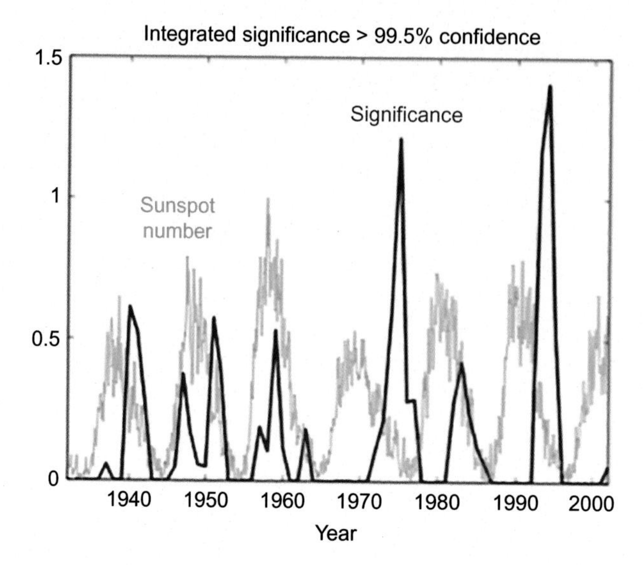

FIG. 4 Significance of differences between $D_{\mathcal{M}}$ obtained from an analysis of the time series of K_p compared with surrogate datasets that share the same distribution and linear correlation function. The results are obtained by integrating all significances that exceed three standard deviations during the 3-year period centered on the year. It is clear that a regular change in the dynamics of the magnetospheric response occurs in the declining phase of the solar cycle as seen in the comparison with sunspot number. Detailed analysis shows that the nonlinearity peaks around 50 h with a duration of a week similar to storm response timescales (Johnson and Wing, 2005).

Fig. 5A plots the time shifted correlation between aa index and SSN. The red curve shows that $aa(t+\tau)$ is correlated with $SSN(t)$ with a peak correlation at $\tau \sim 30-40$ months and is anticorrelated with $SSN(t)$ a half solar cycle period later. The blue curve shows $SSN(t+\tau)$ is anticorrelated with $aa(t)$ with a minimum at $\tau \sim 30-40$ months and a maximum at half solar cycle period later. Because both the SSN and aa index exhibit solar cycle variations, and the peak $|r|$ in both the blue and red curves are roughly the same, the figure does not provide information on whether the aa index causes SSN or the other way around.

However, we get a different picture, if we examine the transfer entropy. In Fig. 5B, the blue curve shows $\mathcal{T}_{aa \to SSN}$ and the red curve shows $\mathcal{T}_{SSN \to aa}$. Apparently, there is information flow from SSN to the aa index, and vice versa. However, the figure shows that there is more information transferred from the SSN to the aa index than the other way around, which is what we would expect intuitively. Therefore, the aa index is not a good proxy for the solar polar field (Wing et al., 2018). This information cannot be discerned from the correlational analysis in Fig. 5A.

To illustrate how conditional redundancy can be used in space weather studies, we present an analysis that we performed to examine the causal role of solar wind parameters in substorm triggering, which involves a load and release process in which magnetic flux builds up slowly in the magnetotail, followed by a fast release at substorm onset.

Observational studies have led to the question of whether substorms are triggered externally or internally. Lyons (1995) suggested that rapid, sustained northward turning of the IMF following a growth phase is a direct trigger of substorms. A series of quantitative investigations

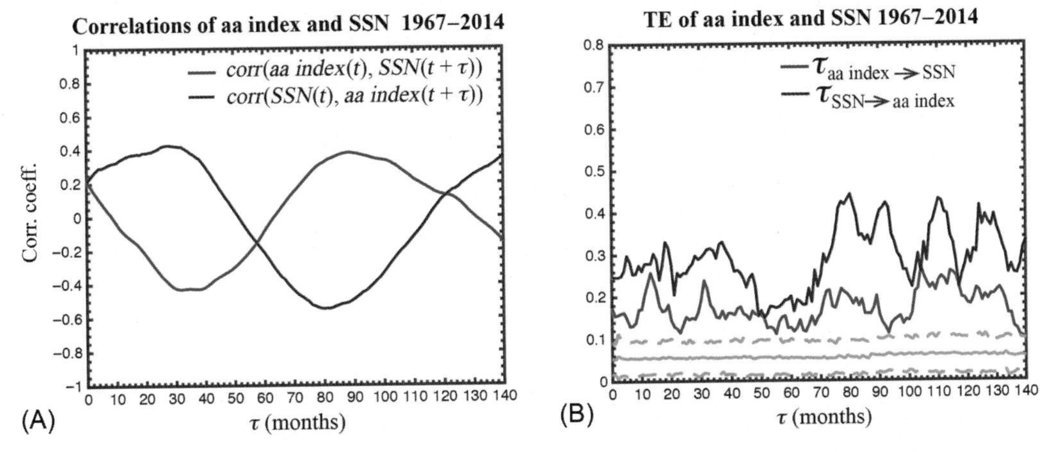

FIG. 5 (A) Shifted correlation $corr[aa\ index(t), SSN(t+\tau)]$ is plotted in *blue* and $corr[SSN(t), aa\ index(t+\tau)]$ is plotted in *red*. The peak $|corr[aa\ index(t), SSN(t+\tau)]|$ is roughly the same as the peak $|corr[SSN(t), aa\ index(t+\tau]|$. (B) $T_{aa \to SSN}$ is plotted in *blue* and $T_{SSN \to aa}$ is plotted in *red*. $T_{SSN \to aa} > T_{aa \to SSN}$ suggesting that more information is transferred from SSN to aa index than the other way around. Such information cannot be discerned from the correlations shown in (A). The *solid and dashed green curves* show the mean and 3σ of the noise. The data are for the period 1967–2014. *(From Wing et al., 2018).*

was subsequently performed by Hsu and McPherron (2002, 2003, 2004) that showed a strong association between external substorm triggers and substorm onsets, suggesting that 60% of substorms are triggered. Morley and Freeman (2007) reevaluated the data set considered by Hsu and McPherron (2002), considering the importance of internal versus external triggering. While their analysis confirmed that there is a high association number between external triggers and substorm onsets, they also considered whether the association was causal or coincidental. Using the same solar wind input, they constructed an alternative dataset of substorm onsets using the integrate and fire MSM model (Freeman and Morley, 2004), which had no requirement of northward turning. Nevertheless, they still found a strong association number between the external triggers and the model substorms, so they concluded that substorms are not causally correlated with northward turning of the IMF. More recently, Newell and Liou (2011) noted that the mean B_z has a northward turning (reversion to the mean) starting 20 min before onset. However, a similar reversion to the mean was found for random elevations of solar wind driving based on several coupling functions, further supporting the concerns of Morley and Freeman (2007).

The conditional redundancy provides a way to quantify the following question. Is there any additional information about substorm onset provided northward turning (external trigger) beyond the requirement of a growth phase (internal trigger)? To answer this question, we used the expanded substorm onset list obtained by Frey et al. (2004) and Frey and Mende (2006), identifying external and internal triggers in the ACE and WIND datasets propagated to 17 R_E using standard methods. The results of our analysis (Johnson and Wing, 2014) are shown in Fig. 6. Panel (A) shows the conditional redundancy of the dependence of substorm onsets (at a given time lag) on external triggers given the internal trigger

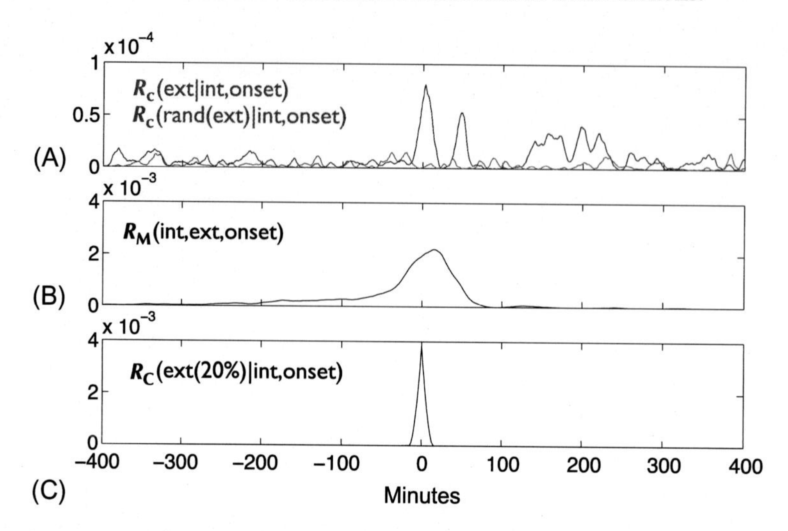

FIG. 6 Conditional redundancy describing how much additional information about onsets, $R_C(\text{ext}(t)|\text{int}(t);\text{onset}(t+\tau))$, is added by knowing external trigger events (ext) given the growth phase requirement (int), as a function of τ for (A) external triggers satisfying Lyons (1995) (*blue*) and random external triggers (*red*); (B) marginal redundancy as a measure of the information in ext and int about onsets; and (C) conditional redundancy if 20% of external triggers artificially coincide with onsets and the remainder are randomly distributed. If 20% of the external triggers were causal, then the information content would be increased by a factor of 50. In reality comparing (A) and (B) only a few percent additional information is added by the northward turning criterion suggesting that northward turning of the IMF is, in general, coincidentally, rather than causally, associated with substorm onset. *(From Johnson, J.R., Wing, S., 2014. External versus internal triggering of substorms: an information-theoretical approach. Geophys. Res. Lett. 41 (16), 5748–5754. ISSN 0094-8276. https://doi.org/10.1002/2014gl060928.)*

(i.e., $R_C(\text{ext}(t)|\text{int}(t);\text{onset}(t+\Delta))$). Panel (B) shows comparison of the marginal redundancy measuring how much information about the onset is contained in the internal and external triggers. Panel (C) shows, for comparison, the conditional redundancy if 20% of the external triggers artificially coincide with the onset list and the remainder are randomized. It is apparent that while R_C is slightly elevated around substorm onset, the fraction of *additional information* about onset added by knowing the external triggers (obtained by comparing $R_C(\text{ext}|\text{int};\text{onsets})$ with $R_M(\text{ext},\text{int};\text{onsets})$) *is about* 2%, suggesting that northward turning of the IMF does not generally have a causal association with substorms.

6 SIGNIFICANCE AS AN INDICATOR OF CHANGES IN UNDERLYING DYNAMICS

Significant deviation from the null hypothesis can be applied to subsets of the original dataset (see Section 4.6). As in a spectrogram, we can consider windowed significance, in which case data is sampled from a window of width N_w and a significance is computed for that dataset. The window is then shifted and the significance recomputed. Time variations in the significance indicate where the null hypothesis fails. For example, if the null hypothesis assumes a linear versus nonlinear system, then changes in the significance indicate changes in the linear versus nonlinear response. Because the significance is a function of time delay, it

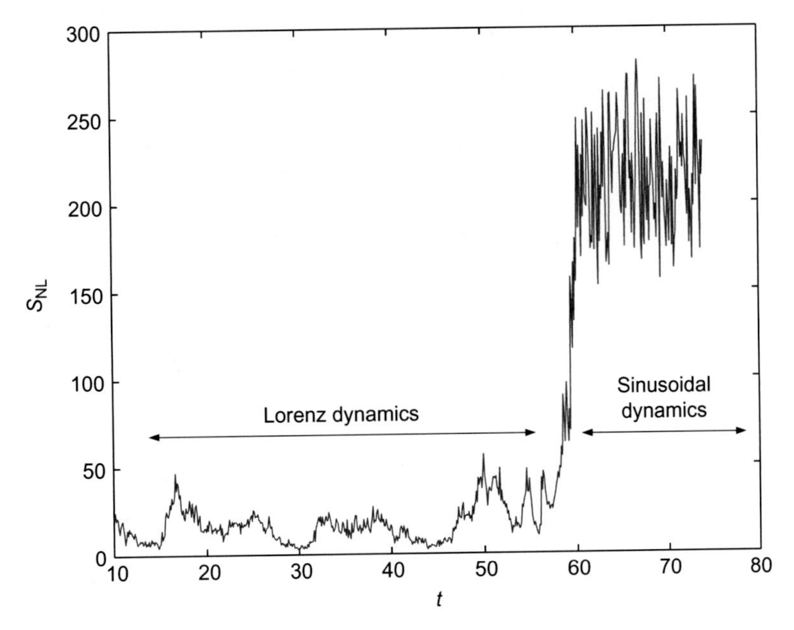

FIG. 7 Windowed significance as a function of time. For $t < 60$ the system is governed by Lorenz dynamics (Lorenz, 1963). Near $t = 60$ there is a transition to sinusoidal dynamics. Note that an abrupt change in the significance clearly indicates the change in the underlying dynamics of the system.

is useful to consider a time delay that is most suitable for the data. While it could be varied, a useful choice is the first minimum of the mutual information of the entire dataset, which is considered to be the best time delay for analyzing the nonlinear dynamics of the system.

For example, in Fig. 7 we examine a system governed by Lorenz dynamics for $t < 60$ with an abrupt change to sinusoidal dynamics for $t > 60$. The Lorenz system (Lorenz, 1963) satisfies the following equations:

$$\frac{dx}{dt} = \sigma(y - x)$$
$$\frac{dy}{dt} = -xz + rx - y$$
$$\frac{dz}{dt} = xy - bz \tag{30}$$

Note that in Fig. 7, the significance remains roughly constant while the system dynamics are stationary, but the significance changes abruptly when the underlying dynamics change. Hence, the significance can be a good indicator of changes in the underlying dynamics of a nonlinear system. The technique of windowed significance has been applied to magnetospheric data to detect changes in the underlying dynamics of the magnetosphere (e.g., Wing et al., 2016). Any predictive model would need be able to account for such changes in the underlying model.

6.1 Detecting Dynamics in a Noisy System

Measurements of physical quantities in the real world usually contain some noise or uncertainties. It is interesting to examine how the cumulant-based significance and mutual information perform in the presence of noise. Therefore, we also performed a similar analysis for the Lorenz system when the signal is artificially contaminated by random noise added to the signal. Fig. 8 plots the cumulant-based significance and mutual information for the Lorentz system. The dashed, dotted, dot-dashed, and gray curves show the results with additive noise with a signal-to-noise ratio of 0.25, 0.50, 0.75, and 1. Surrogate datasets were used in evaluating the significance so that values of $S_\tau > 1.6$ are statistically significant. For the mutual information plot, we have also included three horizontal plots that show the central mean and standard deviation of the mutual information obtained from the surrogate datasets, assuming

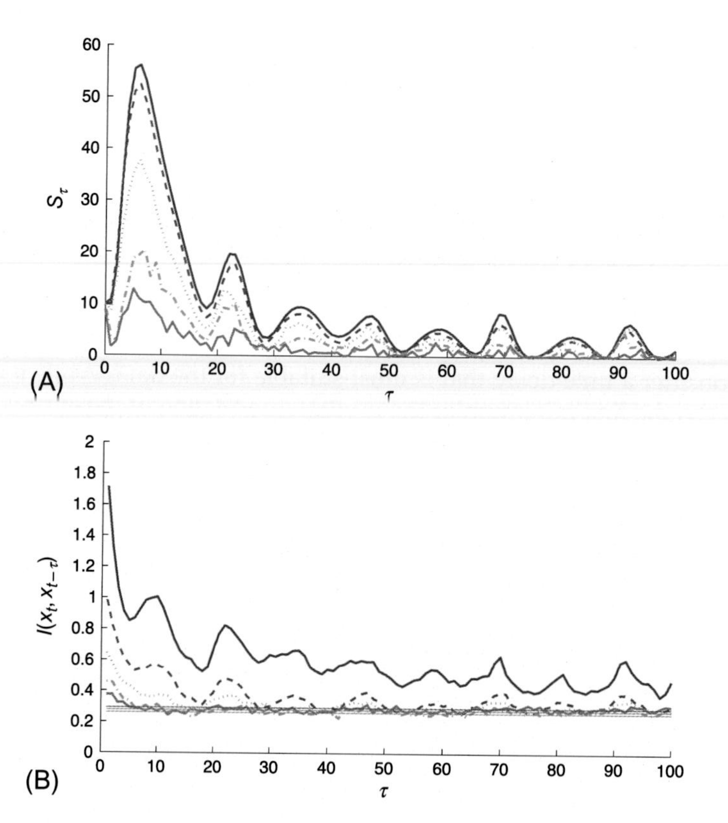

FIG. 8 Comparison of (A) the cumulant-based significance and (B) the mutual information for the Lorenz system (Lorenz, 1963). Shown are the cumulant-based significance, S_τ, and mutual information I_τ verses time lag for one of the Lorenz variables. The *solid curve* is without additive noise. The *dashed, dotted, dot-dashed,* and *gray curves* show the results with additive noise with signal-to-noise ratio of 0.25, 0.50, 0.75, and 1.

no time ordering. The data contaminated by the signal-to-noise ratio 0.75 and 1 cannot be distinguished from the surrogate data with any confidence with the mutual information measure. In contrast, even with a signal-to-noise ratio of 1, the cumulant-based method detects the nonlinearities. Hence, it would appear that the cumulant-based significance can be used to detect the presence of nonlinear dependencies in the underlying dynamics, even when the signal is heavily contaminated by noise.

6.2 Cumulant-Based Information Flow

Information flow deals with changes in the information content of a system. The information flow can be used to detect changes in underlying dynamics and the loss of information in a system.

Unfortunately, it is usually difficult to compute a statistically meaningful information flow for a real system because of the limited size of the dataset. However, cumulants that also carry information about the underlying system dynamics are readily obtained, even from limited data sets. We can therefore introduce a proxy for the information flow which we define as the cumulant-based information flow (Deco and Schürmann, 2000)

$$I_C(p) = \sum_{n=1}^{\infty} \sum_{i_2,\ldots,i_n=1}^{m} (1 - \delta_{1i_2\ldots i_n}) \left\{K^{(p)}_{1i_2\ldots i_n}\right\}^2 \tag{31}$$

where $K^{(p)}_{1i_2\ldots i_n}$ are the cumulants associated with the vector $\{y1,\ldots,y_m\} = \{y(t+p), y(t-\Delta),\ldots,y(t-(m-1)\Delta)\}$. This quantity provides an estimate of how well a predictive model could estimate a future value of the time series p, steps ahead, given the history of the time series. The minimal value of $I_C(p) = 0$ indicates statistical independence, while increasing values of $I_C(p)$ point to increasing dependencies in the time series (Deco and Schürmann, 2000). Ideally, m should be large enough so that $I_C(p)$ becomes a measure of predictability given the entire past of the time series. For practical purposes, we choose the value of m such that the information flow does not change appreciably when m is increased. As for the case of significance, we limit the computation to the fourth-order cumulant.

While the information flow $I_C(p)$ provides an indication of how far into the future one should be able to predict the time series, when examining changes in the information flow, it is more practical to consider an integrated information flow,

$$\widehat{I}_C = \sum_{p=1}^{P} \sum_{n=1}^{\infty} \sum_{i_2,\ldots,i_n=1}^{m} (1 - \delta_{1i_2\ldots i_n}) \left\{K^{(p)}_{1i_2\ldots i_n}\right\}^2 \tag{32}$$

The integrated information flow provides a measure of predictability up to a look ahead, P, which should be taken as large as the desired predictions, keeping in mind that it is only of practical value if it is smaller than value of p, which characterizes the falloff of the information flow $I_C(p)$

As an illustration of this method, in Fig. 9, we compute an information flow based on the multivariate vector $\{x_1, x_2, x_3, x_4\} = \{D_{st}(t+p), D_{st}(t), VBs(t), VBs(t-\Delta)\}$. The integrated

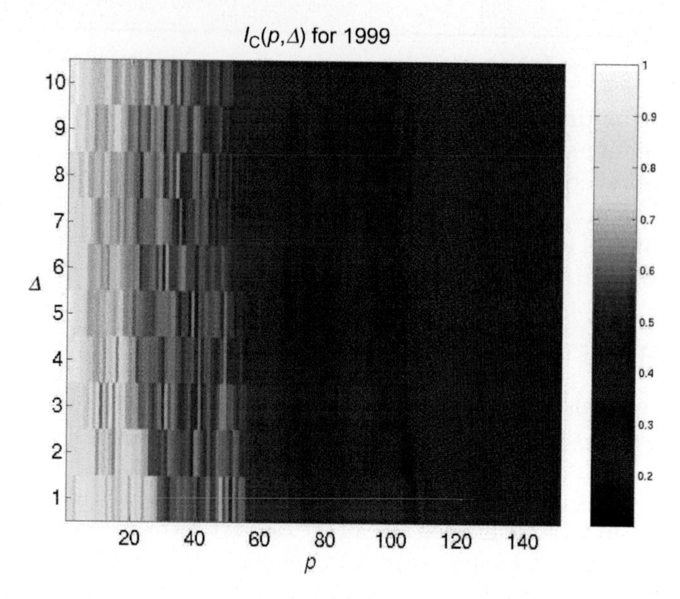

FIG. 9 Information flow, $I_C(p, \Delta)$ as a function of look ahead, p, and solar wind delay time, Δ for 1999. Notice that the information decays on the order of 1 week which is a typical time scale for storm response. The information flow remains relatively high on the order of 20 h and there is a strong nonlinear response at around 40–50 h as detected based on the significance. In considering the integrated information, we will integrate up to 20 h ahead.

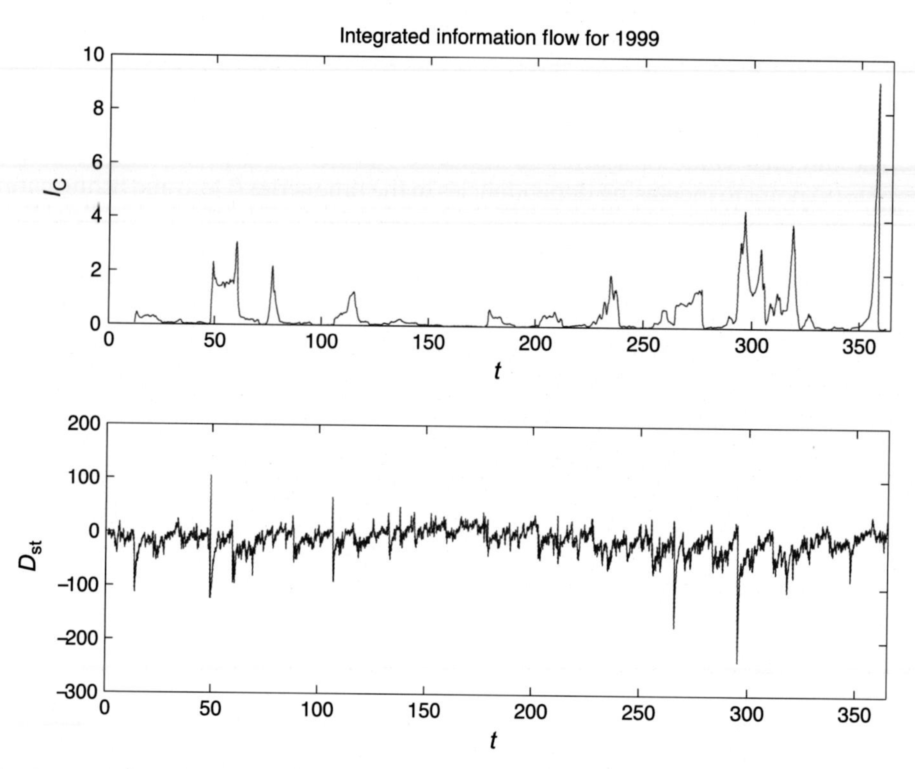

FIG. 10 Integrated information flow verses day of 1999 based on a 20-h look ahead. There are clearly times when there is high predictability. The integrated information flow appears to be largest during times of storm activity indicating increased information flow and better predictability. The most reproducible and predictable behavior of D_{st} appears to be the recovery of the index; however, the information flow does increase significantly prior to the onset of the storm indicating that the storm onset could also be predicted.

information flow, $I_C(p, \Delta)$, is obtained as a function of the look ahead and changes in the solar wind VBs data.

In Fig. 10, we plot the integrated information computed in a sliding window of width of 150 h for the year 1999. Information is integrated up to $P = 20$ h, and a time delay $\Delta = 1$ h is considered in VBs. An important point is that there tend to be times of reasonable predictability. These times are associated with the occurrence of storms.

7 DISCUSSION

This chapter discusses information theoretical tools such as mutual information, conditional mutual information, conditional redundancy, transfer entropy, and their potential benefits to space weather modeling. This nonparametric approach can be used to guide development of parametric models. The output from models depends heavily on the method by which the model is trained. If a training set is noisy, for example, the output is often spoiled or not well constrained. The windowed significance provides a measure of the significance of data within a dataset. Data with significance that does not exceed the surrogate data usually is corrupted by noise and is not useful for training purposes. Such data could result when the method to gather the data is corrupted by noise or by intrinsic noise in the dynamical system. On the other hand, data with high significance exhibits strong dependencies and is therefore more likely to be useful when trying to model those dependencies. Thus, the significance measure can be used to eliminate noisy data from a dataset. One consequence is that models trained with data that best captures the nonlinear dependencies should predict the dynamical evolution of the system better.

Another useful property of the significance is that it can be used to detect changes in the underlying dynamics of a system. When the system dynamics change, the significance detects this change. The magnetosphere is believed to be a multistate system. Magnetospheric behavior is often classified as active or quiescent. The windowed cumulant-based significance is useful for detecting changes in the state of the system. Moreover, it is useful for classifying data according to the state of the system. The dynamics of those states are usually different, and therefore, it would be useful to apply different models to those two different states. By assessing the significance of data and separating it according to its state it is possible to fine tune models to describe the state of the system as illustrated in Fig. 11. It is also possible to dynamically determine when to switch dynamical states using the significance measure for an incoming stream of data.

The cumulant-based information flow (computed by summing over the look-ahead significance) can be used as a proxy for information flow in the system. This information flow can be computed from the data, as well as from the neural network (or other types of models). The network can be trained so as to best approximate the information flow rather than to simply minimize the error in the training set (which does not account for the possibility of noise contamination). The network can be modified (i.e., new nodes added or removed) in a systematic way so as to best approximate the information flow with the minimum number of parameters. By requiring that information flow is well approximated by the parametric model, we are more assured that the underlying dynamics are faithfully modeled than with simple error minimization.

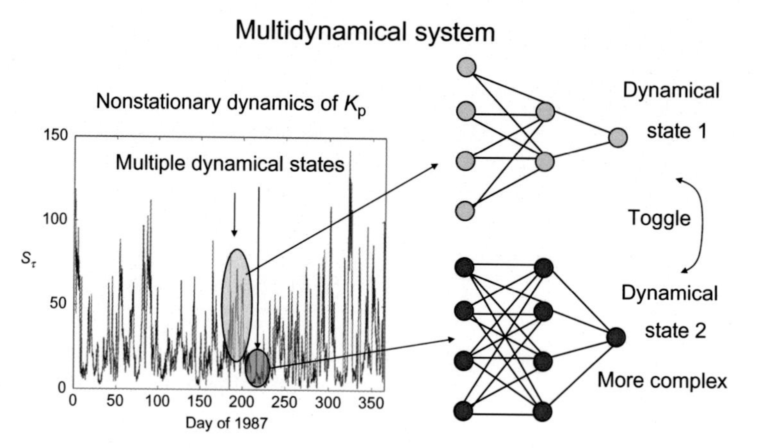

FIG. 11 An illustration of how the windowed significance can be used to identify the state of a system. Larger significance usually indicates a system with simpler underlying dynamics while a state with lower significance is usually more complex. The data can be binned according its windowed significance so that a parametric neural network model can specialize in each state. The instantaneous significance measure can be used to identify whether to toggle between the two networks which predict the state of the system.

8 SUMMARY

In this chapter, we summarized a number of information-theoretical tools that can be used to study complex systems. These tools include mutual information and cumulants which can be used to determine nonlinear dependence among system variables. Causal dependence can be assessed with entropy-based statics such as transfer entropy and conditional redundancy. By looking at the time-variation of dependence, we can also determine an information horizon that gives insight about the predictability of the system dynamics. We also discussed issues related to nonideal datasets contaminated by noise or limited to a small number of observations.

We provided specific examples from space physics showing how these tools can be used to identify timescales of nonlinear interactions in the solar cycle and assess the role of nonlinear dynamics. We also showed how information-theoretical tools can be used to identify causal drivers of dynamical processes such as substorms. In Chapter 6, we show how these tools were used to identify which solar wind variables are causally related to variability in radiation belt fluxes, and we were able to determine that both the solar wind velocity and density play important roles.

Finally, we have applied information dynamical techniques to the Kp and Dst magnetospheric indices. Using the cumulant-based significance, we have established that the underlying dynamics of Kp evolution is, in general, nonlinear, exhibiting a quasiperiodicity that is detectable only if nonlinear correlations are taken into account. As such, one expects that linear auto-regressive moving average (ARMA) models commonly used to predict magneto spheric response should be inaccurate. Local-linear models (which include slow evolution of parameters) are also likely to fail where the dynamics suddenly change, which occurs regularly in Fig. 10. A promising alternative would be to train a neural network with data identified to have a large significance to avoid irrelevant noise that normally spoils the generalization characteristics of the neural network. Additional improvements could be realized by using the fact that magnetospheric dynamics are not stationary (see Chapter 6) to separate the data into "states" and fine tune predictors for those states.

Our analysis of the information flow and integrated information flow provides additional insight. The system appears to have reasonable predictability properties on the time scale of 10 h — a time scale that would be of practical value for protecting vulnerable satellite equipment. However, nonlinear effects already play a significant role, even on that time scale. Longer-term forecasting may also be possible, as evidenced by peaks in the information flow. Finally, improvements to parametric modeling could also be realized by requiring parametric models to approximate the information flow in the real system.

ACKNOWLEDGMENTS

Simon Wing acknowledges support from NASA Grants (NNX13AE12G, NNX15AJ01G, NNX16AR10G, and NNX16AQ87G). Jay R. Johnson acknowledges support from NASA Grants (NNH11AR07I, NNX14AM27G, NNH14AY20I, NNX16AR10G, and NNX16AC39G), NSF Grants (ATM0902730, AGS-1203299, and AGS-1405225), DOE contract DE-AC02-09CH11466 and a faculty research grant from Andrews University.

References

Aldrich, J.H., 1999. Political parties in a critical era. Am. Polit. Q. 27 (1), 9–32.

Babcock, H.W., 1961. The topology of the Sun's magnetic field and the 22-year cycle. Astrophys. J. 133, 572. https://doi.org/10.1086/147060. Available from: http://adsabs.harvard.edu/abs/1961ApJ..133.572B.

Baker, D.N., McPherron, R.L., Cayton, T.E., Klebesadel, R.W., 1990. Linear prediction filter analysis of relativistic electron properties at 6.6 R_E. J. Geophys. Res. 95, 15133–15140. https://doi.org/10.1029/JA095iA09p15133.

Bartels, J., 1949. The standardized index, Ks, and the planetary index, Kp. IATME Bull. 126, 97.

Boozer, A.H., 2012. Theory of tokamak disruptionsa. Phys. Plasmas 19 (5), 058101. https://doi.org/10.1063/1.3703327.

Borovsky, J.E., 2018. Time-integral correlations of multiple variables with the relativistic-electron flux at geosynchronous orbit: the strong roles of the substorm-injected electrons and the ion plasma sheet. J. Geophys. Res. In press.

Borovsky, J.E., Thomsen, M.F., Elphic, R.C., 1998. The driving of the plasma sheet by the solar wind. J. Geophys. Res. 103, 17617–17640. https://doi.org/10.1029/97JA02986.

Darbellay, G.A., Vajda, I., 1999. Estimation of the information by an adaptive partitioning of the observations space. IEEE Trans. Inf. Theory 45, 1315–1321.

De Michelis, P., Consolini, G., Materassi, M., Tozzi, R., 2011. An information theory approach to the storm-substorm relationship. J. Geophys. Res. Space Phys. 116 (A8), A08225.

Deco, G., Schürmann, B., 2000. Information Dynamics. Springer, New York, NY.

Dikpati, M., Gilman, P.A., 2006. Simulating and predicting solar cycles using a flux-transport dynamo. Astrophys. J. 649, 498–514. https://doi.org/10.1086/506314. Available from: http://adsabs.harvard.edu/abs/2006ApJ..649.498D.

Doran, M.F., Pond, G.R., Crowson, C.S., O'Fallon, W.M., Gabriel, S.E., 2002. Trends in incidence and mortality in rheumatoid arthritis in Rochester, Minnesota, over a forty-year period. Arthritis Rheum. 46 (3), 625–631. ISSN 1529-0131. https://doi.org/10.1002/art.509.

Fraser, A.M., 1989. Information and entropy in strange attractors. IEEE Trans. Inf. Theory 35 (2), 245–262. ISSN 0018-9448. https://doi.org/10.1109/18.32121.

Freeman, M.P., Morley, S.K., 2004. A minimal substorm model that explains the observed statistical distribution of times between substorms. Geophys. Res. Lett. 31, L12807. https://doi.org/10.1029/2004GL019989.

Frey, H.U., Mende, S.B., 2006. Substorm onsets as observed by IMAGE-FUV. In: International Conference on Supercomputing.

Frey, H.U., Mende, S.B., Angelopoulos, V., Donovan, E.F., 2004. Substorm onset observations by IMAGE-FUV. J. Geophys. Res. Space Phys. 109, A10304. https://doi.org/10.1029/2004JA010607.

Gershenfeld, N., 1998. The Nature of Mathematical Modeling. Cambridge University Press, Cambridge.

Hathaway, D.H., Wilson, R.M., Reichmann, E.J., 1999. A synthesis of solar cycle prediction techniques. J. Geophys. Res. 104, 22. https://doi.org/10.1029/1999JA900313. Available from: http://adsabs.harvard.edu/abs/1999JGR..10422375H.

Hsu, T.S., McPherron, R.L., 2002. An evaluation of the statistical significance of the association between northward turnings of the interplanetary magnetic field and substorm expansion onsets. J. Geophys. Res. Space Phys. 107, 1398. https://doi.org/10.1029/2000JA000125.

Hsu, T.S., McPherron, R.L., 2003. Occurrence frequencies of IMF triggered and nontriggered substorms. J. Geophys. Res. Space Phys. 108, 1307. https://doi.org/10.1029/2002JA009442.

Hsu, T.S., McPherron, R.L., 2004. Average characteristics of triggered and nontriggered substorms. J. Geophys. Res. Space Phys. 109, A07208. https://doi.org/10.1029/2003JA009933.

Johnson, J.R., Wing, S., 2005. A solar cycle dependence of nonlinearity in magnetospheric activity. J. Geophys. Res. 110, 4211. https://doi.org/10.1029/2004JA010638.

Johnson, J.R., Wing, S., 2014. External versus internal triggering of substorms: an information-theoretical approach. Geophys. Res. Lett. 41 (16), 5748–5754. ISSN 0094-8276. https://doi.org/10.1002/2014gl060928.

Johnson, N.F., Jefferies, P., Hui, P.M., 2003. Financial Market Complexity, OUP Catalogue, 9780198526650. Oxford University Press. ISBN ARRAY(0x53dc0328). Available from: https://ideas.repec.org/b/oxp/obooks/9780198526650.html.

Johnson, J.R., Wing, S., Delamere, P.A., 2014. Kelvin-Helmholtz instability in planetary magnetospheres. Space Sci. Rev. 184, 1–31. https://doi.org/10.1007/s11214-014-0085-z. Available from: http://adsabs.harvard.edu/abs/2014SSRv.184..1J.

Layden, A.C., Fox, P.A., Howard, J.M., Sarajedini, A., Schatten, K.H., 1991. Dynamo-based scheme for forecasting the magnitude of solar activity cycles. Solar. Phys. 132, 1–40. https://doi.org/10.1007/BF00159127. Available from: http://adsabs.harvard.edu/abs/1991SoPh.132..1L.

Leighton, R.B., 1964. Transport of magnetic fields on the Sun. Astrophys. J. 140, 1547. https://doi.org/10.1086/148058. Available from: http://adsabs.harvard.edu/abs/1964ApJ..140.1547L.

Li, W., 1990. Mutual information functions versus correlation functions. J. Stat. Phys. 60, 823–837.

Lorenz, E., [illegible]. Deterministic nonperiodic flows. J. Atmosph. Sci. [illegible].

Lyons, L.R., 1995. A new theory for magnetospheric substorms. J. Geophys. Res. 100, 19069–19082. https://doi.org/10.1029/95JA01344.

Mackey, M.C., Glass, L., et al., 1977. Oscillation and chaos in physiological control systems. Science 197 (4300), 287–289.

March, T.K., Chapman, S.C., Dendy, R.O., 2005. Mutual information between geomagnetic indices and the solar wind as seen by WIND: implications for propagation time estimates. Geophys. Res. Lett. 32, L04101. https://doi.org/10.1029/2004GL021677.

Materassi, M., Ciraolo, L., Consolini, G., Smith, N., 2011. Predictive space weather: an information theory approach. Adv. Space Res. 47, 877–885. https://doi.org/10.1016/j.asr.2010.10.026. Available from: http://adsabs.harvard.edu/abs/2011AdSpR.47.877M.

Materassi, M., Consolini, G., Smith, N., De Marco, R., 2014. Information theory analysis of cascading process in a synthetic model of fluid turbulence. Entropy 16 (3), 1272–1286.

Mayaud, P.N., 1972. The aa indices: a 100-year series characterizing the magnetic activity. J. Geophys. Res. 77, 6870. https://doi.org/10.1029/JA077i034p06870. Available from: http://adsabs.harvard.edu/abs/1972JGR..77.6870M.

Midlarsky, M.I., 1984. Political stability of two-party and multiparty systems: probabilistic bases for the comparison of party systems. Am. Polit. Sci. Rev. 78 (4), 929–951.

Morley, S.K., Freeman, M.P., 2007. On the association between northward turnings of the interplanetary magnetic field and substorm onsets. Gephys. Res. Lett. 34, L08104. https://doi.org/10.1029/2006GL028891.

Newell, P.T., Liou, K., 2011. Solar wind driving and substorm triggering. J. Geophys. Res. Space Phys. 116, A03229. https://doi.org/10.1029/2010JA016139.

Ohl, A., 1966. Forecast of sunspot maximum number of cycle 20. Solice Danie 9, 84.

Prichard, D., Price, C., 1992. Spurious dimension estimates from time series of geomagnetic indices. Geophys. Res. Lett. 19, 1623–1626.

Prichard, D., Theiler, J., 1994. Generating surrogate data for time series with several simultaneously measured variables. Phys. Rev. Lett. 73, 951–954.

Prichard, D., Theiler, J., 1995. Generalized redundancies for time series analysis. Physica D 84 (3–4), 476–493. ISSN 0167-2789. https://doi.org/10.1016/0167-2789(95)00041-2. Available from: http://www.sciencedirect.com/science/article/pii/0167278995000412.

Prokopenko, M., Lizier, J., Price, D., 2013. On thermodynamic interpretation of transfer entropy. Entropy 15, 524–543. https://doi.org/10.3390/e15020524.

Reeves, G.D., 2007. Radiation belt storm probes: a new mission for space weather forecasting. Space Weather 5, S11002. https://doi.org/10.1029/2007SW000341.

Rigler, E.J., Baker, D.N., Weigel, R.S., Vassiliadis, D., Klimas, A.J., 2004. Adaptive linear prediction of radiation belt electrons using the Kalman filter. Space Weather 2, S03003. https://doi.org/10.1029/2003SW000036.

Salvarani, C., Crowson, C.S., O'Fallon, W.M., Hunder, G.G., Gabriel, S.E., 2004. Reappraisal of the epidemiology of giant cell arteritis in Olmsted County, Minnesota, over a fifty-year period. Arthritis Care Res. 51 (2), 264–268. ISSN 1529-0131. https://doi.org/10.1002/art.20227.

Schatten, K.H., Pesnell, W.D., 1993. An early solar dynamo prediction: cycle 23 cycle 22. Geophys. Res. Lett. 20 (20), 2275–2278.

Schatten, K.H., Sofia, S., 1987. Forecast of an exceptionally large even-numbered solar cycle. Geophys. Res. Lett. 14 (6), 632–635.

Schreiber, T., 2000. Measuring information transfer. Phys. Rev. Lett. 85, 461–464. https://doi.org/10.1103/PhysRevLett.85.461.

Theiler, J., Eubank, S., Longtin, A., Galdrikian, B., Doyne Farmer, J., 1992. Testing for nonlinearity in time series: the method of surrogate data. Physica D 58, 77–94.

Theunissen, F.E., Sen, K., Doupe, A.J., 2000. Spectral-temporal receptive fields of nonlinear auditory neurons obtained using natural sounds. J. Neurosci. 20 (6), 2315–2331. ISSN 0270-6474. Available from: http://www.jneurosci.org/content/20/6/2315.

Tóth, G., Sokolov, I.V., Gombosi, T.I., Chesney, D.R., Clauer, C.R., De Zeeuw, D.L., Hansen, K.C., Kane, K.J., Manchester, W.B., Oehmke, R.C., et al., 2005. Space weather modeling framework: a new tool for the space science community. J. Geophys. Res. Space Phys. 110 (A12), 226.

Tsonis, A.A., 2001. Probing the linearity and nonlinearity in the transitions of the atmospheric circulation. Nonlinear Process. Geophys. 8, 341–345.

Vassiliadis, D., Klimas, A.J., Kanekal, S.G., Baker, D.N., Weigel, R.S., 2002. Long-term-average, solar cycle, and seasonal response of magnetospheric energetic electrons to the solar wind speed. J. Geophys. Res. Space Phys. 107, 22–1.

Wing, S., Johnson, J.R., Newell, P.T., Meng, C.I., 2005. Dawn-dusk asymmetries, ion spectra, and sources in the northward interplanetary magnetic field plasma sheet. J. Geophys. Res. Space Phys. 110, A08205. https://doi.org/10.1029/2005JA011086. Available from: http://adsabs.harvard.edu/abs/2005JGRA.110.8205W.

Wing, S., Johnson, J.R., Chaston, C.C., Echim, M., Escoubet, C.P., Lavraud, B., Lemon, C., Nykyri, K., Otto, A., Raeder, J., Wang, C.P., 2014. Review of solar wind entry into and transport within the plasma sheet. Space Sci. Rev. 184, 33–86. https://doi.org/10.1007/s11214-014-0108-9.

Wing, S., Johnson, J.R., Camporeale, E., Reeves, G.D., 2016. Information theoretical approach to discovering solar wind drivers of the outer radiation belt. J. Geophys. Res. Space Phys. ISSN 2169-9380. https://doi.org/10.1002/2016ja022711.

Wing, S., Johnson, J.R., Vourlidas, A., 2018. Information theoretic approach to discovering causalities in the solar cycle. Astrophysi. J. 854 (2), 85. http://stacks.iop.org/0004-637X/854/i=2/a=85.

Zakharov, L.E., Galkin, S.A., Gerasimov, S.N., et al., 2012. Understanding disruptions in tokamaks. Phys. Plasmas 19 (5), 055703. https://doi.org/10.1063/1.4705694.

CHAPTER

4

Regression

Algo Carè, Enrico Camporeale†*

*University of Brescia, Brescia, Italy †Centrum Wiskunde & Informatica, Amsterdam, The Netherlands

CHAPTER OUTLINE

1 WHAT IS REGRESSION?

The term "regression" was introduced in a precise, technical way by Sir Francis Galton who, in 1886, articulated the "law of regression" with the following words:

> *we can define the law of regression very briefly. It is that the height-deviate of the offspring is, on the average, two-thirds of the height-deviate of its mid-parentage.* (Galton, 1886)

Galton inferred this law from 205 families, each of them represented by a couple $(X^{(i)}, Y^{(i)})$, where $X^{(i)}$ was a number representing the height-deviate (i.e., the deviation from the average

Machine Learning Techniques for Space Weather
https://doi.org/10.1016/B978-0-12-811788-0.00004-4

height) of the parents in the ith family, and $Y^{(i)}$ the height-deviate of the corresponding offspring. By denoting by $\mathbb{E}[Y|X]$ the average height-deviate of the offspring whose parents have an height-deviate of X, Galton's regression law can be written as

$$\mathbb{E}[Y|X] = r\,X, \text{ with } r = \frac{2}{3}$$

This is the equation of a line in a 2-dimensional space, linking the value of an "input" variable X to the expected value of an "output" variable Y. One can easily argue that Galton would not have used the word "regression" in his law if $|r|$ had been greater than 1. However, in spite of the noble attempt by G.U. Yule, who suggested calling a relationship of this kind

$$\mathbb{E}[Y|X] = rX + c \tag{1}$$

by the more neutral name of "characteristic line," the term "regression line" soon became predominant, and it is still in use as an abstract concept, much detached from the biological concept that Galton had in mind (Aldrich, 2005).

After Galton, scientists such as G.U. Yule, K. Pearson, and R.A. Fisher, building on the least squares method of Gauss and Legendre (Plackett, 1972) contributed to what is known today as *regression analysis* in statistics. The problem of regression has gone far beyond the task of deriving a linear relationship, such as Eq. (1), from data: it is now an enormous field of study that involves many researchers in statistics and in machine learning, and encompasses a wide variety of approaches and methods. In the limited space of this chapter, we aim to provide an accessible introduction to the regression problem, mainly from a pragmatic, predictive perspective, typical of the machine learning literature. In this perspective, one is not interested in "learning" from data a parameter such as r in Eq. (1) in order to *discover laws* (as Galton wanted to do). Instead, r and c, and even the linear structure of Eq. (1), are considered *tools* for deriving a prediction of an output variable Y, given the observation of an input variable X. Therefore, following a terminology that is well-established in machine learning, we will call *regression problem* the problem of predicting a real-valued output variable Y from an input variable X, given a set of N previous observations of (X, Y)-values. The set of observations, $\mathcal{D} = \{(X^{(1)}, Y^{(1)}), \ldots, (X^{(N)}, Y^{(N)})\}$, is called a *training set*. The idea is that the *training set* is used to "train" a predictor, that is, to identify a function of X with the goal of guessing (at best) the correct value of Y for the corresponding X.

We assume that the input variable X is a d-dimensional vector, $X \in \mathcal{X} \subseteq \mathbb{R}^d$ and the output variable Y takes *real values*, $Y \in \mathcal{Y} \subseteq \mathbb{R}$. For reasons of space, we will not address explicitly any of the many extensions to the case where Y is a vector (multioutput regression).

In machine learning, the word *regression* is reserved for the case where Y can take real values, while the prediction problem where the values of Y are restricted to a *finite* set of possible values is called *classification*.

2 LEARNING FROM NOISY DATA

At a fundamental level, one can assume that the value of Y is completely determined once the values of some other variables, which we denote collectively as $\mathcal{V}$, are assigned, that is

$$Y = F(\mathcal{V})$$

However, even under the nontrivial assumption that $F(\cdot)$ is computable, one rarely has access to *all* the variables $\mathcal{V}$ that would be necessary to compute its value exactly. Denoting by X the variables in $\mathcal{V}$ that we actually measure and by $\tilde{\mathcal{V}}$ the remaining ones, we can write

$$Y = F(X, \tilde{\mathcal{V}}) \tag{2}$$

This simple distinction between observed variables and hidden variables shows clearly that the aim of *exactly* predicting Y based only on X according to a formula of the kind

$$Y = f(X)$$

is doomed to failure as, in most of the cases, Y can take a set of many values for the same value of X. The following toy example should further illustrate the concept.

Example 1. *A scientist wants to predict an output variable Y based on the observation of an input variable. Assume that the variable Y is a function of two variables V_1 and V_2 that take values in the interval $[-1, 1]$, that is,*

$$Y = 0.9V_1 + 0.1V_2 \tag{3}$$

We assume that the experiment that the scientist performs in order to collect data can be described as a repeated and independent uniform sampling of (V_1, V_2) over $[-1, 1]^2$.

Consider now the case where the scientist has access only to Y and V_1, that is, $X = V_1$. Fig. 1 shows the collected data.

Only a very naive scientist might think that data reflects a strong and deterministic relation between X and Y of the kind $Y = f(X)$, such that $Y^{(i)} = f(X^{(i)})$ for all $i = 1, \ldots, N$. The dangers of assuming

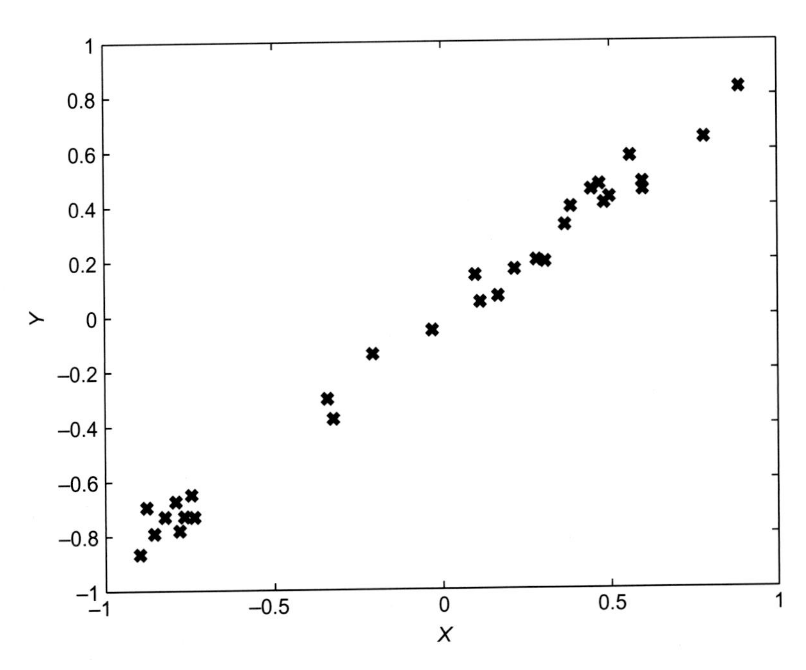

FIG. 1 The data observed by the scientist when $X = V_1$, see Eq. (3).

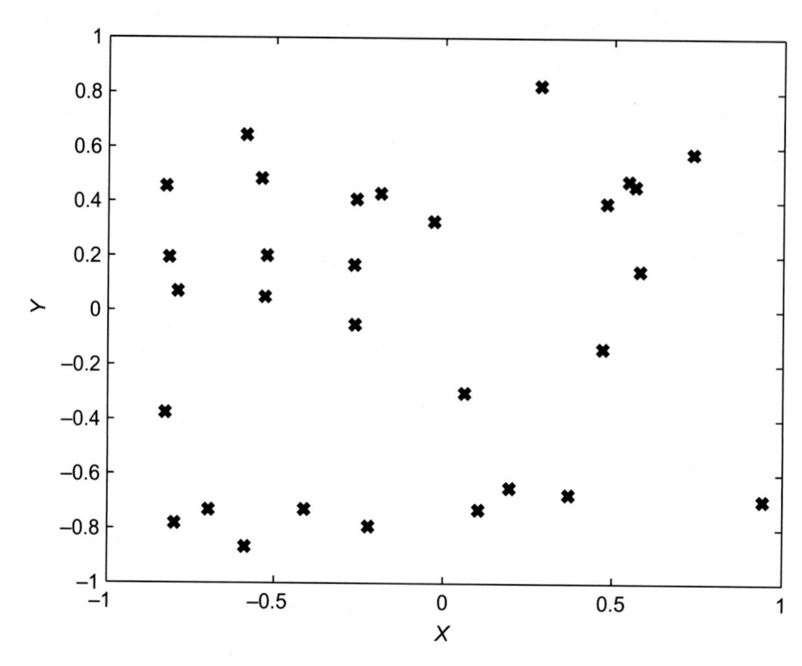

FIG. 2 The data observed by the scientist when $X = V_2$, see Eq. (3).

that a deterministic function exists would be even more apparent in the situation where only V_2 is the observable variable, that is, $X = V_2$. In this case, the same experiment would lead to the data in Fig. 2.

We know instead that such a deterministic relationship does not exist *in our example, unless both V_1 and V_2 are observable, in which case X could be redefined as a vector that includes both of them. In real life, however, it is hardly the case that we have access to* all *the factors that determine the value of an output variable. On the contrary, the number of unobserved variables could be enormous, especially when physical experiments are concerned.*

2.1 Prediction Errors

Although a function of the kind $Y = f(X)$ does not exist in general, we might still want to make an attempt at predicting the value of Y given a value of X by a function $\hat{Y}: \mathcal{X} \to \mathcal{Y}$, which we call a *point predictor*, or just *predictor*. From the discussion so far, we know that our *predictions will be wrong*. However, in many circumstances we can tolerate that $\hat{Y}(X)$ and Y are not perfectly the same, that is, we can tolerate a *prediction error.*

In order to be quantitative in evaluating the prediction errors that we incur, we have to define a function that, given a couple (X, Y) and the predictor $\hat{Y}(X)$, returns a number that quantifies the error in a way that is suitable for our purposes. Such a function, which describes how bad a prediction is for our purposes, is called a *loss function* and will be denoted by $l(\cdot)$.

A very common loss function is the *quadratic error*, which penalizes high deviations from the correct value far more than small ones, and is defined as

$$l(\hat{Y}(X), Y) = (\hat{Y}(X) - Y)^2 \tag{4}$$

A *loss* function allows us to evaluate the error incurred when $\hat{Y}(X)$ is used to predict Y for a given instance of (X, Y). In general, however, a predictor is used to predict the values of Y for more than one instance of (X, Y), and one wants an indicator of its overall performance. The most common ways to define such an indicator prescribe some kind of averaging over the possible values of X and Y. In the following section, we set up the learning problem in a probabilistic framework, which, among the other advantages, provides a natural way to average the loss over infinitely many possible values of X and Y. The probabilistic framework will provide insights also for predictions in nonprobabilistic situations (see also Section 3).

2.2 A Probabilistic Set-Up

A probabilistic description of the input-output relationship is possible under some assumptions about the experiment, such as those made in Example 1. In that case, the input-output relationship can be described through a joint probability distribution, and the observations in $\mathcal{D}$ can be thought of as independent random couples sampled from the same probability distribution. The fact that a probabilistic description exists in theory does not mean that the user has access to that description. In all the interesting cases, in fact, the probability distribution underlying the data-generation mechanism is *not* known to the user and has rather to be thought of as a hidden reality that the user would like to learn. However, postulating its existence allows the user to frame the learning problem in a mathematical set-up where powerful analytical tools are available.

In particular, we have a tool to describe the average loss incurred by a predictor $\hat{Y}(X)$ with respect to all the possible infinite realizations of (X, Y). We can, in fact, define the quantity $\mathbb{E}[l(\hat{Y}(X), Y)]$, which is often called the *risk* of the predictor $\hat{Y}(X)$, where the expected value is with respect to both X and Y. An important caveat is that, in some experiments, the past values of the input variable $X^{(1)}, \ldots, X^{(N)}$ and possibly also the future ones are chosen by an act of will or by some mechanism that might be hard or impossible to describe at a probabilistic level. In this case, the input-output dependency can be described by the set of probability distributions of Y *given* a value of X, say x, with no assumptions on the mechanism that generates X (see Example 2). When no probability distribution on X is available, the definition of risk should be suitably revisited. In general, the suitable definition of the risk can be very problem dependent. For example, one might want to use an integral measure of the loss, such as $\int_{x \in \mathcal{X}} \mathbb{E}[l(\hat{Y}(x), Y)|X = x]dx$, or a worst-case measure such as $\sup_{x \in \mathcal{X}} \mathbb{E}[l(\hat{Y}(x), Y)|X = x]$, and so forth. An extreme (but often practical) choice is neglecting everything that happens outside the available values of X and redefine the risk simply with respect to the variability of the possible output values corresponding to the observed input values $X^{(1)} = x^{(1)}, \ldots, X^{(N)} = x^{(N)}$, as $\mathbb{E}_{Y^{(1)}, \ldots, Y^{(N)}}\left[\frac{1}{N}\sum_{i=1}^{N} l(\hat{Y}(x^{(i)}), Y^{(i)})\right]$. This is called the *in-sample* risk, and it is often used for comparing different predictors, see also Section 2.4.1.

Example 2 (Fixed-X Experiments). *Assume that Y is determined by Eq. (3). The scientist can choose some values of $X = V_1$, and, for each value of X, two independent experiments are performed. The result of an experiment depends on the variation of V_2, which is independent on V_1 and is uniformly distributed on $[-1, 1]$.*

In this case, the input-output relationship can be fully described by the set of probability distributions $\{p_{Y|X=x}: x \in [-1, 1]\}$, where each $p_{Y|X=x}$ is a uniform probability over $[0.9x - 0.1, 0.9x + 0.1]$.

Considering the quadratic error loss function, we can define the predictor $\hat{Y}^*(X)$ as the predictor that minimizes the expected value of the loss for any possible value of X. More formally, the value of the function $\hat{Y}^*(X)$ when $X = x$ is given by the value $\hat{y}^*$ that minimizes $\mathbb{E}[(\hat{y} - Y)^2|X = x]$ with respect to all the possible values $\hat{y} \in \mathbb{R}$.

It is easy to see that

$$\hat{Y}^*(X) = \mathbb{E}[Y|X] \tag{5}$$

Proof. Assuming that the expectations are well-defined and finite, for every fixed $\hat{y} \in \mathbb{R}$ we get

$$\mathbb{E}[(\hat{y} - Y)^2|X = x] = \hat{y}^2 + \mathbb{E}[Y^2|X = x] - 2\hat{y}\mathbb{E}[Y|X = x]$$

The value $\hat{y}^*$ that maximizes this quantity is obtained by setting to zero the derivative with respect to $\hat{y}$: we get

$$2\hat{y} - 2\mathbb{E}[Y|X = x] = 0$$

which is satisfied by $\hat{y}^* = \mathbb{E}[Y|X = x]$. □

Because $\mathbb{E}[(\hat{Y}^*(X) - Y)^2] = \mathbb{E}_X[\mathbb{E}[(\hat{Y}^*(X) - Y)^2|X]]$, and $\hat{Y}^*(X)$ minimizes the average quadratic error over Y for every value of X, $\hat{Y}^*(X)$ is also optimal, on average, over all the possible values of X, that is, $\mathbb{E}[(\hat{Y}^*(X) - Y)^2] \le \mathbb{E}[(\hat{Y}(X) - Y)^2]$ for every possible predictor $\hat{Y}(X)$. In view of these facts, we call $\hat{Y}^*(X)$ the *best predictor*.

The best predictor $\hat{Y}^*(X)$ can be used to represent the input-output relationship in the following way:

$$Y = f(X) + \nu \tag{6}$$

where $f(X) := \hat{Y}^*(X) = \mathbb{E}[Y|X]$ and ν is a stochastic residual $\nu := Y - \mathbb{E}[Y|X]$. In this representation, which is known as a *stochastic data model*, $f(X)$ plays the role of the hidden *structure* that we want to learn from data, while ν accounts for the irreducible variability that depends on the structural lack of determinism, and is often called the *noise*. Starting from Eq. (6) we can formulate the *learning problem* as follows:

> *Starting from the set of (noisy) observations $\mathcal{D}$, the learning problem is the problem of identifying, in a certain class of functions C, the function $f(X)$ in Eq. (6).*

The *ansatz* for many important works in statistics is that the underlying function $f(X)$ has some simple structure, a typical case being that $f(X)$ is a linear function of the elements of X. As for the noise, ν has zero mean and is uncorrelated with $f(X)$ by definition, being $\mathbb{E}[f(X)\nu] = \mathbb{E}[\mathbb{E}[Y|X](Y - \mathbb{E}[Y|X])] = 0$. However, stronger assumptions are usually introduced, for example, that v is *independent* of X (a circumstance known as *homoskedasticity*, as opposed to *heteroskedasticity*), that it is Gaussian, and so forth. Although the approach of working with a stochastic data model under strong limiting assumptions has been criticized for its widespread uncritical use (Breiman et al., 2001), it has been a fruitful source of insights and methods for driving both theory and practical applications.

In the following section, we briefly present the celebrated least squares method for linear regression, which had a very large impact in the regression literature and can be studied in the light of a stochastic data model such as Eq. (6).

2.3 The Least Squares Method for Linear Regression

For $X \in \mathbb{R}^d$, we consider the class of predictors of the kind $\hat{Y}(X) = \theta_0 + \theta_1 X_1 + \cdots + \theta_d X_d$. For brevity, we write $\theta_0 + \theta_1 X_1 + \theta_d X_d$ as $\theta^{\mathrm{T}}\mathbf{X}$, where $\theta = [\theta_0, \ldots, \theta_d]^{\mathrm{T}}$, and $\mathbf{X}$ is defined as $\mathbf{X} = [1, X_1, \ldots, X_d]^{\mathrm{T}}$. Similarly, $\mathbf{X}^{(i)} = [1, X_1^{(i)}, \ldots, X_d^{(i)}]^{\mathrm{T}}$. Starting from a set of data $\mathcal{D} = \{(X^{(i)}, Y^{(i)}) \colon i = 1, \ldots, N\}$, we look for the value of θ that minimizes

$$\sum_{i=1}^{N} (Y^{(i)} - \theta^{\mathrm{T}}\mathbf{X}^{(i)})^2 \tag{7}$$

that is the value of θ that minimizes the average quadratic error *on the training set*. We denote the solution (least squares estimate) by $\hat{\theta}_{\mathrm{LS}}$. Let $\mathbf{X}_o$ be the $(d+1) \times N$ matrix whose ith column is the vector $\mathbf{X}^{(i)}$ and $\mathbf{Y}_0 = [Y^{(1)}, \ldots, Y^{(N)}]^{\mathrm{T}}$, then, if $\mathbf{X}_o\mathbf{X}_o^{\mathrm{T}}$ is invertible, it is easy to check that the minimizer of Eq. (7) is unique, and can be written as

$$\hat{\theta}_{\mathrm{LS}} = (\mathbf{X}_o\mathbf{X}_o^{\mathrm{T}})^{-1}\mathbf{X}_o\mathbf{Y}_o \tag{8}$$

Given $\hat{\theta}_{\mathrm{LS}}$, we define the predictor *trained on* $\mathcal{D}$ as $\hat{Y}_N^{\mathrm{LS}}(X) = \hat{\theta}_{\mathrm{LS}}X$, where the subscript N reminds us that it has been chosen based only on the N input-output couples in our training set. The training set is all we have at our disposal in the real world, all the other (infinite) realizations of (X, Y) being still unobserved. The probabilistic framework provides us with a way to study the behavior of our predictor with respect to all the other infinite possible instances of (X, Y).

2.3.1 *The Least Squares Method and the Best Linear Predictor*

In order to focus on the main ideas, we now assume that X is scalar, and we just consider linear predictors of the kind $\hat{Y}(\theta) = \theta X$. A predictor like $\hat{Y}(\theta)$ is often called a *regression line* (see the historical note in Section 1). Extending the message of this section to linear predictors of the kind $\hat{Y}(X) = \theta_0 + \theta_1 X_1 + \cdots + \theta_d X_d$, $X \in \mathbb{R}^d$, is a matter of technicalities.

Starting from the training set $\mathcal{D} = \{(X^{(i)}, Y^{(i)}) : i = 1, \ldots, N\}$, we find the value of θ that minimizes $\sum_{i=1}^{N}(Y^{(i)} - \theta X^{(i)})^2$, and we get

$$\hat{\theta}_{\mathrm{LS}} = \frac{\sum_{i=1}^{N} X^{(i)}Y^{(i)}}{\sum_{i=1}^{N} X^{(i)2}} \tag{9}$$

Given $\hat{\theta}_{\text{LS}}$, we define the predictor *trained on* $\mathcal{D}$ as $\hat{Y}_N^{\text{LS}}(X) = \hat{\theta}_{\text{LS}}X$. So far, we have not used any assumption on how data are generated. However, if X and Y are stochastic, it holds true under mild assumptions that by increasing N, the sample average $\frac{1}{N}\sum_{i=1}^{N} X^{(i)}Y^{(i)}$ tends to its expected value $\mathbb{E}[YX]$ and $\frac{1}{N}\sum_{i=1}^{N}(X^{(i)})^2$ tends to $\mathbb{E}[X^2]$, so that

$$\lim_{N\to\infty} \hat{\theta}_{\text{LS}} = \frac{\mathbb{E}[YX]}{\mathbb{E}[X^2]}$$

In the limit $N \to \infty$, the obtained predictor is therefore

$$\hat{Y}_\infty^{\text{LS}}(X) = \frac{\mathbb{E}[YX]}{\mathbb{E}[X^2]}X$$

The obtained predictor is a notable one, in fact, for every linear predictor $\hat{Y}(X) = \theta X$ we get $\mathbb{E}[(\hat{Y}(X) - Y)^2] = \theta^2\mathbb{E}[X^2] + \mathbb{E}[Y^2] + 2\theta\mathbb{E}[XY]$, whose minimum point is at $\frac{\mathbb{E}[YX]}{\mathbb{E}[X^2]}$. This means that the performance of $\hat{Y}_\infty^{\text{LS}}(X)$ is the best we can achieve if (i) we measure the performance of a predictor by its average quadratic error over all the possible couples (X, Y) and (ii) we restrict to the class of linear predictors $\hat{Y}(X) = \theta X$. This is why $\hat{Y}_\infty^{\text{LS}}(X)$ is often called the *best linear predictor*.

We can apply the least squares method for finding a predictor in Example 1, in the case where $X = V_1$, Fig. 3. In this case, we have that $\frac{\mathbb{E}[YX]}{\mathbb{E}[X^2]}X = 0.9X = \mathbb{E}[Y|X]$. Therefore, in this special case, the predictor $\hat{Y}_\infty^{\text{LS}}(X)$ coincides with the *best predictor* $\hat{Y}^*(X)$ as defined in

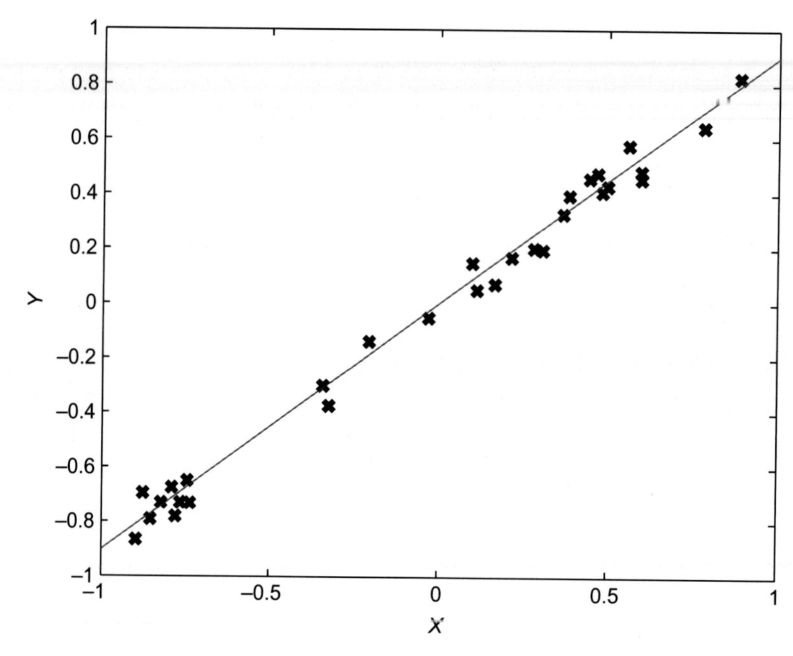

FIG. 3 A regression line obtained with the least squares method.

Section 2.2, see Eq. (5). However, this is *not* always the case. In Example 1 this happens because of the special, that is, linear, data-generation mechanism, which can be represented as a special instance of Eq. (6) where $f(X) = 0.9X$, and because the noise is independent. In general, on the other hand,

$$\frac{\mathbb{E}[YX]}{\mathbb{E}[X^2]}X \neq \mathbb{E}[Y|X]$$

and therefore $\hat{Y}_{\infty}^{LS}(X) \neq \hat{Y}^*(X)$.

When a linear structure is present, and a special kind of noise, namely Gaussian noise, affects the data, the least squares method has another interesting interpretation that is illustrated in the following paragraph.

2.3.2 *The Least Squares Method and the Maximum Likelihood Principle*

Assume now that X takes value in $\mathbb{R}^d$, and the input-output relationship can be described by the stochastic data model (6), where

$$f(X) = \theta_0^* + \theta_1^* X_1 + \cdots + \theta_d^* X_d \tag{10}$$

and ν is an independent Gaussian noise with zero mean and variance σ_ν^2 (in short notation, $\nu \sim \mathcal{N}(0, \sigma_\nu^2)$).

Remark 1. *The Gaussianity of the noise is certainly a limiting assumption; however, it can be quite realistic in situations when the noise arises from an averaged sum of many independent contributions. For example, let us consider an extension of Example 1 where we have $d+1$ observable input variables and an infinite amount of unobserved variables, $\mathcal{V} = \{V_{d+1}, V_{d+2}, \ldots\}$, which are uniformly sampled over $[-1, 1]$, so that*

$$Y = F(X, \mathcal{V}) = \sum_{\ell=0}^{d} \theta_d^* X_d + \lim_{M\to\infty} c_M \sum_{\ell=d+1}^{d+M} \theta_\ell^* V_\ell$$

Assume first that for every $\ell = d+1, d+2, \ldots$, it holds that $|\theta_\ell^*| < B < \infty$, and that the number of variables that contribute significantly to the variance of Y is really infinite, that is, $\sum_{\ell=d+1}^{\infty} \theta_\ell^{*2} = \infty$. Another reasonable assumption is that the variance of the unobservable part, overall, is finite and nonzero, and this can be achieved by assuming that the averaging factor c_M is such that $0 \neq \lim_{M\to\infty} c_M^2 \sum_{\ell=d+1}^{d+M} \theta_\ell^{*2} < \infty$. These assumptions are sufficient to conclude that the unobservable component $\lim_{M\to\infty} c_M \sum_{\ell=d+1}^{M+d} \theta_\ell^* V_\ell$ has a Gaussian distribution in view of the Central Limit Theorem, see, for example, Billingsley (1995, Chapter 5, Example 27.4). Analog conclusions can be drawn for large classes of distributions of the V_ℓ variables, and even relaxing to some extent the independence assumption.

Then, for a fixed value x of X, the probability density of Y given $X = x$ is the Gaussian density

$$p_{Y|X}(y; x) = \frac{1}{\sqrt{2\pi}\sigma_\nu} e^{-\frac{(y - f(x))^2}{2\sigma_\nu^2}}$$

For independent observations, the joint probability distribution of $Y^{(1)}, \ldots, Y^{(N)}$ given the N input values $X^{(1)} = x^{(1)}, \ldots, X^{(N)} = x^{(N)}$ is given by the product of the individual densities, that is,

$$p_{Y^{(1)} \ldots Y^{(N)} | X^{(1)} \ldots X^{(N)}}(y^{(1)}, \ldots, y^{(1)} | x^{(1)}, \ldots, x^{(N)}) = \prod_{i=1}^{N} \frac{1}{\sqrt{2\pi}\sigma_v} e^{-\frac{(y^{(i)} - f(x^{(i)}))^2}{2\sigma_v^2}} \tag{11}$$

Clearly, the function $f(X)$ is unknown in practice. However, under the assumption that it can be expressed as in Eq. (10), learning the correct function boils down to learning the correct parameter $\theta^* = [\theta_0^*, \ldots, \theta_d^*]^{\mathrm{T}}$.

By plugging the values of the N observations in $\mathcal{D}$ into the probability density function (11) we obtain a number, but this number is unknown because it is determined by the unknown θ^*. However, if we replace θ^* in Eq. (11) with a candidate θ, then we get a number that depends on the chosen θ. In this way, we have defined the following function of θ

$$\mathcal{L}(\theta | \mathcal{D}) = \prod_{i=1}^{N} \frac{1}{\sqrt{2\pi}\sigma_v} e^{-\frac{(Y^{(i)} - \theta^{\mathrm{T}} \mathbf{X}^{(i)})^2}{2\sigma_v^2}}$$

which is called "the *likelihood function* of θ given the observations ($\mathcal{D}$)." Among all the possible values of θ, we can choose the one that maximizes this function as an estimate of the true value. This idea, of estimating a parameter by maximizing the likelihood function, is known as the *likelihood principle*.

Note that $\max_\theta \mathcal{L}(\theta|\mathcal{D}) = \max_\theta \log(\mathcal{L}(\theta|\mathcal{D})) = \min_\theta(-\log(\mathcal{L}(\theta|\mathcal{D}))) = \min_\theta \sum_{i=1}^{N} (Y^{(i)} - \theta^{\mathrm{T}} \mathbf{X}^{(i)})^2$. Thus, in the present context, the solution obtained according to the maximum likelihood principle is the *least squares estimate* $\hat{\theta}_{\mathrm{LS}}$, see Eqs. (7), (8).

2.3.3 A More General Approach and Higher-Order Predictors

In the previous sections, we have extended the input vector $X = [X_1, \ldots, X_d]^{\mathrm{T}}$ with a "1" by defining $\mathbf{X} = [1, X_1, \ldots, X_d]^{\mathrm{T}}$. More substantial transformations of the input are possible. For example, starting from a scalar observation $X \in \mathbb{R}$ one can build the vector of its powers up to order n, $[1, X, X^2, X^3, \ldots, X^n]$, and use this vector in place of the original scalar X in the least squares formulae. This leads to the so-called *polynomial regression*, which can be justified in view of the probabilistic set-up whenever it is reasonable to assume that $f(X)$ is a polynomial of order (at most) n. The idea can be pushed further: by using some basis functions, $\Phi_0(X), \ldots, \Phi_{k-1}(X)$, a d-dimensional input $X \in \mathbb{R}^d$ can be mapped into a vector $[\Phi_0(X), \Phi_1(X), \ldots, \Phi_{k-1}(X)]$ in a k-dimensional space, and one can assume that $f(X) = \theta_0 \Phi_0(X) + \cdots + \theta_{k-1} \Phi_{k-1}(X)$ for a suitable θ. The least squares method can be applied in this generalized set-up by writing the predictor as $\hat{Y}(X) = \theta^{\mathrm{T}} \Phi(X)$, where $\Phi(\cdot) = [\Phi_0(\cdot), \ldots, \Phi_{k-1}(\cdot)]^{\mathrm{T}}$, and by minimizing $\sum_{i=1}^{N} (\hat{Y}(X^{(i)}) - Y^{(i)})^2$ with respect to θ, as usual. We get

$$\hat{\theta}_{\mathrm{LS}} = (\Phi_o \Phi_o^{\mathrm{T}})^{-1} \Phi_o \mathbf{Y}_o \tag{12}$$

where Φ_o is the $k \times N$ matrix whose ith column is $\Phi(X^{(i)})$, and letting $\mathbf{Y}_o$ be the column vector $[Y^{(1)}, \ldots, Y^{(N)}]^{\mathrm{T}}$. Finally, the least squares predictor is $\hat{Y}^{\mathrm{LS}}(X) = \theta_{\mathrm{LS}}^{\mathrm{T}} \Phi(X)$.

This powerful idea can be brought to a notable level of sophistication, for example, the number of basis functions can go to infinite. The basis functions can be selected so as to satisfy some good mathematical or practical properties, such as the ability of approximating any possible function $f(X)$ in a very rich set. However, when predictors are selected from a rich class (e.g., high-order polynomials), a fundamental issue that is known as *overfitting* has to be considered. This is the subject of the next section.

2.4 Overfitting

As a motivating example, consider again Fig. 1, and assume that we do not know anything about the data-generation mechanism, so that we try to train two polynomial predictors of different orders. Figs. 4 and 5 show the best polynomial predictors of the kind

$$\hat{Y}_{[3]}(X) = \theta_0 + \theta_1 X + \theta_2 X^2$$

and

$$\hat{Y}_{[30]}(X) = \theta_0 + \theta_1 X + \theta_2 X^2 + \cdots + \theta_{29} X^{29}$$

respectively, trained on the 30 data in Fig. 1 by using the least squares method. We denote by $\hat{Y}^{LS}_{[3]}(X)$ and $\hat{Y}^{LS}_{[30]}(X)$ the trained predictors.

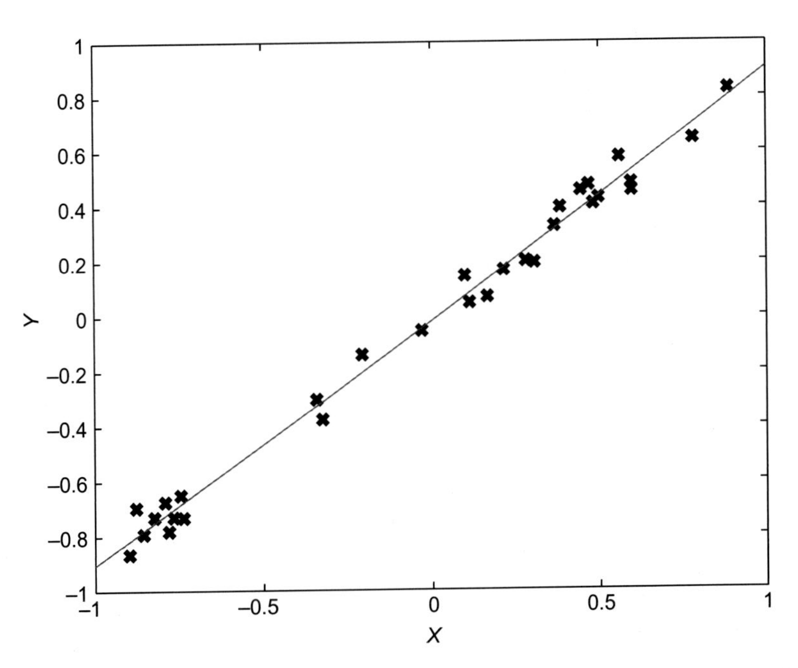

FIG. 4 The polynomial predictor $\hat{Y}^{LS}_{[3]}(X)$ was trained on the data in Fig. 1, by minimizing the quadratic prediction error on the training set.

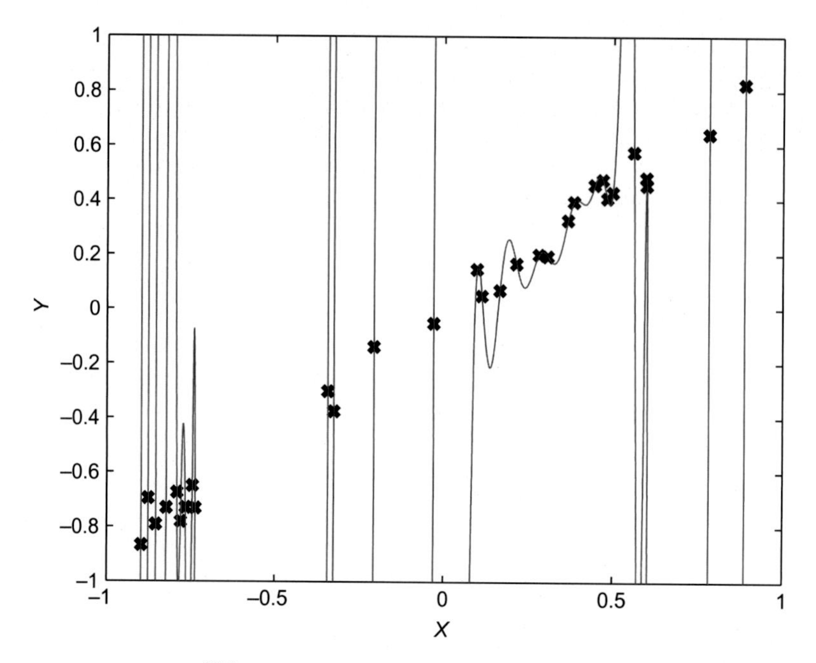

FIG. 5 The polynomial predictor $\hat{Y}^{\mathrm{LS}}_{[30]}(X)$ was trained on the data in Fig. 1, by minimizing the quadratic prediction error on the training set (note that the empirical error here is equal to zero, i.e., the predictor makes no error on the training set). For many values of X, the predictor takes values outside $[-1, 1]$ (not shown).

Which is better? Clearly, the higher is the order of the predictor, the lower is $\frac{1}{N}\sum_{i=1}^{N}(Y^{(i)} - \hat{Y}^{\mathrm{LS}}(X^{(i)}))^2$, which is the average of the quadratic prediction error computed on the training set $\mathcal{D}$. This quantity, which is observable because it depends on the training set, is often called the *empirical risk*. In particular, for $\hat{Y}^{\mathrm{LS}}_{[30]}(X)$, the empirical risk is equal to zero. However, it is clear from Fig. 5 that $\hat{Y}^{\mathrm{LS}}_{[30]}(X)$ will be very bad at predicting Y for new couples (X, Y) that are not among the 30 couples in $\mathcal{D}$. In other words, the empirical risk of $\hat{Y}^{\mathrm{LS}}_{[30]}(X)$ does not reflect its risk. The risk of the predictor $\hat{Y}^{\mathrm{LS}}_{[30]}(X)$ is defined as $\mathbb{E}[(Y - \hat{Y}^{\mathrm{LS}}_{[30]}(X))^2 | \mathcal{D}]$, where the conditioning on $\mathcal{D}$ indicates that the expectation is with respect to (X, Y) only, while $\mathcal{D}$, the observations based on which the predictor $\hat{Y}^{\mathrm{LS}}_{[30]}(X)$ has been selected, are kept fixed. The concepts of empirical risk and risk allow us to distinguish the performance of a predictor on its training set from its performance on new data: this is key to selecting a good predictor.

The probabilistic set-up helps us understand why increasing the order of the predictor can worsen its risk. Intuitively, when too much flexibility is allowed in the choice of the predictor, the predictor will end up describing the "noise" in the observed data rather than "the informative part of the data." Referring to the stochastic model (6), the elusive concepts of "noise" and of "informative part of the data" are well reified into the random variable ν and the function $f(X)$, respectively. In short, there is a trade-off between choosing a predictor $\hat{Y}^{\mathrm{LS}}_{[k]}$

that is simple enough (for the empirical risk to be representative of the risk) and yet complex enough (to approximate sufficiently well the mechanism underlying the generation of data).

Clearly, our example in polynomial regression is just an instance of a more general problem, where predictors can be grouped into classes of increasing complexity $\mathcal{C}_0 \subseteq \mathcal{C}_1 \subseteq \cdots \subseteq \mathcal{C}_k$, and one can choose one predictor, say $\hat{Y}^{l*}_{[j]}(X)$, for each class $\mathcal{C}_j$ by minimizing the value of the empirical risk $\frac{1}{N}\sum_{i=1}^{N} l(\hat{Y}^{l*}_{[j]}(X^{(i)}), Y^{(i)})$ over the training set. In the end, one wants to choose only one predictor, denoted by $\hat{Y}^{l*}_{[0:k]}(X)$, among the trained predictors $\hat{Y}^{l*}_{[0]}(X), \hat{Y}^{l*}_{[1]}(X), \ldots, \hat{Y}^{l*}_{[k]}(X)$. We now discuss some ideas and methods to perform this selection.

2.4.1 The Order Selection Problem

Ideally, we would like to *test* each predictor $\hat{Y}^{l*}_{[j]}$, $j = 0, \ldots, k$, on an infinite amount of new data (*validation data*), and choose $\hat{Y}^{l*}_{[0:k]}(X)$ as the one that performs better. This is clearly unfeasible in practice, and different methods can be used to compensate this impossibility. As a first option, we can resort to a finite validation set $\{(X^{(N+1)}, Y^{(N+1)}), \ldots, (X^{(N+N_V)}, Y^{(N+N_V)})\}$ of N_V new observations. Under the usual assumption that the observed couples (X, Y) are generated independently and according to the same probability distribution,[1] we can estimate the risk

$$\mathbb{E}[l(\hat{Y}^{l*}_{[j]}(X), Y)|\mathcal{D}]$$

by

$$\frac{1}{N_v}\sum_{i=1}^{N_v} l(\hat{Y}^{l*}_{[j]}(X^{(N+i)}), Y^{(N+i)})$$

for every trained predictor $\hat{Y}^{l*}_{[j]}$, $j = 0, \ldots, k$, and choose the predictor that shows a lower prediction error value on the new dataset. One problem with this procedure is that many data are used for validation rather than for improving the choice of a predictor in a given class, and the way in which we split the available data into training set and validation set affects the final outcome.

Cross-validation is an alternative approach to order selection. In cross-validation methods, the set of all the available data is split into m sets $S_1, \ldots, S_m$ of equal size. For every class $\mathcal{C}_j$, m predictors, instead of one, are obtained. Each predictor is trained on $m-1$ sets among $S_1, \ldots, S_m$, while the remaining one is used as a validation set to estimate the risk of the predictor. The average of these risk estimates is an indicator of how well, on average, predictors trained in $\mathcal{C}_j$ perform on new data. Note, however, that this procedure goes in the direction of estimating $\mathbb{E}[l(\hat{Y}^{l*}_{[j]}(X), Y)]$, where the expected value is with respect to the variability of both

[1] This can be critical in many applications, where observations are correlated. See, for example, Burman et al. (1994) and Racine (2000).

the training set (which determines the choice of $\hat{Y}^{l*}_{[j]}(X)$) and the new observation (X, Y), rather than estimating $\mathbb{E}[l(\hat{Y}^{l*}_{[j]}(X), Y)|\mathcal{D}]$ where the variability is only with respect to (X, Y) while the training set $\mathcal{D}$ is kept fixed to the one that is actually at our disposal, and based on which $\hat{Y}^{l*}_{[j]}$ has been chosen (for a discussion on this issue see, e.g., Chapter 7 in Hastie et al., 2009).

A caveat about validation and cross-validation methods is about the correct way of interpreting and *reporting validation results*. Once the predictor $\hat{Y}^{l*}_{[0:k]}(X)$ has been chosen, its performance on the validation set *should not* be reported as a valid estimate of its performance on *new data*. In fact, because the selection of $\hat{Y}^{l*}_{[0:k]}(X)$ is based on the performance of $\hat{Y}^{l*}_{[0]}(X), \hat{Y}^{l*}_{[1]}(X), \ldots, \hat{Y}^{l*}_{[k]}(X)$ on the validation set, the validation set contributes to the choice of $\hat{Y}^{l*}_{[0:k]}(X)$: this means that the validation set is, in fact, part of the training set for $\hat{Y}^{l*}_{[0:k]}(X)$![2] As a safe solution, another, independent, set should be kept for the estimate of the risk of $\hat{Y}^{l*}_{[0:k]}(X)$. This latter set, to be used *after* any selection has been performed, is usually called the *test set* in order to distinguish it from the *validation set*.

ERROR DECOMPOSITION: THE BIAS VERSUS VARIANCE TRADE-OFF

In general, in order to compute the risk of the predictor $\hat{Y}^{l*}_{[j]}(X) \in \mathcal{C}_j$ trained on a given $\mathcal{D}$, that is $\mathbb{E}[(Y - \hat{Y}^{l*}_{[j]}(X))^2|\mathcal{D}]$, we need either some knowledge about the distribution of (X, Y) or some extra observations that can be used for estimation (see the preceding discussion). However, we now show that a simple analysis of the expected value of the risk, $\mathbb{E}[(Y - \hat{Y}^{l*}_{[j]}(X))^2] = \mathbb{E}_{\mathcal{D}}[\mathbb{E}[(Y - \hat{Y}^{l*}_{[j]}(X))^2|\mathcal{D}]]$ provides us with some important insights into the role that the richness of the class $\mathcal{C}_j$ plays in the risk of the obtained predictor. We start by studying $\mathbb{E}[(Y - \hat{Y}^{l*}_{[j]}(X))^2|X = x]$ at a given value x of X. As already observed, $\mathbb{E}[(Y - \hat{Y}^{l*}_{[j]}(X))^2] = \mathbb{E}_X[\mathbb{E}[(Y - \hat{Y}^{l*}_{[j]}(x))^2]|X]$. For the sake of clarity, in what follows, we indicate explicitly when the expectation is with respect to the variability of the training set $\mathcal{D}$ or with respect to the nondeterminism of Y for a given $X = x$, or both.

The expected risk (for $X = x$) can be written as

$$\begin{aligned}\mathbb{E}_{\mathcal{D},Y|X=x}[(Y - \hat{Y}^{l*}_{[j]}(x))^2] &= \mathbb{E}_{\mathcal{D},Y|X=x}[(Y - \mathbb{E}_{\mathcal{D}}[\hat{Y}^{l*}_{[j]}(x)] + \mathbb{E}_{\mathcal{D}}[\hat{Y}^{l*}_{[j]}(x)] - \hat{Y}^{l*}_{[j]}(x))^2] \\ &= \mathbb{E}_{Y|X=x}[(Y - \mathbb{E}_{\mathcal{D}}[\hat{Y}^{l*}_{[j]}(x)])^2] + \mathbb{E}_{\mathcal{D}}[(\mathbb{E}_{\mathcal{D}}[\hat{Y}^{l*}_{[i]}(x)] - \hat{Y}^{l*}_{[i]}(x))^2],\end{aligned}$$

and the first term can be further decomposed as

$$\begin{aligned}\mathbb{E}_{Y|X=x}[(Y - \mathbb{E}_{\mathcal{D}}[\hat{Y}^{l*}_{[j]}(x)])^2] &= \mathbb{E}_{Y|X=x}[(Y - f(x) + f(x) - \mathbb{E}_{\mathcal{D}}[\hat{Y}^{l*}_{[j]}(x)])^2] \\ &= (f(x) - \mathbb{E}_{\mathcal{D}}[\hat{Y}^{l*}_{[j]}(x)])^2 + \mathbb{E}_{Y|X=x}[(Y - f(x))^2]\end{aligned}$$

[2] This phenomenon is often described as "peeking" at the test data (Russell and Norvig, 2009) or as a "data leakage" from the test set into the training set.

where $f(x) = \mathbb{E}[Y|X = x]$, see Eq. (6). So, overall, we have decomposed the quadratic error into three terms (*error decomposition*):

- The *bias* term $(f(x) - \mathbb{E}_{\mathcal{D}}[\hat{Y}^{l^*}_{[j]}(x)])^2$, which accounts for how well the underlying mechanism $f(x)$ is approximated by the predictor on average over all the possible training sets $\mathcal{D}$. Note that if the class C_j is not rich enough and $f(x) \notin C_j$, then this term must be positive for a generic x. Consider, for example, the case where $f(x)$ is a polynomial of order higher than j and C_j is the class of polynomials of order $j-1$. Then, the bias term cannot be zero for all the values $X^{(1)}, \ldots, X^{(N)}$ in the training set.
- The *irreducible error* term $\mathbb{E}_{Y|X=x}[(Y - f(x))^2]$. It accounts for the noise in data (ν in Eq. 6), that is, for the inherent nondeterminism in the data-generation mechanism.
- The *variance* term $\mathbb{E}_{\mathcal{D}}[(\mathbb{E}_{\mathcal{D}}[\hat{Y}^{l^*}_{[j]}(x)] - \hat{Y}^{l^*}_{[j]}(x))^2]$, which accounts for the variability of the predictor with respect to a training set (of a fixed size). It is smaller for simple classes and larger for richer ones. As a very extreme case, consider the case where the class C_0 contains only one predictor: then the predictor does not depend on the training set and the variance term is always zero. On the other hand, consider a polynomial predictor with j coefficients (i.e., of order $j-1$) that is trained over j observations $(X^{(k)}, Y^{(k)})$, $k = 1, \ldots, j$ according to the least squares algorithm. Studying the variance term for every x is not trivial; however, we can get an idea of what happens if we evaluate x at one of the points in the training set $X^{(1)}, \ldots, X^{(N)}$. In other words, assume that $X^{(1)}, \ldots, X^{(N)}$ are given, so that the variability of $\mathcal{D}$ is just with respect to $Y^{(1)}, \ldots, Y^{(N)}$. Then, when $x = X^{(k)}$, we get $\hat{Y}^{l^*}_{[j]}(x) = Y^{(k)}$ and $\mathbb{E}_{\mathcal{D}}[Y^{(k)}] = f(X^{(k)})$, from which we can conclude that the variance term at each point $X^{(k)}$ increases the quadratic error of a value that is as high as the value of the irreducible error at $X^{(k)}$. Intuitively, the variance term will be worse for values of X that are not in the training set (see, e.g., Fig. 5, where the predictor is a polynomial of order 29 trained on 30 data).

Overall, we can conclude that learning schemes that are based on simple predictors lead to a high bias term and a low variance term. Complex predictors, instead, will be more sensitive to the training set (high variance), and their error will have a smaller bias. This is known as the *bias versus variance* trade-off. It is important to remark that what matters here is the complexity of the predictor class *relatively to N, the size of the training set*. At the increasing of N, both the bias term and the variance term can go to zero, even for a complex predictor class (under suitable assumptions). So, if we had an infinite amount of observations, we would be happy to choose a predictor class that contains the true $f(X)$. However, in the real world, we have to deal with a finite and fixed N, and this implies that we will get a high variance term for complex classes of predictors. A consequence of this fact is that, often, it is more convenient to choose a simple class that does not contain the true $f(X)$ (so that we will have a positive bias term) but still minimizes the prediction error because of the lower variance term.

SOME POPULAR ORDER SELECTION CRITERIA

Various criteria to take into account the bias-variance tradeoff in the choice of the predictor order have been proposed. Typically, predictors end up being ranked according to some combination of the value of their empirical risk and a penalization term that depends on the size of their tunable parameters vector θ. The most basic idea is to choose among $Y^{l^*}_{[0]}, \ldots, Y^{l^*}_{[k]}$ the predictor that minimizes some heuristic estimate of the expected risk. For example, estimates can be obtained by resorting to asymptotic arguments or by replacing the risk with its in-sample version. More sophisticated criteria have been proposed. We mention here the Akaike's information criterion (AIC) (Akaike, 1981), which provides a way to adjust the maximum likelihood score (see Section 2.3.2) in order to suitably penalize high-order predictors. AIC aims at minimizing the distance (more precisely, the *Kullback-Leibler divergence*) between the true probability distribution of Y given X and the probability distribution that is obtained by replacing the true θ^* with its maximum likelihood estimate.

A related criterion is known as Bayesian information criterion (BIC), which can be derived in a Bayesian set-up (see also Section 4). BIC usually leads to a stronger penalization of the model complexity (Schwarz, 1978).

Another class of approaches revolves around the minimum description length (MDL) principle (Rissanen, 1978; Grünwald, 2007) and finds its roots in Kolmogorov's attempts of defining rigorously the concept of complexity of data, which led to the definition of *Kolmogorov's complexity* and to an enthralling field of investigations (Li and Vitányi, 1990). Ideally, one would like to define the learning process as the process of finding the most compact representation of the available data $\mathcal{D}$. This can be done in a data-based way, by resorting to information theoretical tools, in particular the theory of codes. Stimulating connections between the MDL principle and probabilistic approaches such as BIC can be established thanks to the well-established relationships between coding and probability.

We conclude this (necessarily incomplete) list of approaches by mentioning the theory of Vapnik and Chervonenkis (VC theory) (Vapnik, 1998). Consider *simultaneously* all the infinite predictors $\hat{Y}(\cdot)$ belonging to a certain class $\mathcal{C}$: is it possible to bound the difference between the empirical risk and the actual risk of *every* predictor $\hat{Y}(\cdot)$ in the class? Conditions for the existence of such bounds, that hold true for finite samples of training data, have been discovered, and are essentially related to the complexity of the class of predictors that one wants to consider. The VC theory does not assume any specific data-generation model, and relies only on the fact that data are sampled independently and according to a fixed, unknown, probability distribution. An order selection approach based on this theory is known as structural risk minimization (SRM). Besides being tightly related to the development of the celebrated support vector machines (Cortes and Vapnik, 1995), connections of the VC theory with learning based on compression schemes (Littlestone and Warmuth, 1986; Floyd and Warmuth, 1995), MDL, PAC learning (Valiant, 1984, 2013), and even merely philosophical concepts such as Karl Popper's Falsificationism (Popper, 1935; Corfield et al., 2009) have been investigated. Traditionally, bounds on the risk that are delivered by the VC theory are data- and algorithm-independent in the sense that they do not depend on the observations in the training set (i.e., only the size of the training set counts) or on the specific algorithm that is used to select a predictor in its own class $\mathcal{C}$. Research on *algorithm-aware* conditions for a learning process to be effective (see, e.g., Bousquet and Elisseeff, 2002; Mukherjee et al., 2006; Shalev-Shwartz et al., 2010), and on *data-dependent* bounds (see, e.g., Koltchinskii, 2001; Herbrich and Williamson, 2002; Boucheron et al., 2004; Bartlett et al., 2005) has been carried out in more recent years.

2.4.2 Regularization

So far, we have considered a sequential approach to order selection. According to this approach, (i) different classes of predictors, $\mathcal{C}_1 \subseteq \mathcal{C}_2 \subseteq \cdots \subseteq \mathcal{C}_k$, are considered; (ii) k predictors, each for its class, are chosen by minimizing some prediction error; (iii) the best predictor among the k obtained predictor is selected according to some score that should take into account the trade-off between variance and bias. Variations on the theme are possible and sometimes necessary for computational issues. For example, one can start from a simple predictor and then increase the complexity of its structure in an incremental manner, without computing explicitly all the alternative candidate predictors, see, for example, *stepwise regression schemes* (Draper and Smith, 1998).

An alternative approach is starting from a very rich class of predictors, say $\mathcal{C}_\infty$, and directly selecting one predictor by minimizing a cost function that combines (e.g., by weighted summation) a component that penalizes the prediction error in the training set and a component that penalizes some measure of "complexity." This latter component is usually called a *regularization* term.

Example 3. *Consider Example 1, with* $X = V_1$, *and assume that we want to find a predictor in the class of polynomial predictors of order* 50. *For this purpose, we define* $\mathbf{X} = [1, X, X^2, \ldots, X^{49}]^{\mathrm{T}}$ *and* $\mathbf{X}^{(i)} = [1, X^{(i)}, (X^{(i)})^2, \ldots, (X^{(i)})^{49}]^{\mathrm{T}}$. *Instead of finding the predictor that minimizes the sum of the quadratic errors*

$$\frac{1}{N}\sum_{i=1}^{N}(\theta^{\mathrm{T}}\mathbf{X}^{(i)} - Y^{(i)})^2$$

which, however, is not unique because we only have 30 *data, we look for the minimizer of*

$$\frac{1}{N}\sum_{i=1}^{N}(\theta^{\mathrm{T}}\mathbf{X}^{(i)} - Y^{(i)})^2 + \alpha\frac{1}{k}\sum_{i=0}^{k-1}\theta_i^2 \tag{13}$$

which is represented in Fig. 6 when $\alpha = 1$. *It can be shown that an equivalent way for finding a solution to Eq. (13) is minimizing* $\frac{1}{N}\sum_{i=1}^{N}(\theta^{\mathrm{T}}\mathbf{X}^{(i)} - Y^{(i)})^2$ *subject to the constraint*

$$\frac{1}{k}\sum_{i=0}^{k-1}\theta_i^2 \leq \alpha_c(\alpha)$$

where $\alpha_c(\alpha)$ *is a nonincreasing function of* α *(this can be seen by writing the Karush-Kuhn-Tucker conditions, see, e.g., Chapter 3 in Hastie et al. (2009) and in Ivanov et al. (2002)). This shows clearly that increasing* α *boils down to restricting the class of candidate predictors.*

The difficulty of selecting the model order is here moved to the problem of selecting α. *Various options for selecting* α *are available. We mention here the* discrepancy principle *and the* L-curve principle, *see, for example, Tveito et al. (2010).*

The approach followed in Example 3, Eq. (13), is an instance of the so-called *ridge regression* or *Tykhonov regularization*. Many variations on the theme are possible. For example, it is easy to see that if the regularization term uses an L1 norm rather than a Euclidean norm, sparse predictors tend to be selected, that is, predictors with many coefficients set to zero (this

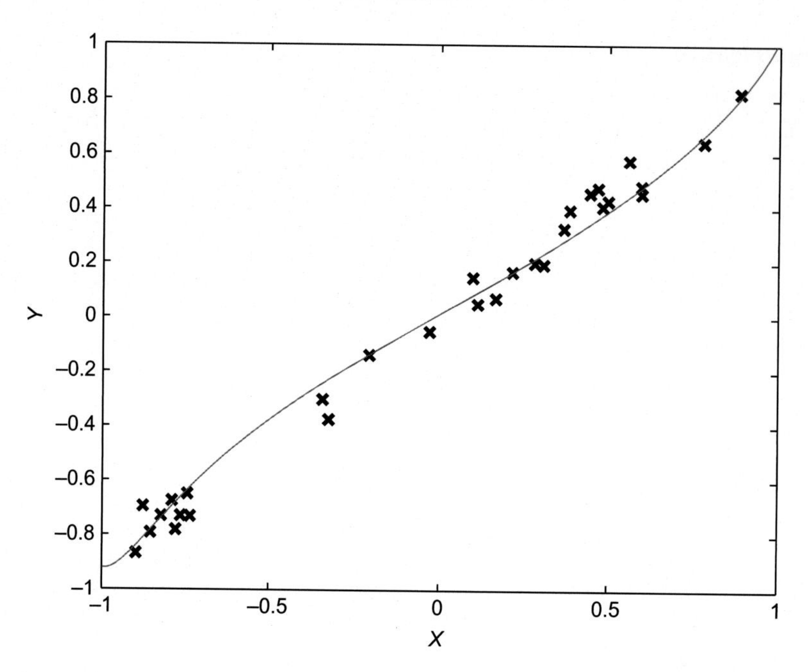

FIG. 6 The polynomial predictor of order 50 trained on the data in Fig. 1 by minimizing the regularized cost function (13) with $\alpha = 1$.

phenomenon sometimes is called *L1 magic*; Tibshirani, 1996; Candes et al., 2006). A notable fact is that regularization is able to turn ill-posed problems into well-posed ones, as in Example 3, where the solution to the plain least squares problem was not unique. Regularization can be justified from first principles in a probabilistic, Bayesian set-up, where one can use a probability measure in order to express the prior belief that the function $f(X)$ is simple (see also Section 4, Example 4).

2.5 From Point Predictors to Interval Predictors

So far, we have discussed how to train a point predictor $\hat{Y}(X)$ with the aim of minimizing its prediction error. We might instead look for a predictor whose output is a *prediction interval* rather than a single value. Prediction intervals can be built starting from point predictors, see, for example, Geisser (1993). Assume, for example, that a stochastic data model such as Eq. (6) is postulated, and that the structure of $f(X)$ is known, for example, $f(X) = \theta^* X$, with $\theta^* \in \mathbb{R}$ unknown. Then, we can aim at building a confidence interval such that $\theta^* \in [\hat{\theta}_l, \hat{\theta}_u]$ with high probability. From this, in fact, a prediction interval of the kind $[\hat{Y}_{\max}(x), \hat{Y}_{\min}(x)]$ for the new output value Y at any given value x of X can be obtained. However, constructing intervals that are guaranteed and practically useful (i.e., not too large) requires, other than the knowledge of the structure of $f(X)$, very strong assumptions on the noise, such as Gaussianity or sub-Gaussianity (Ledoux and Talagrand, 2013). Often, one relies on the asymptotic analysis of prediction intervals, which can be made rigorous under much milder assumptions, but can deliver very wrong results in the real world where only finite training sets are available.

It is remarkable, however, that by following a different line of reasoning, one can build prediction intervals for Y with strong probabilistic guarantees without any assumption on the function $f(X)$ and the noise ν. This is possible by pursuing a direct approach, where a high-probability region is built directly in the (X, Y) space. All one has to know is that couples (X, Y) are independently generated from the same, *unknown*, probability distribution, which we denote by $\mathbb{P}$. Examples of these interesting constructions are provided in what follows.

2.5.1 Distribution-Free Interval Predictors

For $X \in \mathbb{R}$ and $Y \in \mathbb{R}$, given a training set $\mathcal{D} = \{(X^{(1)}, Y^{(1)}), \ldots, (X^{(N)}, Y^{(N)})\}$, we can find the value of θ that minimizes the function $h(\theta)$ defined as

$$h(\theta) = \max_{i=1,\ldots,N} |Y^{(i)} - \theta X^{(i)}|$$

and denote the minimizer (min-max solution) by $\hat{\theta}$.[3] Now, for every X, consider the prediction interval $I(X; \mathcal{D}) = [\hat{\theta}X - h(\hat{\theta}), \hat{\theta}X + h(\hat{\theta})]$. In this context, we call the risk, and denote by $R(\mathcal{D})$, the probability that a new couple (X, Y) is observed and $Y \notin I(X; \mathcal{D})$. Under the assumption that (X, Y) couples are independently and identically distributed (i.i.d.) according to $\mathbb{P}$, it holds true that

$$\mathbb{P}^N\{R(\mathcal{D}) > \epsilon\} \leq (1-\epsilon)^N + N\epsilon(1-\epsilon)^{N-1} \tag{14}$$

where $\mathbb{P}^N$ is the product probability over the N couples in the training set $\mathcal{D}$ based on which we train the interval predictor $I(X; \mathcal{D})$. The proof of this fact can be found, for example, in Garatti and Campi (2009) and Carè et al. (2015). Moreover, under the assumption that $\mathbb{P}$ has a density, it holds true that

$$\mathbb{P}^N\{R(\mathcal{D}) > \epsilon\} = (1-\epsilon)^N + N\epsilon(1-\epsilon)^{N-1} \tag{15}$$

Note that, because the right-hand side of Eqs. (14), (15) does not depend on $\mathbb{P}$, the probability that the constructed interval fails to predict the correct value can be kept under control even if $\mathbb{P}$ is unknown. Note also that in this approach the structure of the predictor (which is given by θX) is just a tool for prediction purposes and is nowhere assumed that it reflects the "true" structure $f(X)$. Indeed, $f(X)$ can be arbitrarily complicated, and this does not impact on the reliability of the predictor.

Fig. 7 shows a prediction band for a training set of $N = 30$ i.i.d. points generated by some unknown mechanism.[4] Using Eq. (15), we can conclude that the probability that the band will not contain the next point (X, Y) is at most $\epsilon \approx 25\%$, unless a very unlucky event, that is,

[3] As we shall see, defining here $\hat{\theta}$ as the min-max solution rather than the more traditional least squares solution allows us to apply very strong mathematical results. For a study about extending the theory of this Section 2.5.1 to least squares solutions, see Carè et al. (2017).

[4] For the record, the hidden generation mechanism was $Y = X + 0.1X^2 \sin(100 \cdot X)$, with X sampled from a uniform distribution over $[-1, 1]$. We remark again that the proposed algorithm builds guaranteed bands independently of the underlying generation mechanism, which in real life is unknown to the user.

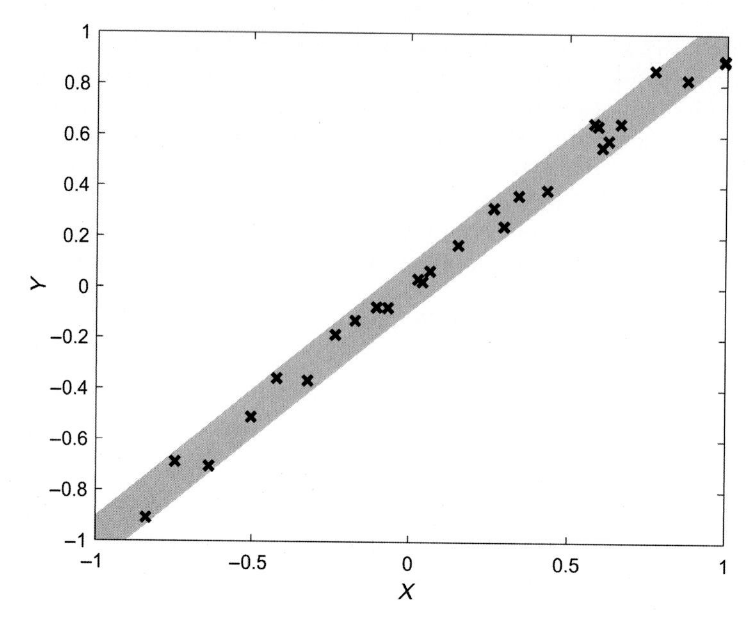

FIG. 7 A guaranteed (min-max based) prediction band from a training set of 30 i.i.d. points. The band will contain a new point (X, Y) with at least 75% probability (unless an unlucky, 0.2%-probability event has happened).

an event of probability $(1-\epsilon)^N + N\epsilon(1-\epsilon)^{N-1} \approx 0.2\%$, has happened when the nature has generated our training set. Because the distribution of $R(\mathcal{D})$ is known and is given by Eq. (15), we can also compute its expected value:

$$\mathbb{E}_{\mathcal{D}}[R(\mathcal{D})] = \frac{2}{N+1} \tag{16}$$

So, the expected probability of leaving a next point out decreases linearly with N, and turns out to be about 6.5% in the example where we only have $N = 30$ data.

The construction in this section is a simple instance of generalization results for *data-driven convex problems*, see Campi et al. (2009), Garatti and Campi (2009), Calafiore (2010), Carè et al. (2015), and Campi and Garatti (2016). Min-max interval predictors around generic linearly parameterized models can be constructed as illustrated in Fig. 7, and the probability of missing a new couple (X, Y) turns out to depend only on N and the number of parameters k. Precisely, with k parameters we get that $R(\mathcal{D})$ is distributed as a beta distribution with parameters $k+1$ and $N-k$, and $\mathbb{E}_{\mathcal{D}}[R(\mathcal{D})] = \frac{k+1}{N+1}$. When many parameters are used for training the predictor, guaranteed regularization techniques can be employed in order to get good prediction properties in spite of a large k (Campi and Carè, 2013).

Overall, this shows that studying a prediction procedure at an *appropriate level of abstraction* can be very rewarding: the goal of delivering a guaranteed interval for a new point in the (X, Y) space turns out to be much easier than trying to approximate an elusive and possibly very complicated function $f(X)$. Clearly, if the structure of the interval predictor is too simplistic, the *accuracy* of the predictor will be poor, that is, the amplitude of the prediction band will be large: this relationship between accuracy and reliability reflects the traditional trade-off between underfitting and overfitting (bias and variance).

Guarantees on prediction intervals that are obtained from convex optimization as in Fig. 7 are very strong, but one might be interested in more flexible constructions, for example, such that the amplitude of the band is not constant for all values of X. The theory of *conformal prediction* offers a powerful and flexible framework for finding predictor intervals with guarantees on the expectation of the risk, in line with Eq. (16), and the reader is referred to Vovk et al. (2005) and Shafer and Vovk (2008) for an introduction to this topic, and Lei et al. (2013); Lei and Wasserman (2014) for more recent advancements.

In concluding, we remark that prediction intervals that we have considered in Section 2.5.1 are *not* guaranteed *for a fixed*, user-chosen value x of X: the probability $R(\mathcal{D})$ that is kept under control by these constructions is the probability of observing *a couple* (X, Y) such that Y is not included in an X-dependent interval, where X is also chosen by the probabilistic generation mechanism. The reader that is interested in turning these methods into methods for predicting Y conditioning on a *given* value x of X is referred to Vovk (2013) and Lei and Wasserman (2014).

2.6 Probability Density Estimation

A radical way of dealing with the regression problem is recasting the learning task as the task of learning from data the probability $\mathbb{P}$ according to which the (X, Y) couples are distributed: once a description of $\mathbb{P}$ is available, prediction on Y can be obtained from the estimated probability distribution. In the so-called *parametric approaches*, the distribution is known, except for a set of parameters, and the problem is reduced to the estimation of the unknown parameters. This can be done, for example, by resorting to the maximum likelihood principle (see Section 2.3.2). In so-called *nonparametric approaches*, the correct distribution (or the best one in terms of some loss function) has to be identified from within a class, typically a very large class. The distribution estimation problem is very hard in general, and becomes easily ill-posed without suitable restrictions on the candidate set of distributions or on training procedures, such as those introduced by regularization approaches, see, for example, Vapnik and Stefanyuk (1978). Standard methods for nonparametric density estimation include *histograms, kernel methods*, and *nearest-neighbor methods*, see, for example, Silverman (1986) and Scott (2015).

3 PREDICTIONS WITHOUT PROBABILITIES

In this section, we reinterpret the regression problem as the problem of approximating a function, and we refer to some popular approximation methods. Although the formulation and the analysis of these methods might benefit from probabilistic concepts such as those introduced in the previous section, in this section we stick to a functional and deterministic point of view.

3.1 Approximation Theory

The aim of *approximation theory* (Powell, 1981; Trefethen, 2013) is that of studying general concepts and conditions under which it is possible to approximate a function $f(\cdot)$ belonging to certain class $\mathcal{F}$ by means of some other functions (simpler, in some sense), while keeping under control suitably defined approximation errors. The mathematical foundations of approximation theory lie in functional analysis and operator theory (Bollobás, 1990; Rudin, 1991; Kreyszig, 1989).

We here recall only a few fundamental ideas, in particular, the concept of *dense sets*, and the concept of the *best approximator* of a function.

DENSE SETS

A well-known theorem developed by Weierstrass states that if f: $[0,1] \to \mathbb{R}$ is a continuous function, then, for every $\epsilon > 0$, there exists a polynomial of a suitable order d, $\wp_d(x) = c_0 + c_1 x + \cdots + c_d x^d$, such that for all $x \in [0,1]$, $|f(x) - \wp_d(x)| < \epsilon$. This fact, which is expressed in technical terms by saying that *the set of polynomials $\mathcal{P}$ is dense in $\mathcal{F} = C[0,1]$ with respect to the supremum norm*, can be informally summarized by saying that continuous functions (on compact sets) can be approximated arbitrarily well by polynomials, that is, sums of simple monomials. Similar results are available for various classes of functions $\mathcal{F}$ other than $C[0,1]$ and approximators $\mathcal{A}$ other then polynomials $\mathcal{P}$. For example, from harmonic analysis, we know under which conditions a periodic function can be approximated arbitrarily well by summing up a finite number of sinusoidal functions at increasing frequencies; wavelet analysis does the same by means of sums of orthonormal wavelets, and so forth.

BEST APPROXIMATOR

Once a set of functions $\mathcal{A}$ that can be used to represent all the elements of $\mathcal{F}$ is available, the quest for good approximators can be restricted to a subset of $\mathcal{A}$, and the best approximator of $f \in \mathcal{F}$ can be sought in this restricted set. Clearly, the problem has to be formulated in such a way that it is well-posed, that is, the approximator exists, is unique and possibly also enjoys the desirable property of not being too "sensitive" with respect to "irrelevant" variations in the target function f (where all these concepts have to be made precise in terms of suitable norm operators). For example, given a function $f \in C[0,1]$ and a polynomial order d, it is well-known that there is a unique polynomial of order (at most) d that minimizes the difference $\max_{x \in [0,1]} |f(x) - \wp_d(x)|$. It was also proved that the first $d+1$ terms of a Fourier series define the best possible approximator of a function in terms of trigonometric polynomials of the same order in the squared error norm. These, and similar facts, are interesting in applications whenever constraining the set of the approximators can be interpreted as focusing on some particularly remarkable characteristics of f. For example, if one aims at approximating a periodic function f at best in terms of squared error, a low-order Fourier series might be enough. For example, the 0-order Fourier approximation of f is given by the average of f over its period, which might be a sufficient approximation for some purposes, while, if more information about the variability of the function f is required, more and more coefficients can be added one after another. On the other hand, when it is more important to detect differences in f (e.g., in image analysis when the boundaries of an object have to be detected) rather than average values, a wavelet representation might be preferred, so that a few coefficients (the first ones) might be enough for obtaining an appropriate approximation of f.

In conclusion, under the point of view of approximation theory,

> *given a* known *function* $f : \mathcal{X} \to \mathcal{Y}$, *the learning process can be stated as the process of choosing, from a set of approximators* $\mathcal{A}$, *a function* $\hat{f}(\cdot)$ *that represents sufficiently well the most important features of* $f(\cdot)$.

In reality, we do not know the function f, but we only have a finite set of points where the value of f is known, that is, the training set $\mathcal{D}$.

The complexity of the data-based approximator is then kept under control by sufficiently restricting $\mathcal{A}$, while its accuracy is kept under control by suitably penalizing the differences between $\hat{f}(X^{(i)})$ and $Y^{(i)}$, that is, its error on the training set. *Regularization approaches* such as the one in Example 3 fit this framework, which also include techniques such as *smoothing splines*, *kernel smoothers*, generic *curve fitting* techniques, *local regression methods*, *random forests* (Breiman, 2001), and *neural networks*. Neural networks, and related concepts, unquestionably played an important role in the history of machine learning (e.g., Novikoff's 1962 analysis of the learning process for perceptrons (Novikoff, 1962) is often credited to have started learning theory (Vapnik, 1998)), and have recently contributed to some extraordinary application results, see, for example, Krizhevsky et al. (2012). The next section provides a (necessarily very limited) introduction to this classic topic (Haykin, 2009).

3.1.1 *Neural Networks*

Let $X \in \mathcal{X} \subseteq \mathbb{R}^d$. For a given natural number q and a vector $\mathbf{p}$ of $q + q(d+1)$ parameters, $\mathbf{p} = (w_1, \ldots, w_q, c_{0,1}, c_{1,1}, \ldots, c_{d,1}, c_{0,2}, c_{1,2}, \ldots, c_{d,2}, \ldots, c_{0,q}, c_{1,q}, \ldots, c_{d,q})$, consider the functions of the kind

$$\hat{f}_q(X) = \sum_{i=1}^{q} w_i \sigma\left(\sum_{j=1}^{d} c_{j,i} X_j + c_{0,i}\right) \tag{17}$$

where $\sigma(\cdot)$ is a continuous function such that $\lim_{s\to\infty} \sigma(s) = +1$ and $\lim_{s\to-\infty} \sigma(s) = 0$. The function (17) is called a *neural network*, namely, a *single hidden layer feed-forward neural network*.[5]

It is a fact (Cybenko, 1989) that every continuous real function f over a *compact* $\mathcal{X}$ can be approximated arbitrarily well by a function such as $\hat{f}_q(X)$ for q large enough. By making the

[5] The evocative name comes from the fact that a function such as Eq. (17) with $q = 1$ is often called an artificial *neuron* (McCulloch and Pitts, 1943) and all that can be obtained by taking linear combinations and compositions of neurons tends to be called a *neural network*. The *single hidden layer* part of the name is related to the fact that in the argument of $\sigma(\cdot)$ there are no other $\sigma(\cdot)$ functions. The *feed-forward* part accounts for the absence of feed-backs, that is, loops between neurons. Neural networks with feed-backs are called *recurrent* neural networks (Bishop, 1995; Ripley, 1996; Haykin, 2009), see also Section 5.

dependence of $\hat{f}_q(X)$ on $\mathbf{p}$ explicit, we can say that the set $\mathcal{A} = \{\hat{f}_{q;\mathbf{p}}(\cdot) : \mathbf{p} \in \mathbb{R}^{q+q(d+1)}, q = 1, 2, \ldots\}$ is dense in $\mathcal{F} = \mathcal{C}(\mathcal{X})$, the set of continuous functions over the compact set $\mathcal{X}$. Restricting q to small values is a way to control the expressive power of the network, that is, to restrict $\mathcal{A}$. Given data $\mathcal{D}$, a suitable approximator could be obtained by fixing q and then selecting the function $\hat{f}_{q;\mathbf{p}}$ that minimizes the quadratic error on the training set, that is,

$$\frac{1}{N}\sum_{i=1}^{N}(\hat{f}_{q;\mathbf{p}}(X^{(i)}) - Y^{(i)})^2 \tag{18}$$

Unfortunately, function (18) is nonlinear and nonconvex in general, so that the minimization problem is typically solved by resorting to iterative methods that come with no guarantees of finding a global optimum. However, in practice, iterative methods have been used with success in a number of situations. We now describe the simple idea behind the most popular training algorithms for neural networks.

THE BACKPROPAGATION ALGORITHM: HIGH-LEVEL IDEA

Initialize the network by initializing the vector $\mathbf{p}$ of parameters (the "weights" of the network) to a certain candidate value $\mathbf{p}_0 \in \mathbb{R}^d$. Consider the first input-output couple $(X^{(1)}, Y^{(1)})$ and the corresponding error $e_1(\mathbf{p}) = (\hat{f}_{q;\mathbf{p}}(X^{(1)}) - Y^{(1)})^2$ as a function of $\mathbf{p}$. Denote the gradient function of the error function $e_1(\mathbf{p})$ with respect to $\mathbf{p}$ by $\nabla_{\mathbf{p}}e_1$, and $\nabla_{\mathbf{p}}e_1(\mathbf{p}_0)$ is its value at $\mathbf{p}_0$. The neural network is updated from $\hat{f}_{q;\mathbf{p}_0}$ to $\hat{f}_{q;\mathbf{p}_1}$, by defining

$$\mathbf{p}_1 = \mathbf{p}_0 - \rho\nabla_{\mathbf{p}}e_1(\mathbf{p}_0) \tag{19}$$

where ρ is a suitably chosen, small step size. Now the second input-output couple is considered and $\mathbf{p}_1$ is updated as $\mathbf{p}_2 = \mathbf{p}_1 - \rho\nabla_{\mathbf{p}}e_2(\mathbf{p}_1)$, and so on until all data in $\mathcal{D}$ have been considered and $\mathbf{p}$ has been updated N times. At this point, the procedure can be repeated: the parameter vector after N updates, $\mathbf{p}_N$, can be updated N more times as $(X^{(1)}, Y^{(1)}), \ldots, (X^{(N)}, Y^{(N)})$ are again considered one after another. In the neural network jargon, each of these macroiterations (which involve updating the parameter vector N times, i.e., one for each input-output couple) is called an *epoch*. The procedure goes on until a predetermined number of epochs is reached, or as soon as the training error (Eq. 18) is small enough. The training process can also be willingly stopped if there is evidence of *overfitting*: for example, the prediction error on a separate validation set can be tested at each cycle, and the learning algorithm can be stopped when a degradation of the validation error is detected. This procedure is known as *early stopping*, and is another source of "regularization," other than fixing q and loosening the requirements on the value of Eq. (18). Another common method for regularizing the solution and hedging against overfitting is setting an upper bound on the norm of $\mathbf{p}$.

We have tacitly assumed that $\sigma(\cdot)$ is differentiable. Indeed, $\sigma(\cdot)$ is traditionally chosen as a smooth function whose derivative can be computed in closed form, so that the gradient in Eq. (19) can be easily computed element by element. Traditional choices for the $\sigma(\cdot)$ functions are the logistic function $\sigma(u) = \frac{1}{1+e^{-u}}$, the arctangent function, the hyperbolic tangent function, and so forth.[6] More recently, functions such as $\sigma(u) = \max(0, u)$, called *rectifiers*,[7] have become very popular, especially in deep neural networks (see the following).

The name of "backpropagation algorithm" for the iterative algorithm outlined herein can be better understood when generalizations of Eq. (17), known as *multiple layers* feed-forward networks, are considered.

MULTIPLE LAYERS NETWORKS (DEEP NETWORKS)

Referring back to the definition (17) of $\hat{f}_{q;\mathbf{p}}(X)$, we can recognize two different kinds of parameters in $\mathbf{p}$. The $q(d+1)$ parameters $\{c_{j,i}\}$ define a transformation from d input values to q values; the remaining q parameters $\{w_i\}$ define a transformation from the resulting q values to a single output value. So, the function $\hat{f}_{q;\mathbf{p}}(X)$ is often thought of as a network where there is an *input layer* (where the observable values $X_1, \ldots, X_d$ enter the network), followed by a *hidden layer*, where the transformation from d to q values is operated, and an *output layer*, where the q values are finally mapped into a scalar observable value. Neural networks with more than one hidden layer can be defined. In general the kth hidden layer is defined by a vector of q_k functions $[\Phi_1^{(k)}(Z), \ldots, \Phi_{q_k}^{(k)}(Z)]^T$ that maps a vector $Z \in \mathbb{R}^{q_{k-1}}$ of q_{k-1} values from the previous layer into a vector of q_k values. Each function $\Phi_i^{(k)}(Z)$, $i = 1, \ldots, q_k$, is defined by $q_{k-1} + 1$ parameters $\{c_{j,i}^{(k)}\}$ as follows

$$\Phi_i^{(k)}(Z) = \sigma\left(\sum_{j=1}^{q_{k-1}} c_{j,i}^{(k)} Z_j + c_{0,i}^{(k)}\right)$$

With this notation, a neural network with two hidden layers can be written as

$$\hat{f}_{q_1,q_2}(X) = \sum_{j=1}^{q_2} w_j \Phi_j^{(2)}([\Phi_1^{(1)}(X)\ \Phi_2^{(1)}(X)\ \cdots\ \Phi_{q_1}^{(1)}(X)]^T) \tag{20}$$

see Fig. 8 for a network with $q_1 = q_2 = 2$.

In a deep neural network, parameters *with different roles* are available as tuning knobs of the sought approximator. The quest for explanations of the different roles played by the parameters in the approximation of functions is an interesting and open research topic (see, e.g., Zeiler and Fergus, 2014 for a study in image recognition).

Going back to the *backpropagation algorithm*, the gradient in Eq. (19) can be computed in a neat way if the hierarchy of the parameters in a deep network is taken into account. In fact, a simple application of the chain rule for the derivative of the composition of functions suggests

[6] The fact that the hyperbolic tangent function and the arctangent function do not satisfy condition $\lim_{u\to-\infty}\sigma(u) = 0$ is not an issue. Indeed, the theorem stating that functions (Eq. 17) are dense still holds true for more general classes of $\sigma(\cdot)$ functions; for example, it holds true for all continuous, bounded, and nonconstant functions $\sigma(\cdot)$ (Hornik, 1991).

[7] See, for example, Sonoda and Murata (2017) for their approximation properties.

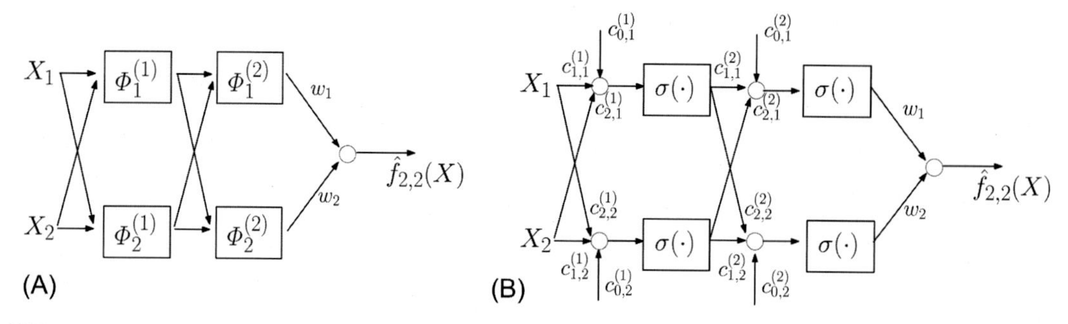

FIG. 8 A neural network with two hidden layers, and $d = q_1 = q_2 = 2$. In (A) we have represented the hidden layers by using the $\Phi_i^{(k)}$ functions, whose structure is made explicit in (B).

starting by computing the derivative of the error with respect to output-layer parameters first, and then "back-propagating" suitably modified versions of the measured error toward the previous layer. At each layer, the partial derivative of the error with respect to a parameter in the layer is carried out according to the same pattern, by exploiting the information that comes from the next layer, together with information that is available given the input $X^{(i)}$ and the current values of the network parameters, until the input-layer is reached. In this way, all the terms that are needed to update all the parameters according to rule (19) are obtained. This way of proceeding applies and is computationally convenient, also in the presence of many layers (Haykin, 2009).

4 PROBABILITIES EVERYWHERE: BAYESIAN REGRESSION

In this section, we introduce regression in a Bayesian framework. The ingredients of a Bayesian framework are two. The first is the choice of modeling *unknown quantities* as *random variables* in order to account for the user's uncertainty about them.[8] The second is the use of Bayes' rule as an inference tool, in order to exploit information coming from data and reduce the uncertainty on the unknown quantities. For the sake of simplicity, we will first consider the case where X is kept fixed, say $X = 1$, and we want to predict Y based on the observations $Y^{(1)}, \ldots, Y^{(N)}$ for the same input $X^{(1)} = \cdots = X^{(N)} = 1$, see also Example 2.

The distribution of Y is given by the following probability density function

$$p_Y^\theta(y)$$

where θ is a parameter that determines the correct shape of the distribution. That is, for each value of θ we have a distribution over Y, and we do not know which value of θ is the one that

[8] Clearly, different modeling choices stem from different views about what probability *is* and whether it is meant to describe a relative frequency with which something actually occurs (frequentist interpretation) or expectations about virtually everything, as in the Bayesian interpretation.

determines the true probability distribution of Y. We model our uncertainty about the true value of θ by a probability distribution over a random variable Θ. The density

$$p_\Theta(\theta) \tag{21}$$

is called *prior probability* and allows us to make statements such as "a priori (i.e., before data are observed), we expect that the value of Θ belongs to the interval $[a,b]$ with probability $\int_{[a,b]} p_\Theta(\theta)\mathrm{d}\theta$." Once the constant but unknown value of the uncertain parameter is modeled as a random variable Θ, the distribution of Y given a certain value θ of Θ can be thought of as a conditional probability of Y given $\Theta = \theta$, that is,

$$p_{Y|\Theta}(y;\theta) \quad = \quad p^\theta(y) \tag{22}$$

Distributions (21), (22) together define a joint probability distribution for Y and Θ, that is, $p_{Y,\Theta}(y,\theta) = p_{Y|\Theta}(y;\theta)p_\Theta(\theta)$. If we want to predict Y before any observation is available, the best that we can do is integrating $p_{Y,\Theta}(y,\theta)$ with respect to the uncertainty about Θ, that is, we can weight each predictive distribution $p_{Y|\Theta}(y;\theta)$ by our prior $p_\Theta(\theta)$ and then sum up, as follows:

$$p_Y(y) = \int_{-\infty}^{+\infty} p_{Y|\Theta}(y;\theta)p_\Theta(\theta)\mathrm{d}\theta$$

For independent observations, conditionally to a given value θ of Θ, the sequence $Y^{(1)},\ldots,Y^{(N)}$ is distributed according to the probability density function

$$p_{\mathcal{D}|\Theta}(y^{(1)},\ldots,y^{(N)};\theta) = \prod_{i=1}^{N} p_{Y|\Theta}(y^{(i)};\theta) \tag{23}$$

If the function (23) is evaluated with respect to θ while keeping fixed the values of the observations $\mathcal{D} = (Y^{(1)},\ldots,Y^{(N)})$, then it is called the *likelihood function* of θ given the observations $\mathcal{D}$, see also Section 2.3.2. Then, we can update our distribution over Θ by using the Bayes' rule:[9]

$$p_{\Theta|\mathcal{D}}(\theta;y^{(1)},\ldots,y^{(N)}) = \frac{p_{\mathcal{D}|\Theta}(y^{(1)},\ldots,y^{(N)};\theta)p_\Theta(\theta)}{p_{\mathcal{D}}(y^{(1)},\ldots,y^{(N)})} \tag{24}$$

Note that in Eq. (24) the value of the likelihood function of θ given the observations is weighted by our prior $p_\Theta(\theta)$. The denominator is $p_{\mathcal{D}}(y^{(1)},\ldots,y^{(N)}) = \int_{-\infty}^{\infty} p_{\mathcal{D}|\Theta}(y^{(1)},\ldots,y^{(N)};\theta)p_\Theta(\theta)$ $\mathrm{d}\theta = \int_{-\infty}^{\infty}\prod_{i=1}^{N} p_{Y|\Theta}(y^{(i)};\theta)p_\Theta(\theta)\mathrm{d}\theta$, and does not depend on θ: it just plays the role of a

[9] The Bayes' rule for two events A and B such that $p(B) \neq 0$ is commonly stated as

$$p(A|B) = \frac{p(B|A)p(A)}{p(B)}$$

Formula (24) is an application of the Bayes' rule to continuous random variables for which the probability density functions and conditional densities can be properly defined. Bayes' rule follows easily as a theorem from the Kolmogorov's axioms of probability. Besides Kolmogorov's axioms, other foundations have been proposed for Bayesian inference, for example, the Cox-Jaynes framework (Cox, 1946; Jaynes, 2003).

normalizing constant that ensures that $\int_{-\infty}^{\infty} p_{\Theta|\mathcal{D}}(\theta; y^{(1)}, \ldots, y^{(N)}) \mathrm{d}\theta = 1$. Eq. (24) defines a distribution of Θ given the observations, which is usually called the *a posteriori distribution*.

Now, if we want to predict a new value of Y, we can weight all the candidate distributions of Y given θ, $p_{Y|\Theta}(y;\theta)$, by our a posteriori distribution, $p_{\Theta|\mathcal{D}}(\theta; y^{(1)}, \ldots, y^{(N)})$, namely, we predict Y using the distribution

$$p_{Y|\mathcal{D}}(y; y^{(1)}, \ldots, y^{(N)}) = \int_{-\infty}^{\infty} p_{Y|\Theta}(y;\theta) p_{\Theta|\mathcal{D}}(\theta; y^{(1)}, \ldots, y^{(N)}) \mathrm{d}\theta \tag{25}$$

The extension to nonscalar parameters θ is immediate. In the case where $X^{(1)}, \ldots, X^{(N)}$ and the new X are allowed to take different values, we need to parameterize the probabilities considered so far with respect to the corresponding input values. For example, by defining $p^{x}_{Y|\Theta}(y)$ as the distribution that describes the variability of Y in the couple (X, Y) when $X = x$, formula (23) becomes $p^{x^{(1)},\ldots,x^{(n)}}_{\mathcal{D}|\Theta}(y^{(1)}, \ldots, y^{(N)}; \theta) = \prod_{i=1}^{N} p^{x^{(i)}}_{Y|\Theta}(y^{(i)}; \theta)$, and so on. In most of the Bayesian literature, the dependence on the input is expressed in an equivalent way by conditioning on the variable X. However, no probability distribution is normally defined explicitly on the X variable, and the value of X at which a certain Y is observed can be simply considered as an indexing parameter that is available to us, together with the values of $X^{(1)}, \ldots, X^{(N)}$ at which we have collected our training data.

Now, some important practical issues remain open. How to parameterize the distribution of Y given certain values of Θ and X? How to define the prior probability over Θ? And how to do that in such a way that inferences according to Eq. (24) and integrals such as Eq. (25) can be efficiently computed? Due to space limits we do not even try to answer these questions in a general way, and the reader is referred to, for example, Gelman et al. (2013) for more information. In what follows, we introduce Gaussian process regression methods, which are popular and powerful methods to address these issues in a way that is often reasonable and computationally convenient.

4.1 Gaussian Process Regression

Assume that the input-output relationship is of the kind

$$Y = f_{\Theta}(X) + \nu \tag{26}$$

where $f_{\Theta}(X) = \Theta_0 + \Theta_1 X_1 + \cdots + \Theta_d X_d$. If ν is independent of X and Θ, the structure (26) defines a parametrization of the distribution of Y given the values of X and of Θ. A common choice for ν is a Gaussian noise with zero mean and variance σ_ν^2 (an attempt to motivate the Gaussian noise assumption was made in Remark 1). Outputs $Y^{(1)}, \ldots, Y^{(N)}$, for a fixed value of Θ and given the inputs $X^{(1)}, \ldots, X^{(N)}$, are independent and jointly distributed as a product of Gaussians, with individual means equal to $\Theta_0 + \Theta_1 X_1^{(i)} + \cdots + \Theta_d X_d^{(i)}$, $i = 1, \ldots, N$, and with the same variance σ_ν^2. Now, we are left to define a prior probability over Θ. If the prior distribution of Θ is chosen as a multidimensional Gaussian, then all the distributions that are involved in the inference process are Gaussian, and can be computed by updating their mean and variance according to the well-known formulae for Gaussian distributions. For example, in the case where Θ is zero mean with nonsingular co-variance matrix C_Θ, it can be easily verified that the expected value of Θ given $Y^{(1)}, \ldots, Y^{(N)}$ (observed at $X^{(1)}, \ldots, X^{(N)}$) is

$$\bar{\Theta}_\mathcal{D} = C_\Theta \mathbf{X}_o (\mathbf{X}_o^\mathrm{T} C_\Theta \mathbf{X}_o + I\sigma_v^2)^{-1} \mathbf{Y}_o \tag{27}$$

where $\mathbf{X}_o$ is the $(d+1) \times N$ matrix whose ith column is $[1, X_1^{(i)}, \ldots, X_d^{(i)}]^\mathrm{T}$ and $\mathbf{Y}_o = [Y^{(1)}, \ldots, Y^{(N)}]^\mathrm{T}$. By applying some basic matrix algebra one can verify that an equivalent formula is

$$\bar{\Theta}_\mathcal{D} = \left(\frac{1}{\sigma_v^2} \mathbf{X}_o \mathbf{X}_o^\mathrm{T} + C_\Theta^{-1} \right)^{-1} \frac{1}{\sigma_v^2} \mathbf{X}_o \mathbf{Y}_o \tag{28}$$

which can be computationally convenient as it requires the inversion of a smaller $(d+1) \times (d+1)$ matrix in place of an $N \times N$ matrix. The covariance of Θ given the observations turns out to be

$$C_{\Theta|\mathcal{D}} = \left(\frac{1}{\sigma_v^2} \mathbf{X}_o \mathbf{X}_o^\mathrm{T} + C_\Theta^{-1} \right)^{-1} \tag{29}$$

or, equivalently,

$$C_{\Theta|\mathcal{D}} = C_\Theta - C_\Theta \mathbf{X}_o (\mathbf{X}_o^\mathrm{T} C_\Theta \mathbf{X}_o + I\sigma_v^2)^{-1} \mathbf{X}_o^\mathrm{T} C_\Theta \tag{30}$$

$\bar{\Theta}_\mathcal{D}$ and $C_{\Theta|\mathcal{D}}$ together define the a posteriori distribution of Θ given the observations, and we indicate this fact by writing

$$\Theta | \mathcal{D} \sim \mathcal{N}(\bar{\Theta}_\mathcal{D}, C_{\Theta|\mathcal{D}}) \tag{31}$$

Moreover, for a new input $X = [X_1, \ldots, X_d]^\mathrm{T}$, and defining $\mathbf{X} = [1, X_1, \ldots, X_d]^\mathrm{T}$, we have that

$$f_\Theta(X) | \mathcal{D} \sim \mathcal{N}(\mathbf{X}^\mathrm{T} \bar{\Theta}_\mathcal{D}, \mathbf{X}^\mathrm{T} C_{\Theta|\mathcal{D}} \mathbf{X}) \tag{32}$$

and

$$Y | \mathcal{D} \sim \mathcal{N}(\mathbf{X}^\mathrm{T} \bar{\Theta}_\mathcal{D}, \mathbf{X}^\mathrm{T} C_{\Theta|\mathcal{D}} \mathbf{X} + \sigma_v^2)$$

The following example illustrates some important aspects of how the distribution over Θ is updated based on data according to Eq. (31).

Example 4. *We want to get a bit more intuition about formulae (28), (29) that allow us to update our knowledge of Θ (see Eq. 31) and consequently to make predictions about Y. Let us start by simplifying Eq. (26) by considering the scalar case where*

$$Y = X\Theta + v$$

with $X \in \mathbb{R}$ and Θ distributed as $\mathcal{N}(0, \sigma_\Theta^2)$. We can think of σ_Θ^2 as a parameter that describes our degree of uncertainty about Θ: the more uncertain we are about the value of Θ, and in particular about its distance from 0, the bigger σ_Θ^2 should be. Assume that we observe $Y^{(1)}, \ldots, Y^{(N)}$ at $X^{(1)}, \ldots, X^{(N)}$, respectively. The a posteriori distribution on Θ given the observations $\mathcal{D}$ is a Gaussian distribution with mean $\bar{\Theta}_\mathcal{D}$ and variance $C_{\Theta|\mathcal{D}}$, with

$$\bar{\Theta}_\mathcal{D} = \left(\frac{\sum_{i=1}^N X^{(i)2}}{\sigma_v^2} + \frac{1}{\sigma_\Theta^2} \right)^{-1} \frac{1}{\sigma_v^2} \sum_{i=1}^N X^{(i)} Y^{(i)} \tag{33}$$

and

$$C_{\Theta|\mathcal{D}} = \left(\frac{\sum_{i=1}^{N} X^{(i)2}}{\sigma_v^2} + \frac{1}{\sigma_\Theta^2} \right)^{-1}$$

cf. Eqs. (28), (29). Let us focus on $\bar{\Theta}_\mathcal{D}$*. Although its main role is defining the mean of the a posteriori probability distribution of* Θ *given* $\mathcal{D}$*,* $\bar{\Theta}_\mathcal{D}$ *can be used also per se, as an estimate of the true value of* Θ *(in which case, it is called a* Bayesian (point) estimate*). It is useful to compare Eq. (33) with the standard least squares estimate Eq. (9), which is recovered in the limit when* $\sigma_\Theta^2 \to \infty$*, that is, the more uncertain we are about* Θ*, the closer our estimate will be to the least squares estimate.*

To understand Eq. (33) even better, let us focus on the case where $N = 1$*, that is, we want to update our distribution over* Θ *based on one observation* $\mathcal{D} = (X^{(1)}, Y^{(1)})$ *only. We get* $\bar{\Theta}_\mathcal{D} = \frac{\sigma_\Theta^2}{X^{(i)2}\sigma_\Theta^2 + \sigma_v^2} X^{(1)} Y^{(1)}$*. Obviously, if* $X^{(1)} = 0$ *the observation does not change our prior, and indeed it does not carry any information on* Θ *as we are just measuring an instance of the noise* v*. Assume, instead, that* $X^{(1)} \neq 0$*, say* $X^{(1)} = 1$*. In this case, the effect of* $Y^{(1)}$ *on* $\bar{\Theta}_\mathcal{D}$ *depends on the ratio between the magnitude of the noise, represented by* σ_v^2*, and our a priori uncertainty on* Θ*, represented by* σ_Θ^2*. Indeed, if* σ_v^2 *is dominated by* σ_Θ^2*, then the Bayesian estimate will "trust" the data* $Y^{(1)}$*, to the point that if* $\frac{\sigma_v^2}{\sigma_\Theta^2} \approx 0$*, then* $\bar{\Theta}_\mathcal{D} \approx Y^{(1)}$*. On the other hand, if* σ_v^2 *is larger than* σ_Θ^2*, then the Bayesian estimate will compensate the uncertainty due to the noise by sticking to our prior belief that the parameter is close to zero, to the point that if* $\frac{\sigma_\Theta^2}{\sigma_v^2} \approx 0$*, then* $\bar{\Theta}_\mathcal{D} \approx 0$*. When more data are observed, a similar balance between information gained from data and our prior belief is carried out by the more general formula* (33).

Given the preceding discussion, it comes as no surprise that the value of θ *that minimizes*

$$\frac{1}{N} \sum_{i=1}^{N} (\theta X^{(i)} - Y^{(i)})^2 + \frac{1}{N} \frac{\sigma_v^2}{\sigma_\Theta^2} \theta^2$$

is exactly $\hat{\theta} = \left(\frac{\sum_{i=1}^{N} X^{(i)2}}{\sigma_v^2} + \frac{1}{\sigma_\Theta^2} \right)^{-1} \frac{1}{\sigma_v^2} \sum_{i=1}^{N} X^{(i)} Y^{(i)} = \bar{\Theta}_\mathcal{D}$*. So, the Bayesian estimate can be seen as a regularized least squares estimate (see Section 2.4.2), where the regularization parameter depends on the magnitude of the noise with respect to our uncertainty about* Θ*. Although we will not discuss this topic any further, it is worth remarking that the connection between Bayesian regression and regularization is not limited to our simple example: very elegant results have been established, see, for example, Rasmussen and Williams (2006) and the references therein, also when* $f_\Theta(X)$ *is nonlinear and* Θ *is infinite-dimensional, as in the framework that we shall introduce as follows.*

The prediction mechanism developed so far for Eq. (26) can be immediately extended to the case where $f_\Theta(X)$ is nonlinear in X, by defining

$$f_\Theta(X) = \sum_{i=0}^{M} \Theta_i \Phi_i(X) \tag{34}$$

for a suitable sequence of basis functions $\Phi_0, \Phi_1, \ldots, \Phi_M$. An important generalization to this scheme is when $M \to \infty$. The step from finite M to infinite M is not trivial: for example, it becomes important to choose the basis functions Φ_i and the priori distribution over Θ (an infinite dimensional vector!) in such a way that $\sum_{i=0}^{M} \Theta_i \Phi_i(X)$ converges with probability 1 to a reasonable function as $M \to \infty$. However, one can avoid dealing with the problem of defining the basis functions Φ_i and the distribution over Θ explicitly, by resorting to a more synthetic point of view, which is provided by the theory of Gaussian processes. A Gaussian process is a random function $f(\cdot)$ such that the distribution of every finite sequence of the kind $[f(X^{(1)}),\ldots,f(X^{(n)})]$, for arbitrary n and $X^{(1)},\ldots,X^{(n)}$, is Gaussian. A Gaussian process can be defined by defining a function $m\colon \mathcal{X} \to \mathbb{R}$, which represents the mean of the process $m(X) = \mathbb{E}_f[f(X)]$ (the subscript denotes that the expected value is taken with respect to all the realizations of the process $f(\cdot)$) and a positive definite function $K\colon \mathcal{X} \times \mathcal{X} \to \mathbb{R}$, which represents the covariance of the process, that is, $K(X^{(i)}, X^{(j)}) = \mathbb{E}_f[(f(X^{(i)}) - m)(f(X^{(j)}) - m)]$, and is usually called the *kernel* of the Gaussian process.[10] So, by defining the functions m and K, we are defining a probability distribution over an infinite space of functions $f\colon \mathcal{X} \to \mathbb{R}$, see also Fig. 9. In the light of Mercer's theorem, these functions can be reinterpreted as the limit functions $f_\Theta(X) = \lim_{M\to\infty} \sum_{i=1}^{M} \Theta_i \Phi_i(X)$ for the various realizations of the Gaussian coefficients $\Theta_0, \Theta_1, \ldots$, see, for example, Rasmussen and Williams (2006).

Remarkably, this considerable generalization can be achieved by small modifications of the formulae for the linear case. To begin with, consider again a linear and zero mean $f_\Theta(X)$. If we

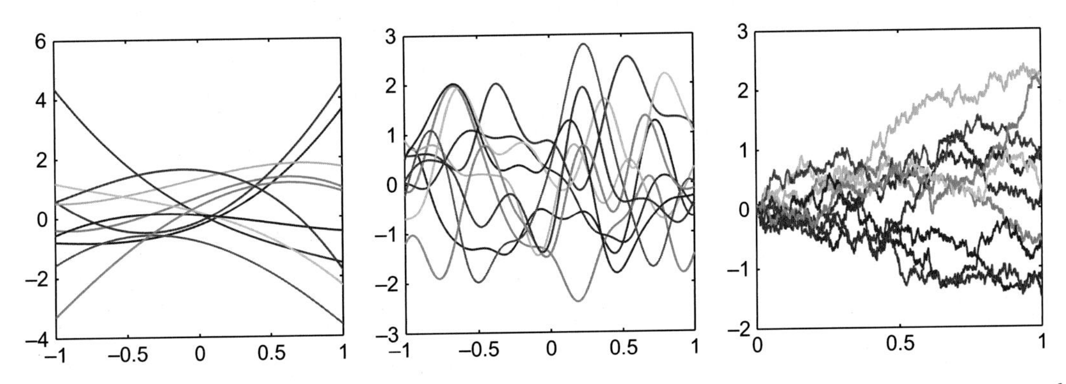

FIG. 9 Each *box* represents 10 randomly drawn realizations of a Gaussian process with scalar input, zero mean and kernel $K(X^{(i)}, X^{(j)})$ defined as *left*: $(1 + X^{(i)}X^{(j)})^3$ [inhomogeneous polynomial (cubic) kernel]; *center*: $e^{-20(X^{(i)}-X^{(j)})^2}$ [exponential kernel]; *right*: $\min(X^{(i)}, X^{(j)})$. Note that the last function is a legitimate kernel (positive definite) only for $X \geq 0$.

[10] The fact that for the functions m and K a Gaussian process exists is a consequence of an abstract formulation of the Daniell-Kolmogorov existence theorem (see Dudley, 2002, Chapter 12). This theorem by itself, however, does not imply that the realizations of the corresponding Gaussian process have desirable properties, such as boundedness and continuity. The reader interested in a discussion on these rather technical details is referred for example to the first chapter of Adler and Taylor (2009) and to Rasmussen and Williams (2006).

rewrite more explicitly formula (32) by using Eqs. (27), (30) we get that at a new input X the random variable $f_\Theta(X)$ is distributed as

$$\mathcal{N}\left(\mathbf{X}^\mathrm{T}C_\Theta\mathbf{X}_o(\mathbf{X}_o^\mathrm{T}C_\Theta\mathbf{X}_o+I\sigma_v^2)^{-1}\mathbf{Y}_o,\mathbf{X}^\mathrm{T}C_\Theta\mathbf{X}-\mathbf{X}^\mathrm{T}C_\Theta\mathbf{X}_o(\mathbf{X}_o^\mathrm{T}C_\Theta\mathbf{X}_o+I\sigma_v^2)^{-1}\mathbf{X}_o^\mathrm{T}C_\Theta\mathbf{X}\right) \quad (35)$$

where $\mathbf{X}=[1,X_1,\ldots,X_d]$. Denoting the new input X as $X^{(N+1)}$, and $\mathbf{X}^{(i)}=[1,X_1^{(i)},\ldots,X_d^{(i)}]$, it is interesting to note that $X^{(1)},\ldots,X^{(N)},X^{(N+1)}$ enter Eq. (35) only through scalar products of the kind $\mathbf{X}^{(i)\mathrm{T}}C_\Theta\mathbf{X}^{(j)}, i,j=1,\ldots,N+1$. It is easy to check that when $f_\Theta(X)=\Theta_0+\Theta_1X_1+\cdots+\Theta_dX_d$, with $\Theta\sim\mathcal{N}(0,C_\Theta)$, it holds that $\mathbb{E}[f_\Theta(X^{(i)})f_\Theta(X^{(j)})]=\mathbf{X}^{(i)\mathrm{T}}C_\Theta\mathbf{X}^{(j)}$. On the other hand, setting $m(X)=0$, the function $K(X^{(i)},X^{(j)})=\mathbf{X}^{(i)\mathrm{T}}C_\Theta\mathbf{X}^{(j)}$ defines the linear Gaussian process $f_\Theta(X)=\Theta_0+\Theta_1X_1+\cdots+\Theta_dX_d$, with $\Theta\sim\mathcal{N}(0,C_\Theta)$. So, we conclude that, in the linear case, the crucial updating formulae depend on the inputs only through the kernel function $K(\cdot,\cdot)$. This is a general fact: when a (zero mean) Gaussian process defined by a function $K(X^{(i)},X^{(j)})\neq\mathbf{X}^{(i)\mathrm{T}}C_\Theta\mathbf{X}^{(j)}$ is in use, one has just to replace the scalar products $\mathbf{X}^{(i)\mathrm{T}}C_\Theta\mathbf{X}^{(j)}$ with $K(X^{(i)},X^{(j)})$ in formula (35). Clearly, in the case of nonzero mean processes, formulae have to be slightly changed to accommodate the nonzero mean function m, but the extension is straightforward.

In conclusion, a suitable definition of the kernel function $K(\cdot,\cdot)$ and the ability of computing it on a *finite* number of points is all we need to define both a Gaussian process on a possibly infinite-dimensional space of possibly nonlinear functions *and* to carry out all the Bayesian inferences and predictions in an efficient way. To be precise, when N is large, computing all the values $K(X^{(i)},X^{(j)})$ and the matrix inversion in the formulae can still be practically unfeasible in the general case, while in the linear case the computation is easier (in fact, Eq. 28 simplifies Eq. 27, and Eq. 29 simplifies Eq. 30). As a consequence of this, tractable methods have been studied and proposed that use online updating formulae and rely on sparsity, see, for example, Csató and Opper (2002).

Many tools and ideas have been developed to design kernels that are suitable for specific applications, see, for example, Shawe-Taylor and Cristianini (2004). Once a class of kernel functions has been chosen, some parameters defining a specific kernel in the class are often left undecided and are chosen based on data. These parameters are called *hyperparameters*. For example, in the linear case, the covariance matrix C_Θ is often taken as $C_\Theta=I\sigma_\Theta^2$, with the hyperparameter σ_Θ^2. The choice of σ_Θ^2 can be made according to the likelihood principle or by maximizing some validation score. In principle, the Bayesian approach can be reapplied at this level, by modeling the hyperparameter as a random variable with a prior.

For some recent applications of Gaussian processes to space weather, see also Chandorkar et al. (2017) and Camporeale et al. (2017).

5 LEARNING IN THE PRESENCE OF TIME: IDENTIFICATION OF DYNAMICAL SYSTEMS

In this section, we discuss some methods to deal with the presence of *time*, that is, of predictions when the relation between input and output is *dynamical*.

Assume that $u(t)$ is a physical quantity that varies in time and determines the evolution of another physical quantity $y(t)$. For simplicity, assume that $u(t)$ and $y(t)$ are real valued, discrete-time functions, that is, with $t\in\mathbb{Z}$, so that they can be thought of as sequences obtained by sampling continuous time functions. In order to avoid confusion, we reserve symbols as $u(t)$ and $y(t)$ for denoting the entire *sequences* while, when we want to refer to the value of the sequences at a certain time step $t = k$, we will use the symbols $u[k]$ and $y[k]$. From an engineering point of view, sequences such as $u(t)$ and $y(t)$ are often called *signals*.

Using one of the techniques discussed in this chapter, one can try to predict the value of $Y = y[k]$ given $X = u[k]$ by using a predictor that is a function of $u[k]$ only and that has been trained using, for example, $(X^{(1)}, Y^{(1)}), \ldots, (X^{(N)}, Y^{(N)})$, with $X^{(i)} = u[i]$ and $Y^{(i)} = y[i]$. However, in the presence of a physical relation between $u(t)$ and $y(t)$, this attempt is often doomed to failure. In fact, if $u(t)$ and $y(t)$ are connected by a *functional* relation of the kind

$$y(t) = \mathbf{F}(u(t)) \tag{36}$$

the value of $y(t)$ at a specific point in time can depend on the *entire evolution* in time of $u(t)$. For example, if $u(t)$ is a physical acceleration and $y(t)$ a velocity, knowing $u[k]$, the acceleration at time k, is not enough to guess the velocity $y[k]$, which is determined also by the *past*, in particular by value $y[k-1]$, or, equivalently, by all the inputs $u[k-1], u[k-2], \ldots$ until a time for which the velocity is known (initial condition) is reached. A relation such as Eq. (36) is an *input-output* representation of the relation between $u(t)$ and $y(t)$. It is a general fact that physical signals are part of *systems*, that is, they are subject to constraints on their joint behaviors: the fact that some of these signals are labeled as inputs and some as outputs depends much on our aims and our practical possibilities of controlling and measuring them, see, for example, Willems and Polderman (2013). Although we will refer to Eq. (36) as a *system*, the study of systems extends far beyond input-output relations, and even beyond the traditional study of physical systems (Bertalanffy, 1950; Wiener, 1961).

An accurate knowledge of system (36) can be derived from the laws of physics (*white-box modeling* or *analytical modeling*). However, this is not always possible or practical, and often one wants to approximate a relation such as Eq. (36) by modeling it through some predefined class of models, which is good for practical purposes. The art and science of reconstructing (approximating) relation $\mathbf{F}$ given a finite sample of input $(u[1], \ldots, u[N])$ and output $(y[1], \ldots, y[N])$ values is called *system identification* (Ljung, 2007, 2010).

In the following, we briefly introduce some basic facts about the most studied class of systems, which is certainly the class of linear and time-invariant systems (LTI).

5.1 Linear Time-Invariant Systems

A system $y(t) = \mathbf{F}(u(t))$ is linear if for every $\alpha, \beta \in \mathbb{R}$ and every two input sequences $u_1(t)$, $u_2(t)$ it holds true that $\mathbf{F}(\alpha u_1(t) + \beta u_2(t)) = \alpha \mathbf{F}(u_1(t)) + \beta \mathbf{F}(u_2(t))$ (superposition property). It is time-invariant if for every couple of signals $u_0(t), y_0(t)$ such that $\mathbf{F}(u_0(t)) = y_0(t)$ and every time shift $t_d \in \mathbb{Z}$, it holds that $\mathbf{F}(u_0(t - t_d)) = y_0(t - t_d)$. Linear and time-invariant systems are fully described by their impulse response, that is, by the sequence $h(t) = \mathbf{F}(\delta(t))$, where $\delta[0] = 1$ and $\delta[k] = 0$ for all $k \neq 0$. In fact, it can be easily shown that the output of the linear system $\mathbf{F}(u(t))$ can be written explicitly as a convolution of $u(t)$ and $h(t)$, that is the output value at time t is

$$\sum_{k=-\infty}^{\infty} h[k]u[t-k] \tag{37}$$

Thus, the behavior of an LTI system with respect to any other input signal can be deduced from the impulse response, and therefore the study of LTI systems can be reduced to the study of the properties of $h(t)$. For example, if, for all $k < 0, h[k] = 0$, then the system is called *causal* and past output values are not affected by future input values; if $\sum_{k=-\infty}^{\infty} |h[k]| < \infty$, then the system is *stable* and it will tend to "forget the past," and so forth; see, for example, Kailath (1980) and Oppenheim and Schafer (2010). LTI systems are *frequency filters*, that is, if a sinusoidal input enters the system, the output will still be a sinusoid at the same frequency but with a different phase and module. Plenty of systems identification tools exploit these powerful properties, both in time and frequency domain (Jenkins and Watts, 1968; Pintelon and Schoukens, 2012) to reconstruct **F** from input and output data.

It is probably worth clarifying the extent to which the common feeling that linear systems are "simple systems" is correct. LTI systems, by virtue of Eq. (37), are certainly "simple" as far as homogeneity of behavior with respect to very different inputs is concerned. However, one should not forget that linearity does not entail any special regularity or simplicity of the impulse response, so that the output of a linear system to a very simple input can be very complex and unpredictable. See also Rolewicz (1969) and Grosse-Erdmann and Peris Manguillot (2011).

In applications, linear systems are often represented by difference equations. For example, consider the linear difference equation

$$y(t) = a_1 y(t-1) + b_1 u(t-1) \tag{38}$$

If an initial condition $y[0]$ is known, Eq. (38) can be solved in a forward way in order to derive $y[k]$ for every $k > 0$ given $u[k], k \geq 0$. If $y[0] = 0$, the impulse response of Eq. (38) can be easily computed and it turns out to be $h[k] = 0$ for $k \leq 0$ and $h[k] = a_1^{k-1} b_1$ for $k \geq 1$. Therefore, Eq. (38) can be used to represent a class of LTI system with infinite impulse response by means of only two parameters, that is, in a very *concise way*. The response of system (38) to any input signal from $k = 0$ on is then obtained by convolution (37). If $y[0] \neq 0$, the effect of the initial conditions at any time $k \geq 1$ consists of an additive output term given by $a_1^{k-1} y[0]$. Note that, if $|a_1| < 1$, the effect of the initial conditions vanishes in time (stable system).

Disturbances are often added to equations like Eq. (38) for two reasons. First, they model the uncertainty in the measurements of the output variable $y(t)$. Second, they can be used as a modeling device to account for the fact that the mathematical formulation of the system does not reflect exactly the "true" (physical) system. Overall, a model such as

$$y(t) = a_1 y(t-1) + b_1 u(t-1) + w(t) \tag{39}$$

where $w(t)$ is a noise sequence, is likely to be more useful than Eq. (38). Eq. (39) can be written in a more compact way by using the delay operator q^{-1} as follows

$$(1 - a_1 q^{-1}) y(t) = b_1 q^{-1} u(t) + w(t)$$

A more general representation of a linear difference equation involving the output, the input, and an unmeasurable disturbance is the following one (see Söderström and Stoica, 1988)

$$A(q^{-1}) y(t) = \frac{B(q^{-1})}{F(q^{-1})} u(t) + \frac{C(q^{-1})}{D(q^{-1})} w(t) \tag{40}$$

where $A(q^{-1}) = 1 + a_1 q^{-1} + \cdots + a_{n_A} q^{-n_A}$, $B(q^{-1}) = b_1 q^{-1} + \cdots + b_{n_B} q^{-n_B}$, $C(q^{-1}) = 1 + c_1 q^{-1} + \cdots + c_{n_C} q^{-n_C}$, $F(q^{-1}) = 1 + f_1 q^{-1} + \cdots + f_{n_F} q^{-n_F}$, $D(q^{-1}) = 1 + d_1 q^{-1} + \cdots + d_{n_D} q^{-n_D}$ are formally treated as polynomials in the delay operator q^{-1}. By multiplying both sides of Eq. (40) by $F(q^{-1})D(q^{-1})$, one gets an equation relating the signals $y(t)$, $u(t)$, $w(t)$ and their delayed versions. By putting to zero some coefficients, common models of linear systems are obtained from Eq. (40), such as finite impulse response (FIR, when $1 = A(q^{-1}) = D(q^{-1}) = F(q^{-1})$ and $C(q^{-1}) = 0$), AutoRegressive with eXogenous input (ARX, when $1 = D(q^{-1}) = F(q^{-1})$ and $C(q^{-1}) = 1$: Eq. 39 is an example), and the *ARMAX* model, which is as follows

$$A(q^{-1})y(t) = B(q^{-1})u(t) + C(q^{-1})w(t) \tag{41}$$

The characterization of the noise $w(t)$ plays an important role. For reasons that are rooted in the Wiener-Kolmogorov theory of stationary processes (Kailath, 1970), the noise $w(t)$ is most often assumed to be white. Given this restrictive assumption, polynomials $C(q^{-1})$ and $D(q^{-1})$ provide us with useful degrees of freedom in shaping the stochastic properties of the disturbances that affect the system.

The *validation* of the model class is often made by resorting to statistical tests on the prediction residuals, such as whiteness and symmetry tests, see, for example, Chapter 11 in Söderström and Stoica (1988), while the model *order selection* can be done by resorting to methods and principles such as those discussed in Section 2.4.1. We mention, in particular, that regularization techniques in the framework of Bayesian estimation in line with the contents of Section 4 but adapted to the dynamical context are available, see, for example, Pillonetto et al. (2014) for a recent survey.

Once a model class has been chosen, the correct parameters have to be estimated from input-output data. The difficulty of the identification process depends on the model class adopted. For example, the parameters of an ARX system such as Eq. (39) can be easily estimated by the least squares method: the couple of parameters $(\hat{a}_1, \hat{b}_1)$ that minimizes the function

$$\sum_{k=1}^{N} (y[k] - y[k-1]\hat{a}_1 - u[k-1]\hat{b}_1)^2$$

is obtained from data $(u[k-1], y[k-1], y[k])$, $k = 1, \ldots, N$, and, under suitable assumptions on the input and the noise, the estimate will converge to the true parameters (Söderström and Stoica, 1988; Ljung, 1999). The case of ARMAX systems (41) is more difficult, because, in order to apply the least squares method, one should know the values of the unknown disturbances. Common solutions to this problem include the *extended least squares* algorithm, where the unknown noise terms are replaced with estimates (Söderström and Stoica, 1988), and *instrumental variables* methods (Söderström and Stoica, 2002). A very general class of robust methods for estimating the parameters of a system is the class of *prediction error methods* (PEMs). The idea is as follows. First, we pretend that the parameters of the system, say θ, are known and, based on them, we find the *optimal predictor* of $y[k]$ given the available inputs and the measured outputs until time $k-1$. Then, the predictor is written as a function of the unknown parameters $\hat{y}[k;\theta]$, so that the value of $\hat{y}[k;\theta]$ at every time step k given the available data becomes a function of the parameters θ only. Finally, one chooses the parameters that minimize the prediction error over a certain time horizon. For example, if the loss function is chosen as $(\hat{y}[k] - y[k])^2$, then, for ARX systems, PEM coincides with the least squares method, as the optimal predictor of $y[k]$ given the past is $\hat{y}[k] = a_1 y[k-1] + b_1 u[k-1]$ (assuming Eq. 39 is the correct model and $w(t)$ is white noise). On the other hand, for more general classes of systems, PEMs lead easily to nonconvex optimization problems, which are computationally challenging due to the presence of local minima, and are usually solved by resorting to heuristics.

An important class of identification algorithm is that of *online (or recursive) algorithms*, where data are processed sequentially rather than in a batch way. Recursive algorithms range from basic recursive least squares methods, where the least squares estimate is updated online for every new datum, to far more sophisticated stochastic algorithms; see also the backpropagation algorithm in Section 3.1.1. Other than the computation advantages, these algorithms, often with minor modifications, lend themselves to be employed as *adaptive* algorithms that are able to track *time-varying* parameters in relations such as Eq. (41) (Benveniste et al., 2012). Therefore, they are key in dealing with *time-varying systems*.

Traditionally, methods for systems identification are backed by theoretical analyses that, under suitable conditions, guarantee some good *asymptotic properties*. Results for finite samples of data ($N < \infty$, as in real life) are available only in the presence of strong assumptions on the distribution of $w(t)$, for example, Gaussianity. However, it is important to recognize that asymptotic results can lead to deceiving conclusions when they are applied to finite samples of data, see, for example, Garatti et al. (2004). *Set-membership approaches* (Milanese et al., 2013) allow one to bound the estimation error starting from a description of the noise range, while methods that deliver guaranteed confidence regions for the parameters estimated from *finite samples* of data under mild statistical assumptions on the noise have been recently proposed, see, for example, Carè et al. (2018).

Equations like Eq. (40) model the dynamical input-output relationship, and are often very useful for prediction purposes. Another common model uses additional state variables as follows:

$$\begin{aligned} z[k+1] &= Az[k] + Bu[k] + v[k] \\ y[k+1] &= Cz[k] + e[k] \end{aligned} \tag{42}$$

where $z[k] \in \mathbb{R}^{n_z}$ is a vector of state variables, $u[k] \in \mathbb{R}^{n_u}$ is the input variable, A is a $d \times d$ matrix that defines the state dynamic, C is a $1 \times d$ vector that defines the output transformation, that is, the relation between the state and the output, $v[k]$ is a disturbance on the state dynamic and $e[k] \in \mathbb{R}$ is an output noise. Eq. (42) is a *state space model*. From the input-output point of view, the expressive power of Eqs. (40), (42) is the same. However, a model such as Eq. (42) is often closer to the way in which physical systems are represented, so that physical insights can be more easily exploited. The study of state space models, with the notions of observability and controllability (Kálmán, 1959), is at the basis of modern linear control theory, and provides the background for the celebrated *Kalman predictor* (Kálmán, 1960). The identification of models in the form of Eq. (42) can be carried out by *subspace identification* techniques (Van Overschee and De Moor, 2012). Subspace identification methods provide natural ways to determine the order of the system from input-output data. Indeed, they are related to techniques such as principal component analysis and can be used for *dimensionality reduction*, that is, for finding *reduced model order* representations of a system that preserve its fundamental dynamical properties.

Overall, linear models are a powerful modeling tool. It is well known that they approximate very well the behavior of nonlinear systems in the neighborhood of equilibrium points, and linearized models of nonlinear systems are ubiquitous in physics and engineering applications, a notable example being small-signal modeling in electronic circuits. Many practitioners and engineers can also witness that models such as Eqs. (39), (41), used as *black-box* models and trained on input-output data without any physical insight, provide practically acceptable and sometimes surprisingly good predictive performances. However, we now briefly consider some more general models that can be considered when a linear model does not capture the behavior of the system sufficiently well.

5.2 Nonlinear Systems

The operating conditions of a nonlinear system can often be partitioned into an acceptable number of sets where the system behaves in an approximately linear way, and where the identification problem can be reduced to a linear identification problem. Then the global behavior can be approximated by suitably merging the behaviors of the *locally linear systems*. This can be done in a smooth way by representing the output of the global system as a weighted sums of the outputs of the linear systems, where weights depend on the operating conditions. Various methods for combining locally linear systems, and even partitioning their operating conditions, have been proposed based on, for example, *fuzzy logic* and *neural networks* (Takagi and Sugeno, 1985; Nelles, 2013). Locally linear systems are a viable options when a reasonable number of operating conditions can be identified, and under suitable constraints on the input signal (Billings and Voon, 1984). This is related to a fundamental issue in nonlinear systems identification. We have seen that linear systems treat all the inputs on a plane of equality, to the point that LTI systems are fully characterized by their response to a single signal, Eq. (37). In the presence of unmeasured disturbances, a linear system can still be identified by choosing inputs that, in some sense, cancel the effect of the disturbances and are sufficiently exciting to squeeze structural information out of the system (Ljung, 1971; Godfrey, 1993). The behavior of a nonlinear system, instead, can be very input-dependent as no uniform description such as Eq. (37) is available in general.

In practice, some kind of uniformity in the response for "similar" input sequences is always assumed, and input is designed so as to excite the system and reveal its fundamental behaviors at a certain range of bandwidth and amplitudes (Leontaritis and Billings, 1987).

Linear models such as Eqs. (40), (42) can be extended in order to model nonlinear systems by rewriting the constant parameters as *signal-dependent* functions, that is, functions that depend on the inputs, on the outputs or on other external signals, see, for example, Chapter 3 in Haber and Keviczky (1999) for some examples in applications.

Nonlinear systems are often modeled by interconnecting linear subsystems with nonlinear transformations (*block structured models*). Among the most studied structures there are *Hammerstein systems*, which consist of a static nonlinearity followed by an LTI system, and *Wiener systems*, where an LTI system is followed by a static nonlinearity. Powerful results are available from the system identification perspective. For example, in the case of Gaussian noise, the identification of the dynamical linear part and the static nonlinear part can be separated, as a consequence of Bussgang's theorem (Bussgang, 1952; Giri and Bai, 2010).

Volterra series (Volterra, 1887) can represent a large variety of nonlinear systems. They can be thought of as a generalization of relation (37): convolutions such as Eq. (37) stands to Volterra series as first-order approximations stand to Taylor expansions. Indeed, Volterra series (for causal systems) include terms such as $\sum_{m_1=1}^{M} h_1[m_1]u[k-m_1]$, but also $\sum_{m_1=1}^{M}\sum_{m_2=1}^{M} h_2[m_1,m_2]u[k-m_1]u[k-m_2]$, $\sum_{m_1=1}^{M}\sum_{m_2=1}^{M}\sum_{m_3=1}^{M} h_3[m_1,m_2,m_3]u[k-m_1]$ $u[k-m_2]u[k-m_3]$, and so on. *Wiener series* (Wiener, 1959) are an alternative representation in the same spirit (although not exactly equivalent), see, for example, Rugh (1981). The literature on the Volterra and Wiener series is extensive, as they are certainly a landmark in nonlinear system analysis and nonlinear identification. The main practical problem with using them directly for identification purposes is the lack of parsimony, that is, the proliferation of terms and parameters to be estimated. Therefore, identification can be carried out by considering implicit, more concise representations, see, for example, Franz and Schölkopf (2006). Just as a linear system can be modeled by selecting the correct parameters in an ARX model (39), in the nonlinear case one tries to reconstruct relations of the kind

$$y[k]=f(y[k-1],y[k-2],\dots,y[k-n_A],u[k-1],u[k-2],\dots,u[k-n_B])+w[k] \quad (43)$$

where f is some nonlinear function of a finite number of inputs and outputs. Eq. (43) is called a *NARX* model (nonlinear ARX), and the function f can be identified by using function approximation techniques, see Section 3. For example, a neural network (17) can be trained with data sequentially defined as $X^{(i)}=[y[k-i],y[k-i-1],\dots,y[k-i-(n_A-1)],u[k-i],u[k-i-1],\dots,u[k-i-(n_B-1)]]^{\mathrm{T}}$ and $Y^{(i)}=y[k-i+1]$, for $i=1,\dots,N$. Then, the same network can be used to predict future values of $y[k]$ by feeding it with the past values $y[k-1],y[k-2],\dots,y[k-n_A],u[k-1],u[k-2],\dots,u[k-n_B]$ grouped in an individual (n_A+n_B)-dimensional input vector. What we have described is a simple instance of a *dynamic neural network*, as its operations are synchronized with the time steps of the signals $u[k]$ and $y[k]$. We can also make more-steps predictions by estimating $y[k]$ at time k, and then re-using the estimate $\hat{y}[k]$ in the input vector to predict $y[k+1]$ in the absence of the true $y[k]$. In this way, we are building a *recurrent neural network*, and an input will keep affecting the output of the network for possibly infinitely many time steps.[11] Similarly, a Gaussian process approach can be used to reconstruct the function f in Eq. (43) in a Bayesian framework (Section 4), see, for example, Kocijan et al. (2005).

Eq. (43) can be extended by including in the argument of the nonlinear function some disturbance terms, thus obtaining a *NARMAX* model:

$$y[k]=f(y[k-1],\dots,y[k-n_A],u[k-1],\dots,u[k-n_B],w[k-1],\dots,w[k-n_C])+w[k] \quad (44)$$

cf. Eq. (41).

We recall here a few facts drawn from Billings (2013), which provides an extensive treatment of techniques both for the identification and the analysis of NARX and NARMAX systems, with applications to real data and also to space weather (for space weather, see Chapter 14.4 in Billings, 2013). (i) From the identification point of view, by using suitable orthogonalization techniques, f can be built in an incremental way as a weighted sum of linear and nonlinear

[11] For some recent successful case in using dynamic neural networks, see Hinton et al. (2012) and Graves et al. (2013).

terms (typically monomials), by selecting the most promising model terms from a large set of candidate model terms, while pursuing parsimony. (ii) In Eq. (44), the presence of unknown disturbance terms in f can be dealt with in the spirit of the extended least squares algorithm for ARMAX systems. (iii) Methods that work well for validating models in the linear set-up can fail in the nonlinear set-up. For example, correlation methods must be enriched by considering high-order correlations. When the real system exhibit complex, for example, chaotic behaviors, qualitative validation criteria can be useful (Aguirre and Billings, 1994). (iv) Tools for analyzing the contribution of the terms in f also in the frequency domain have been developed.

References

Adler, R.J., Taylor, J.E., 2009. Random Fields and Geometry. Springer Science & Business Media, New York, NY.

Aguirre, L.A., Billings, S.A., 1994. Validating identified nonlinear models with chaotic dynamics. Int. J. Bif. Chaos 4 (1), 109–125.

Akaike, H., 1981. Likelihood of a model and information criteria. J. Econ. 16 (1), 3–14.

Aldrich, J., 2005. Fisher and regression. Stat. Sci. 20 (4), 401–417.

Bartlett, P.L., Bousquet, O., Mendelson, S., et al., 2005. Local Rademacher complexities. Ann. Stat. 33 (4), 1497–1537.

Benveniste, A., Métivier, M., Priouret, P., 2012. Adaptive Algorithms and Stochastic Approximations, vol. 22. Springer Science & Business Media, Berlin.

Bertalanffy, L.V., 1950. An outline of general system theory. Br. J. Philos. Sci. 1 (2), 134.

Billings, S.A., 2013. Nonlinear System Identification: NARMAX Methods in the Time, Frequency, and Spatio-Temporal Domains. John Wiley & Sons, London.

Billings, S.A., Voon, W.S.F., 1984. Least squares parameter estimation algorithms for non-linear systems. Int. J. Syst. Sci. 15 (6), 601–615. https://doi.org/10.1080/00207728408547198.

Billingsley, P., 1995. Probability and Measure, third ed. Wiley-Interscience, London.

Bishop, C.M., 1995. Neural Networks for Pattern Recognition. Oxford University Press, Inc., New York, NY.

Bollobás, B., 1990. Linear Analysis, vol. 199. Cambridge University Press, Cambridge.

Boucheron, S., Lugosi, G., Bousquet, O., 2004. Concentration inequalities. In: Advanced Lectures on Machine Learning. Springer, New York, NY, pp. 208–240.

Bousquet, O., Elisseeff, A., 2002. Stability and generalization. J. Mach. Learn. Res. 2, 499–526.

Breiman, L., 2001. Random forests. Mach. Learn. 45 (1), 5–32.

Breiman, L., et al., 2001. Statistical modeling: the two cultures (with comments and a rejoinder by the author). Stat. Sci. 16 (3), 199–231.

Burman, P., Chow, E., Nolan, D., 1994. A cross-validatory method for dependent data. Biometrika 81 (2), 351–358.

Bussgang, J.J., 1952. Crosscorrelation functions of amplitude-distorted Gaussian signals. Technical Report 216.

Calafiore, G.C., 2010. Learning noisy functions via interval models. Syst. Control Lett. 59 (7), 404–413.

Campi, M.C., Carè, A., 2013. Random convex programs with L_1-regularization: sparsity and generalization. SIAM J. Control. Optim. 51 (5), 3532–3557.

Campi, M.C., Garatti, S., 2016. Wait-and-judge scenario optimization. Mathematical Programming, July.

Campi, M.C., Calafiore, G.C., Garatti, S., 2009. Interval predictor models: identification and reliability. Automatica 45 (2), 382–392.

Camporeale, E., Carè, A., Borovsky, J.E., 2017. Classification of solar wind with machine learning. J. Geophys. Res. Space Phys. 122 (11), 10910–10920. https://doi.org/10.1002/2017JA024383.

Candes, E.J., Romberg, J.K., Tao, T., 2006. Stable signal recovery from incomplete and inaccurate measurements. Commun. Pure Appl. Math. 59 (8), 1207–1223.

Carè, A., Garatti, S., Campi, M.C., 2015. Scenario min-max optimization and the risk of empirical costs. SIAM J. Optim. 25 (4), 2061–2080.

Carè, A., Garatti, S., Campi, M.C., 2017. A coverage theory for least squares. J. R. Stat. Soc. Ser. B Stat. Methodol. 79 (5), 1367–1389.

Carè, A., Csáji, B., Campi, M.C., Weyer, E., 2018. Finite-sample system identification: an overview and a new correlation method. IEEE Control Syst. Lett. 2 (1), 61–66.

Chandorkar, M., Camporeale, E., Wing, S., 2017. Probabilistic forecasting of the disturbance storm time index: an autoregressive Gaussian process approach. Space Weather 15 (8), 1004–1019.

Corfield, D., Schölkopf, B., Vapnik, V.N., 2009. Falsificationism and statistical learning theory: comparing the Popper and Vapnik-Chervonenkis dimensions. J. Gen. Philos. Sci. 40 (1), 51–58.

Cortes, C., Vapnik, V.N., 1995. Support-vector networks. Mach. Learn. 20 (3), 273–297.

Cox, R.T., 1946. Probability, frequency and reasonable expectation. Am. J. Phys. 14 (1), 1–13.

Csató, L., Opper, M., 2002. Sparse on-line Gaussian processes. Neural Comput. 14 (3), 641–668.

Cybenko, G., 1989. Approximation by superpositions of a Sigmoidal function. Math. Control Signals Syst. 2 (4), 303–314.

Draper, N.R., Smith, H., 1998. Selecting the "Best" Regression Equation. John Wiley & Sons, New York, NY, pp. 327–368.

Dudley, R.M., 2002. Real Analysis and Probability, vol. 74. Cambridge University Press, Cambridge.

Floyd, S., Warmuth, M., 1995. Sample compression, learnability, and the Vapnik-Chervonenkis dimension. Mach. Learn. 21 (3), 269–304.

Franz, M.O., Schölkopf, B., 2006. A unifying view of Wiener and Volterra theory and polynomial kernel regression. Neural Comput. 18 (12), 3097–3118.

Galton, F., 1886. Regression towards mediocrity in hereditary stature. J. Anthropol. Inst. Great Britain Ireland 15, 246–263.

Garatti, S., Campi, M.C., 2009. L_∞ layers and the probability of false prediction. IFAC Proc. 42 (10), 1187–1192.

Garatti, S., Campi, M.C., Bittanti, S., 2004. Assessing the quality of identified models through the asymptotic theory—when is the result reliable? Automatica 40 (8), 1319–1332.

Geisser, S., 1993. Predictive Inference, vol. 55. CRC Press, Boca Raton, FL.

Gelman, A., Stern, H., Carlin, J., Dunson, D., Vehtari, A., Rubin, D., et al., 2013. Bayesian Data Analysis, third ed. Chapman and Hall/CRC, New York, NY.

Giri, F., Bai, E.W., 2010. Block-Oriented Nonlinear System Identification, vol. 1. Springer, New York, NY.

Godfrey, K., 1993. Introduction to Perturbation Signals for Time-Domain Systems Identification. Prentice-Hall International (UK) Ltd., Englewood Cliffs, NJ.

Graves, A., Mohamed, A.R., Hinton, G., 2013. Speech recognition with deep recurrent neural networks. In: 2013 IEEE International Conference on Acoustics, Speech and Signal Processing, May, pp. 6645–6649.

Grosse-Erdmann, K., Peris Manguillot, A., 2011. Linear Chaos. Springer Science & Business Media, New York, NY.

Grünwald, P.D., 2007. The Minimum Description Length Principle. MIT Press, Cambridge, MA.

Haber, R., Keviczky, L., 1999. Nonlinear System Identification. Input Output Modeling Approach. Kluwer Academic Publishers, Dordrecht.

Hastie, T., Tibshirani, R., Friedman, J., 2009. The Elements of Statistical Learning, second ed. Springer, New York, NY.

Haykin, S., 2009. Neural Networks and Learning Machines, third ed. Pearson Education, Upper Saddle River, NJ.

Herbrich, R., Williamson, R.C., 2002. Algorithmic luckiness. J. Mach. Learn. Res. 3, 175–212.

Hinton, G., Deng, L., Yu, D., Dahl, G.E., Mohamed, A.R., Jaitly, N., Senior, A., Vanhoucke, V., Nguyen, P., Sainath, T.N., et al., 2012. Deep neural networks for acoustic modeling in speech recognition: the shared views of four research groups. IEEE Signal Process. Mag. 29 (6), 82–97.

Hornik, K., 1991. Approximation capabilities of multilayer feedforward networks. Neural Netw. 4 (2), 251–257.

Ivanov, V.K., Vasin, V.V., Tanana, V.P., 2002. Theory of Linear Ill-Posed Problems and Its Applications, vol. 36. Walter de Gruyter. Berlin, Boston.

Jaynes, E.T., 2003. Probability Theory: The Logic of Science. Cambridge University Press, Cambridge.

Jenkins, G.M., Watts, D.G., 1968. Spectral Analysis and Its Applications. Holden-Day, San Francisco, CA, pp. 243–238.

Kailath, T., 1970. The innovations approach to detection and estimation theory. Proc. IEEE 58 (5), 680–695.

Kailath, T., 1980. Linear Systems, vol. 156. Prentice-Hall, Englewood Cliffs, NJ.

Kálmán, E.R., 1959. On the general theory of control systems. IRE Trans. Autom. Control 4 (3), 110.

Kálmán, E.R., 1960. A new approach to linear filtering and prediction problems. J. Basic Eng. 82 (1), 35–45.

Kocijan, J., Girard, A., Banko, B., Murray-Smith, R., 2005. Dynamic systems identification with Gaussian processes. Math. Comput. Model. Dyn. Syst. 11, 411–424.

Koltchinskii, V., 2001. Rademacher penalties and structural risk minimization. IEEE Trans. Inf. Theory 47 (5), 1902–1914.

Kreyszig, E., 1989. Introductory to Functional Analysis with Applications. Wiley Classics Library. Wiley, New York, NY.

Krizhevsky, A., Sutskever, I., Hinton, G.E., 2012. Imagenet classification with deep convolutional neural networks. In: Advances in Neural Information Processing Systems, pp. 1097–1105.
Ledoux, M., Talagrand, M., 2013. Probability in Banach Spaces: Isoperimetry and Processes. Springer Science & Business Media, New York, NY.
Lei, J., Wasserman, L., 2014. Distribution-free prediction bands for non-parametric regression. J. R. Stat. Soc. Ser. B 76, 71–96.
Lei, J., Robins, J., Wasserman, L., 2013. Distribution-free prediction sets. J. Am. Stat. Assoc. 108 (501), 278–287.
Leontaritis, I.J., Billings, S.A., 1987. Experimental design and identifiability for non-linear systems. Int. J. Syst. Sci. 18 (1), 189–202.
Li, M., Vitányi, P.M.B. 1990, Kolmogorov complexity and its applications (Chapter 4), Leeuwen, J., (Ed.), Algorithms and Complexity, Handbook of Theoretical Computer Science, Elsevier, Amsterdam, pp. 187–254, Available from: https://www.sciencedirect.com/science/article/pii/B9780444880710500096.
Littlestone, N., Warmuth, M., 1986. Relating data compression and learnability. Technical Report. University of California, Santa Cruz, CA.
Ljung, L., 1971. Characterization of the concept of "persistently exciting" in the frequency domain. Technical Report 7119. Lund Institute of Technology.
Ljung, L., 1999. System Identification: Theory for the User, Information and System Sciences Series. Prentice-Hall, Upper Saddle River, NJ.
Ljung, L., 2007. The System Identification Toolbox: the manual. Natick, MA, USA: The MathWorks Inc., (1st ed. 1986, 7th ed. 2007).
Ljung, L., 2010. Perspectives on system identification. Annu. Rev. Control. 34 (1), 1–12.
McCulloch, W.S., Pitts, W., 1943. A logical calculus of the ideas immanent in nervous activity. Bull. Math. Biophys. 5 (4), 115–133.
Milanese, M., Norton, J., Piet-Lahanier, H., Walter, É., 2013. Bounding Approaches to System Identification. Springer Science & Business Media, New York, NY.
Mukherjee, S., Niyogi, P., Poggio, T., Rifkin, R., 2006. Learning theory: stability is sufficient for generalization and necessary and sufficient for consistency of empirical risk minimization. Adv. Comput. Math. 25 (1), 161–193.
Nelles, O., 2013. Nonlinear System Identification: From Classical Approaches to Neural Networks and Fuzzy Models. Springer Science & Business Media, New York, NY.
Novikoff, A.B.J., 1962. On convergence proofs for perceptrons. In: Proceedings of the Symposium on the Mathematical Theory of Automata, pp. 615–620.
Oppenheim, A.V., Schafer, R.W., 2010. Discrete-Time Signal Processing. Pearson, Upper Saddle River, NJ.
Pillonetto, G., Dinuzzo, F., Chen, T., De Nicolao, G., Ljung, L., 2014. Kernel methods in system identification, machine learning and function estimation: a survey. Automatica 50 (3), 657–682.
Pintelon, R., Schoukens, J., 2012. System Identification: A Frequency Domain Approach. John Wiley & Sons, London.
Plackett, R.L., 1972. Studies in the history of probability and statistics. XXIX: the discovery of the method of least squares. Biometrika 59, 239–251.
Popper, K., 1935. Logik der Forschung. Verlag von Julius Springer, Vienna, Austria.
Powell, M.J.D., 1981. Approximation Theory and Methods. Cambridge University Press, Cambridge.
Racine, J., 2000. Consistent cross-validatory model-selection for dependent data: HV-block cross-validation. J. Econ. 99 (1), 39–61.
Rasmussen, C.E., Williams C.K.I., 2006. Gaussian Processes for Machine Learning. The MIT Press, Cambridge, MA, USA.
Ripley, B.D., 1996. Pattern Recognition and Neural Networks. Cambridge University Press, Cambridge.
Rissanen, J., 1978. Modeling by shortest data description. Automatica 14 (5), 465–471.
Rolewicz, S., 1969. On orbits of elements. Stud. Math. 32 (1), 17–22.
Rudin, W., 1991. Functional Analysis, International Series in Pure and Applied Mathematics. McGraw-Hill, Inc., New York, NY.
Rugh, W.J., 1981. Nonlinear System Theory. Johns Hopkins University Press, Baltimore, MD.
Russell, S.J., Norvig, P., 2009. Artificial Intelligence: A Modern Approach. third ed. Prentice Hall, Upper Saddle River, NJ, USA.
Schwarz, G., 1978. Estimating the dimension of a model. Ann. Stat. 6 (2), 461–464.
Scott, D.W., 2015. Multivariate Density Estimation: Theory, Practice, and Visualization. John Wiley & Sons, London.
Shafer, G., Vovk, V., 2008. A tutorial on conformal prediction. J. Mach. Learn. Res. 9, 371–421.

Shalev-Shwartz, S., Shamir, O., Srebro, N., Sridharan, K., 2010. Learnability, stability and uniform convergence. J. Mach. Learn. Res. 11, 2635–2670.

Shawe-Taylor, J., Cristianini, N., 2004. Kernel Methods for Pattern Analysis. Cambridge University Press, Cambridge.

Silverman, B.W., 1986. Density Estimation for Statistics and Data Analysis, vol. 26. CRC Press, Boca Raton, FL.

Söderström, T., Stoica, P., 1988. System Identification. Prentice-Hall, Inc., Englewood Cliffs, NJ.

Söderström, T., Stoica, P., 2002. Instrumental variable methods for system identification. Circ. Syst. Signal Process. 21 (1), 1–9.

Sonoda, S., Murata, N., 2017. Neural network with unbounded activation functions is universal approximator. Appl. Comput. Harmon. Anal. 43 (2), 233–268.

Takagi, T., Sugeno, M., 1985. Fuzzy identification of systems and its applications to modeling and control. IEEE Trans. Syst. Man Cybern. SMC-15 (1), 116–132.

Tibshirani, R., 1996. Regression shrinkage and selection via the lasso. J. R. Stat. Soc. Ser. B Methodol. 58, 267–288.

Trefethen, L.N., 2013. Approximation Theory and Approximation Practice. SIAM. Philadelphia, Pennsylvania, USA.

Tveito, A., Langtangen, H.P., Nielsen, B.F., Cai, X., 2010. Parameter estimation and inverse problems. In: Elements of Scientific Computing. Springer, pp. 411–421.

Valiant, L.G., 1984. A theory of the learnable. Commun. ACM 27 (11), 1134–1142.

Valiant, L.G., 2013. Probably Approximately Correct: Nature's Algorithms for Learning and Prospering in a Complex World. Basic Books, Inc., New York, NY.

Van Overschee, P., De Moor, B., 2012. Subspace Identification for Linear Systems: Theory-Implementation-Applications. Springer Science & Business Media, New York, NY.

Vapnik, V.N., 1998. Statistical Learning Theory, vol. 1. Wiley, New York, NY.

Vapnik, V.N., Stefanyuk, A.R., 1978. Nonparametric methods for restoring the probability densities. Avtomatika i Telemekhanika 8, 38–52.

Volterra, V., 1887. Sopra le funzioni che dipendono da altre funzioni. Atti della Reale Accademia dei Lincei 3, 97–105.

Vovk, V., 2013. Conditional validity of inductive conformal predictors. Mach. Learn. 92 (2–3), 349–376.

Vovk, V., Gammerman, A., Shafer, G., 2005. Algorithmic Learning in a Random World. Springer, New York, NY.

Wiener, N., 1959. Nonlinear problems in random theory. Phys. Today 12 (8), 52–54.

Wiener, N., 1961. Cybernetics or Control and Communication in the Animal and the Machine, vol. 25. MIT Press, Cambridge, MA.

Willems, J.C., Polderman, J.W., 2013. Introduction to Mathematical Systems Theory: A Behavioral Approach, vol. 26. Springer Science & Business Media, New York, NY.

Zeiler, M.D., Fergus, R., 2014. Visualizing and Understanding Convolutional Networks. Springer, New York, NY, pp. 818–833.

CHAPTER

5

Supervised Classification: Quite a Brief Overview

Marco Loog

Delft University of Technology, Delft, The Netherlands; University of Copenhagen, Copenhagen, Denmark

CHAPTER OUTLINE

Machine Learning Techniques for Space Weather
https://doi.org/10.1016/B978-0-12-811788-0.00005-6

1 INTRODUCTION

Consider the playful, yet in a way realistic problem of comparing apples and oranges. More precisely, let us consider the goal of telling them apart in an automated way, for example, by means of a machine or, more specifically, a computer. How does one go about building such a machine? One approach could be to try to construct an accurate physiological model of both types of fruit. We gather what is known about their appearances, their cell structure, their chemical compounds, and so forth, together with any type of physiological laws that relate these quantities and their innerprocesses. Being presented by a new piece of fruit, we can then measure the quantities we think matter most and check these with our two models. A new piece of fruit for which we want to predict whether it is an apple or an orange is then best assigned to the class for which the model fits best. Clearly, the performance of this approach critically depends on issues such as how well we can build such models, how good we are at deciding what the quantities are that matter, and how accurately we can measure these, how we actually decide whether we have good model fit, and so on. Such a model may, for example, not be adequate when dealing with cases that are pathological from a physiological point of view, for example, pieces of fruit that suffer from deformations or rot. Nevertheless, having enough understanding of the problem at hand, we may be able to tackle it in the way it is described here.

1.1 Learning, Not Modeling

Now, let us consider the, in a way, even more challenging problem of comparing foos and bars. Again, let us consider the goal of telling them apart in an automated way on the basis of particular measurements taken from the individual foos and bars. How are we going to go about our problem now? We might not even know exactly what we are dealing with here. Foos? Bars? So, where for the fruits we could try and build, for example, a physiological model, we now do not even know whether physiology at all applies. Or is it physics that we need in this case to describe our objects? Chemistry maybe? Economics? Linguistics?

In the absence of any precise knowledge of our problem at hand,[1] another approach to construct the asked-for machine that tells two or more object classes apart is by means of learning from examples. With this, we move away from any precisely interpretable model based on more or less factual knowledge about the object classes with which we are dealing. As a substitute, we consider a different type of model—in a sense more general purpose—that can learn from examples. This is one of the premier learning settings that is studied in the fields of pattern recognition and machine learning.[2] The particular task is referred to as

[1] Unfortunately, people may often not be able to recognize such absence. Work by Tversky, Kahneman, and others tells us, for example, that people can suffer from systematic deviations from rationality or good judgment, that is, cognitive biases (Kahneman et al., 1982).

[2] We do not care to elaborate on the possible distinctions between machine learning and pattern recognition. For this chapter, it is perfectly fine to consider them one and the same and we, indeed, are going to treat them as such.

the supervised classification task: given a number of example input-output relations, can a general mapping be learned that takes any new and unseen feature vector to its correct class? That is, Can we generalize from finite examples?

1.2 An Outline

The next section starts out with a basic introduction into classification technology and some of the underlying ideas. Classifiers are the aforementioned mappings, or functions, that take the measurements made on our objects that need assigning and aim to map these values to the correct corresponding output, or label, as it is often called. Section 3 discusses a matter that typically warrants consideration prior even to the actual learning, or training, of the classifier: what measurement is the classifier that we are about to construct actually going to rely on when predicting the labels for new and unseen examples? In the same section, we more broadly cover the issue of object representation, which generally considers in what way to present our objects to the classifier. At this point, we can build various classifiers and have different possibilities to represent our objects, but really, how good is the actual classifier built at our prediction task? Section 4 delves into the question of classifier evaluation, providing some tools to get insight into the behavior and performance of the machines that we have constructed. Two of the chapter's main conclusions are covered in this section: (1) there is no single best classifier and (2) there is an inherent tradeoff between the number of examples from which to generalize and classifier complexity. Section 5 then quickly introduces elementary regularization, that is, another way of controlling classifier complexity. In conclusion, Section 6 mentions some variations to basic supervised classification.

2 CLASSIFIERS

Before we get to our selection of classifiers, we introduce some notation, make mathematically more precise what the scope is within which we operate, and describe the ultimate aim of a classifier in equally unambiguous terms.

2.1 Preliminaries

In the general classification setting considered, there are a total of N labeled objects o_i, with $i \in \{1, \ldots, N\}$, to learn from. Every object o_i is represented by a d-dimensional feature vector $x_i \in \mathbb{R}^d$ and has corresponding labels $y_i \in \{-1, +1\}$. So we assume we actually already have a numerical representation that makes up the d measurements in every vector x_i. In addition, with the choice of $y_i \in \{-1, +1\}$, we limit ourselves here to two-class classification problems for simplicity of exposition.[3] Although typically not stated explicitly, we should be aware that the pairs (x_i, y_i) are assumed to be random yet i.i.d. draws from some underlying true

[3] Every two-class classifier can, however, in a more or less principled way, be extended to the general multiclass case. One way is through combining multiple two-class classifiers (see, e.g., Galar et al., 2011). Other classifiers, in particular probabilistic ones, adapt more naturally to the multiclass case.

distribution p_{XY}. For notational convenience, we also define N_+ and N_- ($N_+ + N_- = N$), being the number of positive and negative samples, respectively. In general, we may use shorthand $+$ and $-$, rather than the slightly more elaborate $+1$ and -1, for example, $N_+ = N_{+1}$.

A classifier C is a function or mapping from the space of feature vectors $\mathbb{R}^d$ to the set of labels $\{-1, +1\}$. The input space on which C is defined can be taken smaller than $\mathbb{R}^d$, for example, if we know that some measurements can only take on continuous values between 0 and 1 or if they can only be integer. At the least, the input space is typically larger than the set of N training points as the aim is for our classifier to generalize to new and unseen input vectors. Here we generally consider all of $\mathbb{R}^d$ as possible inputs. One way or the other, a classifier C splits up the space in two sets. One in which points are assigned to $+1$ and one in which the class assignment is -1. The boundary between these two sets is generally referred to as the classification or decision boundary of C.

Finally, we define the objective that every classifier sets out to optimize. In a sense, this will also act as the ultimate measure to decide which of two or more classifiers performs the best. The de facto standard is the measure ε that determines the fraction of misclassified objects when applying classifier C to the problem described by p_{XY}.[4] Using Iverson brackets, we have

$$\varepsilon = \int_{\mathbb{R}^d} \sum_{y \in \{-1,+1\}} [C(x) \neq y] p_{XY}(x, y) dx \tag{1}$$

This measure goes under various names such as (true) error rate, classification error, 0–1 risk, and probability of misclassification. Then again, some may rather refer to it in terms of accuracy, which equals 1 minus the error rate. The lower ε for a particular classifier the better we would say it is for the particular problem p_{XY}, as it will perform better in expectation.[5]

[4] This is not to say that other measures have not been studied. Depending on the actual goal, measures such as AUC, F score, or the H measure have been considered as well. For both general and critical coverage of such measures refer to Bradley (1997), Fawcett (2006), Hand and Anagnostopoulos (2014), Hand and Christen (2017), Lavrač et al. (1999), Landgrebe et al. (2006), and Provost et al. (1997).

[5] Expanding a bit further on Footnote 4, we remark that one of the more important settings, in which another performance measure or another way of evaluating may be appropriate, is in the case where the classification cost per class is different. Eq. (1) tacitly assumes that predicting the wrong class incurs a cost of one, while predicting the right class comes at no cost. In many real-world settings, however, making the one error is not as costly as making the other. For instance, building a rotten fruit detector, classifying a fresh piece of fruit as rotten could turn out less costly than classifying a bad piece of fruit as good. When building an actual classifier, life often is even worse as one may not even know what the cost really is that will be incurred by a misclassification. This is one reason to resort to an analysis of the so-called receiver operating characteristic (ROC) curve and its related measure: the area under the ROC curve (AUC, an abbreviation mention already in the previous footnote). This curve and the related area provide, in some sense, tools to study the behavior of classifiers over all possible misclassification costs simultaneously.

Another important classification setting is the one in which there is a strong imbalance in class sizes, for example, where we expect the one class to occur significantly much more often than the other class—a situation easily imagined in various applications. Also here analyses through ROC, AUC, and related techniques are advisable. For more on this topic, the reader is kindly referred to Fawcett (2006), Hand and Anagnostopoulos (2014), and related work.

2.2 The Bayes Classifier

One should understand that if we know p_{XY}, we are done. In that case, we can construct an optimal classifier C that attains the minimum risk ε^*. But how do we get to that classifier? Let C^* refer to this optimal classifier, fix the feature vector that we want to label to x, and consider the corresponding term within the integral of Eq. (1):

$$\sum_{y\in\{-1,+1\}} [C^*(x) \neq y] p_{XY}(x,y) = [C^*(x) \neq +1] p_{XY}(x,+1) + [C^*(x) \neq -1] p_{XY}(x,-1) \quad (2)$$

As C^* should assign x to $+1$ or -1, we see that the choice that adds the least to the integral at this feature vector value is the assignment to that class for which p_{XY} is largest. We reach an overall minimum if we stick to this optimal choice in every location[6] $x \in \mathbb{R}^d$. In case $p_{XY}(x,+1) = p_{XY}(x,-1)$, where x is actually on the decision boundary, it does not matter what decision the classifier makes, as it will induce an equally large error. In other words, we can define $C^*\colon \mathbb{R}^d \to \{-1,+1\}$ as follows:

$$C^*(x) = \begin{cases} -1 & \text{if } \; p_{XY}(x,-1) > p_{XY}(x,+1) \\ +1 & \text{otherwise} \end{cases} \quad (3)$$

Again using Iverson brackets, we could equally well write this as $C^*(x) = 1 - 2[p_{XY}(x,-1) > p_{XY}(x,+1)]$.

A possibly more instructive reformulation is by considering the conditional probabilities $p_{Y|X}$, often referred to as the posterior probabilities or simply the posteriors, instead of the full probabilities. Equivalent to checking $p_{XY}(x,-1) > p_{XY}(x,+1)$, we can verify whether $p_{Y|X}(-1|x) > p_{Y|X}(+1|x)$ and in the same vein as in Eq. (3) decide to assign x to -1 if this is indeed the case and assign it to $+1$ otherwise. The latter basically states that, given the observations made, one should assign the corresponding object to the class with the largest probability conditioned on those observations. Especially formulated like this, it seems like the obviously optimal assignment strategy.

The theoretical constructs ε^* and C^* are referred to as the Bayes error rate and the Bayes classifier, respectively. The former gives a lower bound on the best error we could ever achieve on the problem at hand. The latter shows us how to make optimal decisions once p_{XY} is known. But these quantities are merely of theoretical importance indeed. In reality, our only hope is to approximate them, as the exact p_{XY} will never be available to us. The objects we can work with are the N draws (x_i, y_i) from that same distribution. Based on these examples, we aim to build a classifier that generalizes well to all of p_{XY}. In all that follows in this chapter, this is the setting considered.

2.3 Generative Probabilistic Classifiers

The previous section showed that if we have p_{XY}, we can compare $p_{XY}(x,-1)$ and $p_{XY}(x,+1)$, and perform an optimal class assignment. One way to get a hold on p_{XY} is by

[6] Of course, if we wish, we generally can decide otherwise on a set of measure 0 without doing any harm to the optimality of the classifier C^*.

making assumptions on the form of the underlying class distributions $p_{X|Y}$ and estimate its free parameters from the training data provided. Such class-conditional probabilities describe the distribution of the underlying individual classes, that is, they consider the distribution of X given Y. Subsequently, these $p_{X|Y}$ can be combined with an estimate of the prior probability p_Y of every class—that is, an estimate of how often every class occurs anyway—to come to an overall estimate of $p_Y p_{X|Y} = p_{XY}$. These models are called generative, because, once they are fitted to the data, they allow us to generate new data from these class-conditional distributions, though the accuracy of this generative process of course heavily depends on the accuracy of the model fit.

A very common, and one of the more simple instantiations of a generative model, is classical linear discriminant analysis (LDA),[7] which assumes the classes to be normally distributed with different means and equal covariance matrices. Denoting the normal distribution with mean μ and covariance Σ by $g(\cdot|\mu, \Sigma)$, the full probability model can be written as

$$p(x, y|\mu_{+1}, \mu_{-1}, \Sigma) = \begin{cases} \pi_+ g(x|\mu_+, \Sigma) & \text{if} \quad y = +1 \\ \pi_- g(x|\mu_-, \Sigma) & \text{if} \quad y = -1 \end{cases} \tag{4}$$

where π_+ and $\pi_- = 1 - \pi_+$ are the class priors, which are in the $[0, 1]$ interval.

The preceding merely specifies the class of models that we consider, but it does not tell us how we fit it to the data that we have. A classical way to come to such parameters is to determine the maximum likelihood estimates. Other approaches, like maximum a posteriori and proper Bayesian estimation, are possible as well. Relying on the log-likelihood, the objective function we find equals

$$\begin{aligned} L(\pi_+, \pi_-, \mu_+, \mu_-, \Sigma) &= \sum_{i=1}^{N} \log p(x_i, y_i|\mu_+, \mu_-, \Sigma) \\ &= \sum_{i:y_i=+1} \left(\log \pi_+ + \log g(x_i|\mu_+, \Sigma)\right) \\ &\quad + \sum_{i:y_i=-1} \left(\log \pi_- + \log g(x_i|\mu_-, \Sigma)\right) \end{aligned} \tag{5}$$

Maximizing this leads to the well-known closed-form estimators

$$\hat{\pi}_+ = \frac{N_+}{N} \tag{6}$$

$$\hat{\pi}_- = \frac{N_-}{N} \tag{7}$$

$$\hat{\mu}_+ = \frac{1}{N_+} \sum_{i:y_i=+1} x_i \tag{8}$$

[7] Some authors take the term "linear discriminant analysis" to refer to a more broadly defined class of classifiers. The normality-based classifier discussed here is also referred to by some as Fisher's linear discriminant or classical discriminant analysis.

$$\hat{\mu}_{-} = \frac{1}{N_{-}} \sum_{i:y_i=-1} x_i \tag{9}$$

$$\hat{\Sigma} = \frac{1}{N} \sum_{i=1}^{N} (x_i - \hat{\mu}_{y_i})(x_i - \hat{\mu}_{y_i})^{\top} \tag{10}$$

We note that, no matter what estimates one chooses, the L in the abbreviation LDA refers to the fact that the decision boundary forms a $(d-1)$-dimensional hyperplane, as can be checked by explicitly solving $\pi_{+}g(x|\mu_{+}, \Sigma) = \pi_{-}g(x|\mu_{-}, \Sigma)$ for x and finding that it takes on the form of a linear equation[8]:

$$(\hat{\mu}_{+} - \hat{\mu}_{-})^{\top} \hat{\Sigma}^{-1} x + c = 0 \tag{11}$$

with c a constant offset that depends on the dimensionality d, the determinant of $\hat{\Sigma}$, and the prior probabilities $\hat{\pi}_{+}$ and $\hat{\pi}_{-}$.

Again, these estimates are just one way to specify the free parameters in the model. For instance, Σ could also have been estimated by an estimator such as $\frac{N}{N-2}\hat{\Sigma}$ or we could have applied Laplace smoothing to our prior estimates and, for example, consider $(N_{+}+1)/(N+2)$ instead of N_{+}/N. Of course, probably of a more dramatic influence is the actual choice of normal class-conditional models, which just may not be realistic. Even though having a misspecified model, as such, does not mean that the classifier will not perform well, more advanced, and complex choices for these class-conditional distributions have been introduced that can potentially improve upon the use of normal distributions (see, e.g., Bishop, 1995; Hastie et al., 2001; McLachlan, 2004; Rasmussen and Williams, 2006; Ripley, 2007, but let us also refer already to Section 4.3 for some important notes regarding classifier complexity in relation to training set size). As an example, we may substitute our relatively rigid parametric choice for a highly flexible nonparametric kernel density estimator per class, which leads to a classifier that is often referred to as the Parzen classifier.

2.4 Discriminative Probabilistic Classifiers

In the previous section, we decided to model p_{XY}, based on which we can then come to a decision on whether to assign o_i to the $+$ or the $-$ class considering its corresponding feature vector x_i. Section 2.2, however, showed that we might as well use a model of $p_{Y|X}$ to reach a decision. Of course, from the full model p_{XY}, we can get to the conditional $p_{Y|X}$, while going into the other direction is not possible. For classification, however, we merely need to know $p_{Y|X}$ and so we can save the trouble of building a full model. In fact, if we are unsure about the true form of the underlying class-conditionals or the marginal p_X that describes the feature vector distribution, directly modeling $p_{Y|X}$ may be wise, as we can avoid potential problems due to such model misspecification. On the other hand, if the full model is accurate enough this may have a positive effect on the classifier's performance (McLachlan, 2004; Rubinstein

[8] It actually is a nonhomogeneous linear or affine one, to be more precise.

and Hastie, 1997). Approaches that directly model $p_{Y|X}$ are called discriminative as they aim to get straightaway to the information that matters to tell the one class apart from the other.

The classical model in this setting, and in a sense a counterpart of LDA, is called logistic regression.[9] One way to get to this model is to assume that the logarithm of the so-called posterior odds ratio takes on a linear form in x, that is,

$$\log \frac{p_{Y|X}(+1|x)}{p_{Y|X}(-1|x)} = w^\top x + w_\circ \tag{12}$$

with $w \in \mathbb{R}^d$ and $w_\circ \in \mathbb{R}$. From this we derive that the posterior for the positive class takes on the following form:

$$p_{Y|X}(+1|x) = 1 - p_{Y|X}(-1|x) = \frac{\exp(w^\top x + w_\circ)}{1 + \exp(w^\top x + w_\circ)} = 1 - \frac{1}{1 + \exp(w^\top x + w_\circ)} \tag{13}$$

The parameters w and $w_\circ$ again are typically estimated by maximizing the log-likelihood. Formally, we have to consider the likelihood of the full model and not only of its posterior, but the choice of the necessary additional marginal model for p_X is of no influence on the optimum of the parameters we are interested in McLachlan (2004), Minka (2005), and so we may just consider

$$(\hat{w}, \hat{w}_\circ) = \underset{(w,w_\circ)\in\mathbb{R}^{d+1}}{\operatorname{argmax}} \sum_{i=1}^{N} \log\left(\frac{y_i + 1}{2} - \frac{y_i}{1 + \exp(w^\top x_i + w_\circ)}\right) \tag{14}$$

Note that, similar to LDA, this classifier is linear as well. Generally, the decision boundary is located at the x for which $p_{XY}(x, +1) = p_{XY}(x, -1)$ or, similarly, for which $p_{Y|X}(+1|x) = p_{Y|X}(-1|x)$. But the latter case implies that the log-odds equals 0 and so the decision boundary takes on the form

$$\hat{w}^\top x_i + \hat{w}_\circ = 0 \tag{15}$$

As for generative models, discriminative probabilistic ones come in all different kinds of flavors. A particularly popular and fairly general form of turning linear classifiers into nonlinear ones is discussed in Section 3.1. These and more variations can be found, among others, in Bishop (1995), Hastie et al. (2001), Hinton (1989), McLachlan (2004), Rasmussen and Williams (2006), and Ripley (2007).

2.5 Losses and Hypothesis Spaces

As made explicit in Eqs. (11), (15), both LDA and logistic regression lead to linear decision boundaries, that is, $(d-1)$-dimensional hyperplanes in a d-dimensional feature space. The functional form of these decision boundaries is indirectly fixed through the choice of

[9] Again, there are various other names that refer to, at least more or less, the same approach, for example, the logistic classifier, logistic discrimination, or logistic discriminant analysis.

probabilistic model. Ultimately, they can be seen as mapping a feature vector x in a linear way to a value on the real line. A subsequent decision to assign this point to the + or − class depends on whether the obtained value is smaller or larger than 0. This last operation, which basically turns the linear mapping into an actual classifier, can in essence be carried out by applying the sign function following the linear transform. This maps every number to a point in the set $\{-1, +1\}$, except for the points on the decision boundary that are mapped to 0.[10] The main difference between LDA and logistic regression is the way the free parameters of their respective linear transformations have been obtained.

Especially within some areas of machine learning, it is common to altogether forget about the potentially probabilistic interpretation of a classifier. Instead, one explicitly defines the set H of functions that transform any feature vector into a single number that can subsequently be turned into actual classifiers by "signing" them. Next to that, one defines a measure $\ell\colon \mathbb{R} \times \{-1, +1\} \to \mathbb{R}$ of how good or bad a particular function fits to any point in the training data. The set of functions is often referred to as the hypothesis space, while the measure of fit is typically referred to as the loss (which is something that one minimizes). More specifically, the loss function ℓ takes in a predicted value $h(x)$, with $h \in H$, and decides how good this value matches the desired output y. Once, H and ℓ are fixed, we can search for a best fitting $h^* \in H$,

$$h^* = \underset{h \in H}{\operatorname{argmin}} \sum_{i=1}^{N} \ell(h(x_i), y_i) \tag{16}$$

which can then be used to classify new and unseen object x by assigning it to class $\operatorname{sign}(h^*(x))$.

The foregoing is a fairly general way of formulating the training of a classifier and not every choice for H and ℓ may be equally convenient or suitable to be used on a classification problem. In the following, we present some of the better-known choices and briefly introduce some of the classifiers related to the particular options.

2.5.1 0–1 Loss

In a way, the obvious choice is to take the loss that we are actually interested in: the fraction of misclassified observations. Eq. (1) defines this fraction, that is, the classification error or the 0–1 risk, under the true distribution. Considering our finite number N of training data, the best we can do is just count the number of incorrectly assigned samples:

$$L_{0\text{-}1}(h) := \sum_{i=1}^{N} \ell_{0\text{-}1}(h(x_i), y_i) \tag{17}$$

[10] As this is typically a set of measure zero, where the decision boundary gets mapped to does not really influence the performance of the classifier. Nevertheless, one may of course decide to have the boundary points mapped to −1 or +1, rather than 0.

with

$$\ell_{0\text{-}1}(a,b) := [\operatorname{sign}(a) \neq b] \tag{18}$$

being the 0–1 loss.[11]

A major problem with this loss is that finding the optimal h^* is for many cases computationally very hard. Take for H, for example, all linear functions, then finding our h^* turns out to be NP-hard and even settling for approximate solutions does not necessarily help (Ben-David et al., 2003; Hoffgen et al., 1995).

2.5.2 *Convex Surrogate Losses*

The most broadly used approach to get around the problem of the computational complexity that the 0–1 loss poses is to consider a relaxation of the original problem, which turns it into a convex optimization problem that is relatively easy to solve, for example, by gradient descent or variations of this technique.

To achieve this, first of all, one chooses H to be convex. The classical choice would be the space of linear functions, but more complex choices are possible as we will see later. The second step in the relaxation of the optimization problem is to turn to so-called surrogate losses. These are losses that aim to approximate $\ell_{0\text{-}1}$ as well as possible, but have some additional benefits. For instance, one could choose a loss that is everywhere differentiable, something that does not hold for the 0–1 loss. To turn the optimization for h into a convex optimization, the function in Eq. (17) needs to be convex as well. To generally do so, we approximate $\ell_{0\text{-}1}$ in Eq. (18) by a convex function that upper bounds this loss everywhere. The idea of the upper bound is that if we find the minimizer to it, we know that 0–1 loss one would achieve on the training set is certainly not larger than the surrogate risk the minimizer attains.

2.5.3 *Particular Surrogate Losses*

As we would decide on the label of a sample x based on the output $h(x)$ that a trained classifier $h \in H$ provides, one typically only needs to consider what the loss function does for the value $yh(x)$. For instance, we can rewrite $\ell_{0\text{-}1}(a,b)$ as $\ell_{0\text{-}1}(a,b) = \ell_{0\text{-}1}(ba) = [b \operatorname{sign}(a) \neq 1]$ and achieve the same loss. In Fig. 1 the shape of the 0–1 loss is plotted in these terms. The same figure shows various widely used upper bounds for $\ell_{0\text{-}1}$.

Maybe the first one to note is the logistic loss, which is defined as

$$\ell_{\text{logistic}}(a,b) := \log_2\left(1 + e^{-ba}\right) \tag{19}$$

The figure displays it as the solid light gray curve. Using this loss, in combination with a linear hypothesis class, leads to standard logistic regression as introduced in Section 2.4. So in this case, we have both a probabilistic view of the resulting classifier as well as an interpretation of logistic regression as minimizer of a specific surrogate loss. A second well-known classifier, or at least a basic form of it, is obtained by using the so-called hinge loss:

$$\ell_{\text{hinge}}(a,b) := \max(1 - ba, 0) \tag{20}$$

[11] We note here that the expected loss is often referred to as the risk, which explains why the error defined in Eq. (1) is also referred to as the 0–1 risk.

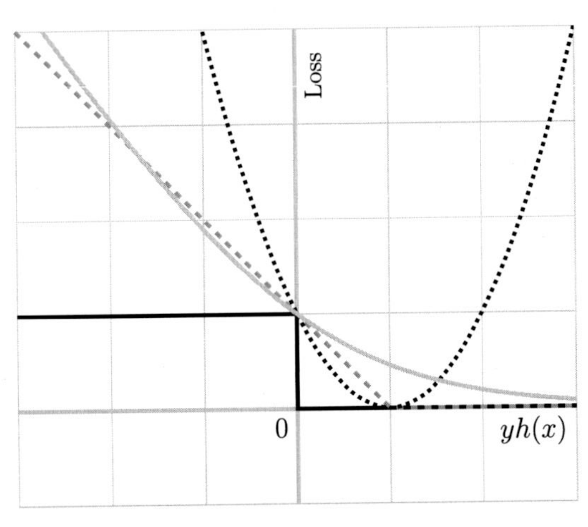

FIG. 1 Plot of the 0–1 loss in terms y times the predicted output $h(x)$ (in *solid black*) and three examples of surrogate losses that upper bound the 0–1 loss. Logistic loss is in *solid light gray*, hinge loss is in *dashed dark gray*, and squared loss is *dotted black*.

This loss is at the basis of the support vector machine[12] (SVM) (Boser et al., 1992; Christianini and Shawe-Taylor, 2000; Cortes and Vapnik, 1995; Vapnik, 1998). A third classifier that fits the general formalism and is widely employed is obtained by using the squared loss function

$$\ell_{\text{squared}}(a, b) := (1 - ba)^2 \tag{21}$$

Using again the set of linear hypotheses, we get basically what is, among others, referred to as the linear regression classifier, the least squares classifier, the least squares SVM, the Fisher classifier, or Fisher's linear discriminant (Duda and Hart, 1973; Hastie et al., 2001; Poggio and Smale, 2003; Suykens and Vandewalle, 1999). Indeed, this classifier is a reinterpretation of the classical decision function introduced by Fisher (1936) in the language of losses and hypotheses.

Finally, other losses one may encounter in the literature are the exponential loss $\exp(-ba)$, the truncated squared loss $\max(1 - ba, 0)^2$, and the absolute loss $|1 - ba|$. In Section 3.1, we introduce ways of designing nonlinear classifiers, which often rely on the same formalism as presented in this section.

12 Together with the underlying theory, SVMs caused all the furor in the late 1990s and early 2000s. To many, the development of the SVM may still be one of the prime achievements of the mathematical field of statistical learning theory that started with Vapnik and Chervonenkis in the early 1970s. At least, SVMs are still one of the most widely known and used classifiers within the fields of pattern recognition and machine learning. Possibly one of the main feats of statistical learning theory was that it broke with the statistical tradition of merely studying the theory of the asymptotic behavior of estimators. Statistical learning theory is also concerned with the finite sample setting and makes, for example, statements on the expected performance on unseen data for classifiers that have been trained on a limited set of examples (Christianini and Shawe-Taylor, 2000; Devroye et al., 1996; Vapnik, 1998).

2.6 Neural Networks

The use of artificial neural networks for supervised learning can be traced back at least to 1958. In that year, the perceptron was introduced (Rosenblatt, 1958), providing a linear classifier that could be trained using a basic iterative updating scheme for its parameters. Currently, neural networks are the dominant technique in many applications and application-related areas when massively sized datasets to train from are available.

Even though the original formulation of the perceptron does not give a direct interpretation in terms of the solution to the optimizing of a particular loss within its hypothesis space of linear classifiers, such formulations are possible (Devroye et al., 1996). Neural networks that are employed nowadays readily follow the earlier sketched loss-hypothesis space paradigm. Possibly the most characteristic feature of neural networks is that the hypotheses considered are built up of relatively simple computational units called neurons or nodes. Such a unit is a function $g: \mathbb{R}^q \to \mathbb{R}$ that takes in q inputs and maps these to a single numerical output. Typically, g takes on the form of a linear mapping followed by a nonlinear transfer or activation function $\sigma: \mathbb{R} \to \mathbb{R}$:

$$g(x) = \sigma(w^\top x + w_\circ) \tag{22}$$

Often σ is taken to be of sigmoidal shape, like the logistic function in Eq. (13) used in logistic regression. This function squeezes the real line onto the interval $[0, 1]$, resembling a smooth threshold function. Other choices are possible, however. A choice popularized more recently, with a clearly different characteristic is the so-called rectified linear unit, which is defined as $\sigma(x) = \max(0, x)$ (Nair and Hinton, 2010). As for various of the previously mentioned classifiers, the free parameters are tuned by measuring how well g fits the given training data. A widely used choice is the squared loss, but also likelihood-based methods have been considered and links with probabilistic models have been studied (Bishop, 1995; Hinton, 1989; Ripley, 2007).

One should realize that, whatever the choice of activation function, as long as it is monotonic using g for classification will lead to a linear classifier. Nonlinear classifiers are constructed by combining various gs, both in parallel and in sequence. In this way, one can build arbitrarily large networks that can perform arbitrarily complex input-output mappings. This means that we are dealing with large and diverse hypothesis classes H. The general construction is that multiple nodes, connected in parallel, provide the inputs to subsequent nodes. Consider, for example, the nonlinear extension where, instead of a single node g, as a first step, we have multiple nodes $g_1, \ldots, g_D$ that all receive the same feature vectors x as input. In a second step, these D outputs are collected by yet another node, $g: \mathbb{R}^D \to \mathbb{R}$ and transformed in a similar kind of way. So, all in all, we get a function G of the form

$$G(x) = g\left(g_1(x), \ldots, g_D(x)\right) = \sigma\left(\sum_{i=1}^{D} w_i \sigma(w_{ij}^\top x + w_{i\circ}) + w_\circ\right) \tag{23}$$

To fully specify a particular G, one needs to set all the parameters in all $D + 1$ nodes. Once these are set, we can again use it to classify any x to the sign of $G(x)$.

Of course, one does not have to stop at two steps. The network can have an arbitrary number of steps, or layers as they are typically called. Nowadays, so-called deep[13] networks are being employed with hundreds of layers and millions of parameters. In addition, there are generally many different variations to the basic scheme we have sketched here (Schmidhuber, 2015). By making different choices for the transfer function, by using multiple transfer functions, by changing the structure of the network, the number of nodes per layer, and so forth, one basically changes the hypothesis class that is considered. In addition, where in Section 2.5 the choice of H and ℓ would typically be such that we end up with a convex optimization problem, using neural networks, we typically move away from optimization problems for which one can reasonably expect to be able to find the global optimum. As a result, to fully define the classifier, we should not only specify the loss and the hypothesis class, but also the exact optimization procedure that is employed. There are many choices possible to carry out the optimization, but most approaches rely on gradient descent or variations to this basic scheme (Bishop, 1995; Bottou, 1991; Hinton, 1989; White, 1989).

2.7 Neighbors, Trees, Ensembles, and All that

There are a few approaches to classification that do not really fit the aforementioned settings, but that we feel do need brief mentioning either for historical reasons or for completeness.

2.7.1 *k Nearest Neighbors*

The nearest neighbor rule (Fix and Hodges, 1951; Cover and Hart, 1967) is possibly the classifier with the most intuitive appeal. It is a widely used and classical decision rule and one of the earliest nonparametric classifiers proposed. In order to classify a new and unseen object, one simply determines the distances between its describing feature vector and the feature vectors in the training set and decides to assign the object to the same class the closest feature vector that training set has. Most often, the Euclidean distance is used to determine the nearest neighbor in the training dataset, but in principle any other, possibly even expert-designed or learned, distance measure can be employed.

A direct, yet worthwhile extension is to not only consider the closest sample in the training set, that is, the first nearest neighbor, but to consider the closest k and assign any unseen object to the class that occurs most often among these k nearest neighbors. The k nearest neighbor classifier has various interesting properties (Fix and Hodges, 1951; Cover and Hart, 1967; Devroye et al., 1996). One of the more interesting ones may be the result that roughly states that with increasing numbers of training data, the k nearest neighbor classifier converges to the Bayes classifier C^*, given k increases at the appropriate rate.

2.7.2 *Decision Trees*

Classification trees, or decision trees, are classifiers that can be visualized by a hierarchical tree structure (Breiman et al., 1984; Quinlan, 1986). Every observation that is classified traverses

[13] There is, of course, not a clear-cut definition of what number of layers makes a network deep. This is similar to wondering what amount of data makes it big or what number of features makes a problem high-dimensional.

the tree's nodes to arrive at a leave that contains the final decision, that is, assign label $+1$ or -1. In every node, a basic operation decides what next node, at a level deeper, will be visited.

Probably the simplest and most extensively used form of decisions that are made at every node are those that just check whether or not a single particular feature value is larger than a chosen threshold value. As there are only two outcomes possible at every step, there will only be two nodes available at every next level that can be reached. Using training data, there are various ways to come to an automated construction of the exact tree hierarchy, together with the decisions to make at the different nodes.

One of the possible benefits of these kinds of classifiers, especially of the last mentioned type of decision tree, given the tree is not too deep, is that one can easily trace back and interpret how the decision to assign a sample to class -1 or $+1$ was reached. One can retrace the steps through the tree and see what subsequent decisions led to the class label provided.

2.7.3 *Multiple Classifier Systems*

The terms multiple classifier systems, classifier combining, and ensemble methods all refer to roughly the same idea: potentially more powerful classifiers can be built by combining two or more of them (kit, 2000; Kuncheva, 2004; Polikar, 2006). The latter are often referred to as base[14] classifiers. So, these techniques are not classifiers as such, but ways to compile base classifiers into classifiers that in some sense befit the data better. There can be various reasons to combine classifiers.

Sometimes a classifier turns out to be overly flexible and one may wish to stabilize the base classifier (see also Section 5). One way to do so is by a well-known combining technique called bagging (Breiman, 1996), which trains various classifiers based on bootstrap samples of the same dataset and assigns any new sample based on the average output of this often large set of base classifiers. Another way to construct different base classifiers is to consider random subspaces by sampling a set of different features for every base learner. This technique has been extensively exploited in random forests and the like (Ho, 1995, 1998).

Combining classifiers can also be exploited when dealing with a problem where, in some sense, essentially different sets of features play a role. For instance, in the analysis of patient data, one might want to use different classifiers for high-dimensional genetic measurements and low-dimensional clinical data, as these sets may behave rather differently from each other. Once the two or more specialized classifiers have been trained, various forms of so-called fixed and trained combiners can be applied to come to a final decision rule (Duin, 2002; Kuncheva, 2004).

At times, the base classifiers can already be quite complex, possibly being a multiple classifier system in itself. Examples are available from medical image analysis (Niemeijer et al., 2011) and recommender systems (Jahrer et al., 2010).[15] In these cases, advanced systems have been developed independently from each other. As a result, there is a fair chance that every system has its own strengths and weaknesses, and even the best performing individual system cannot be expected to perform the best in every part of feature space. Hence combining such systems can result in significantly improved overall performance.

[14] That is, \ 'bās \ rather than \ 'bāz-\.

[15] Though illustrative, strictly speaking, this work does not report on a classification task.

Another reason to employ classifier combining is to integrate contextual features into the classification process. Such approaches can especially be beneficial when integrating contextual information into image and signal analysis tasks (Cohen and Carvalho, 2005; Loog and van Ginneken, 2002; Loog, 2004). These techniques can be seen as a specific form of stacked generalization, or stacking (Wolpert, 1992b), and are becoming relevant again these days in connection with deep learning (see, e.g., Fu et al., 2016; Li et al., 2016; Shrivastava and Gupta, 2016).

Finally, we should mention boosting approaches to multiple classifier systems and in particular adaboost (Freund and Schapire, 1995). Boosting was initially studied in a more theoretical setting to show that so-called weak learners, that is, classifiers that barely reach better performance than an error rate equal to the a priori probability of the largest class, could be combined into a strong learner to significantly improve performance over the weak ones. This research culminated in the development of a combining technique that sequentially adds base classifiers to the ensemble that has already been constructed, where the next base classifier focuses especially on samples that previous base learners were unable to correctly classify. This last feature is the adaptive characteristic of this particular combining scheme that warrants the prefix ada-.

3 REPRESENTATIONS AND CLASSIFIER COMPLEXITY

In all of the foregoing, we tacitly assumed that we already had decided which feature set to use for every object or, more generally, on how to represent every object before constructing our classification rule. In actual applications, there are, however, choices to be made.

To start with, we may be lucky and even have influence on what kind of raw measurements are going to be taken. Are we making color images with a standard camera of our oranges? Do we make a CT or an MR scan? Do we measure weight, diameter, plasticity? Will we sequence tissue samples? Clearly, the actual measurements that are most informative depend on the classification task at hand. More importantly, one should realize that once key measurements have been omitted,[16] one can never recover the lost information through the building of a classifier, no matter how advanced or clever the classifier one designs is.

In reality, the practitioner often has little influence on what precise measurements are going to be made and s/he has to work with a predetermined set of initial features. But even in this setting there are considerations to be made. Are we going to use all of these features? Are we using them as is, or do we construct derivative features. Such decisions can hinge, say, on one's own insight into the problem or on possible external expert knowledge that one has acquired. Do we really need the RGB values of every pixel when describing our oranges, or is an average value per channel enough? Do we at all need RGB or is a measure of orangeness sufficient? And do we need that weight measurement or does our consulting expert suggest that it does not contain any relevant information and can we do without?

[16] Such a situation may of course arise from the fact that particular measurements simply cannot be realized for a number of reasons. For instance, measurements can be too expensive to consider as part of a realistic solution or they can be invasive to an unacceptable degree.

We give a brief overview of ways to represent objects and some tools with which we can partially automate the process of selecting features and create derivatives. In addition, we discuss some of the—initially maybe counterintuitive—effects of using more and more measurements and how this roughly relates to the complexity of the classifier. We also introduce a tool possibly valuable in the analysis of the effect of the number of features: the so-called feature curve.

3.1 Feature Transformations

In the previous section, we saw various linear classifiers such as logistic regression, SVMs, and Fisher's discriminant. At first sight, the linearity of a classifier may seem like a limitation, but this is actually easily removed. Nonlinear classifiers can be created by including nonlinear transformations of the original features into the original feature vector and subsequently train a linear classifier in this extended space. This essentially extends the hypothesis space H. For example, if $x_i \in \mathbb{R}^2$, then we can form new and nonlinear features by transforming the original individual features, for example, we can transform the first feature x_{i1} into x_{i1}^3 or $\sin x_{i1}$, and by combining individual features, for example, we can form the product feature $x_{i1}x_{i2}$. Clearly, the possibilities are endless.

Let $\varphi: \mathbb{R}^d \to \mathbb{R}^D$ be the transformation that maps the original feature vector to its extended representation. Once we learned a classifier C_D in the extended space, it induces a classifier in the original space through $C_d = C_D \circ \varphi$: we simply take every feature vector to be classified, transform it by φ, and apply the classifier trained in the higher-dimensional space. Using this construct, even if we would limit ourselves to linear classifiers in $\mathbb{R}^D$, we would typically find nonlinear decision boundaries in the original d-dimensional feature space.

3.1.1 The Kernel Trick

SVMs not only received a lot of attention as a result of statistical learning theory, the SVM literature also introduced what has become widely known as the kernel trick or kernel method (Boser et al., 1992), which has its roots in the 1960s (Aizerman et al., 1964). The kernel trick allows one to extend many inherently linear approaches to nonlinear ones in a computationally simple way. At its basis is that—following the representer theorem (Schölkopf et al., 2001a; Wahba, 1990)—many solutions for the type of optimization problems for linear classifiers that we have considered in Section 2.5 can be expressed in terms of a weighted combination of inner products of training feature vectors and the x that is being classified, that is,

$$h^*(x) = w^\top x + w_\circ = \sum_{i=1}^{N} a_i x_i^\top x + a_\circ \tag{24}$$

with $a_i \in \mathbb{R}$. Therefore, finding h^* becomes equivalent to finding the optimal coefficients a_i.

After mapping the original feature vectors with φ, we would be optimizing the equivalent in the D-dimensional space to get to a possibly nonlinear classifier:

$$h^*(x) = w^\top x + w_\circ = \sum_{i=1}^{N} a_i \varphi(x_i)^\top \varphi(x) + a_\circ \tag{25}$$

It becomes apparent that the only thing that matters in these settings is that we know how to compute inner products $k(z, x) := \varphi(z)^\top \varphi(x)$ between any two mapped feature vectors x and z. The function k is also referred to as a kernel function or simply a kernel. Of course, once we have explicitly defined φ, we can always construct the corresponding kernel functions, but the power of the kernel trick is that in many settings this can be avoided. This is interesting in at least two ways.

The first one is that if one wants to construct highly nonlinear classifiers, the explicit expansion φ could grow inconveniently large. Take a simple expansion in which we consider all (unique) second degree monomials, which number equals $\binom{d}{2}$. So the dimensionality D of the feature space in which we have to take the inner product grows as $O(d^2)$. By a direct calculation, one can, however, show that the inner product in this larger space can be expressed in terms of a much simpler k. In this case particularly, we have that[17]

$$\varphi(z)^\top \varphi(x) = (z^\top x)^2 \tag{26}$$

As one can imagine, moving to nonlinearities of even an higher degree, the effect becomes more pronounced.[18] At some point, explicitly expanding the feature vector nonlinearly becomes prohibitive, while calculating the induced inner product may still be easy to do. An extreme example is the radial basis function or Gaussian kernel defined by

$$k(z, x) = \exp\left(-\frac{1}{\sigma^2}||x - z||^2\right) \tag{27}$$

which corresponds to a mapping that takes the original d-dimensional space to an infinite dimensional expansion (Vapnik, 1998).

A second reason why the formulation in terms of inner products is of interest is that it, in principle, allows us to forget about an explicit feature representation altogether. Going back to our original objects o_i, if we can construct a function $k(\cdot, \cdot) \mapsto \mathbb{R}_0^+$ that takes in two objects and fulfils all the criteria of an kernel, we can directly use $k(o_i, o)$ (with o the object that we want to classify) as a substitute for $\varphi(x_i)^\top \varphi(x)$ in Eq. (25). Once such a kernel function k has been constructed—whether it is through an explicit feature space or not, one can use it to build classifiers.

All in all, kernel methods define a very general, powerful, and flexible formalism, which allows the design of problem-specific kernels. Research into this direction has spawned a massive amount of publications about such approaches (see, e.g., Christianini and Shawe-Taylor, 2000; Schölkopf and Smola, 2002).

3.2 Dissimilarity Representation

Any kernel k provides, in a way, a similarity measure between two feature observations x and z (or possible directly between two objects): the larger the value is, the more similar the two observations are. As k has to act like an inner product that, at least implicitly, corresponds

[17] This can be demonstrated by explicitly writing out both sides of the equation.

[18] We note that first degree monomials can also be included, either by explicitly including an additional feature to the original feature vector that is constant, say c, or implicitly by defining the inner product as $(z^\top x + c^2)^2$.

to some underlying feature space, limitations apply. In many settings, one might actually have an idea of a proper way to measure the similarity or the, in some sense equivalent, dissimilarity between two objects.[19] Possibly, such measure is provided by an expert that is working in the field where you are asked to build your classifier. It therefore may be expected to be a well thought-through quantity that captures the essential resemblance of or difference between two objects.

The dissimilarity approach (Duin et al., 1997; Pękalska et al., 2001; Pękalska and Duin, 2005) allows one to build classifiers similar to kernel-based classifiers, but without some of the restrictions. One of the core ideas is that every object can be represented, not by what one can see as absolute measurements that can be performed on every individual object, but rather by relative measurements that tell us how (dis)similar the object of interest is with a set of D representative objects. These representative objects are also referred to as the prototypes. In particular, having such a set of prototypes p_i with $i \in \{1,\dots,D\}$, and having our favorite dissimilarity measure δ, every object o can be represented by the D-dimensional dissimilarity vector

$$x = (\delta(p_1,o),\dots,\delta(p_D,o))^\top \tag{28}$$

Training, for example, a linear classifier in this space leads to a hypothesis of the form

$$w^\top x + w_\circ = \sum_{i=1}^{N} a_i \delta(p_i,o) + w_\circ \tag{29}$$

which should be compared with Eq. (25).

The linear classifier is just one example of a classifier one can use in these D dimensions. In this dissimilarity space, one can of course use the full range of classifiers that have been introduced in this chapter.

3.3 Feature Curves and the Curse of Dimensionality

Measuring more and more features on every object seems to imply that we gather more and more useful information on them. The worst that can happen is that we measure features that are partly or completely redundant, for example, measuring the density, while we already have measured the mass and the volume. But once we have the information present in the features, it cannot vanish anymore. In a sense this is indeed true, but the question is whether we can still extract the information relevant with growing feature numbers. All classifiers rely on some form of estimation, which is basically what we do when we train a classifier, but estimation typically becomes less and less reliable when the space in which we carry it out grows.[20] The net result of this is that, while we typically would get improved performance with every additional feature in the very beginning, this effect will gradually wear off, and in the long run even leads to a deterioration in performance as soon as the estimates become

[19] Depending on the requirements one imposes upon dissimilarities (or, proximities, distances, etc.), similarities s can be turned into dissimilarities δ. For instance, by taking $\delta = \frac{1}{s}$ or $\delta = -s$. Next to these very basic transforms, there are various more advanced possibilities to construct such conversions (Pękalska and Duin, 2005).

[20] One can say that, in this specific sense, the complexity of the classifier increases.

unreliable enough. This behavior is what is often referred to as the curse of dimensionality (Bishop, 1995; Duin and Pękalska, 2015; Hastie et al., 2001; Jain et al., 2000).

A curve that plots the performance of a classifier against an increasing number of features is called a feature curve (Duin and Pękalska, 2015). It can be used as a simple analytic tool to get an idea of how sensitive our classifier is to the number of measurements with which each object is described. Possibly of equal importance is that such curves can be used to compare two or more classifiers with each other. The forms feature curves take on depends heavily on the specific problem that we are dealing with, on the complexity of the classification method, the way this complexity relates to the specific problem, and on the number N of training samples we have to train our classifier. As far as it is at all possible, the exact mathematical quantification of these quantities is a real challenge. Very roughly, one can state that the more complex a classifier is, the quicker its performance starts deteriorating with an increasing number of features. On the other hand: the more training data that is available, the later the deterioration in performance sets in. Also, the one classification technique is more complex than the other if the possible decision boundaries the former can model are more flexible or, similarly, less smooth. Another way to think about this is that the hypothesis of the former classification method is larger than the latter one.[21]

3.4 Feature Extraction and Selection

The curse of dimensionality indicates that in particular cases it can be beneficial for the performance of our classifier to lower the feature dimensionality. This may be applicable, for example, if one has little insight into the classification problem at hand, in which case one tends to define lots of potentially useful features and/or dissimilarities in the hope that at least some of them pick up what is important to discriminate between the two classes. Carrying out a more or less systematic reduction of the dimensionality after defining such large class of features can lead to acceptable classification results.

Roughly speaking, there are two main approaches (Guyon and Elisseeff, 2003; Hastie et al., 2001; Jain and Zongker, 1997; Ripley, 2007). The first one is feature selection and the second one is feature extraction. The former reduces the dimensionality by picking a subset from the original feature set, while the latter allows the combination of two or more features into fewer new features. This combination is often restricted to linear transformations, that is, weighted sums, of original features, meaning that one considers linear subspaces of the original feature space. In principle, however, feature extraction also encompasses nonlinear dimensionality reductions. Feature selection is, by construction, linear, where the possible subspace is even further limited to linear spaces that are parallel to the feature space axes. Lowering the feature dimensionality by feature selection can also aid in interpreting classification results. At least it can shed some light on which features seem to matter most and possibly we can gain some insight into their interdependencies, for example, by studying the coefficients of a trained linear classifier. Aiming for a more interpretable classifier, we might even sacrifice some of the performance for the sake of a really limited feature set size. Feature selection can also

[21] Vapnik was one of the first to concern himself with quantifying classifier complexity in a mathematically rigorous fashion. Details can be found in his book on learning theory: (Vapnik, 1998).

be used to select the right prototypes when employing the dissimilarity approach (Pękalska et al., 2006), as in this case every feature basically corresponds to all distances to a particular prototype.

4 EVALUATION

So how good are all these classification approaches really? How can we decide on one feature set over the other? How to compare two classifiers? What prototypes work and which do not? In Section 2.1, we already stated that the (true) error rate as defined in Eq. (1) is, in our context, the ultimate measure that decides which procedure is best: the lower the error rate, the better. A first problem we face, the same one we encountered in dealing with the Bayes error and classifier—ε^* and C^*, is that p_{XY} is not available to us and so we are unable to determine the exact classification error for any procedure. Like in building a classifier, we have to rely on the N samples from p_{XY} that we have and we can merely come to an estimated error rate. But the situation is worse even. Not only do we have to give an estimate of our performance measure based on these N samples—the simplest of ways would be to just count the number of misclassifications on that set and divide that number by N. Typically, we also have to use these same N samples to train our classifier. In many real-world settings there simply is not more data available or there is no time or money left to gather more.

We consider some alternatives to estimate an error rate (see also Schiavo and Hand, 2000), introduce so-called learning curves that give some basic insight into classifier behavior, mention overtraining and in what sense there is no single best classifier, and offer some further considerations when it comes to developing a complete classifier system.

4.1 Apparent Error and Holdout Set

A major mistake, which is still being made among users and practitioners of pattern recognition and machine learning, is that one simply uses all available samples to build a classifier and then estimates the error on these same N samples. This estimate is called the resubstitution, or apparent error, denoted ε^{A}. The problem with this approach is that one gets an overly optimistic estimate. The classifier has been adapting to these specific points with these specific labels and therefore performs particularly well on this set. To more faithfully estimate the actual generalization performance of a classifier, one would need a training set to train the classifier and a completely independent so-called test set[22] to estimate its performance. The latter is also referred to as the holdout set.

In reality, we often have only a single set at our disposal, in which case we can construct a training and a holdout set by splitting the initial set in two. But how do we decide on the sizes of these two sets? We are dealing with two conflicting goals here. We would like to train

[22] In particular settings, when dealing with transductive learning for example Vapnik (1998), the inputs may be available and exploited in the training phase. In that case, the training is still performed independent of the labels in the test set, which is the primary requirement. The related setting of semisupervised learning is briefly covered in Section 6.4.

on as much data as possible, as this would typically give us the best performing classifier.[23] So the final classifier we would deliver, say, to a client, would be trained on the full initial set available. But to get an idea of the true performance of this classifier—a possible selling point if low, we at least need some independent samples. The smaller we take this set, however, the less trustworthy this estimate will be. In the extreme case of one test sample, for example, the estimate for the error rate will always be equal to 0 or 1. But adding data to the test set will reduce the amount of data in the training set, which removes us further from the setting in which we train our final classifier on the full set. The following approaches, relying on resampling the training data, provide a resolution.

4.2 Resampling Techniques

We consider some resampling techniques that provide us with some possibilities to evaluate classifiers that, in real-world settings, are often more practical then using a holdout set.

4.2.1 *Leave-One-Out and k-Fold Cross-Validation*

Cross-validation is an evaluation technique that offers the best of both worlds and allows us to both train and test on large datasets. Moreover, when it comes to estimation accuracy, so-called leave-one-out cross-validation is probably one of the best options we have.

The latter approach loops through the whole dataset for all $i \in \{1, \ldots, N\}$. At step i the pair (x_i, y_i) is used to evaluate the classifier that has been trained on all examples from the full set, except for that single sample (x_i, y_i). So we have a training set of size $N - 1$ and a test set size of 1. Given that we want at least some data to test, this is the best training set size we can have. The test set size is almost the worst we can have, but this is just for this single step in our loop. Going through all data available, every single sample will at some point act as a test set, giving us an estimated error rate ε_i (all of value 0 or 1), which we can subsequently average to get to a better overall estimate

$$\varepsilon^{\mathrm{LOO}} = \frac{1}{N} \sum_{i=1}^{N} \varepsilon_i \tag{30}$$

This procedure is called leave-one-out cross-validation and its resulting estimate the leave-one-out estimate (Efron, 1982; Lachenbruch and Mickey, 1968; McLachlan, 2004).

For computational reasons, for example, when dealing with rather large datasets or classifiers that take long to train, one can consider to settle for so-called k-fold cross-validation instead of its leave-one-out variant. In that case, the original dataset is split in k, preferably, equal sized sets or k-folds. After this, the procedure is basically the same as with leave-one-out: we loop over the k-folds, which we consecutively leave out during training and which we then test. Leave-one-out is then the same as N-fold cross-validation.

[23] Typically, yes, but classifiers can act counterintuitively and perform structurally worse with increasing numbers of labeled data in certain settings (Loog and Duin, 2012).

4.2.2 Bootstrap Estimators

Bootstrapping is a common resampling technique in statistics, the basic version of which samples from the observed empirical distribution with replacement. Various bootstrap estimators of the error rate aim to correct the bias—that is, the overoptimism, in the apparent error.

One of the more simple approaches proceeds as follows (Efron, 1982, 1983). From our dataset of N training samples, we generate M bootstrap samples of size N and calculate the M corresponding apparent error rates $\varepsilon_i^{\mathrm{A}}$ for our particular choice of classifier. Using every time that same classifier, we also calculate the error $\varepsilon_i^{\mathrm{T}}$ rate on the total dataset. An estimate of the bias is now given by their averaged difference.

$$\beta = \frac{1}{M}\sum_{i=1}^{M} \varepsilon_i^{\mathrm{A}} - \varepsilon_i^{\mathrm{T}} \tag{31}$$

The bias corrected version of the apparent error, and as such an improved estimate of the true error, is now given by $\varepsilon^{\mathrm{A}} - \beta$.

Various improvements upon and alternatives to this scheme have been suggested and studied (Efron, 1982, 1983; Efron and Tibshirani, 1997). Possibly the best-known is the 0.632 estimator $\varepsilon^{0.632} = 0.368\varepsilon^{\mathrm{A}} + 0.632\varepsilon^{\mathrm{O}}$. With the first term on the right-hand side the apparent error and the second term the out-of-bootstrap error. The latter is determined by counting all the samples from the original dataset that are misclassified and that are not part of the current bootstrap sample based on which the classifier is built. Adding all these mistakes over all M rounds and dividing this number by the total number of out-of-bootstrap samples, gives us ε^{O}.

4.2.3 Tests of Significance

In all of the foregoing, it is of course important to remember that our estimates are based on random samples and so we will only observe an instantiation of a random variable. As a result, to decide on anything like a significant difference in performances of two or more classifiers, we have to resort to statistical tests (Dietterich, 1998). To get a rough idea of the standard deviation of any of our error estimates ε^{X} based on a test size of size N_{test}, we can, for example, use a simple estimator that can be derived as the result of an averaging of N_{test} Bernoulli trials:

$$\mathrm{std}[\varepsilon^{\mathrm{X}}] = \sqrt{\frac{\varepsilon^{\mathrm{X}}(1-\varepsilon^{\mathrm{X}})}{N_{\mathrm{test}}}} \tag{32}$$

4.3 Learning Curves and the Single Best Classifier

In Section 3.3, we briefly introduced feature curves, which give us an impression of how the error rate evolves with an increasing numbers of features. We discussed the curse of dimensionality in this context. It should be clear by now that also these feature curves can, like the error rate, only be estimated and to do so, one would typically apply the estimation techniques described in the foregoing. Another, possibly more important curve that provides us with insight into the behavior of a classification method is the so-called learning curve (Cortes et al., 1993; Duin and Pękalska, 2015; Langley, 1988). The learning curve plots the (estimated) true error rate against the number of training samples. To complete the picture, one typically also plots the apparent error in the same figure.

Fig. 2 displays stylized learning curves for two classifiers of different complexity. There are various characteristics of interest that we can observe in these plots and that reflect the typical behavior for many a classifier on many classification problems. To start with, with growing sample size, the classifier is expected to perform better in terms of the error rate (see, however, Footnote 23). In addition, for the apparent error we would typically observe the opposite behavior: the curve increases as it becomes more and more difficult to solve the classification problem for the growing training set.[24] In the limit of an infinite amount of data points, both curves come together[25]: the more training data one has, the better it describes the general data that we may encounter at test time and the closer to each other true error and apparent error get. In fact, the gap that we see is an indication that the trained classifier focuses too much on specifics that are in the training but not in the test set. This is called overtraining or overfitting. The larger the gap between true and apparent error is, the more overtraining has occurred.

From the way that both learning curves for one classifier come together, one can also glean some insight. Classifiers that are less complex typically drop off more quickly, but also level out earlier than more complex ones. In addition, the former converges to an error rate that is most often above the limiting error rate of the latter: given enough data, one can get closer to the Bayes error when employing a more complex classifier.[26] As a result, it often is the

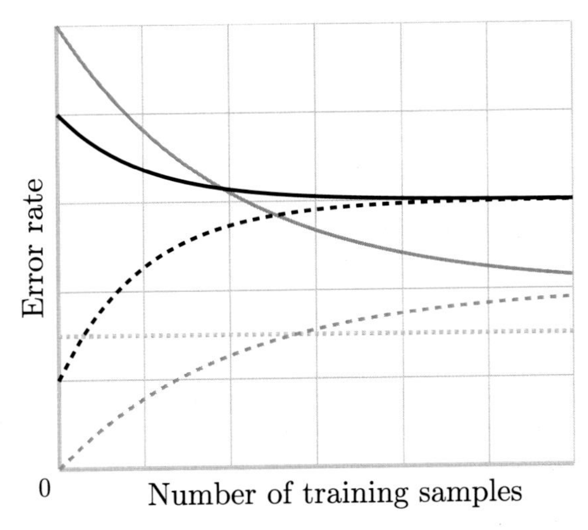

FIG. 2 Stylized plots of the learning curves of two different classifiers. *Solid lines* are the (estimated) true errors for different training set sizes, while the *dashed lines* sketch the apparent errors. *Black lines* below to a classifier with relatively low complexity, while the *gray lines* illustrate the behavior of a more complex classifier. The *light gray, dotted, horizontal line* is the Bayes error for the problem considered.

[24] Clearly, there are exceptions. For instance, the nearest neighbor classifier generally gives a zero error rate on the training set.

[25] Again exceptions apply. Again the nearest neighbor is an example. Statistical learning theory formally studies settings for which such consistency does apply.

[26] This also shows that complex classifiers typically have a smaller bias than simple classifiers.

case that one classifier is not uniformly better than another, even if we consider the same classification problem. It really matters what training set size we are dealing with and, when benchmarking the one classification method against the other, this should really be taken into account. Generally, the smaller the training dataset is, the better it is to stick with simple classifiers, for example, using a linear hypothesis class and few features.

The fact that the best choice of classifier may depend not only on the type of classification problem we need to tackle, but also on the number of training samples that we have at our disposal, may lead one to wonder what generally can be said about the superiority of one classifier over the other. Wolpert (1992a, 1996) (see also Duda et al., 2001) made this question mathematically precise and demonstrated that for this and several variations of this question the answer is that, maybe surprisingly, such distinctions between learning algorithms do not exist. This so-called no free lunch theorem states, very roughly, that averaged over all classification problems possible, there is no one classification method that outperforms any other. Although the result is certainly gripping, one should interpret it with some care. It should be realized, for example, that among all possible classification problems that one can construct, there probably are many that do not reflect any kind of realistic setting. What we can say, nevertheless, is that generally there is no single best classifier.

Finally, a learning curve may give us an idea of whether gathering more training data may improve the performance. In Fig. 2, the classifier corresponding to the black curves can hardly be improved, even if we add enormous amounts of additional data. The other classifier, the gray curves, can probably improve a bit, reaching a slightly lower error rate when enlarging the training set.

4.4 Some Words About More Realistic Scenarios

In real-world applications, designing and building a full classifier system will often be a process in which one may consider many feature representations, in which one will try various feature reduction schemes, and in which one will compare many different types of classifiers. On top of all that, there might be all kinds of preprocessing steps that are applied to the data (and that are not explicitly covered in this chapter). Working with images or signals, for example, one can perform various types of enhancement, smoothing, and normalization techniques that may have positive or negative effects on the performance of our final classifier.

A real problem in all this is that it is difficult to dispose of a truly independent test set. Unless one has a massive amount of labeled training data, one easily gets into the situation that data that is also going to be used for evaluation leaks into the training phase. The estimated test errors are therefore overly optimistic, and more so for complex classifiers than for simple ones. In the end, the result of this is that we may end up with a wrongly trained classifier, together with an overly optimistic estimate of its performance.

Let us consider some examples where things go wrong.

- A very simple instance is where one has decided, at some point, to use the k-nearest-neighbor classifier. The only thing that remains to be done is finding the best value for k and one decides to determine it on the basis of the performance for every k on the test set. It may seem like a minor offense, but often there are many such choices: the best number of features, the number of nodes in a layer of a neural network, the free parameters in some of the kernels, etc. (cf. Levine et al., 2009 and, in particular point 7 in the list).

- Here is an example where it is maybe more difficult to see that one may have gone wrong. We decide to set up everything in a seemingly clean way. We prefix all classifiers that we want to study, all the feature selection schemes that we want to try, decide beforehand on all the kernels we want to consider, and all classifier combining schemes that we may want to employ. This gives a finite number of different classification schemes that we then compare based on cross-validation. In the end, we then pick the scheme that provides the best performance. Even though this approach is actually fairly standard, again something does go wrong here. If the number of different classification schemes that we try out in this way gets out of hand, and it easily does, we still run the risk that we pick an overtrained solution with a badly biased estimate for its true error rate, especially when dealing with small training sets (cf. Braga-Neto and Dougherty, 2004; Isaksson et al., 2008; Schaffer, 1993).
- Even more complicated issues arise when multiple groups work on a large and challenging classification task. Nowadays, there are various research initiatives in which labeled data is provided publicly by a third party on which researchers can work simultaneously and collaboratively, but also in competition with each other. The flow of information and, in particular, the possibly indirect leakage of test data becomes difficult to oversee, let alone that we can easily correct for it when providing error estimates and corresponding confidence intervals or the like. How does one, for example, correct for the fact that one's own method is inspired by some of the results by another group one has read about in the research literature? Though some statistical approaches are available that can alleviate particular problems (Dwork et al., 2015a,b), it is safe to say that there currently is no generally applicable solution—if such at all exists.

The preceding scenarios primarily pertain to evaluation. In real scenarios, we of course also have to worry about the reproducibility and replicability of our findings. Otherwise, what kind of science would this be? Clearly, these are all issues that in one way or the other also play a significant role in other areas of research. In general, it turns out, however, that it is difficult to control all of these aspects and that mistakes are made, mostly unwittingly, but in some cased possibly even knowingly. For some potential, more or less dramatic consequences, we refer to the following references: Duin (1994), Fanelli (2011), Ioannidis (2005), Leek and Peng (2015), Moonesinghe et al. (2007), and Nissen et al. (2016).

5 REGULARIZATION

Regularization is actually a rather important yet relatively advanced topic in supervised learning (Christianini and Shawe-Taylor, 2000; Poggio and Smale, 2003; Schölkopf and Smola, 2002; Vapnik, 1998; Wahba, 1981), and unfortunately we are going to be fairly brief about it here.

The main idea of regularization is to have a means of performing complexity control. As we have seen already, classifier complexity can be controlled by the number of features that are used or through the complexity of the hypothesis class and, in a way, regularization is related to both of these. One of the well-known ways of regularizing a linear classifier is by constraining the already limited hypothesis space further. This is typically done by restricting the admissible weights w of the linear classifier to a sphere with radius $t > 0$ around the origin of the hypothesis space, which means we solve the constraint optimization problem

$$\begin{aligned} h^* = & \underset{(w,w_\circ)\in\mathbb{R}^{d+1}}{\operatorname{argmin}} \sum_{i=1}^{N} \ell(w^\top x_i + w_\circ, y_i) \\ & \text{subject to } ||w||^2 \leq t \end{aligned} \tag{33}$$

A formulation that is essentially equivalent is constructed by including the constraint directly into the objective function:

$$h^* = \underset{(w,w_\circ)\in\mathbb{R}^{d+1}}{\operatorname{argmin}} \sum_{i=1}^{N} \ell(w^\top x_i + w_\circ, y_i) + \lambda||w||^2 \tag{34}$$

where $\lambda > 0$ is known as the regularization parameter. The regularization is stronger with larger λ.

This procedure is the same as the one used in classical ridge regression (Hoerl and Kennard, 1970) and effectively stabilizes the solution that is obtained. The effect of regularization is that the bias of our classification method increases, as we cannot reach certain linear classifiers anymore due to the added constraint. At the same time, the variance in our classifier estimates decreases due to the constraint (which is another way of saying that the classifier becomes more stable). In the average, with a small to moderate parameter λ, the worsening in performance we may get because of the increased bias is offset with an improvement in performance due to the reduced variance, in which case, regularization will lead to an improved classifier. If, however, we regularize too strongly, the bias will start to dominate and pull our model too far away from any reasonable solution, at which point the true error rate will start to increase again.

A basic explanation of the effects of this so-called bias-variance tradeoff can be found in the earlier mentioned work of Hoerl and Kennard (1970). The phenomenon can be seen in various guises, and its importance has been acknowledged early on in statistics and data analysis (Wahba, 1979, 1981). A more explicit dissection of the bias-variance tradeoff, in the context of learning methods, was published in Geman et al. (1992). The more complex a classifier is, the higher the variance we are faced with when training such model, and the more important some form of regularization becomes.

Eqs. (33), (34) only consider the most basic form of regularization. There are many more variations on this theme. Among others, there are regularizers with built-in feature selectors (Tibshirani, 1996) and regularizers that have deep connections to our earlier discussed kernels (Girosi et al., 1995; Smola and Schölkopf, 1998).

6 VARIATIONS ON STANDARD CLASSIFICATION

We have introduced and discussed the main aspects of supervised classification. In this section, we like to review some slight variations in this basic learning problem. Although basic, there are very many decision problems that can actually be cast into a classification problem. Nevertheless, in reality, one may often be confronted with problems that still do not completely fit this restricted setting. We cover some of these here.

6.1 Multiple Instance Learning

In particular settings, it is more appropriate, or it simply is easier, to describe every object o_i, not with a single feature vector x_i, but with a set of such feature vectors. This approach is, for example, common in various image analysis tasks, in which a set of so-called descriptors, that is, feature vectors that capture the local image content at various locations in the image, act as the representation of that image. Every image, in both the training and the test set, is represented by such a set of descriptors and the goal is to construct a classifier for such sets. The research area that studies approaches applicable to this setting, in which every object can be described with sets of feature vectors having different sizes, but where the feature vectors are from the same measurement space, is called multiple instance learning. A large number of classification routines have been developed for this specific problem, which range from basic extensions of classifiers from the supervised classification domain by means of combining techniques, via dissimilarity-based approaches, to approaches specifically designed for the purpose of set classification (Carbonneau et al., 2016; Cheplygina et al., 2015; Li et al., 2013; Maron and Lozano-Pérez, 1998; Zhou, 2004). The classical reference, in which the initial problem has been formalized, is Dietterich et al. (1997).

6.2 One-Class Classification, Outliers, and Reject Options

There are various problems in which it is difficult to find sufficient examples of one of the classes, because they are very difficult to find or simply occur very seldom. In that case, one-class classification might be of use. Instead of trying to solve the two-class problem straightaway, it aims to model the distribution or support of the oft-occurring class accurately, and based on that, decides which points really do not belong to that class and, therefore, will be assigned to the class of which little is known (Schölkopf et al., 2001b; Tax and Duin, 1999). Such techniques have direct relations to approaches that perform outlier or novelty detection in data and data streams (Chandola et al., 2009; Markou and Singh, 2003), in which one aims to identify objects that are, in some sense, far away from the bulk of the data.

The more a test data point is an outlier, the less training data will be present in its vicinity and, therefore, the less certain a classifier will be in assigning the corresponding object to one or the other class. Consequently, outlier detection and related techniques are also used to implement so-called reject options (Chow, 1970). These aim to identify points for which, say, p_X is small and any automated decision by the classifier at hand is probably unreliable. In such case, the ultimate decision may be better left to a human expert. We might, for example, be dealing with a sample from a third class; something that our classifier never encountered. This kind of rejection is also referred to as the distance reject option (Dubuisson and Masson, 1993). A second option is ambiguity rejection, in which case the classifier rather looks at $p_{Y|X}$ and leaves the final decision to a human expert if (in the two-class case) the two posteriors are very close to each other, that is, approximately $\frac{1}{2}$ (Dubuisson and Masson, 1993). For ambiguity and distance rejection, one should realize that both an erroneous decision by the classifier and deploying a human expert come with their own costs. One of the main challenges in the use of a reject option is then to trade these two costs off in an optimal way.

6.3 Contextual Classification

Contextual classification has already been mentioned in Section 2.7.3 on multiple classifier systems. In these contextual approaches, samples are not classified in isolation, but they may have various types of neighborhood relations that can be exploited to improve the overall performance. The classical approach to this employs Markov random fields (Besag, 1986; Li, 2009) and specific variations to those techniques such as conditional random fields (Lafferty et al., 2001). The earlier mentioned methods using classifier combining techniques (Cohen and Carvalho, 2005; Loog and van Ginneken, 2002; Loog, 2004) are often more easily applicable and can leverage the full potential of more general classification methodologies. As already indicated, in Section 2.7.3 as well, the latter class of techniques seems to become relevant again in the context of deep learning approaches.

6.4 Missing Data and Semisupervised Learning

In many real-world setting, missing data is a considerable and reoccurring problem. In the classification setting this means that particular features and/or class labels have not been observed. Missing features can occur because of the failure of a measurement apparatus, because of human nonresponse, or because the data was not recorded or was accidentally erased. There are various ways to deal with such deletions, which is a topic thoroughly studied in statistics (Little and Rubin, 2014).

The case of missing labels can have additional causes. It may simply have been too expensive to label more data, or additional input data has been collected afterward to extend the already available data, but the collector is not a specialist that can provide the necessary annotation. The case of missing label data is known within pattern recognition and machine learning as semisupervised learning (Chapelle et al., 2006; Zhu, 2008). Also for this problem, which has been studied for more than 50 years, many different techniques have been developed. Although in a theoretical sense, there is still no completely satisfactory and practicable solution to the problem.[27] One of the major issues is the question to what extent one can guarantee that a supervised classifier can indeed be improved by taking all unlabeled data into account as well (Krijthe and Loog, 2017; Li and Zhou, 2011; Loog, 2010, 2016).

6.5 Transfer Learning and Domain Adaptation

For various reasons, the distribution of the data at training time can be different from that at test time. Examples are medical devices that are trained on samples (subjects) from one country, while the machine is also deployed in another country. More generally, machines and sensors suffer from wear and tear and, as a result, measurement statistics at a later

[27] In a way, this can probably also be said about many of the other problems discussed.

point may not really match their distribution at the time of training. Depending on what can be assumed about the difference of the two domains or, as they are often referred to more specifically, the source and target, particular approaches can be employed that can alleviate the discrepancy between them (Pan and Yang, 2010; Quiñonero-Candela et al., 2009). The areas of domain adaptation and transfer learning study techniques for these challenging settings. Depending on the actual transfer that has to be learned, more or less successful approaches can be identified to tackle these problems.

6.6 Active Learning

The final variation on supervised classification is actually concerned with regular supervised classification. The difference, however, with the main setting discussed throughout this chapter is that active learning sets out to improve the data collection process. It tries to answer various related questions, one of which is as follows. Given that we have a large number of unlabeled samples and a budget to label N of these samples, what instances should we consider for labeling to enable us to train a better classifier than we would be able to in case we would rely on random sampling?[28] So can we, in a more systematic way, collect data to be labeled, such that we more quickly come to a well-trained classifier?

The problem formulation has direct relations to sequential analysis (Wald, 1973) and optimal experimental design (Fedorov, 1972). Overviews of current techniques can be found in Cohn et al. (1996), Settles (2010), and Yang and Loog (2016). One of the major issues in active learning is that the systematic collection of labeled training data typically leads to a systematic bias as well. Correcting for this seems essential (Beygelzimer et al., 2009) (see also Loog and Yang, 2016). In a way, it points to a problem one will more generally encounter in practical settings and which directly relates to some of the issues indicated in Section 6.5: one of the key assumptions in supervised classification is that the training and test set consist of i.i.d. samples from the same underlying problem defined by the density p_{XY}. In reality, this assumption is probably violated, and care should be taken.

ACKNOWLEDGMENTS

Many thanks go to Mariusz Flasiński (Jagiellonian University, Poland) and a second, anonymous reviewer for their critical, yet encouraging appraisal of this overview. Their comments helped to improve the exposition of the chapter and to drastically reduce the number of spelling and grammar errors.

References

Aizerman, A., Braverman, E.M., Rozoner, L.I., 1964. Theoretical foundations of the potential function method in pattern recognition learning. Autom. Remote. Control. 25, 821–837.

Ben-David, S., Eiron, N., Long, P.M., 2003. On the difficulty of approximately maximizing agreements. J. Comput. Syst. Sci. 66 (3), 496–514.

[28] The latter of which is the de facto standard.

Besag, J., 1986. On the statistical analysis of dirty pictures. J. R. Stat. Soc. Ser. B Methodol. 48, 259–302.

Beygelzimer, A., Dasgupta, S., Langford, J., 2009. Importance weighted active learning. In: Proceedings of the 26th Annual International Conference on Machine Learning. ACM, pp. 49–56.

Bishop, C.M., 1995. Neural Networks for Pattern Recognition. Oxford University Press, Oxford.

Boser, B.E., Guyon, I.M., Vapnik, V.N., 1992. A training algorithm for optimal margin classifiers. In: Proceedings of the Fifth Annual Workshop on Computational Learning Theory. ACM, pp. 144–152.

Bottou, L., 1991. Stochastic gradient learning in neural networks. In: Proceedings of Neuro-Nîmes 91. Nimes, France.

Bradley, A.P., 1997. The use of the area under the ROC curve in the evaluation of machine learning algorithms. Pattern Recogn. 30 (7), 1145–1159.

Braga-Neto, U.M., Dougherty, E.R., 2004. Is cross-validation valid for small-sample microarray classification? Bioinformatics 20 (3), 374–380.

Breiman, L., 1996. Bagging predictors. Mach. Learn. 24 (2), 123–140.

Breiman, L., Friedman, J., Stone, C.J., Olshen, R.A., 1984. Classification and Regression Trees. CRC Press, Boca Raton, FL.

Carbonneau, M.A., Cheplygina, V., Granger, E., Gagnon, G., 2016. Multiple instance learning: a survey of problem characteristics and applications. ArXiv preprint arXiv:1612.03365.

Chandola, V., Banerjee, A., Kumar, V., 2009. Anomaly detection: a survey. ACM Comput. Surv. (CSUR) 41 (3), 15.

Chapelle, O., Schölkopf, B., Zien, A., 2006. Semi-Supervised Learning. MIT Press, Cambridge, MA.

Cheplygina, V., Tax, D.M.J., Loog, M., 2015. Multiple instance learning with bag dissimilarities. Pattern Recogn. 48 (1), 264–275.

Chow, C., 1970. On optimum recognition error and reject tradeoff. IEEE Trans. Inf. Theory 16 (1), 41–46.

Christianini, N., Shawe-Taylor, J., 2000. An Introduction to Support Vector Machines and Other Kernel-Based Learning Methods. Cambridge University Press, Cambridge, United Kingdom.

Cohen, W.W., Carvalho, V.R., 2005. Stacked sequential learning. In: Proceedings of the 19th International Joint Conference on Artificial Intelligence, pp. 671–676.

Cohn, D.A., Ghahramani, Z., Jordan, M.I., 1996. Active learning with statistical models. J. Artif. Intell. Res. 4 (1), 129–145.

Cortes, C., Vapnik, V., 1995. Support-vector networks. Mach. Learn. 20 (3), 273–297.

Cortes, C., Jackel, L.D., Solla, S.A., Vapnik, V., Denker, J.S., 1993. Learning curves: asymptotic values and rate of convergence. In: Proceedings of the 6th International Conference on Neural Information Processing Systems. Morgan Kaufmann Publishers Inc., pp. 327–334.

Cover, T., Hart, P., 1967. Nearest neighbor pattern classification. IEEE Trans. Inf. Theory 13 (1), 21–27.

Devroye, L., Györfi, L., Lugosi, G., 1996. A Probabilistic Theory of Pattern Recognition. Springer-Verlag, New York, NY.

Dietterich, T.G., 1998. Approximate statistical tests for comparing supervised classification learning algorithms. Neural Comput. 10 (7), 1895–1923.

Dietterich, T.G., Lathrop, R.H., Lozano-Pérez, T., 1997. Solving the multiple instance problem with axis-parallel rectangles. Artif. Intell. 89 (1), 31–71.

Dubuisson, B., Masson, M., 1993. A statistical decision rule with incomplete knowledge about classes. Pattern Recogn. 26 (1), 155–165.

Duda, R.O., Hart, P.E., 1973. Pattern Classification and Scene Analysis. John Wiley, London.

Duda, R.O., Hart, P.E., Stork, D.G., 2001. Pattern Classification. Wiley-Interscience, London.

Duin, R.P.W., 1994. Superlearning and neural network magic. Pattern Recogn. Lett. 15 (3), 215–217.

Duin, R.P.W., 2002. The combining classifier: to train or not to train? In: Proceedings of the 16th International Conference on Pattern Recognition, vol. 2. IEEE, pp. 765–770.

Duin, R.P.W., Pękalska, E., 2015. Pattern recognition: Introduction and terminology. 37 Steps. http://37steps.com/documents/printro/.

Duin, R.P.W., de Ridder, D., Tax, D.M.J., 1997. Experiments with a featureless approach to pattern recognition. Pattern Recogn. Lett. 18 (11), 1159–1166.

Dwork, C., Feldman, V., Hardt, M., Pitassi, T., Reingold, O., Roth, A., 2015a, Generalization in adaptive data analysis and holdout reuse. In: Advances in Neural Information Processing Systems, pp. 2350–2358.

Dwork, C., Feldman, V., Hardt, M., Pitassi, T., Reingold, O., Roth, A., 2015b. The reusable holdout: preserving validity in adaptive data analysis. Science 349 (6248), 636–638.

Efron, B., 1982. The Jackknife, the Bootstrap and Other Resampling Plans. SIAM, Philadelphia, PA.

Efron, B., 1983. Estimating the error rate of a prediction rule: improvement on cross-validation. J. Am. Stat. Assoc. 78 (382), 316–331.
Efron, B., Tibshirani, R., 1997. Improvements on cross-validation: the 632+ bootstrap method. J. Am. Stat. Assoc. 92 (438), 548–560.
Fanelli, D., 2011. Negative results are disappearing from most disciplines and countries. Scientometrics 90 (3), 891–904.
Fawcett, T., 2006. An introduction to ROC analysis. Pattern Recogn. Lett. 27 (8), 861–874.
Fedorov, V.V., 1972. Theory of Optimal Experiments. Elsevier, Amsterdam.
Fisher, R.A., 1936. The use of multiple measurements in taxonomic problems. Ann. Eugen. 7 (2), 179–188.
Fix, E., Hodges Jr., J.L., 1951. Discriminatory analysis-nonparametric discrimination: consistency properties. Technical report, DTIC Document.
Freund, Y., Schapire, R.E., 1995. A decision-theoretic generalization of on-line learning and an application to boosting. In: European Conference on Computational Learning Theory. Springer, pp. 23–37.
Fu, H., Wang, C., Tao, D., Black, M.J., 2016. Occlusion boundary detection via deep exploration of context. In: Proceedings of the IEEE Conference on Computer Vision and Pattern Recognition, pp. 241–250.
Galar, M., Fernández, A., Barrenechea, E., Bustince, H., Herrera, F., 2011. An overview of ensemble methods for binary classifiers in multi-class problems: experimental study on one-vs-one and one-vs-all schemes. Pattern Recogn. 44 (8), 1761–1776.
Geman, S., Bienenstock, E., Doursat, R., 1992. Neural networks and the bias/variance dilemma. Neural Comput. 4 (1), 1–58.
Girosi, F., Jones, M., Poggio, T., 1995. Regularization theory and neural networks architectures. Neural Comput. 7 (2), 219–269.
Guyon, I., Elisseeff, A., 2003. An introduction to variable and feature selection. J. Mach. Learn. Res. 3, 1157–1182.
Hand, D.J., Anagnostopoulos, C., 2014. A better Beta for the H measure of classification performance. Pattern Recogn. Lett. 40, 41–46.
Hand, D., Christen, P., 2017. A note on using the F-measure for evaluating record linkage algorithms. In: Statistics and Computing, pp. 1–9.
Hastie, T., Tibshirani, R., Friedman, J., 2001. The Elements of Statistical Learning. Springer, New York, NY.
Hinton, G.E., 1989. Connectionist learning procedures. Artif. Intell. 40 (1–3), 185–234.
Ho, T.K., 1995. Random decision forests. In: Proceedings of the Third International Conference on Document Analysis and Recognition, vol. 1. IEEE, pp. 278–282.
Ho, T.K., 1998. The random subspace method for constructing decision forests. IEEE Trans. Pattern Anal. Mach. Intell. 20 (8), 832–844.
Hoerl, A.E., Kennard, R.W., 1970. Ridge regression: biased estimation for nonorthogonal problems. Technometrics 12 (1), 55–67.
Hoffgen, K.U., Simon, H.U., Vanhorn, K.S., 1995. Robust trainability of single neurons. J. Comput. Syst. Sci. 50 (1), 114–125.
Ioannidis, J.P.A., 2005. Why most published research findings are false. PLoS Med. 2 (8), e124.
Isaksson, A., Wallman, M., Göransson, H., Gustafsson, M.G., 2008. Cross-validation and bootstrapping are unreliable in small sample classification. Pattern Recogn. Lett. 29 (14), 1960–1965.
Jahrer, M., Töscher, A., Legenstein, R., 2010. Combining predictions for accurate recommender systems. In: Proceedings of the 16th ACM SIGKDD International Conference on Knowledge Discovery and Data Mining. ACM, pp. 693–702.
Jain, A., Zongker, D., 1997. Feature selection: evaluation, application, and small sample performance. IEEE Trans. Pattern Anal. Mach. Intell. 19 (2), 153–158.
Jain, A.K., Duin, R.P.W., Mao, J., 2000. Statistical pattern recognition: a review. IEEE Trans. Pattern Anal. Mach. Intell. 22 (1), 4–37.
Kahneman, D., Slovic, P., Tversky, A. (Eds.) 1982. Judgment Under Uncertainty: Heuristics and Biases. Cambridge University Press, Cambridge.
Kittler, J., Roli, F. (Eds.) 2000. Multiple classifier systems: first international workshop. In: Kittler, J., Roli, F. (Eds.), Volume 1857 of Lecture Notes in Computer Science. Springer, Cagliari.
Krijthe, J.H., Loog, M., 2017. Projected estimators for robust semi-supervised classification. Mach. Learn. 106 (7), 993–1008.

Kuncheva, L.I., 2004. Combining Pattern Classifiers: Methods and Algorithms. John Wiley & Sons, New York, NY.
Lachenbruch, P.A., Mickey, M.R., 1968. Estimation of error rates in discriminant analysis. Technometrics 10 (1), 1–11.
Lafferty, J., McCallum, A., Pereira, F., et al., 2001. Conditional random fields: probabilistic models for segmenting and labeling sequence data. In: Proceedings of the Eighteenth International Conference on Machine Learning, vol. 1, pp. 282–289.
Landgrebe, T.C.W., Paclik, P., Duin, R.P.W., 2006. Precision-recall operating characteristic (P-ROC) curves in imprecise environments. In: 18th International Conference on Pattern Recognition, vol. 4. IEEE, pp. 123–127.
Langley, P., 1988. Machine learning as an experimental science. Mach. Learn. 3 (1), 5–8.
Lavrač, N., Flach, P., Zupan, B., 1999. Rule evaluation measures: a unifying view. In: International Conference on Inductive Logic Programming. Springer, pp. 174–185.
Leek, J.T., Peng, R.D., 2015. Statistics: P values are just the tip of the iceberg. Nature 520 (7549), 612.
Levine, D., Bankier, A.A., Halpern, E.F., 2009. Submissions to radiology: our top 10 list of statistical errors. Radiology 253 (2), 288–290.
Li, S.Z., 2009. Markov Random Field Modeling in Image Analysis. Springer Science & Business Media, New York, NY.
Li, Y.F., Zhou, Z.H., 2011. Towards making unlabeled data never hurt. In: Proceedings of the 28th ICML, pp. 1081–1088.
Li, Y., Tax, D.M.J., Duin, R.P.W., Loog, M., 2013. Multiple-instance learning as a classifier combining problem. Pattern Recogn. 46 (3), 865–874.
Li, K., Hariharan, B., Malik, J., 2016. Iterative instance segmentation. In: Proceedings of the IEEE Conference on Computer Vision and Pattern Recognition, pp. 3659–3667.
Little, R.J.A., Rubin, D.B., 2014. Statistical Analysis With Missing Data. John Wiley & Sons, New York, NY.
Loog, M., 2004. Supervised Dimensionality Reduction and Contextual Pattern Recognition in Medical Image Processing. Ph.D. thesis. Utrecht University.
Loog, M., 2010. Constrained parameter estimation for semi-supervised learning: the case of the nearest mean classifier. In: European Conference on Machine Learning, pp. 291–304.
Loog, M., 2016. Contrastive pessimistic likelihood estimation for semi-supervised classification. IEEE Trans. Pattern Anal. Mach. Intell. 38 (3), 462–475.
Loog, M., Duin, R.P.W., 2012. The dipping phenomenon. In: Proceedings of the 2012 Joint IAPR International Conference on Structural, Syntactic, and Statistical Pattern Recognition. Springer-Verlag, pp. 310–317.
Loog, M., van Ginneken, B., 2002. Supervised segmentation by iterated contextual pixel classification. In: 16th International Conference on Pattern Recognition, vol. 2. IEEE, pp. 925–928.
Loog, M., Yang, Y., 2016. An empirical investigation into the inconsistency of sequential active learning. In: 23rd International Conference on Pattern Recognition. IEEE, pp. 210–215.
Markou, M., Singh, S., 2003. Novelty detection: a review—part 1: statistical approaches. Signal Process. 83 (12), 2481–2497.
Maron, O., Lozano-Pérez, T., 1998. A framework for multiple-instance learning. In: Advances in Neural Information Processing Systems, pp. 570–576.
McLachlan, G., 2004. Discriminant Analysis and Statistical Pattern Recognition. John Wiley & Sons, New York, NY.
Minka, T., 2005. Discriminative models, not discriminative training. Technical Report MSR-TR-2005-144, Microsoft Research.
Moonesinghe, R., Khoury, M.J., Cecile, A., Janssens, J.W., 2007. Most published research findings are false—but a little replication goes a long way. PLoS Med. 4 (2), e28.
Nair, V., Hinton, G.E., 2010. Rectified linear units improve restricted Boltzmann machines. In: Proceedings of the 27th International Conference on Machine Learning, pp. 807–814.
Niemeijer, M., Loog, M., Abramoff, M.D., Viergever, M.A., Prokop, M., van Ginneken, B., 2011. On combining computer-aided detection systems. IEEE Trans. Med. Imaging 30 (2), 215–223.
Nissen, S.B., Magidson, T., Gross, K., Bergstrom, C.T., 2016. Publication bias and the canonization of false facts. Elife 5, e21451.
Pan, S.J., Yang, Q., 2010. A survey on transfer learning. IEEE Trans. Knowl. Data Eng. 22 (10), 1345–1359.
Pękalska, E., Duin, R.P.W., 2005. The Dissimilarity Representation for Pattern Recognition: Foundations and Applications, vol. 64. World Scientific, River Edge, NJ.
Pękalska, E., Paclik, P., Duin, R.P.W., 2001. A generalized kernel approach to dissimilarity-based classification. J. Mach. Learn. Res. 2, 175–211.
Pękalska, E., Duin, R.P.W., Paclík, P., 2006. Prototype selection for dissimilarity-based classifiers. Pattern Recogn. 39 (2), 189–208.

Poggio, T., Smale, S., 2003. The mathematics of learning: dealing with data. Not. AMS 50 (5), 537–544.
Polikar, R., 2006. Ensemble based systems in decision making. IEEE Circuits Syst. Mag. 6 (3), 21–45.
Provost, F., Fawcett, T., Kohavi, R., 1997. The case against accuracy estimation for comparing induction algorithms. In: Fifteenth International Conference on Machine Learning.
Quiñonero-Candela, J., Suyiyama, M., Schwaighofer, A., Lawrence, N.D., 2009. Dataset Shift in Machine Learning. MIT Press, Cambridge, MA.
Quinlan, J.R., 1986. Induction of decision trees. Mach. Learn. 1 (1), 81–106.
Rasmussen, C.E., Williams, C.K.I., 2006. Gaussian Processes for Machine Learning. The MIT Press, Cambridge, MA.
Ripley, B.D., 2007. Pattern Recognition and Neural Networks. Cambridge University Press, Cambridge.
Rosenblatt, F., 1958. The perceptron: a probabilistic model for information storage and organization in the brain. Psychol. Rev. 65 (6), 386.
Rubinstein, Y.D., Hastie, T., 1997. Discriminative vs informative learning. In: Third International Conference on Knowledge Discovery and Data Mining, pp. 49–53.
Schaffer, C., 1993. Selecting a classification method by cross-validation. Mach. Learn. 13 (1), 135–143.
Schiavo, R.A., Hand, D.J., 2000. Ten more years of error rate research. Int. Stat. Rev. 68 (3), 295–310.
Schmidhuber, J., 2015. Deep learning in neural networks: an overview. Neural Netw. 61, 85–117.
Schölkopf, B., Smola, A.J., 2002. Learning With Kernels: Support Vector Machines, Regularization, Optimization, and Beyond. MIT Press, Cambridge, MA.
Schölkopf, B., Herbrich, R., Smola, A.J., 2001a, A generalized representer theorem. In: International Conference on Computational Learning Theory. Springer, pp. 416–426.
Schölkopf, B., Platt, J.C., Shawe-Taylor, J., Smola, A.J., Williamson, R.C., 2001b. Estimating the support of a high-dimensional distribution. Neural Comput. 13 (7), 1443–1471.
Settles, B., 2010. Active Learning Literature Survey, Computer Sciences TR 1648. University of Wisconsin.
Shrivastava, A., Gupta, A., 2016. Contextual priming and feedback for faster R-CNN. In: European Conference on Computer Vision. Springer, pp. 330–348.
Smola, A.J., Schölkopf, B., 1998. On a kernel-based method for pattern recognition, regression, approximation, and operator inversion. Algorithmica 22 (1), 211–231.
Suykens, J.A.K., Vandewalle, J., 1999. Least squares support vector machine classifiers. Neural. Process. Lett. 9 (3), 293–300.
Tax, D.M.J., Duin, R.P.W., 1999. Support vector domain description. Pattern Recogn. Lett. 20 (11), 1191–1199.
Tibshirani, R., 1996. Regression shrinkage and selection via the Lasso. J. R. Stat. Soc. Ser. B Methodol. 58 (1), 267–288.
Vapnik, V.N., 1998. Statistical Learning Theory. Wiley, London.
Wahba, G., 1979. Smoothing and ill-posed problems. In: Solution Methods for Integral Equations. Springer, pp. 183–194.
Wahba, G., 1981. Constrained regularization for ill posed linear operator equations, with applications in meteorology and medicine. Technical report, DTIC Document.
Wahba, G., 1990. Spline Models for Observational Data. SIAM, Philadelphia, PA.
Wald, A., 1973. Sequential Analysis. Courier Corporation. Dover Publications, Mineola, NY.
White, H., 1989. Learning in artificial neural networks: a statistical perspective. Neural Comput. 1 (4), 425–464.
Wolpert, D.H., 1992a. On the connection between in-sample testing and generalization error. Complex Syst. 6 (1), 47–94.
Wolpert, D.H., 1992b. Stacked generalization. Neural Netw. 5 (2), 241–259.
Wolpert, D.H., 1996. The lack of a priori distinctions between learning algorithms. Neural Comput. 8 (7), 1341–1390.
Yang, Y., Loog, M., 2016. A benchmark and comparison of active learning for logistic regression. arXiv preprint arXiv:1611.08618.
Zhou, Z.H., 2004. Multi-instance learning: a survey. Technical report. Department of Computer Science & Technology, Nanjing University.
Zhu, X., 2008. Semi-supervised learning literature survey, Computer Sciences TR 1530. University of Wisconsin.

PART III

APPLICATIONS

CHAPTER

6

Untangling the Solar Wind Drivers of the Radiation Belt: An Information Theoretical Approach

Simon Wing[*], *Jay R. Johnson*[†], *Enrico Camporeale*[‡], *Geoffrey D. Reeves*[§]

[*]Johns Hopkins University, Laurel, MD, United States [†]Andrews University, Berrien Springs, MI, United States [‡]Centrum Wiskunde & Informatica, Amsterdam, The Netherlands [§]Los Alamos National Laboratory; Space Science and Applications Group, Los Alamos, NM, United States

CHAPTER OUTLINE

Machine Learning Techniques for Space Weather
https://doi.org/10.1016/B978-0-12-811788-0.00006-8

1 INTRODUCTION

The Earth's radiation belts refer to a region in space that is inhabited by trapped energetic particles, electrons, and ions. Typically, there are two radiation belts: the inner belt and outer belt. The inner belt is located at equatorial distance approximately between 1.2 and 3 R_E (R_E = radius of the Earth ~6371 km) from the center of the Earth and is inhabited by electrons having energies of hundreds of keVs and ions having hundreds of MeVs. The outer belt is located at equatorial distance approximately between 4 and 8 R_E and is inhabited by mostly electrons having energies ranging from a few hundred keVs to tens of MeVs. Sometimes, a third belt appears between the inner and outer belts.

This chapter deals only with the outer radiation belt electron population. These radiation belt electrons are often referred to as "killer electrons" because they can cause serious damage to satellites. For example, the radiation belt electrons with energies of a few MeVs or higher can penetrate deep into spacecraft components, and those with energies lower than 1 MeV can lodge on the surface of the spacecraft bodies, leading to devastating electrical discharges. Hence, radiation belts are quite relevant to the studies of space weather.

The existence of radiation belt MeV electrons is usually explained by some acceleration mechanisms that can accelerate electrons from a few keVs to tens of MeVs. There have been several acceleration mechanisms proposed, but most studies generally suggest either local acceleration or global acceleration. In the local acceleration, the storm and substorm plasma injection from plasma sheet into the innermagnetosphere accelerates low energy (e.g., a few keV) electrons to a few hundred keVs. Once in the innermagnetosphere, electrons interact with locally grown ultra low-frequency (ULF) waves (e.g., Elkington et al., 1999; Rostoker et al., 1998; Ukhorskiy et al., 2005; Mathie and Mann, 2000, 2001), very low-frequency (VLF) waves (e.g., Summers et al., 1998; Omura et al., 2007; Thorne, 2010; Simms et al., 2015; Camporeale and Zimbardo, 2015; Camporeale, 2015), or magnetosonic waves (e.g., Horne et al., 2007; Shprits et al., 2008), which can energize electrons to the MeV energy range.

The global acceleration mechanism also invokes ULF waves for electron acceleration, but here the ULF waves are generated globally by Kelvin-Helmholtz instability (KHI) along the magnetopause flanks due to large solar wind velocity (V_{sw}) (e.g., Johnson et al., 2014; Engebretson et al., 1998; Vennerstrøm, 1999). Indeed, studies have shown that V_{sw} is a dominant, if not the most dominant, driver of relativistic electron fluxes (herein J_e refers to geosynchronous MeV electron energy flux) at geosynchronous orbit (6.6R_E) (e.g., Paulikas and Blake, 1979; Baker et al., 1990; Li et al., 2001, 2005; Vassiliadis et al., 2005; Ukhorskiy et al., 2004; Rigler et al., 2007; Kellerman and Shprits, 2012; Reeves et al., 2011). However, the geosynchronous electron response to V_{sw} has a lag time. For MeV electrons, a lag time of about 2 days has been consistently observed in many studies (e.g., Baker et al., 1990; Vassiliadis et al., 2005; Reeves et al., 2011; Balikhin et al., 2011; Lyatsky and Khazanov, 2008), suggesting the time scale needed to accelerate electrons to MeV energy range is about 2 days.

In contrast to V_{sw}, which correlates with J_e, solar wind density (n_{sw}) anticorrelates with J_e for reasons that are not entirely clear (e.g., Lyatsky and Khazanov, 2008; Kellerman and Shprits, 2012). Li et al. (2005) suggests that an increase in n_{sw} increases solar wind dynamic pressure (P_{dyn}), which, in turn, pushes the magnetopause inward, leading to electron losses. However, Lyatsky and Khazanov (2008) argues that the poor correlation between P_{dyn} and J_e suggests that compression of the magnetosphere is probably not the main factor. Moreover, the effectiveness of n_{sw} at influencing J_e is also not clear. Some studies found that n_{sw} has weaker effects than V_{sw} on J_e (e.g., Vassiliadis et al., 2005; Rigler et al., 2007; Kellerman and Shprits, 2012). However, Balikhin et al. (2011) finds that J_e has the strongest dependence on n_{sw} with a lag of 1 day.

The interpretation of the relationship between n_{sw} and J_e is complicated by the anticorrelation between V_{sw} and n_{sw} (e.g., Hundhausen et al., 1970). Because J_e and V_{sw} are correlated, the anticorrelation between J_e and n_{sw} could simply be coincidence.

Other studies showed that the interplanetary magnetic field (IMF) and other solar wind parameters can also contribute to J_e variations (e.g., Balikhin et al., 2011; Rigler et al., 2007; Vassiliadis et al., 2005; Li et al., 2005; Onsager et al., 2007; Simms et al., 2014), but it is not entirely clear, quantitatively, given the main driver, for example, V_{sw} (or n_{sw}), how much additional information these parameters provide to J_e. This knowledge can help radiation belt modelers decide what input parameters to consider for their models.

The solar wind-magnetosphere and solar wind-radiation belt systems have been shown to be nonlinear (e.g., Wing et al., 2005a; Johnson and Wing, 2005; Reeves et al., 2011; Kellerman and Shprits, 2012). An example is presented in Fig. 1, which plots $\log J_e(t + \tau)$ versus $V_{sw}(t)$ for $\tau = 0, 1, 2$, and 7 days. The figure, which is similar to Fig. 9 in Reeves et al. (2011), shows that the relationship between J_e and V_{sw} is nonlinear. For nonlinear systems, qualitative linear correlational analysis can be misleading (e.g., Balikhin et al., 2010, 2011). Moreover, correlational analysis cannot establish causalities.

Information theory can help identify nonlinearities in the system and information transfer from input to output parameters. Hence, it can be a useful tool to study the solar wind-radiation belt system. In particular, we apply mutual information (MI) (e.g., Li, 1990; Tsonis, 2001), conditional mutual information (CMI) (e.g., Wyner, 1978), and transfer entropy (TE) (e.g., Schreiber, 2000; Wing et al., 2018) to determine the solar wind drivers of J_e and to quantify how much information is transferred from solar wind parameters to J_e.

The scatter plots of J_e versus V_{sw} in Fig. 1A–C look like a triangle, which Reeves et al. (2011) refers to as the triangle distribution. The mystifying part of the triangle distribution is that there is a large variability of J_e when $V_{sw} < 500$ km s^{-1}. The present study probes deeper into this triangle distribution using information theory.

2 DATA SET

Most studies of geosynchronous MeV electrons have been performed with data having 1-day resolution (e.g., Reeves et al., 2011; Balikhin et al., 2011; Kellerman and Shprits, 2012; Wing et al., 2016). As stated by Reeves et al. (2011), because of the asymmetry of the geomagnetic field along the geosynchronous orbit, geosynchronous electron fluxes exhibit a diurnal, or magnetic local time (MLT) variation as well as latitude-longitude dependence. However, these effects are reduced in daily resolution data.

The present study uses the same dataset in Reeves et al. (2011). The data and format description can be found at ft://ftop.agu.org/apend/ja/2010ja015735. This dataset contains daily averages of electron fluxes obtained from energetic sensors for particles (ESP) (Meier et al., 1996) and synchronous orbit particle analyzers (SOPA) (Belian et al., 1992) on board of all seven Los Alamos National Laboratory (LANL) geosynchronous satellites from September 22, 1989 to December 31, 2009. The present study only examines the fluxes of electrons with an energy range of 1.8–3.5 MeVs (which is referred to herein as J_e). A detailed description of the dataset and its processing are given in Reeves et al. (2011). The daily and hourly averaged solar wind data 1989–2009 come from OMNI dataset provided by NASA (http://omniweb.gsfc.nasa.gov/). The LANL and solar wind data are merged. The LANL dataset has 7187 data points (days of data), out of which, 6438 data points have simultaneous solar wind observations.

3 MUTUAL INFORMATION, CONDITIONAL MUTUAL INFORMATION, AND TRANSFER ENTROPY

Dependency is a key discriminating statistic that is commonly used to understand how systems operate. The standard tool used to identify dependency is cross-correlation. Considering two variables, x and y, the correlation analysis essentially tries to fit the data to a 2D Gaussian cloud, where the nature of the correlation is determined by the slope and the strength of correlation is determined by the width of the cloud perpendicular to the slope.

By nature, the response of the radiation belts to solar wind variables is nonlinear (Reeves et al., 2011; Kellerman and Shprits, 2012) as evidenced by the triangle distribution in J_e versus V_{sw} seen in Fig. 1A–C. Such a distribution is not well described by a Gaussian cloud of points and is not well characterized by a slope. For such distributions, it is better to use a statistical based measure such as MI (Tsonis, 2001; Li, 1990; Darbellay and Vajda, 1999). MI between two variables, x and y, compares the uncertainty of measuring variables jointly with the uncertainty of measuring the two variables independently. The uncertainty is measured by the entropy. In order to construct the entropies, it is necessary to obtain the probability distribution functions, which in this study are obtained from histograms of the data based on discretization of the variables (i.e., bins).

Suppose that two variables, x and y, are binned so that they take on discrete values, $\hat{x}$ and $\hat{y}$, where

$$x \in \{\hat{x}_1, \hat{x}_2, \ldots, \hat{x}_n\} \equiv \aleph_1; \quad y \in \{\hat{y}_1, \hat{y}_2, \ldots, \hat{y}_m\} \equiv \aleph_2 \tag{1}$$

The variables may be thought of as letters in alphabets $\aleph_1$ and $\aleph_2$, which have n and m letters, respectively. The extracted data can be considered sequences of letters. The entropy associated with each of the variables is defined as

$$H(x) = -\sum_{\aleph_1} p(\hat{x}) \log p(\hat{x}); \quad H(y) = -\sum_{\aleph_2} p(\hat{y}) \log p(\hat{y}) \tag{2}$$

where $p(\hat{x})$ is the probability of finding the word $\hat{x}$ in the set of x-data and $p(\hat{y})$ is the probability of finding word $\hat{y}$ in the set of y-data. To examine the relationship between the variables, we extract the word combinations $(\hat{x}, \hat{y})$ from the dataset. The joint entropy is defined by

$$H(x,y) = -\sum_{\aleph_1 \aleph_2} p(\hat{x}, \hat{y}) \log p(\hat{x}, \hat{y}) \tag{3}$$

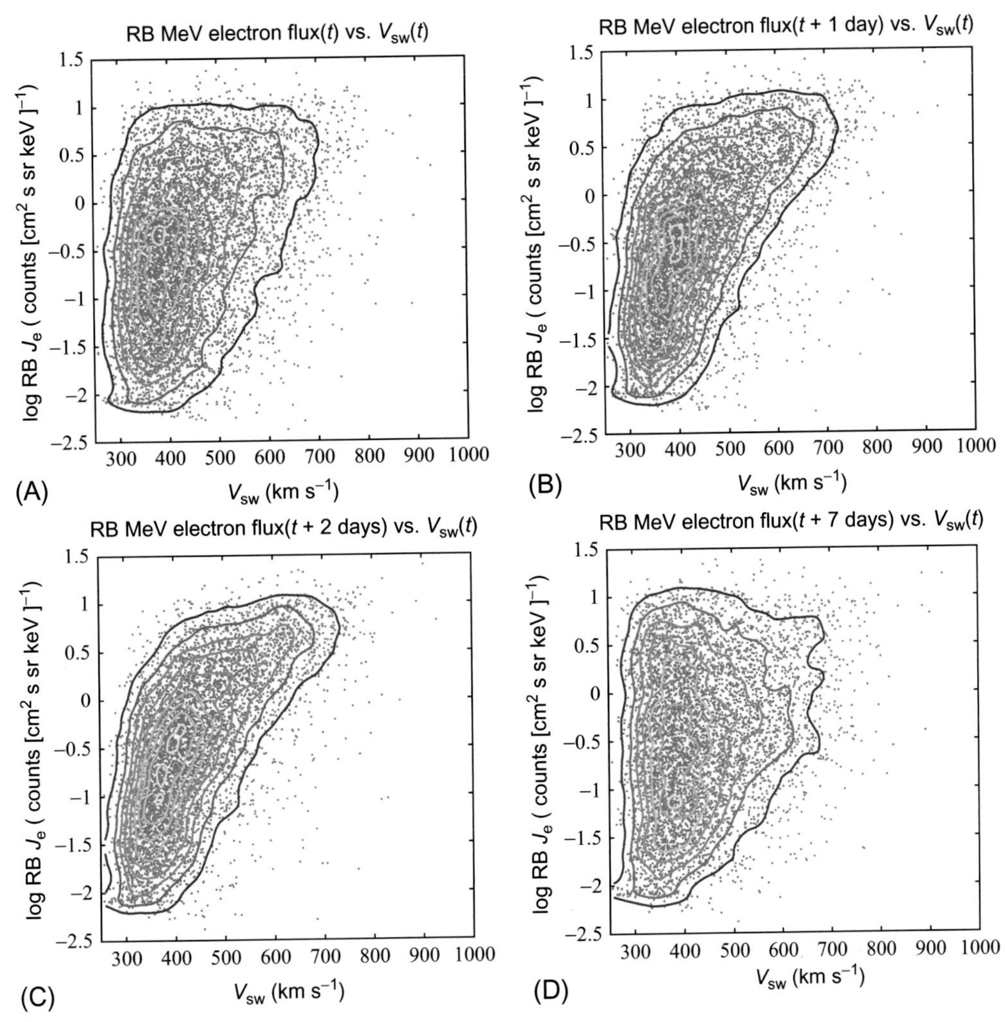

FIG. 1 Scatter plots of $\log J_e(t+\tau)$ versus $V_{SW}(t)$ for $\tau = 0, 1, 2$, and 7 days in panels (A), (B), (C), and (D), respectively. The data points are overlain with density contours showing the nonlinear trends. The panels show that J_e has dependence on V_{SW} for $\tau = 0, 1$, and 2 days and the dependence is strongest for $\tau = 2$ days. (D) At large τ, for example, $\tau = 7$ days, J_e dependence on V_{SW} is very weak. The *triangle* distribution (Reeves et al., 2011) can be seen in panels (A), (B), and (C). This is essentially the same as Fig. 9 in Reeves et al. (2011), except that no density contours are drawn and Fig. 1D plots $\tau = 7$ days instead of $\tau = 3$ days.

where $p(\hat{x}, \hat{y})$ is the probability of finding the word combination $(\hat{x}, \hat{y})$ in the set of (x, y) data.

The MI is then defined as

$$\text{MI}(x, y) = \text{H}(x) + \text{H}(y) - \text{H}(x, y) \tag{4}$$

In the case of Gaussian distributed data, the MI can be related to the correlation function; however, it also includes higher-order correlations that are not detected by the correlation function. Hence, MI is a better measure of dependency for variables having a nonlinear relationship (Johnson and Wing, 2005).

Although MI is useful to identify nonlinear dependence between two variables, it does not provide information about whether the dependence is causal or coincidental. Herein, we use the working definition that if there is a transfer of information from x to y, then x causes y. In this case, it is useful to consider conditional dependency with respect to a conditioner variable z that takes on discrete values, $\hat{z} \in \{z_1, z_2, \ldots, z_k\} \equiv \aleph_3$. The CMI (Wyner, 1978)

$$\mathrm{CMI}(x,y|z) = \sum_{\aleph_1\aleph_2\aleph_3} p(\hat{x},\hat{y},\hat{z}) \log \frac{p(\hat{x},\hat{y}|\hat{z})}{p(\hat{x}|\hat{z})p(\hat{y}|\hat{z})} = \mathrm{H}(x,z) + \mathrm{H}(y,z) - \mathrm{H}(x,y,z) - \mathrm{H}(z) \quad (5)$$

determines the MI between x and y given that z is known. In the case where z is unrelated, $\mathrm{CMI}(x,y|z) = \mathrm{MI}(x,y)$, but in the case that x or y is known based on z, then $\mathrm{CMI}(x,y|z) = 0$. CMI therefore provides a way to determine how much additional information is provided, given another variable. CMI can be seen as a special case of the more general conditional redundancy that allows the variable z to be a vector (e.g., Prichard and Theiler, 1995; Johnson and Wing, 2014).

A common method to establish causal relationships between two time series (e.g., $[x_t]$ and $[y_t]$) is to use a time-shifted correlation function

$$r(\tau) = \frac{\langle x_t y_{t+\tau}\rangle - \langle x\rangle\langle y\rangle}{\sqrt{\langle x^2\rangle - \langle x\rangle^2}\sqrt{\langle y^2\rangle - \langle y\rangle^2}} \quad (6)$$

where r is the correlation coefficient and τ is the lag time. The results of this type of analysis may not be particularly clear when the correlation function has multiple peaks or there is not an obvious asymmetry. Additionally, correlational analysis only detects linear correlations. If the feedback involves nonlinear processes, its usefulness may be seriously limited.

Alternatively, time shifted MI, $\mathrm{MI}(x(t), y(t+\tau))$, can be used to detect causality in nonlinear systems, but this too suffers from the same problems as time-shifted correlation when it has multiple peaks and long-range correlations.

A better choice for studying causality is the one-sided TE (Schreiber, 2000)

$$TE_{x\rightarrow y}(\tau) = \forall t \sum_{\aleph_1\aleph_2} p(y_{t+\tau}, yp_t, x_t) \log\left(\frac{p(y_{t+\tau}|yp_t, x_t)}{p(y_{t+\tau}|yp_t)}\right) \quad (7)$$

where $yp_t = [y_t, y_{t-\Delta}, \ldots, y_{t-k\Delta}]$, $k+1$ is the dimensionality of the system, and Δ is the first minimum in MI. TE can be considered a specialized case of CMI:

$$TE_{x\rightarrow y}(\tau) = CMI(y(t+\tau), x(t)|yp(t)) \quad (8)$$

where $yp(t) = [y(t), y(t-\Delta), \ldots, y(t-k\Delta)]$. The TE can be considered CMI that detects how much average information is contained in an input, x, about the next state of a system, y, that is not contained in the history, yp, of the system (Prokopenko et al., 2013). In the absence of information flow from x to y, $\mathrm{TE}(x \rightarrow y)$ vanishes. Also, unlike correlational analysis and MI, TE is directional, $\mathrm{TE}(x \rightarrow y) \neq \mathrm{TE}(y \rightarrow x)$. The TE accounts for static internal correlations, which can be used to determine whether x and y are driven by a common driver or whether x drives y or y drives x.

4 APPLYING INFORMATION THEORY TO RADIATION BELT MeV ELECTRON DATA

4.1 Radiation Belt MeV Electron Flux Versus V_{sw}

A good starting point for our analysis is Fig. 9 in Reeves et al. (2011), which is replotted in Fig. 1 with some modifications. As in Reeves et al. (2011), this chapter uses the convention that V_{sw} is positive in the anti-Sunward direction. Consistent with Reeves et al. (2011), Fig. 1 shows that (1) the correlation is best at $\tau = 2$ days (Fig. 1C); (2) the relationship between $\log J_e$ and V_{sw} is nonlinear, which can be seen more clearly in the data density contours; and (3) the data point distribution looks like a triangle. This so-called triangle distribution is discussed further in Section 5.4.

The blue curve in Fig. 2A shows the correlation coefficient of $[\log J_e(t + \tau), V_{sw}(t)]$. Note that herein, unless otherwise stated, all linear and nonlinear analyses performed with J_e uses $\log J_e$ values. Fig. 2A shows that the linear correlation coefficient peaks at $\tau_{max} = 2$ days with $r = 0.63$. There is a smaller peak at $\tau = 29$ days ($r = 0.42$), which can be attributed to the 27-day synodic solar rotation. Because of the large number of data points ($n > 5772$), the two peak correlation coefficients are highly significant with $P < 0.01$ (the probability of two random variables giving a correlation coefficients as large as r is <0.01).

However, the relationship between J_e and V_{sw} is nonlinear, and hence, linear cross-correlation may not capture the full extent of the relationship, as described in Section 3. Although the correlation coefficient may give some indication about the sign and strength of the relationship, it is not quantitatively precise (Reeves et al., 2011). In order to take into account the nonlinearities in the relationship, we apply MI and TE.

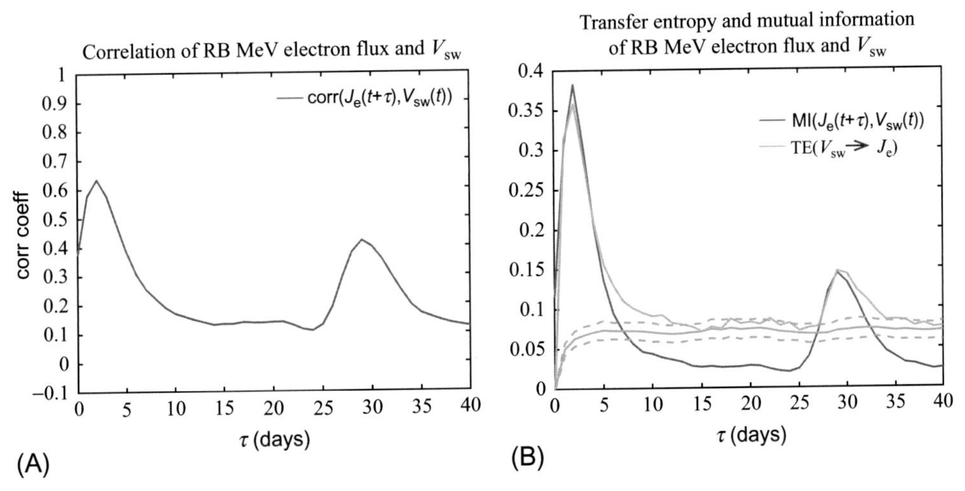

FIG. 2 (A) Correlation coefficient of $[J_e(t+\tau), V_{sw}(t)]$. (B) MI$[J_e(t+\tau), V_{sw}(t)]$ (*blue*) and TE$[J_e(t+\tau), V_{sw}(t)]$ (*yellow*). The *solid and dashed green curves* are the mean and the 3σ from the mean of the noise. The transfer of information from V_{sw} to J_e[TE($V_{sw} \rightarrow J_e$)] peaks at $\tau_{max} = 2$ days.

Fig. 2B plots the MI of $[J_e(t + \tau), V_{sw}(t)]$ (blue curve) and TE of $[J_e(t + \tau), V_{sw}(t)]$ (yellow curve). For simplicity, we assume $k = 0$ in Eqs. (7), (8). The TE from V_{sw} to J_e, TE($V_{sw} \rightarrow J_e$), peaks at $\tau_{max} = 2$ days (yellow curve), suggesting that the transfer of information from V_{sw} to geosynchronous MeV electrons has a 2-day delay. Similar to MI and correlational analysis, TE($V_{sw} \rightarrow J_e$) has a small peak at $t = 29$ days.

To get a measure of the significance of TE($V_{sw} \rightarrow J_e$), we calculate noise = TE[sur(V_{sw}) $\rightarrow J_e$] where sur(V_{sw}) is the surrogate data of V_{sw}, which is obtained by randomly permuting the order of the time series array V_{sw}. The mean and standard deviation of the noise are calculated from an ensemble of 100 random permutations of TE[sur(V_{sw}) $\rightarrow J_e$]. The mean noise and 3σ (standard deviation) from the mean noise are plotted with solid and dashed green curves, respectively, in Fig. 2B. The maximum TE, TE$[J_e(t + 2 \text{ days}), V_{sw}(t)]$ has a peak information transfer (it_{max}) = 0.30, signal-to-noise ratio (snr) = 5.7 and significance = 94σ where it_{max} = peak − mean noise, snr = peak/mean noise, and significance = it_{max}/σ(noise). From the snr, it_{max}, and significance, we conclude that there is a significant transfer of information from V_{sw} to J_e with a 2-day delay. Note that the linear correlation, MI, and TE analyses are consistent with the previous studies (e.g., Baker et al., 1990; Vassiliadis et al., 2005; Reeves et al., 2011; Balikhin et al., 2011; Lyatsky and Khazanov, 2008).

The TE($V_{sw} \rightarrow J_e$) (yellow) curve shows that V_{sw} has little influence on the geosynchronous MeV electrons after a delay of 7–10 days, which is essentially the prediction or information horizon. This result is consistent with Fig. 1D, which shows poor correlation in $\log J_e(t{+}7 \text{ days})$ versus V_{sw} distribution.

In applying our information theoretical tools, the number of bins (n_b) need to be chosen appropriately. Sturges (1926) proposes that for a normal distribution, optimal $n_b = \log_2(n) + 1$ and bin width (w) = range/n_b, where n = number of points in the dataset, range = maximum value − minimum value of the points. In practice, there is usually a range of n_b that would work. Using Sturges (1926) formula, with roughly 6400 points, $n_b \sim 13.6$. For the present study, we find that $n_b = 10$ to 15 would work well. Having too few bins would lump too many points into the same bin, leading to loss of information. Conversely, having too many bins would leave many bins with 0 or fewer points, which also leads to loss of information. For the present study, we choose $n_b = 10$.

4.2 Radiation Belt MeV Electron Flux Versus n_{sw}

We repeat the preceding analyses for J_e versus n_{sw}. Fig. 3 plots $J_e(t + \tau)$ versus n_{sw} for $\tau = 0, 1, 2$, and 7 days. It shows that (1) J_e anticorrelates with n_{sw} and (2) $J_e(t + 1 \text{ day})$ versus n_{sw} (panel B) has narrowest cloud, suggesting the largest anticorrelation. The anticorrelation is shown more clearly in Fig. 4A, which plots corr$[J_e(t + \tau), n_{sw}(t)]$. The blue curve shows $\tau_{min} = 1$ day ($r = -0.40$) and a secondary minimum at $\tau = 28$ days ($r = -0.23$). The latter can be attributed to solar rotation. The two peaks are highly significant ($P < 0.01$) due to the large number of data points.

Fig. 4B is similar to Fig. 2B, except that it shows MI and TE for (J_e, n_{sw}). MI and TE, both have $\tau_{max} = 1$ day, which is consistent with Fig. 2A. Note that unlike correlational analysis, MI and TE only give positive values for both correlations and anticorrelations. TE($n_{sw} \rightarrow J_e$) is not as large as TE($V_{sw} \rightarrow J_e$) shown in Fig. 2B. *TE*$[n_{sw}(t) \rightarrow J_e(t + 1 \text{ day})]$ has $it_{max} = 0.13$, $snr = 4.4$,

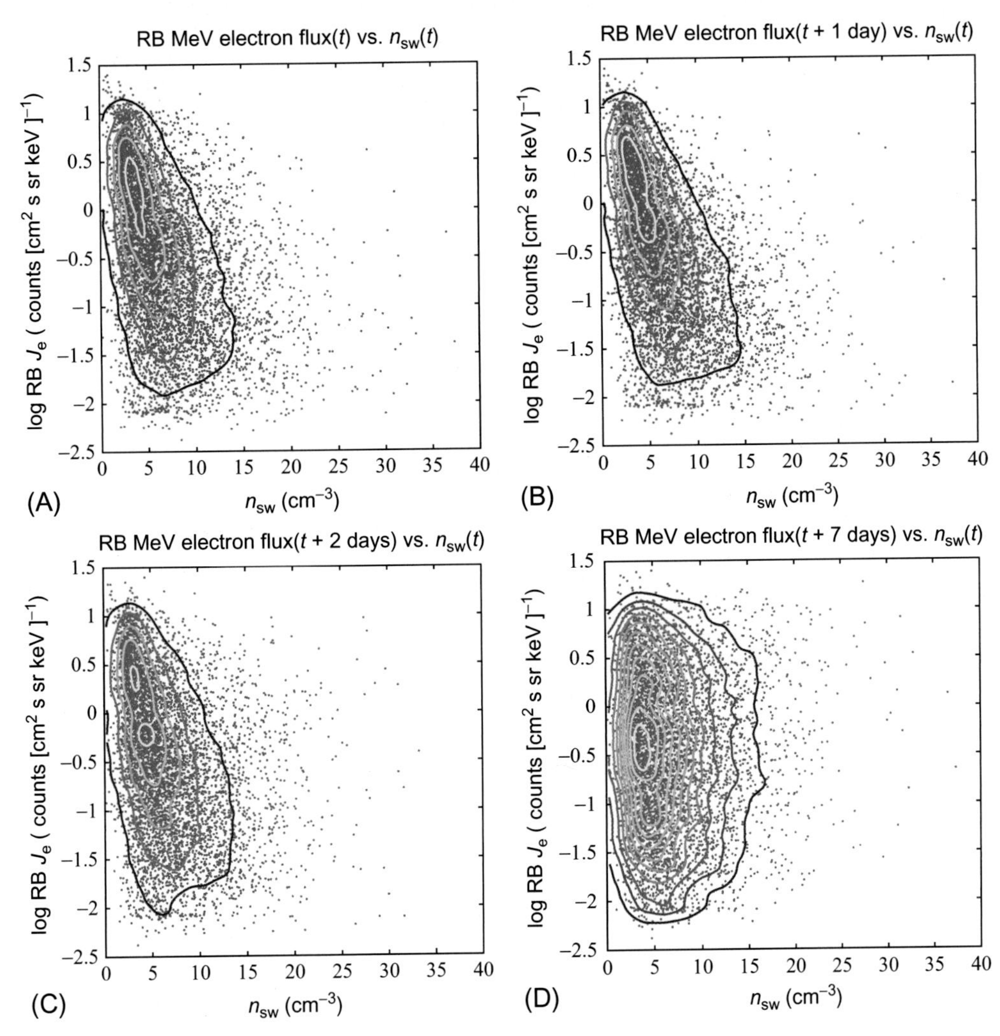

FIG. 3 J_e anticorrelates with n_{sw}. Scatter plots of $\log J_e(t+\tau)$ versus $n_{sw}(t)$ for $\tau = 0, 1, 2,$ and 7 days in panels (A), (B), (C), and (D), respectively. The data points are overlain with density contours, which show the trends. The panels show that J_e has dependence on n_{sw} for $\tau = 0, 1,$ and 2 days. The slope of the contours is most negative for $\tau = 1$ day, suggesting strongest dependence on n_{sw} at $\tau = 1$ day. (D) At large τ, for example, $\tau = 7$ days, J_e dependence on n_{sw} vanishes.

and significance = 42σ. This result suggests that there is a transfer of information from n_{sw} to geosynchronous MeV electrons with a 1-day delay. There is little information transfer from n_{sw} to J_e after 4 days. That is, $\mathrm{TE}[J_e(t+\tau), n_{sw}(t)]$ for $\tau > 4$ days is in the noise level, which is consistent with Fig. 3D for $\tau = 7$ days.

4.3 Anticorrelation of V_{sw} and n_{sw} and Its Effect on Radiation Belt

In Section 4.1, we show that $J_e(t + 2\text{ days})$ linearly and nonlinearly correlate with $V_{sw}(t)$. It is well known that n_{sw} anticorrelates with V_{sw} (e.g., Hundhausen et al., 1970). However, if the

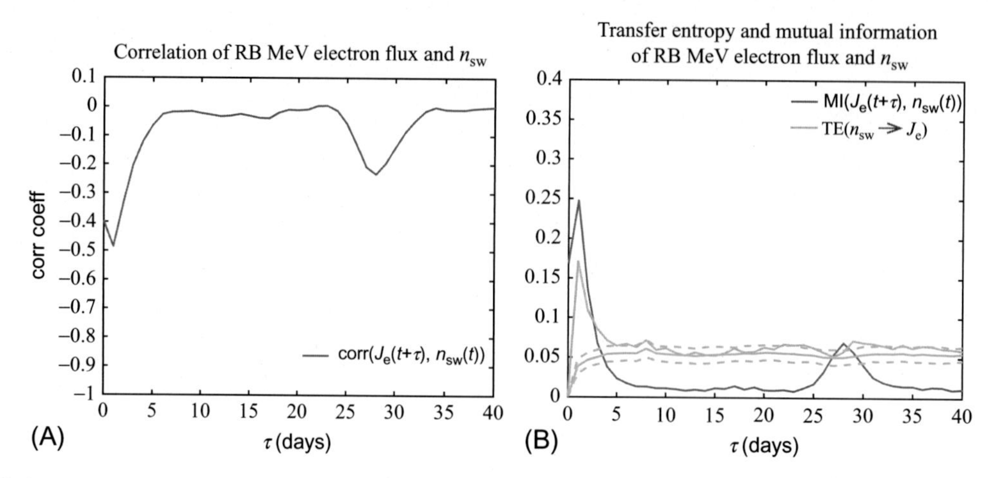

FIG. 4 (A) Correlation coefficient of $[J_e(t+\tau), n_{sw}(t)]$. (B) MI$[J_e(t+\tau), n_{sw}(t)]$ (*blue*) and TE$[J_e(t+\tau), n_{sw}(t)]$ (*yellow*). The *solid and dashed green curves* are the mean and 3σ from the mean of the noise. The transfer of information from n_{sw} to J_e[TE($n_{sw} \rightarrow J_e$)] peaks at $\tau_{max} = 1$ day.

anticorrelation was instantaneous, that is, $n_{sw}(t + 0 \text{ day})$ anticorrelates with $V_{sw}(t)$, then we would expect $J_e(t + 2 \text{ days})$ to anticorrelate with $n_{sw}(t)$. However, in Section 4.2, we show that $J_e(t + 1 \text{ day})$ linearly and nonlinearly anticorrelate with $n_{sw}(t)$, suggesting that other factors may be involved.

To investigate this, we plot in Fig. 5 $n_{sw}(t + \tau)$ versus $V_{sw}(t)$ in the same format as in Figs. 1 and 3. Fig. 5 suggests that n_{sw} anticorrelates with V_{sw} and the relationship is not linear.

Fig. 6A plots corr$[n_{sw}(t + \tau), V_{sw}(t)]$ (blue curve). The blue curve has a minimum at $\tau_{min} = 1$ day ($r = -0.56$) and a secondary minimum at $\tau = 28$ days ($r = -0.32$).

Fig. 6B plots TE and MI for (n_{sw}, V_{sw}) in a similar manner as in Figs. 2B and 4B. Both, MI and TE for $[n_{sw}(t + \tau), V_{sw}(t)]$, blue and yellow curves, respectively, show peaks at $\tau_{max} = 1$ day, which is consistent with the linear correlational analysis. TE$[n_{sw}(t + 1 \text{ day}), V_{sw}(t)]$ has $it_{max} = 0.20$, $snr = 7.4$, and significance $= 95\sigma$.

From the considerations of the lag times, it is entirely possible that anticorrelation of $[J_e(t + 1 \text{ day}), n_{sw}(t)]$ is caused by $[J_e(t + 2 \text{ days}), V_{sw}(t)]$ correlation and the anticorrelation of $[n_{sw}(t + 1 \text{ day}), V_{sw}(t)]$. Note by correlation here we mean both linear and nonlinear correlations. However, we cannot rule out that n_{sw} may also influence J_e independently of V_{sw}. To investigate this, we perform CMI calculation as described in Section 4.4.

So far, we have determined the lag times at daily resolution because we use daily solar wind and LANL data. The LANL higher time resolution data are not yet available, but the OMNI solar wind data are available at hourly resolution. Hence, we can investigate the corr(n_{sw}, V_{sw}) at hourly resolution.

Fig. 7A plots the corr$[n_{sw}(t + \tau), V_{sw}(t)]$ for $\tau = 0\text{–}100$ h (solid curve). It shows that the correlation reaches a minimum at $\tau_{min} = 14$ h for the data interval used in the present study 1989–2009. It also shows that n_{sw} anticorrelation with V_{sw} has a broad minimum. To quantify

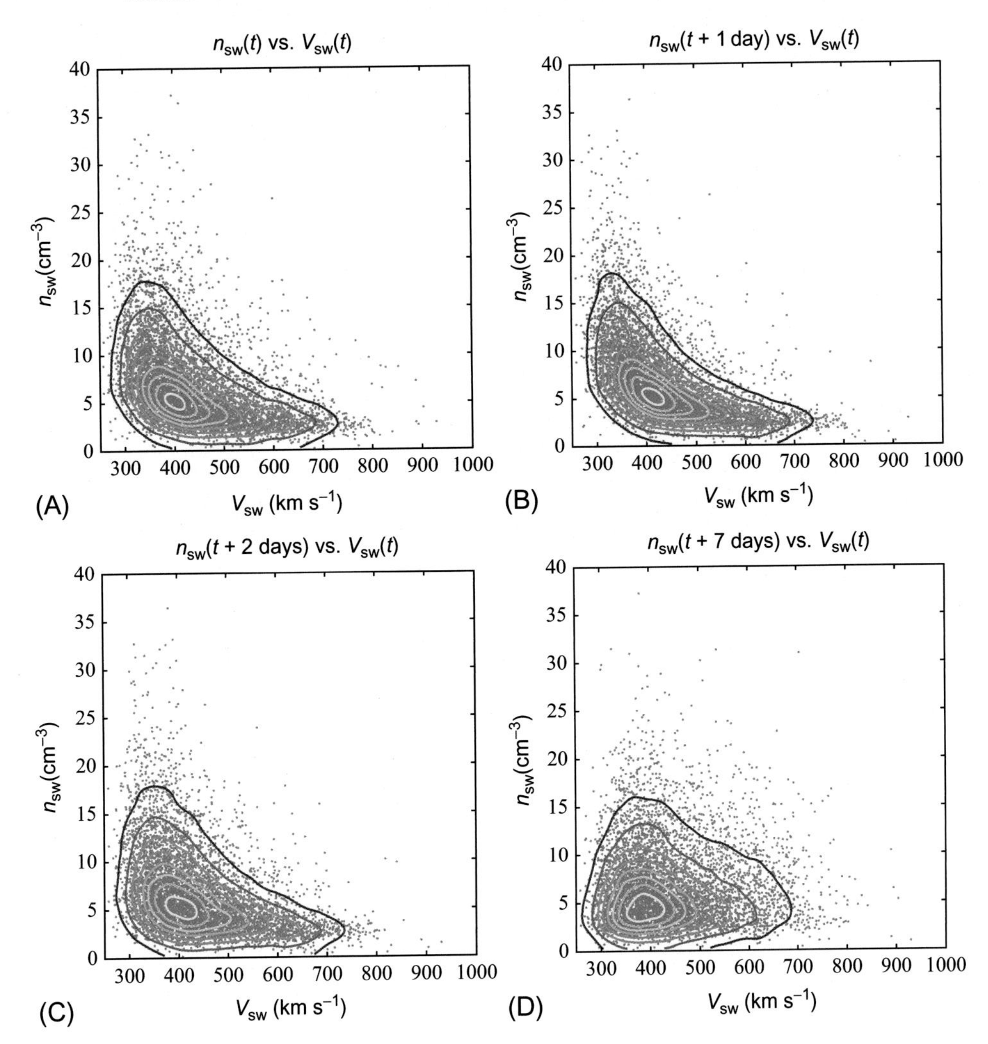

FIG. 5 n_{SW} anticorrelates with V_{SW}. Scatter plots of $n_{SW}(t+\tau)$ versus $V_{SW}(t)$ for $\tau = 0, 1, 2$, and 7 days in panels (A), (B), (C), and (D), respectively. The data points are overlain with density contours showing strongly nonlinear the trends. The panels show that n_{SW} has dependence on V_{SW} for $\tau = 0, 1$, and 2 days and the dependence is strongest for $\tau = 1$ day. (D) At large τ, for example, $\tau = 7$ days, n_{SW} dependence on V_{SW} is very weak.

the width of the minimum, we draw a dashed horizontal line that intersects the solid curve at $\tau = 0$ h and show that the anticorrelation does not worsen more than that at $\tau = 0$ h until $\tau = 36$ h. However, τ_{min}, the correlation coefficient at τ_{min}, and the width of the minimum are time dependent. As an example, Fig. 7B shows that for the period 2000–14, $\tau_{min} = 17$ h and the width of the minimum using the preceding criterion is about 46 h. Moreover, Fig. 6B shows that TE$[n_{SW}(t+\tau), V_{SW}(t)]$ does not reach the noise level until $\tau > 3$ days, suggesting a rather long period when V_{SW} affects the trailing density, n_{SW}.

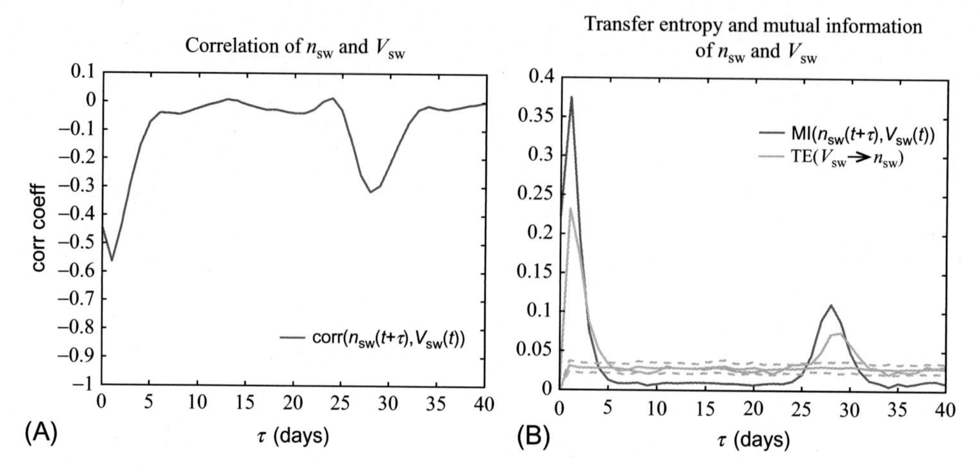

FIG. 6 (A) Correlation coefficient of $[n_{SW}(t+\tau), V_{SW}(t)]$. (B) MI$[n_{SW}(t+\tau), V_{SW}(t)]$ (*blue*) and TE$[n_{SW}(t+\tau), V_{SW}(t)]$ (*yellow*). The *solid and dashed green curves* are the mean and 3σ from the mean of the noise. The transfer of information from V_{SW} to n_{SW}[TE($V_{SW} \rightarrow n_{SW}$)] peaks at $\tau_{max} = 1$ day.

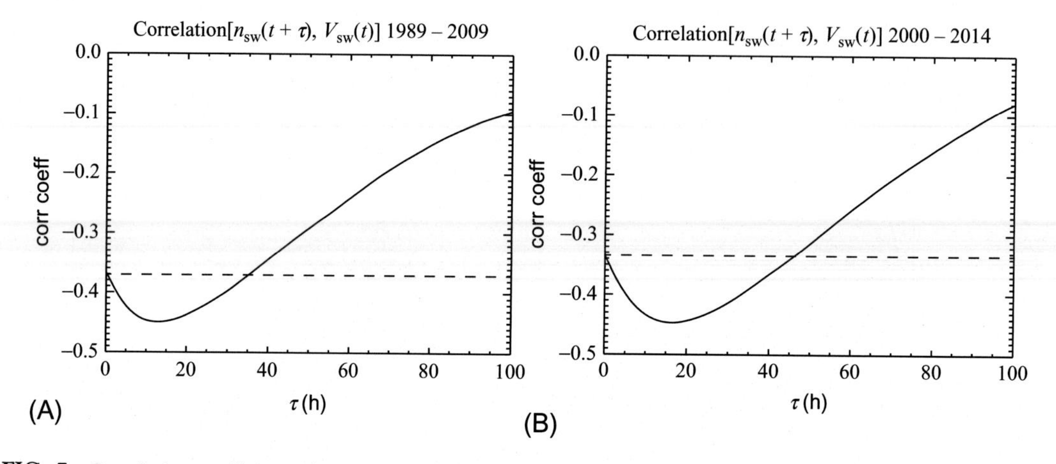

FIG. 7 Correlation coefficient of $[n_{SW}(t+\tau), V_{SW}(t)]$ for (A) 1989–2009 and (B) 2000–14 at hourly resolution. The anticorrelation improves with increasing τ, reaching minimum at $\tau_{min} = 14$ and 16 h in (A) and (B), respectively. The correlation coefficient finally reaches the same value as that at $\tau = 0$ h (the *dashed line*) at $\tau = 36$ and 46 h in (A) and (B), respectively.

4.4 Ranking of Solar Wind Parameters Based on Information Transfer to Radiation Belt Electrons

From our preceding analysis, V_{SW} is a stronger driver of J_e than n_{SW}, that is, V_{SW} transfers more information to J_e than n_{SW}. For example, TE$[V_{SW}(t) \rightarrow J_e(t + 2 \text{ days})]$ has $it_{max} = 0.30$ and $snr = 5.7$ while TE$[n_{SW}(t) \rightarrow J_e(t + 1 \text{ day})]$ has $it_{max} = 0.13$ and $snr = 4.4$. Because V_{SW} anticorrelates with n_{SW}, there is some embedded dependence. So, it is necessary to use CMI to determine how much information passes from n_{SW} to J_e, given V_{SW} and vice versa.

To calculate how much information flows from n_{sw} to J_e, given V_{sw}, we calculate $\text{CMI}[J_e(t+\tau), n_{sw}(t)|V_{sw}(t)]$, which is plotted as a blue curve in Fig. 8A. Using a similar approach as for TE, we determine the noise level of the surrogates: $\text{CMI}[J_e(t+\tau), \text{sur}[n_{sw}(t)]|V_{sw}(t)]$. The mean and σ of the noise are calculated in the same manner as TE (described in Section 4.1) and used to determine the significance of the results. The mean noise and 3σ are plotted as solid and dashed green curves, respectively. Fig. 8A shows that $\text{CMI}[J_e(t+\tau), n_{sw}(t) \mid V_{sw}(t)]$ peaks at $\tau_{max} = 0$ day with $it_{max} = 0.091$ and $snr = 3.2$. The $\tau_{max} = 0$ day suggests that the J_e response lag time to n_{sw} is less than 24 h.

We can now revisit the J_e response lag times to V_{sw} and n_{sw}. Earlier we established that $J_e(t+2 \text{ days})$ correlates with $V_{sw}(t)$ (Fig. 2), $J_e(t+1 \text{ day})$ anticorrelates with $n_{sw}(t)$ (Fig. 4), but $n_{sw}(t+1 \text{ day})$ anticorrelates with $V_{sw}(t)$ (Fig. 6). However, our CMI analysis (Fig. 8A) shows that given V_{sw}, J_e response lag time to n_{sw} is 0 day (<24 h). This suggests that the $J_e(t+1 \text{ day})$ anticorrelation with $n_{sw}(t)$ seen in Fig. 4 mainly comes from $J_e(t+2 \text{ days})$ correlation with $V_{sw}(t)$ and $V_{sw}(t)$ anticorrelation with $n_{sw}(t+1 \text{ day})$. However, Fig. 8A shows that the peak is rather broad, suggesting that the J_e response is still significant at $\tau = 1$ day.

We also calculate $\text{CMI}[J_e(t+\tau), V_{sw}(t)|n_{sw}(t)]$, which is plotted in Fig. 8B as a solid blue curve. The blue curve peaks at $\tau = 2$ days with $it_{max} = 0.25$, which is about 2.7 times larger than the it_{max} of 0.091 for $\text{CMI}[J_e(t+\tau), n_{sw}(t)| V_{sw}(t)]$. Thus, V_{sw} transfers more information to J_e than n_{sw} does.

Interestingly, the peak in $\text{CMI}[J_e(t+\tau), V_{sw}(t)|n_{sw}(t)]$ (Fig. 8B) is broader than the peak in $\text{TE}[J_e(t+\tau), V_{sw}(t)]$ (Fig. 2B). The former also has slightly higher snr (6.6) than the latter (5.7). Removing the effect of n_{sw}, which anticorrelates with J_e, has the effects of broadening the peak, lowering the noise, and increasing the snr. These effects are discussed in Section 5.3.

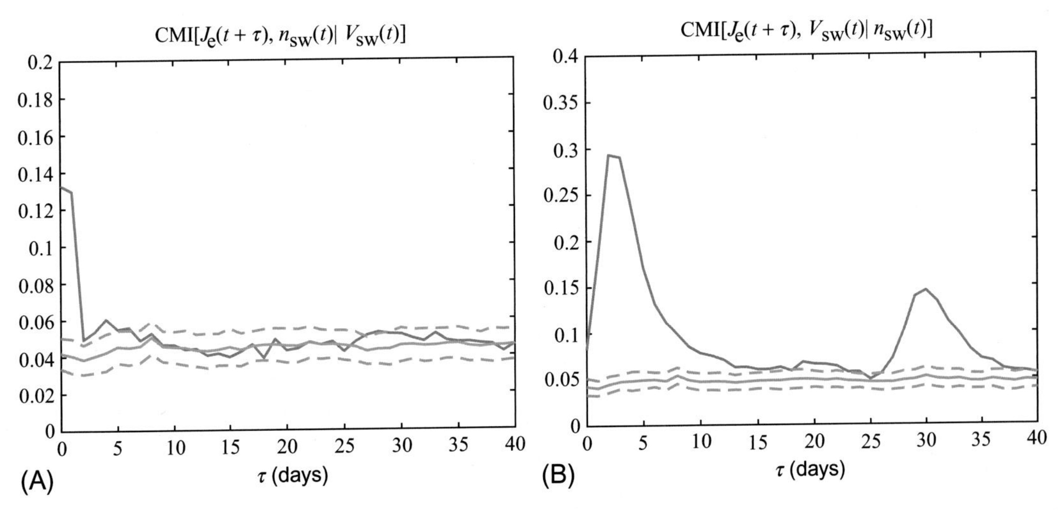

FIG. 8 *Blue curve* showing (A) $\text{CMI}[J_e(t+\tau), n_{sw}(t) \mid V_{sw}(t)]$ and (B) $\text{CMI}[J_e(t+\tau), V_{sw}(t) \mid n_{sw}(t)]$. The *solid and dashed green curves* are the mean and 3σ from the mean of the noise. (A) Unlike $\text{TE}[J_e(t+\tau), n_{sw}(t)]$, which peaks at $\tau_{max} = 1$ day, $\text{CMI}[J_e(t+\tau), n_{sw}(t) \mid V_{sw}(t)]$ peaks at $\tau_{max} = 0$ day ($it_{max} = 0.091$). The smaller τ_{max} comes about because CMI removes the effect of V_{sw} on J_e (see text). (B) The peak in $\text{CMI}[J_e(t+\tau), V_{sw}(t) \mid n_{sw}(t)]$($it_{max} = 0.25$) is broader and has slightly higher snr than that of $\text{TE}[J_e(t+\tau), V_{sw}(t)]$ in Fig. 2B because CMI removes the effect of n_{sw}, which anticorrelates with J_e. V_{sw} transfers about 2.7 times more information to J_e than n_{sw}.

The preceding analysis suggests that V_{sw} is the major driver of J_e. Next, we investigate whether other solar wind parameters also contribute to J_e. We calculate the information transfer from |IMF **B**|, P_{dyn}, σ(IMF B), southward IMF B_Z, northward IMF B_Z, IMF B_y, IMF B_X, and solar wind electric field (E_{sw}) to J_e, given V_{sw}. The northward (southward) IMF Bz is calculated from the daily average of the hourly IMF B_z when IMF $B_Z > 0$ (IMF $B_Z < 0$). The results are tabulated in Table 1, which ranks various solar wind parameters based on the it_{max}. Thus, the ranking gives the importance of each solar wind parameter based on the information transfer to J_e. Table 1 also lists τ_{max}, which signifies the lag time when information transfer to J_e maximizes.

Note that the ranking in Table 1 is obtained with daily resolution data. It is possible that the ranking of some parameters may change if the data are analyzed at higher time resolution. For example, some studies showed that southward IMF B_z can influence J_e (e.g., Li et al., 2005; Onsager et al., 2007; Miyoshi and Kataoka, 2008), but southward IMF B_z is only ranked number 5 in Table 1. IMF fluctuates with periods of northward and southward IMF at minutes or tens of minutes timescale. Thus, the low ranking of the southward IMF B_z most likely result from the fluctuations of IMF B_z within a 1-day period (e.g., Li et al., 2001; Balikhin et al., 2011; Reeves et al., 2011). Consistent with our result, Li et al. (2001) found IMF B_z is poorly correlated with J_e at daily resolution. Interestingly, although southward IMF B_z has higher it_{max} than northward IMF B_z, northward IMF B_z has a lower noise level, and hence a

TABLE 1 Ranking of the Importance of the Solar Wind Parameters Based on Information Transfer to Geosynchronous Mev Electron Flux (J$_e$) at τ_{max}, Where τ_{max} Is the Lag Time When the Information Transfer Peaks

Rank	Solar Wind Parameters	Peak Information Transfer (it_{max})	Signal-to-Noise Ratio at τ_{max}	Significance at $\tau_{max}(\sigma)$	τ_{max} (Days)	Prediction Horizon (Days)
1	V_{sw}	0.25	6.6	94	2	10[a]
2	IMF |**B**|	0.12	3.9	48	0	2
3	P_{dyn}	0.092	3.4	35	0	2
3	n_{sw}	0.091	3.2	34	0	2
4	σ(IMF B)	0.075	3.9	48	0	2
5	IMF $B_Z < 0$	0.064	2.7	26	0	2
6	E_{sw}	0.056	2.9	22	1	5
7	IMF B_y	0.052	2.3	20	0	2
8	IMF $B_z > 0$	0.048	3.1	22	0	2
9	IMF B_X	0.044	2.2	19	0	2

Notes: Parameters 2–9 are calculated from CMI[$J_e(t+\tau), x(t)|V_{sw}(t)$] where as parameter 1 is calculated from CMI[$J_e(t+\tau), V_{sw}(t)\,|\,n_{sw}(t)$], where x = parameters 2–9. The peak information transfer (it_{max}) = peak − mean noise, the signal-to-noise ratio = peak/noise, and significance = it_{max}/σ(noise). Noise is calculated from surrogate data (see Section 4.1). The prediction horizon gives a measure of how far ahead J_e can be predicted. Note that n_{sw} and P_{dyn} are both ranked at number 3 because they have similar it_{max} (the effect of V_{sw} has been removed (see Section 5.3)). Northward IMF has slightly higher snr than southward IMF because northward IMF has lower noise level than southward IMF.
[a]*Excluding the effect of solar rotation.*

higher *snr* than southward IMF B_z. The τ_{max} for E_{sw} is 1 day, which may be the average of $\tau_{max} = 2$ days for V_{sw} and $\tau_{max} = 0$ day for IMF B_z or IMF $|\mathbf{B}|$.

4.5 Detecting Changes in the System Dynamics

As described in Section 3, TE from x to y, TE($x \rightarrow y$), gives a measure of information transfer from variable x to y. In the solar wind-magnetosphere system, the solar wind driving of the magnetosphere is not constant, depending on the strength of the driver and internal dynamics (e.g., Wing et al., 2005a; Johnson and Wing, 2005). So, the system dynamics may not be stationary. The dynamics of the system can be detected by applying TE to a sliding window of data. Fig. 9 shows the behavior of windowed TE[$V_{sw}(t) \rightarrow J_e(t + 2$ days) over the course of 0–2500 days since January 1, 1989 (a sliding 50-day window is used). One of the key features of the figure is the variation in TE over the course of 7 years, indicative of nonstationary dynamics. There are periods when TE has higher values, suggesting stronger solar wind-radiation belt coupling and vice versa.

With windowed TE, we are limited to a small number of points, which can pose difficulties for noise calculation from an ensemble of surrogates. A small number points can lead to a rather large noise and uncertainty (large σ(noise)/mean noise), rendering the noise to be unreliable. Therefore, rather than using *it* (TE—mean noise), we just use TE to characterize the dynamics of the system. Generally, periods of low TE can be considered a baseline and deviations from the baseline may indicate the presence of significant information transfer in the dynamics. The value of this approach can ultimately be measured by how well it detects dependencies and changes in system dynamics. In Section 5.4, we show that even with just using TE, we are able to dissect the triangle distribution (Reeves et al., 2011) quite well, and in Section 5.5, we discuss applications to modeling.

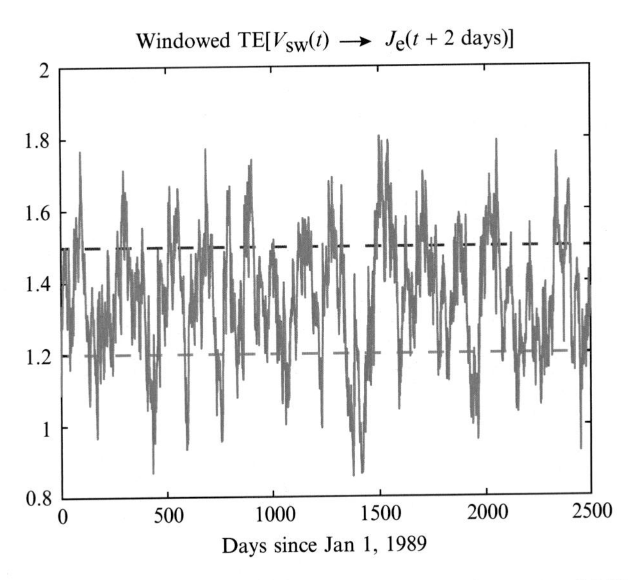

FIG. 9 Blue curve showing windowed TE[$J_e(t + 2$ days), $V_{sw}(t)$] over the course of 0–2500 days after January 1, 1989. The dynamics in the solar wind-outer radiation belt system changes with time, showing periods of high and low TEs.

5 DISCUSSION

5.1 Geo-Effectiveness of Solar Wind Velocity

Studies suggest that the substorm or storm injection process from plasma sheet into the innermagnetosphere accelerate low-energy electrons from a few keV to ~100 keV, and once in the innermagnetosphere, wave-electron interactions accelerate the electrons further to several MeV (e.g., Lyatsky and Khazanov, 2008; Baker and Kanekal, 2008). The mechanisms for accelerating the electrons to the MeV energy range generally fall into two categories. In the first mechanism, electron interactions with ULF waves can increase radial diffusion, or nonadiabatic transport, of electrons, resulting in acceleration (e.g., Baker and Kanekal, 2008; Li and Temerin, 2001; Li et al., 2005; Elkington et al., 1999; Rostoker et al., 1998; Ukhorskiy et al., 2005; Mathie and Mann, 2000, 2001; Reeves, 2007; Shprits et al., 2009; Green and Kivelson, 2004; Kellerman and Shprits, 2012). The second mechanism is often referred to as local acceleration, where acceleration can occur when low-energy electrons interact with locally grown waves such as VLF whistler mode waves (Summers et al., 1998, 2007; Omura et al., 2007; Horne et al., 2005), or fast magnetosonic waves (e.g., Horne et al., 2007; Shprits et al., 2008). These two mechanisms are not necessarily mutually exclusive.

Fig. 1 shows that the relationship between J_e and V_{sw} is nonlinear, and hence it is necessary to use information theoretical tools to discover the full extent of the relationships between these two parameters. Previous correlational analyses show that $J_e(t + 2 \text{ days})$ correlates best with $V_{sw}(t)$ (e.g., Reeves et al., 2011; Lyatsky and Khazanov, 2008), but correlational analysis only establishes linear correlation. The present study establishes that $J_e(t+2 \text{ days})$ and $V_{sw}(t)$ are nonlinearly correlated. Using TE, we establish that indeed, information transfer from V_{sw} to J_e, peaks at $\tau = 2$ days. However, n_{sw} anticorrelates with V_{sw}. Removing the effects of n_{sw}, $\text{CMI}[J_e(t + \tau), V_{sw}(t)|n_{sw}(t)]$ shows that the information transfer peaks still at $\tau = 2$ days, but the peak is broader. For example, the information transfer at $\tau = 3$ days is only slightly lower than that at $\tau = 2$ days, as shown in Fig. 8B.

Our result is consistent, at least with the first electron acceleration mechanism mentioned previously. Large V_{sw} can increase the occurrences of KHI along the magnetopause flanks (e.g., Fairfield et al., 2000; Johnson et al., 2014; Wing et al., 2005b), leading to enhancements of ULF waves within the magnetosphere (e.g., Engebretson et al., 1998; Vennerstrøm, 1999) and electron acceleration. Thus, the process to accelerate the electrons to the MeV energy range takes 2 days, as previously suggested (e.g., Kellerman and Shprits, 2012; Reeves et al., 2011). V_{sw} may also be tied to the local acceleration mechanism through substorm particle injections (e.g., Baker and Kanekal, 2008; Kissinger et al., 2011; Tanskanen, 2009; Kellerman and Shprits, 2012; Newell et al., 2016).

5.2 n_{sw} and V_{sw} Anticorrelation

The anticorrelation of n_{sw} and V_{sw} is well known (e.g., Hundhausen et al., 1970), but the long lag time for this anticorrelation is relatively unknown. The anticorrelation may result from the solar wind high-speed streams that originate from the coronal holes, which have higher velocities and lower densities than the background solar wind. Surprisingly, the anticorrelation peaks at $\tau = 14\text{–}16$ h, depending on the year. It is not clear what causes the lag time to peak at 14–16 h. The solar wind high-speed stream may originate from the coronal holes at high latitudes of the Sun, whereas the background slower solar wind may originate from lower latitudes (e.g., Kreiger et al., 1973; Schwenn et al., 1978). This lag time may result from the compression of the leading edge of the high-speed stream structure when it encounters the denser background solar wind. Such compression may create slower and denser structure at the leading edge of the high-speed stream (e.g., Gosling et al., 1972). As a result, the anticorrelation at the leading edge of the high-speed stream is not as good as that at the trailing edge, which may better better preserve the high-speed–low-density structure. Fig. 6B suggests that there is information transfer from V_{SW} to n_{SW} up to 3–5 days, suggesting perhaps the longevity of the high-speed stream structure is about 3–5 days. This is consistent with the Gosling et al. (1972) study that reports that the widths of high-speed streams are about 4 days. This property needs to be further investigated.

The correlational analyses of $[n_{SW}(t+1\text{ day}), V_{SW}(t)]$ and $[J_e(t+2\text{ days}), V_{SW}(t)]$ return correlation coefficients of −0.56 and 0.63, respectively. So, the former has a slightly lower correlation than the latter (the two set of data are a similar size). However, the scatter plots in Figs. 1C and 5B show that both sets of data exhibit nonlinear behaviors. Hence, the linear correlational analysis may not capture the full extent of their relationships. Indeed, analyses with TE reveal that $\text{TE}[n_{SW}(t+1\text{ day}), V_{SW}(t)]$ has comparable significance (95σ) to that of $\text{TE}[J_e(t+2\text{ days}), V_{SW}(t)](94\sigma)$.

The significant transfer of information from V_{SW} to n_{SW} has implications on the studies of solar wind driving of the magnetosphere that involve n_{SW} and V_{SW}. These studies should take into account the strong anticorrelation between n_{SW} and V_{SW} that can persist, even at large lag times. For example, any attempt to isolate the effects of $n_{SW}(V_{SW})$ on the magnetosphere would need to effectively remove the effects of $V_{SW}(n_{SW})$ using CMI or similar methods.

5.3 Geo-Effectiveness of Solar Wind Density

Fig. 4 shows that $J_e(t+1\text{ day})$ anticorrelates with $n_{SW}(t)$. The lag time of 1 day is consistent with that found in Balikhin et al. (2011). However, $n_{SW}(t+1\text{ day})$ anticorrelates with $V_{SW}(t)$. Moreover, $\text{CMI}[J_e(t+\tau), n_{SW}(t) \mid V_{SW}(t)]$ peaks at $\tau = 0$ day, suggesting that given V_{SW}, J_e responds to n_{SW} in <24 h. Hence, J_e response lag time of 1 day to n_{SW} in Fig. 4 and in Balikhin et al. (2011) can be attributed at least partly to $J_e(t+2\text{ days})$ correlation with $V_{SW}(t)$, and $V_{SW}(t)$ anticorrelation with $n_{SW}(t+1\text{ day})$. Fig. 8 and Table 1 show that V_{SW} is by far the dominant driver of J_e, transferring 2.7 times more information to J_e than n_{SW} does.

Figs. 6 and 7 show that the anticorrelation of V_{SW} and n_{SW} is not trivial, and any attempt to interpret the effects of n_{SW} and V_{SW} on J_e should take into account the anticorrelation of V_{SW} and n_{SW}. To illustrate, $\text{TE}[n_{SW}(t) \rightarrow J_e(t+1\text{ day})]$ has $it_{max} = 0.13$, but removing the effects of V_{SW}, the it_{max} drops ~30% to 0.091 [it_{max} of $(\text{CMI}[J_e(t+0\text{ day}), n_{SW}(t)|V_{SW}(t)]$ is 0.091)].

Conversely, some of the effects attributed to V_{SW} may be due to n_{SW}, but the effects of n_{SW} are smaller. For example, $\text{TE}[V_{SW}(t) \rightarrow J_e(t+2\text{ days})]$ has $it_{max} = 0.30$, but removing the effects of n_{SW}, the it_{max} drops only ~17% to 0.25 [it_{max} of $\text{CMI}[J_e(t+2\text{ days})V_{SW}(t)|n_{SW}(t)] = 0.25$]. Interestingly, Fig. 8B shows that the $\text{CMI}[J_e(t+\tau), V_{SW}(t)|n_{SW}(t)]$ peak is broader than that of $\text{TE}[V_{SW}(t) \rightarrow J_e(t+\tau)]$ as shown in Fig. 2B. For example, the CMI at $\tau = 3$ days is only slightly smaller than that at $\tau = 2$ days. The reason for the broader peak is that as shown in Fig. 8A, there is a significant information transfer from n_{sw} to J_e at $\tau = 0-1$ day, but it falls off rapidly at larger τ. Because the anticorrelation between V_{sw} and n_{sw} has a 1-day lag, removing the effects of n_{sw} would lower information transfer from V_{sw} to J_e (i.e., $TE[V_{sw}(t) \rightarrow J_e(t+\tau)]$) at $\tau = 1-2$ days. So, *it* for $TE[V_{sw}(t) \rightarrow J_e(t+\tau)]$ at $\tau = 1, 2$, and 3 are 0.26, 0.30, and 0.23, respectively, whereas the corresponding values for $CMI[J_e(t+\tau), V_{sw}(t)|n_{sw}(t)]$ are 0.14, 0.25, and 0.24, respectively. Note that, at $\tau = 1$ and 2, there are reductions in information transfer, while at $\tau = 3$, the information transfer is more or less the same (the difference is within one σ [$\sim$0.01]). This leads to a broader peak in the $CMI[J_e(t+\tau), V_{sw}(t)|n_{sw}(t)]$ curve in Fig. 8B than that in the $TE[V_{sw}(t) \rightarrow J_e(t+\tau)]$ curve in Fig. 2B.

An increase in n_{sw} would increase solar wind dynamic pressure (P_{dyn}), which, in turn, would push the magnetopause inward, leading to electron losses at the high L shell (e.g., Li et al., 2001). Furthermore, the magnetopause compression would drive ULF waves (e.g., Korotova and Sibeck, 1995; Kepko and Spence, 2003; Claudepierre et al., 2010) leading to fast radial diffusion, which redistributes the losses to the magnetopause to lower L shells, including at geosynchronous orbit (Shprits et al., 2006; Kellerman and Shprits, 2012; Turner et al., 2012). Ukhorskiy et al. (2006) uses a test particle simulation to demonstrate this scenario, which is known as the magnetopause shadowing.

Our result suggests that based on information transfer from n_{sw} to J_e, any mechanism for n_{sw} anticorrelation with J_e has to operate or start operating within <24 h (Fig. 8A). We note that the magnetospheric compression due to an increase in n_{sw} or P_{dyn} would be nearly instantaneous, although it is not clear how long it would take for the electron losses to redistribute radially and how long the loss process would continue. Although the J_e response to n_{sw} peaks at $\tau = 0$, at $\tau = 1$ day, there is still a significant amount of information transferred from n_{sw} to J_e, as shown in Fig. 8A.

5.4 Revisiting the Triangle Distribution

Reeves et al. (2011) is the first to note the right triangle distribution exhibited in Fig. 1A. Fig. 1A plots $J_e(t+\tau)$ versus $V_{sw}(t)$ with no delay, $\tau = 0$. However, we note that as shown in Fig. 1B and C, even with $\tau = 1$ or 2 days, respectively, the triangle distribution is still evident, albeit not as prominently as for $\tau = 0$. For example, the triangle distribution can still be seen in Fig. 1C, which is replotted in Fig. 10A without the contour overlays. Reeves et al. (2011) notes that the left-hand side of the triangle forms because V_{sw} rarely goes below 300 km s^{-1}. The hypotenuse of the triangle suggests that the lower limit of J_e more or less increases with V_{sw}. The top side of the triangle suggests that J_e saturates, which can be attributed to local instabilities (Kennel and Petschek, 1966). Reeves et al. (2011) considers this and other possible explanations for the J_e saturation. As noted by Reeves et al. (2011), the most interesting and perhaps mystifying aspect of the triangle distribution is that high J_e is observed for all V_{sw} conditions and the variability of J_e at lower V_{sw} is much larger than that at higher V_{sw}.

Reeves et al. (2011) notes that the triangle distribution appears in the declining phase of the solar maximum, but it vanishes during solar maximum (but high J_e and low V_{sw} points still appear during solar maximum). This dependence of the solar cycle suggests that perhaps the mode in which the solar wind couples to the magnetosphere/radiation belt can be a factor.

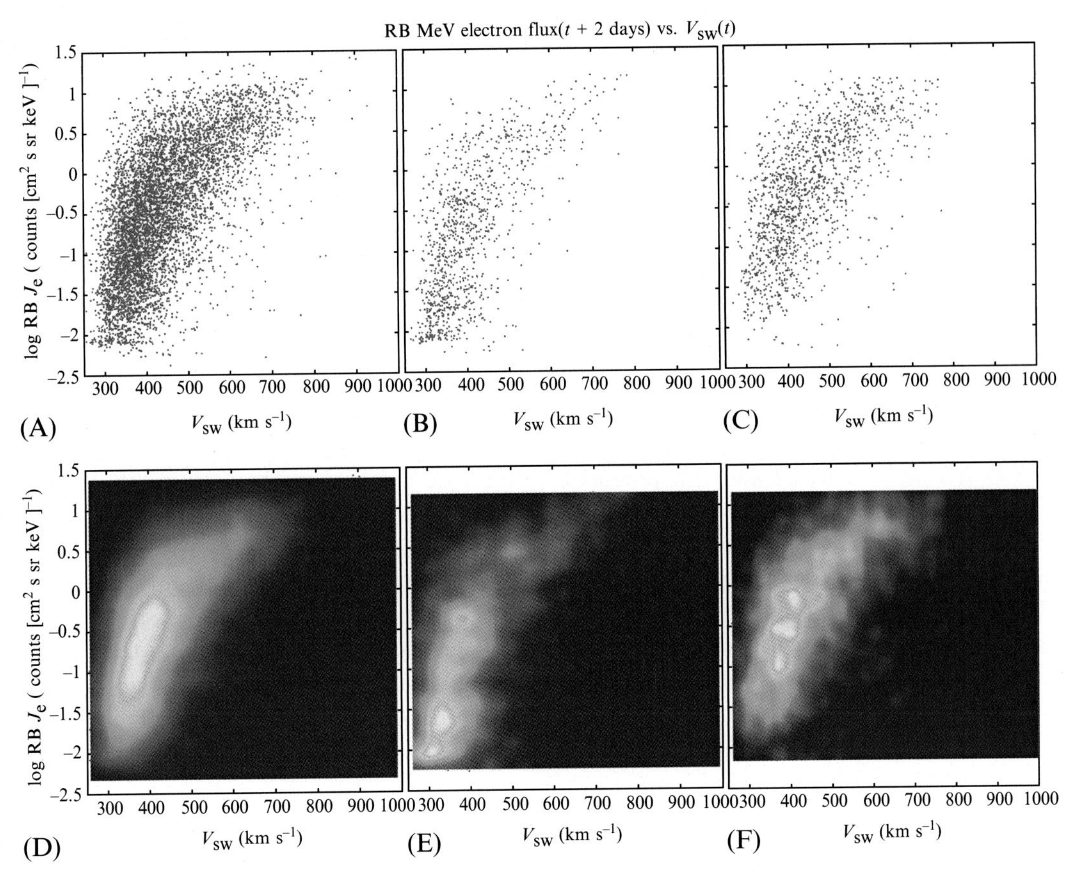

FIG. 10 (A) Scatter plot of $\log J_e(t+2 \text{ days})$ versus $V_{SW}(t)$ showing the triangle distribution. This is similar to Fig. 1C, but without the density contours. (B) The points in (A) are plotted when $\text{TE}[J_e(t+2 \text{ days}), V_{SW}(t)] < 1.2$ (below the *green dashed line* in Fig. 9). (C) The points in (A) are plotted when $\text{TE}[J_e(t+2 \text{ days}), V_{SW}(t)] > 1.5$ (above the *red dashed line* in Fig. 9). The distributions in (B) and (C) differ and both do not have the same triangle distribution as in (A). Panels (D), (E), and (F) show the data density maps of the data in panels (A), (B), and (C), respectively. Panels (E) and (F) reinforce the differences in the data distributions in panels (B) and (C).

We probe the possible effects of solar wind-radiation belt coupling further with information theoretical tools. Particularly, we separate the points in Fig. 10A based on the information transfer from V_{SW} to J_e.

Fig. 10B shows $J_e(t+2 \text{ days})$ versus $V_{SW}(t)$ when $\text{TE}[V_{SW}(t) \rightarrow J_e(t+2 \text{ days})]$ is below 1.2, below the dashed green line in Fig. 9. Fig. 10C plots the points when TE is above 1.5, above the dashed red line in Fig. 9. It is clear that the data distribution in Fig. 10B looks different than that in Fig. 10C. Also, the triangle distribution in Fig. 10A is not reproduced in Fig. 10B nor in Fig. 10C. Although Fig. 10C still shows a triangle, the triangle, which is not a right triangle, shows a different characteristic than that in Fig. 10A.

Fig. 10B and C contrasts the differences between low and high TE cases. Fig. 10B shows that (1) most of the points tend to have $V_{SW} < 500 \text{ km s}^{-1}$ and (2) excluding points with $V_{SW} >$

$500\,\mathrm{km\,s^{-1}}$, J_e tends to have only weak dependency on V_{sw} and the distribution looks more like a rectangle than a triangle. On the other hand, Fig. 10C shows that (1) the points tend to have more uniform distribution in velocity; (2) the hypotenuse of the triangle in Fig. 10A that shows the lower limit of J_e increases with V_{sw} can still be seen, (3) unlike in Fig. 10A where the left side of the triangle is nearly parallel to the y-axis, the left side of the triangle now has a positive slope, suggesting that J_e increases with V_{sw}; and (4) for $V_{sw} \geq 600\,\mathrm{km\,s^{-1}}$, the higher limit of J_e saturates as in Fig. 10A. In general, Fig. 10C shows stronger dependence of J_e on V_{sw} than Fig. 10A or B. Thus, during the periods when TE is large, there is a large information transfer from V_{sw} to J_e, and we can indeed see that there is a stronger dependence of J_e on V_{sw}.

Fig. 10D, E, and F shows the data density maps of Fig. 10A, B, and C, respectively. Fig. 10E and F helps draw sharper contrasts between the distributions in Fig. 10B and C. There are proportionally more points with higher J_e and stronger dependency of J_e on V_{sw} in Fig. 10F than in Fig. 10E.

We investigate further the large spread of J_e at lower V_{sw} that can be seen in Fig. 10A. The $\log J_e$ versus V_{sw} data in Fig. 10A are binned in 0.3 counts $(\mathrm{cm^2\,s\,sr\,keV})^{-1} \times 30\,\mathrm{km\,s^{-1}}$ bins. From Fig. 9, we have calculated windowed TE for each point in the dataset. We then assign the windowed TE for each point in the $\log J_e$ versus V_{sw} bins. Fig. 11 shows the mean TE in each bin. Bins with fewer than 15 points are not displayed. The figure shows that for $V_{sw} < 500\,\mathrm{km\,s^{-1}}$, there is a large spread of J_e. However, these J_es are well ordered by TE. Large TE corresponds to large J_e, and conversely, small TE corresponds to small J_e. This suggests for $V_{sw} < 500\,\mathrm{km\,s^{-1}}$, when there is small information transfer from V_{sw} to J_e, J_e is small, and vice versa. We have also binned the data in Fig. 10B in a similar manner and obtained a similar result, albeit with higher noise due to lower statistics in the bins.

Balikhin et al. (2011) suggests that the triangle distribution can be attributed to n_{sw} and Kellerman and Shprits (2012) suggests the saturation of J_e in the triangle distribution can be attributed to n_{sw}. Our analysis in Section 4.4 certainly supports the argument that n_{sw} has a significant effect on J_e. We investigate further the effect of n_{sw} on the triangle distribution. We assign $n_{sw}(t)$ for each point in the $\log J_e(t+2\text{ days})$ versus $V_{sw}(t)$ scatter plot in Fig. 10A. These points are then binned using the same bin size as in Fig. 11. Fig. 12A shows the mean n_{sw} of each bin. As in Fig. 11, bins with fewer than 15 points are not displayed. The most prominent trend in Fig. 12A is a strong density gradient in the x-direction because n_{sw} anticorrelates with V_{sw}.

However, our analysis and Fig. 8A suggest that the maximum transfer of information from $n_{sw}(t)$ to $J_e(t+\tau)$ occurs at $\tau = 0$ day (<24 h). Hence, instead of assigning $n_{sw}(t)$ to each point in the $J_e(t+2\text{ days})$ versus $V_{sw}(t)$ plot, we assign $n_{sw}(t+2\text{ days})$ so that J_e is not time shifted with respect to n_{sw}. We repeat the same procedure done for Fig. 12A and the result is shown in Fig. 12B. Now, there are density gradients in both x- and y-directions. As in Fig. 12A, the density gradient in the x-direction is due to the anticorrelation of n_{sw} with V_{sw}. Fig. 12B clearly shows that for $V_{sw} < 500\,\mathrm{km\,s^{-1}}$, larger n_{sw} hence larger P_{dyn} can be associated with lower J_e and vice versa. This density gradient in the y-direction may be attributed to the magnetopause shadowing effect, which rapidly depletes radiation belt fluxes when solar wind pressure is increased, as discussed in Section 5.3.

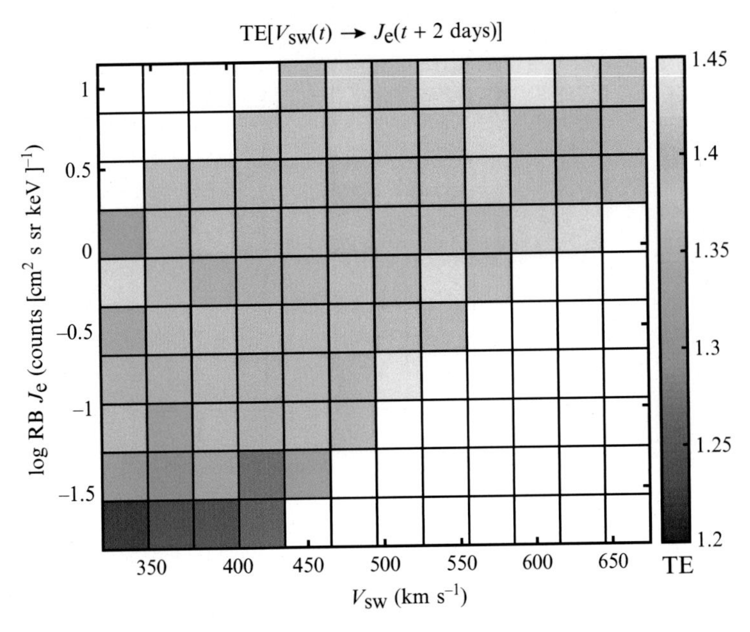

FIG. 11 Mean TE[$V_{sw}(t) \rightarrow J_e(t+2\ \text{days})$] of each bin in $J_e(t+2$ days) versus $V_{sw}(t)$ distribution shown in Fig. 10A. The bin size is 0.3 counts $(\text{cm}^2\ \text{s}\ \text{sr}\ \text{keV})^{-1} \times 30\ \text{km}\ \text{s}^{-1}$. Fig. 10A shows that at $V_{sw} < 500\ \text{km}\ \text{s}^1$, J_e has a large variance as previously shown, but it turns out that these points are well ordered by TE. Low J_e corresponds to low TE and vice versa.

Fig. 12B shows that large V_{sw} can be associated with large J_e and small n_{sw}. The latter can be mostly attributed to the anticorrelation of n_{sw} with V_{sw}. Fig. 12B also shows that large n_{sw} can decrease J_e, consistent with our analysis in Section 5.3, but it is not clear if n_{sw} alone can explain why small $V_{sw}(<500\ \text{km}\ \text{s}^{-1})$ can lead to high J_e and saturation of J_e for small n_{sw}. The high J_e and the saturation of J_e when $V_{sw} < 500\ \text{km}\ \text{s}^{-1}$ can probably be attributed to the strong solar wind-radiation belt coupling as suggested by the high TE in Fig. 11.

5.5 Improving Models With Information Theory

Tools based on information theory can be used to improve modeling. Several ideas are discussed as follows.

5.5.1 Selecting Input Parameters

Often the first step in developing a parametric forecasting model is to decide which parameters should be used as inputs to the model. Using TE and CMI, one can determine the ranking of each parameter based on information transfer from the input to the output parameters. For example, Table 1 shows the ranking of solar wind parameters for the solar wind-outer radiation belt system at daily resolution.

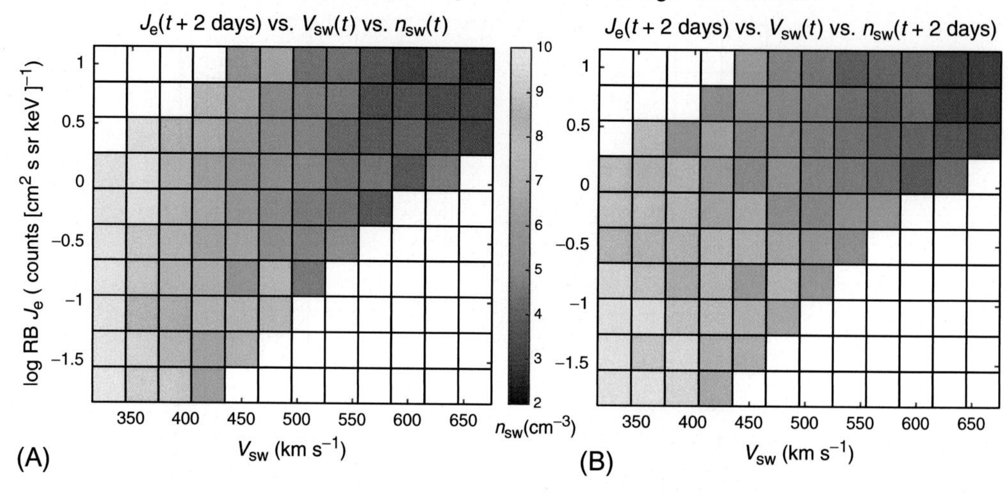

FIG. 12 Points in $J_e(t+2\text{ days})$ versus $V_{SW}(t)$ distribution in Fig. 10A are binned in 0.3 counts $(\text{cm}^2\,\text{s}\,\text{sr}\,\text{keV})^{-1} \times$ 30 km s^{-1} bins. Each point is assigned its $n_{SW}(t)$ and $n_{SW}(t+2\text{ days})$ values. The latter has no time shift with respect to J_e such that information transfer from n_{SW} to J_e maximizes. Panel (A) shows the mean $n_{SW}(t)$ while panel (B) shows the mean $n_{SW}(t+2\text{ days})$ of each bin. In panel (A), the density gradient is mainly in the x-direction due to the anticorrelation between n_{SW} and V_{SW}. However, in panel (B), there are density gradients in x- and y-directions. The latter can be attributed to P_{dyn} and magnetopause shadowing.

5.5.2 Detecting Nonstationarity in System Dynamics

As shown in Fig. 9, TE can be used to detect changes in the system dynamics, for example, nonstationarity of the system dynamics. Higher TE suggests that the solar wind-radiation belt system is more tightly coupled, and vice versa. Figs. 10B, C, and 11 show the differences in the dynamics for low and high TE cases in J_e versus V_{SW}. In this case, TE can help decompose the triangle distribution into something that can be more easily understood. This kind of information can help modelers. For example, modelers may want to create a model that varies the coupling function strength within the model, depending on the value of TE. Alternatively, two separate models may be developed: one for low TE and one for high TE.

5.5.3 Prediction Horizon

TE shows how much information is transferred from the input time series x to output time series y. When TE is significantly above the noise level, it suggests that there is hope for the model to predict parameter y. Conversely, when TE is at noise level, there is not much information transfer from x to y, and hence it can be expected that x would not be able to predict y accurately.

For example, Fig. 8A suggests that n_{SW} transfers the most information to J_e at $\tau_{max} = 0$ day, but the blue curve falls rapidly at $\tau > 2$ days, suggesting that little information is transferred from n_{SW} to J_e after a 2-day lag time. Hence, the prediction horizon for using n_{SW} to predict J_e is about 1 day. Here we use mean noise $+3\sigma$ as the noise threshold, but using different threshold criterion would yield a different prediction horizon. Likewise, Fig. 8B suggests that the prediction horizon for using V_{sw} to predict J_e is about 7–10 days. Table 1 lists the prediction horizon for the parameters considered in the present study.

6 SUMMARY

The present study applies information theoretical tools to investigate the solar wind drivers of the geosynchronous MeV electron fluxes. The following summarizes our results.

1. V_{sw}, n_{sw}, $|\mathbf{B}|$, P_{dyn}, $\sigma(B)$, E_{sw}, IMF B_x, IMF B_y, and IMF B_z (southward and northward) are causally related to J_e, but the amount of information transfer from each of this parameter to J_e differs. The ranking of these 10 parameters in terms of information transfer is given in Table 1.
2. V_{sw} is the most dominant driver and the transfer of information from the time series $V_{sw}(t)$ to $J_e(t+\tau)$ peaks at $\tau_{max} = 2$ days. V_{sw} transfers 2.7 times more information to J_e than n_{sw}.
3. Although the anticorrelation between n_{sw} and V_{sw} is perhaps well known, the large and persistent lag times for this anticorrelation are relatively unknown. $n_{sw}(t+\tau)$ anticorrelates with $V_{sw}(t)$ with $\tau_{max} = 14-16$ h, but the exact τ_{max} has time dependence. It is not clear what causes τ_{max} of 14–16 h. This may be due to the compression of the leading edge of the high-speed stream when it encounters the denser background solar wind. Analyses of solar wind driving of the magnetosphere involving V_{sw} and n_{sw} should take into account this anticorrelation that can persist even at large lag times, up to 3–5 days. For example, the information transfer from n_{sw} to J_e drops 30% after the effects of V_{sw} are removed.
4. $J_e(t+1 \text{ day})$ anticorrelates with $n_{sw}(t)$, but the 1-day lag and the anticorrelation are at least partly due to (1) $J_e(t+2 \text{ days})$ correlation with $V_{sw}(t)$, and (2) $V_{sw}(t)$ anticorrelation with $n_{sw}(t+1 \text{ day})$. Given V_{sw}, the transfer of information from $n_{sw}(t)$ to $J_e(t+\tau)$ peaks at $\tau_{max} = 0$ day (<24 h), suggesting the loss mechanism due to n_{sw} or P_{dyn} has to start operating in <24 h. However, the loss mechanism or mechanisms can operate for a long duration because at $\tau = 1$ day, there is still significant information transfer from n_{sw} to J_e.
5. The triangle distribution in J_e versus V_{sw} plot shows a large variability of J_e for $V_{sw} < 500\,\text{km s}^{-1}$. However, these points are well ordered by their TE values: high TE corresponds to high J_e and vice versa. The triangle distribution can be decomposed to low and high TE cases. In the low TE case, the distribution looks more like a rectangle for $V_{sw} < 500\,\text{km s}^{-1}$, suggesting that V_{sw} has little influence on J_e in these conditions. In the high TE case, the lower and upper limits of J_e increase with V_{sw} for $V_{sw} < 600\,\text{km s}^{-1}$, but for $V_{sw} \geq 600\,\text{km s}^{-1}$, the higher limit of J_e saturates.

TE and CMI can be used effectively to improve modeling by (1) selecting model input parameters, (2) detecting changes in the dynamics of the system, and (3) determining prediction horizon. Table 1 gives this information for the solar wind-outer radiation belt system.

The present study uses daily resolution LANL data. Reeves et al. (2013) investigated longer-term relationships between J_e and V_{sw}. They found that longer-term, 1-month to 1-year, averages show much stronger correlations than 1-day averages. They showed that this is not just because there is more "noise" superposed on a linear distribution. The distribution of log J_e around the baseline (the yearly mean) is very stable throughout the solar cycle. It would be interesting to apply our information theoretical tools to this normalized (rebaselined) data. It would also be interesting to apply our tools to higher-resolution data, for example, hourly resolution data, that are binned by D_{st}, P_{dyn}, and MLT.

ACKNOWLEDGMENTS

All the derived data products in this chapter are available upon request by email (simon.wing@jhuapl.edu). Simon Wing acknowledges support of NASA Grants NNX13AE12G, NNX15AJ01G, NNX16AR10G, and NNX16AQ87G. Jay R. Johnson acknowledges support from NASA Grants (NNH11AR07I, NNX14AM27G, NNH14AY20I, NNX16AC39G), NSF Grants (ATM0902730, AGS-1203299), and DOE contract DE-AC02-09CH11466.

References

Baker, D.N., Kanekal, S.G., 2008. Solar cycle changes, geomagnetic variations, and energetic particle properties in the inner magnetosphere. J. Atmos. Sol. Terr. Phys. 70, 195–206. https://doi.org/10.1016/j.jastp.2007.08.031.

Baker, D.N., McPherron, R.L., Cayton, T.E., Klebesadel, R.W., 1990. Linear prediction filter analysis of relativistic electron properties at 6.6 R_E. J. Geophys. Res. 95 (A9), 15133–15140. https://doi.org/10.1029/JA095iA09p15133.

Balikhin, M.A., Boynton, J.R., Billings, S.A., Gedalin, M., Ganushkina, N., Coca, D., Wei, H., 2010. Data based quest for solar wind-magnetosphere coupling function. Geophys. Res. Lett. 37, L24107. https://doi.org/10.1029/2010GL045733.

Balikhin, M.A., Boynton, R.J., Walker, S.N., Borovsky, J.E., Billings, S.A., Wei, H.L., 2011. Using the NARMAX approach to model the evolution of energetic electrons fluxes at geostationary orbit. Geophys. Res. Lett. 38, L18105. https://doi.org/10.1029/2011GL048980.

Belian, R.D., Gisler, G.R., Cayton, T., Christensen, R., 1992. High-Z energetic particles at geosynchronous orbit during the Great Solar Proton Event Series of October 1989. J. Geophys. Res. 97 (A11), 16897–16906. https://doi.org/10.1029/92JA01139.

Camporeale, E., 2015. Resonant and nonresonant whistlers-particle interaction in the radiation belts. Geophys. Res. Lett. 42, 3114–3121. https://doi.org/10.1002/2015GL063874.

Camporeale, E., Zimbardo, G., 2015. Wave-particle interactions with parallel whistler waves: nonlinear and time-dependent effects revealed by particle-in-cell simulations. Phys. Plasmas 22, 092104. https://doi.org/10.1063/1.4929853.

Claudepierre, S.G., Hudson, M.K., Lotko, W., Lyon, J.G., Denton, R.E., 2010. Solar wind driving of magnetospheric ULF waves: field line resonances driven by dynamic pressure fluctuations. J. Geophys. Res. 115, A11202. https://doi.org/10.1029/2010JA015399.

Darbellay, G.A., Vajda, I., 1999. Estimation of the information by an adaptive partitioning of the observations space. IEEE Trans. Inf. Theory 45, 1315–1321.

Elkington, S.R., Hudson, M.K., Chan, A.A., 1999. Acceleration of relativistic electrons via drift-resonant interaction with toroidal-mode Pc-5 ULF oscillations. Geophys. Res. Lett. 26, 3273.

Engebretson, M., Glassmeier, K.H., Stellmacher, M., Hughes, W.J., Lühr, H., 1998. The dependence of high-latitude PcS wave power on solar wind velocity and on the phase of high-speed solar wind streams. J. Geophys. Res. 103 (A11), 26271–26283. https://doi.org/10.1029/97JA03143.

Fairfield, D.H., Otto, A., Mukai, T., Kokubun, S., Lepping, R.P., Steinberg, J.T., Lazarus, A.J., Yamamoto, T., 2000. Geotail observations of the Kelvin-Helmholtz instability at the equatorial magnetotail boundary for parallel northward fields. J. Geophys. Res. 105 (A9), 21159–21173. https://doi.org/10.1029/1999JA000316.

Gosling, J.T., Hundhausen, A.J., Pizzo, V., Asbridge, J.R., 1972. Compressions and rarefactions in the solar wind: Vela 3. J. Geophys. Res. 77 (28), 5442–5454. https://doi.org/10.1029/JA077i028p05442.

Green, J.C., Kivelson, M.G., 2004. Relativistic electrons in the outer radiation belt: differentiating between acceleration mechanisms. J. Geophys. Res. 109, A03213. https://doi.org/10.1029/2003JA010153.

Horne, R.B., Thorne, R.M., Glauert, S.A., Albert, J.M., Meredith, N.P., Anderson, R.R., 2005. Timescale for radiation belt electron acceleration by whistler mode chorus waves. J. Geophys. Res. 110, A03225. https://doi.org/10.1029/2004JA010811.

Horne, R.B., Thorne, R.M., Glauert, S.A., Meredith, N.P., Pokhotelov, D., Santolík, O., 2007. Electron acceleration in the Van Allen radiation belts by fast magnetosonic waves. Geophys. Res. Lett. 34, L17107. https://doi.org/10.1029/2007GL030267.

Hundhausen, A.J., Bame, S.J., Asbridge, J.R., Sydoriak, S.J., 1970. Solar wind proton properties: Vela 3 observations from July 1965 to June 1967. J. Geophys. Res. 75 (25), 4643–4657. https://doi.org/10.1029/JA075i025p04643.

Johnson, J.R., Wing, S., 2005. A solar cycle dependence of nonlinearity in magnetospheric activity. J. Geophys. Res. 110, A04211. https://doi.org/10.1029/2004JA010638.

Johnson, J.R., Wing, S., 2014. External versus internal triggering of substorms: an information-theoretical approach. Geophys. Res. Lett. 41, 5748–5754. https://doi.org/10.1002/2014GL060928.

Johnson, J.R., Wing, S., Delamere, P.A., 2014. Kelvin Helmholtz instability in planetary magnetospheres. Space Sci. Rev. 184, 1–31. https://doi.org/10.1007/s11214-014-0085-z.

Kellerman, A.C., Shprits, Y.Y., 2012. On the influence of solar wind conditions on the outer-electron radiation belt. J. Geophys. Res. 117, A05217. https://doi.org/10.1029/2011JA017253.

Kennel, C.F., Petschek, H.E., 1966. Limit on stably trapped particle fluxes. J. Geophys. Res. 71 (1), 1–28. https://doi.org/10.1029/JZ071i001p00001.

Kepko, L., Spence, H.E., 2003. Observations of discrete, global magnetospheric oscillations directly driven by solar wind density variations. J. Geophys. Res. 108, 1257. https://doi.org/10.1029/2002JA009676.

Kissinger, J., McPherron, R.L., Hsu, T.S., Angelopoulos, V., 2011. Steady magnetospheric convection and stream interfaces: relationship over a solar cycle. J. Geophys. Res. 116, A00I19. https://doi.org/10.1029/2010JA015763.

Korotova, G.I., Sibeck, D.G., 1995. A case study of transient event motion in the magnetosphere and in the ionosphere. J. Geophys. Res. 100 (A1), 35–46. https://doi.org/10.1029/94JA02296.

Kreiger, A.S., Timothy, A.F., Roelof, E.C., 1973. A coronal hole and its identification as the source of a high velocity solar wind stream. Sol. Phys. 29, 505.

Li, W., 1990. Mutual information functions versus correlation functions. J. Stat. Phys. 60, 823–837.

Li, X., Temerin, M.A., 2001. The electron radiation belt. Space Sci. Rev. 95, 569–580.

Li, X., Temerin, M., Baker, D., Reeves, G., Larson, D., 2001. Quantitative prediction of radiation belt electrons at geostationary orbit based on solar wind measurements. Geophys. Res. Lett. 28 (9), 1887–1890.

Li, X., Baker, D.N., Temerin, M., Reeves, G., Friedel, R., Shen, C., 2005. Energetic electrons, 50 keV to 6 MeV, at geosynchronous orbit: their responses to solar wind variations. Space Weather 3, S04001. https://doi.org/10.1029/2004SW000105.

Lyatsky, W., Khazanov, G.V., 2008. Effect of solar wind density on relativistic electrons at geosynchronous orbit. Geophys. Res. Lett. 35, L03109. https://doi.org/10.1029/2007GL032524.

Mathie, R.A., Mann, I.R., 2000. A correlation between extended intervals of ULF wave power and storm-time geosynchronous relativistic electron flux enhancements. Geophys. Res. Lett. 27, 3621–3264. https://doi.org/10.1029/2000GL003822.

Mathie, R.A., Mann, I.R., 2001. On the solar wind control of Pc5 ULF pulsation power at mid-latitudes: implications for MeV electron acceleration in the outer radiation belt. J. Geophys. Res. 106 (A12), 29783–29796. https://doi.org/10.1029/2001JA000002.

Meier, M.M., Belian, R.D., Cayton, T.E., Christensen, R.A., Garcia, B., Grace, K.M., Ingraham, J.C., Laros, J.G., Reeves, G.D., 1996. The energy spectrometer for particles (ESP): instrument description and orbital performance. AIP Conf. Proc. 383, 203–210.

Miyoshi, Y., Kataoka, R., 2008. Flux enhancement of the outer radiation belt electrons after the arrival of stream interaction regions. J. Geophys. Res. 113, A03S09. https://doi.org/10.1029/2007JA012506.

Newell, P.T., Liou, K., Gjerloev, J.W., Sotirelis, T., Wing, S., Mitchell, E.J., 2016. Substorm probabilities are best predicted from solar wind speed. J. Atmos. Sol. Terr. Phys. 146, 28–37. https://doi.org/10.1016/j.jastp.2016.04.019.

Omura, Y., Furuya, N., Summers, D., 2007. Relativistic turning acceleration of resonant electrons by coherent whistler mode waves in a dipole magnetic field. J. Geophys. Res. 112, A06236. https://doi.org/10.1029/2006JA012243.

Onsager, T.G., Green, J.C., Reeves, G.D., Singer, H.J., 2007. Solar wind and magnetospheric conditions leading to the abrupt loss of outer radiation belt electrons. J. Geophys. Res. 112, A01202. https://doi.org/10.1029/2006JA011708.

Paulikas, G.A., Blake, J.B., 1979. Effects of the solar wind on magnetospheric dynamics: energetic electrons at the synchronous orbit. In: Quantitative Modeling of Magnetospheric Processes. Geophys. Monogr. Ser., vol. 21. AGU, Washington, D.C., pp. 180–202.

Prichard, D., Theiler, J., 1995. Generalized redundancies for time series analysis. Phys. D 84, 476–493. https://doi.org/10.1016/0167-2789(95)00041-2.

Prokopenko, M., Lizier, J.T., Price, D.C., 2013. On thermodynamic interpretation of transfer entropy. Entropy 15 (2), 524–543. https://doi.org/10.3390/e15020524.

Reeves, G.D., 2007. Radiation belt storm probes: a new mission for space weather forecasting. Space Weather 5. https://doi.org/10.1029/2007SW000341.

Reeves, G.D., Morley, S.K., Friedel, R.H.W., Henderson, M.G., Cayton, T.E., Cunningham, G., Blake, J.B., Christensen, R.A., Thomsen, D., 2011. On the relationship between relativistic electron flux and solar wind velocity: Paulikas and Blake revisited. J. Geophys. Res. 116, A02213. https://doi.org/10.1029/2010JA015735.

Reeves, G., Morley, S., Cunningham, G., 2013. Long-term variations in solar wind velocity and radiation belt electrons. J. Geophys. Res. Space Phys. 118, 1040–1048. https://doi.org/10.1002/jgra.50126.

Rigler, E.J., Wiltberger, M., Baker, D.N., 2007. Radiation belt electrons respond to multiple solar wind inputs. J. Geophys. Res. 112, A06208. https://doi.org/10.1029/2006JA012181.

Rostoker, G., Skone, S., Baker, D.N., 1998. On the origin of relativistic electrons in the magnetosphere associated with some geomagnetic storms. Geophys. Res. Lett. 25, 3701–3704. https://doi.org/10.1029/98GL02801.

Schreiber, T., 2000. Measuring information transfer. Phys. Rev. Lett. 85, 461–464. https://doi.org/10.1103/PhysRevLett.85.461.

Schwenn, R., Montgomery, M.D., Rosenbauer, H., Miggenrieder, H., Mühlhäuser, K.H., Bame, S.J., Feldman, W.C., Hansen, R.T., 1978. Direct observation of the latitudinal extent of a high-speed stream in the solar wind. J. Geophys. Res. 83 (A3), 1011–1017. https://doi.org/10.1029/JA083iA03p01011.

Shprits, Y.Y., Thorne, R.M., Friedel, R., Reeves, G.D., Fennell, J., Baker, D.N., Kanekal, S.G., 2006. Outward radial diffusion driven by losses at magnetopause. J. Geophys. Res. 111, A11214. https://doi.org/10.1029/2006JA011657.

Shprits, Y.Y., Subbotin, D.A., Meredith, N.P., Elkington, S.R., 2008. Review of modeling of losses and sources of relativistic electrons in the outer radiation belt II: local acceleration and loss. J. Atmos. Sol. Terr. Phys. 70, 1694–1713. https://doi.org/10.1016/j.jastp.2008.06.014.

Shprits, Y.Y., Subbotin, D., Ni, B., 2009. Evolution of electron fluxes in the outer radiation belt computed with the VERB code. J. Geophys. Res. 114, A11209. https://doi.org/10.1029/2008JA013784.

Simms, L.E., Pilipenko, V., Engebretson, M.J., Reeves, G.D., Smith, A.J., Clilverd, M., 2014. Prediction of relativistic electron flux at geostationary orbit following storms: multiple regression analysis. J. Geophys. Res. Space Phys. 119, 7297–7318. https://doi.org/10.1002/2014JA019955.

Simms, L.E., Engebretson, M.J., Smith, A.J., Clilverd, M., Pilipenko, V., Reeves, G.D., 2015. Analysis of the effectiveness of ground-based VLF wave observations for predicting or nowcasting relativistic electron flux at geostationary orbit. J. Geophys. Res. Space Phys. 120, 2052–2060. https://doi.org/10.1002/2014JA020337.

Sturges, H.A., 1926. The choice of a class interval. J. Am. Statist. Assoc. 21, 65–66. https://doi.org/10.1080/01621459.1926.10502161.

Summers, D., Thorne, R.M., Xiao, F., 1998. Relativistic theory of wave-particle resonant diffusion with application to electron acceleration in the magnetosphere. J. Geophys. Res. 103 (A9), 20487–20500. https://doi.org/10.1029/98JA01740.

Summers, D., Ni, B., Meredith, N.P., 2007. Timescales for radiation belt electron acceleration and loss due to resonant wave-particle interactions: 1 theory. J. Geophys. Res. 112, A04206. https://doi.org/10.1029/2006JA011801.

Tanskanen, E.I., 2009. A comprehensive high-throughput analysis of substorms observed by IMAGE magnetometer network: years 1993–2003 examined. J. Geophys. Res. 114, A05204. https://doi.org/10.1029/2008JA013682.

Thorne, R.M., 2010. Radiation belt dynamics: the importance of wave-particle interactions. Geophys. Res. Lett. 37, L22107. https://doi.org/10.1029/2010GL044990.

Tsonis, A.A., 2001. Probing the linearity and nonlinearity in the transitions of the atmospheric circulation. Nonlinear Process. Geophys. 8, 341–345.

Turner, D.L., Shprits, Y., Hartinger, M., Angelopoulos, V., 2012. Explaining sudden losses of outer radiation belt electrons during geomagnetic storms. Nat. Phys. 8, 208–212. https://doi.org/10.1038/nphys2185.

Ukhorskiy, A.Y., Sitnov, M.I., Sharma, A.S., Anderson, B.J., Ohtani, S., Lui, A.T.Y., 2004. Data-derived forecasting model for relativistic electron intensity at geosynchronous orbit. Geophys. Res. Lett. 31, L09806. https://doi.org/10.1029/2004GL019616.

Ukhorskiy, A.Y., Takahashi, K., Anderson, B.J., Korth, H., 2005. Impact of toroidal ULF waves on the outer radiation belt electrons. J. Geophys. Res. 110, A10202. https://doi.org/10.1029/2005JA011017.

Ukhorskiy, A.Y., Anderson, B.J., Brandt, P.C., Tsyganenko, N.A., 2006. Storm time evolution of the outer radiation belt: transport and losses. J. Geophys. Res. 111, A11S03. https://doi.org/10.1029/2006JA011690.

Vassiliadis, D., Fung, S.F., Klimas, A.J., 2005. Solar, interplanetary, and magnetospheric parameters for the radiation belt energetic electron flux. J. Geophys. Res. 110, A04201. https://doi.org/10.1029/2004JA010443.

Vennerstrøm, S., 1999. Dayside magnetic ULF power at high latitudes: a possible long-term proxy for the solar wind velocity? J. Geophys. Res. 104 (A5), 10145–10157. https://doi.org/10.1029/1999JA900015.

Wing, S., Johnson, J.R., Jen, J., Meng, C.I., Sibeck, D.G., Bechtold, K., Freeman, J., Costello, K., Balikhin, M., Takahashi, K., 2005a. Kp forecast models. J. Geophys. Res. 110, A04203. https://doi.org/10.1029/2004JA010500.

Wing, S., Johnson, J.R., Newell, P.T., Meng, C.I., 2005b. Dawn-dusk asymmetries, ion spectra, and sources in the northward interplanetary magnetic field plasma sheet. J. Geophys. Res. 110, A08205. https://doi.org/10.1029/2005JA011086.

Wing, S., Johnson, J.R., Camporeale, E., Reeves, G.D., 2016. Information theoretical approach to discovering solar wind drivers of the outer radiation belt. J. Geophys. Res. Space Physics 121, 9378–9399. https://doi.org/10.1002/2016JA022711.

Wing, S., Johnson, J., Vourlidas, A., 2018. Information theoretic approach to discovering causalities in the solar cycle. Ap. J. 854, 85. https://doi.org/10.3847/1538-4357/aaa8e7.

Wyner, A.D., 1978. A definition of conditional mutual information for arbitrary ensembles. Inf. Control 38, 51–59.

CHAPTER

7

Emergence of Dynamical Complexity in the Earth's Magnetosphere

Giuseppe Consolini

National Institute for Astrophysics, Institute for Space Astrophysics and Planetology, Rome, Italy

CHAPTER OUTLINE

1 INTRODUCTION

The Earth's magnetosphere is a highly structured dynamical region where the geomagnetic field is confined by the solar wind flowing in the interplanetary space. This region of the near-Earth space continuously interacts with the solar wind and the Earth's ionosphere by exchanging energy, mass, and momentum and does not passively respond to changes of the solar wind and interplanetary medium conditions. Evidence of the nonpassive dynamics of Earth's magnetosphere dates back to the late 1980s and early 1990s, when it was realized that the

Machine Learning Techniques for Space Weather
https://doi.org/10.1016/B978-0-12-811788-0.00007-X

dynamics of the Earth's magnetosphere is clearly nonlinear, intermittent, and scale-invariant, showing several features common to open, extended stochastic systems, usually in an out-of-equilibrium configuration near criticality (Consolini, 1997, 2002; Consolini et al., 2008; Uritsky and Pudovkin, 1998; Uritsky et al., 2002). These features are especially observed during magneospheric storms and substorms, which are the main manifestations of the interaction between the solar wind and the Earth's magnetosphere.

With the terms "magnetic storms" and "substorms," we indicate a wide variety of phenomena occurring in the Earth's magnetosphere and nearby environment that are detectable in terms of rapid variations in magnetograms recorded in ground-based geomagnetic observatories and/or auroral manifestations in the high-latitude ionosphere (for an extensive description, see Kamide and Chian (2007) and references therein). These two categories of phenomena are generally related to the intensification/activation of different magnetospheric currents flowing in different regions of the Earth's nearby space that affect the high-latitude and low-latitude geomagnetic field. For example, the geomagnetic substorms are generally associated with dipolarization effects occurring in the Earth's geomagnetic tail regions with the successive activation of field-aligned current (FAC) systems and the intensification of the auroral electrojet current systems in the high-latitude polar ionosphere. Conversely, the geomagnetic storms are related to the intensification of the equatorial ring current as a consequence of the increase of plasma convection and advection toward the inner magnetospheric regions due to the occurrence of magnetic reconnection at the Earth's magnetopause for long-standing periods (typically more than 3 h). Different geomagnetic indices are used to monitor these two different phenomena. In the case of geomagnetic storms, the most common index is the *Dst*-index (Sugiura and Poros, 1971), which is a 1-h proxy of the variation of the magnetospheric ring current, although it is also affected by other current systems, such as the magnetopause current and the cross-tail current (Gonzalez et al., 1994). A high-resolution version of the *Dst*-index is the *SYM-H*, which has a 1-min resolution. For geomagnetic substorms, the most commonly used geomagnetic indices are the set of auroral electrojet, *AE*, indices, a set of four indices (*AE*, *AL*, *AU*, and *AO*) with 1-min resolution, that provide information on the variations of the high-latitude eastward and westward electrojets (Davis and Sugiura, 1966).

One of the first papers clearly showing the nonlinear dynamics of the Earth's magnetosphere was from Tsurutani et al. (1990). In this work, comparing the power spectral density of time series from the interplanetary magnetic field (IMF) southward component B_S and auroral electrojet *AE*-index, Tsurutani et al. (1990) clearly showed that the Earth's magnetosphere did not respond linearly to interplanetary conditions' changes for time scales shorter than approximately 200 min. This evidence, along with other results on the highly irregular character of the magnetospheric dynamics during magnetic storms and substorms, suggested that the Earth's magnetosphere might behave as a chaotic system (Baker et al., 1990; Roberts et al., 1991; Vassiliadis et al., 1990). For example, the analyses of *AE*-index times series during disturbed conditions (see, e.g., Vassiliadis et al., 1990) found a typical correlation dimension in the range from 2.4 to 4.2, indicating that the global dynamics of the Earth's magnetosphere might be described in terms of a low-order system of equations.

The possible occurrence of chaos, and, in particular, of low-dimensional chaos in the magnetospheric response to interplanetary changes stimulated further studies in the solar-wind magnetosphere-ionosphere coupling. These further studies evidenced how there were some features of the magnetospheric dynamics that could not be simply explained in terms of low-dimensional dynamics, suggesting that a more appropriate scenario was that of nonlinear input-output dynamical systems, instead of autonomous attractor dynamics (see, e.g., Klimas et al., 1996). For example, Consolini et al. (1996) found the evidence of multifractal features in *AE*-index time series, similar to those of turbulent signals, which was interpreted as the fingerprint of multiscale and high-dimensional dynamics (see also, e.g., Consolini and De Michelis, 1998, 2011). In other words, the magnetospheric dynamics in response to solar wind changes at certain time scales (typically less than 200 min) was characterized by many degrees of freedom and/or by stochastic and turbulent fluctuations.

In the 1990s a different scenario for the nearly low-dimensional features of the Earth's magnetospheric dynamics was proposed by Chang (1992), and successively supported by analyzing *AE*-index time series and observations of auroral displays in the polar regions (Consolini, 1997, 2002; Lui et al., 2000; Uritsky and Pudovkin, 1998; Uritsky et al., 2002). In 1992, Chang stated that stochastic systems, driven out-of-equilibrium, might exhibit a low-dimensional dynamic when operating near "forced and/or self-organized criticality" (FSOC). In this case the emergence of a pseudo low-dimensional evolution was the counterpart of the limited number of relevant eigenoperators near the critical point.

The emergence of scale-invariant and near-criticality features in the Earth's magnetospheric dynamics was later confirmed by means of cellular automata, coupled-map, and magnetohydrodynamic (MHD) numerical simulations of the dynamics of the magnetotail central plasma sheet (CPS) region (see, e.g., Klimas et al., 2000).

In the first decades of the 21st century, several works revealed that the global magnetospheric dynamics are characterized by an inherent multiscale dynamic that could be treated in terms of nonequilibrium-phase transitions (Sharma et al., 2001; Sitnov et al., 2000, 2001). These features have been observed in coincidence with the occurrence of geomagnetic substorms, which are characterized by both external driving features and internal dynamics. The fractal nature of several geomagnetic indices (such as *AE*, *Dst*, and *SYM-H*) has been shown to undergo significative changes during magnetic storms and substorms (Uritsky and Pudovkin, 1998; Wanliss, 2005; Balasis et al., 2006; Wanliss and Dobias, 2007; Dobias and Wanliss, 2009), suggesting the occurrence of dynamical phase transitions. For example, Balasis et al. (2006) reported clear evidence of a transition in the persistency character of *Dst* fluctuations in correspondence with the transition from quiet to disturbed magnetospheric activity during intense geomagnetic storms.

Furthermore, the role that stochastic fluctuations play in the magnetospheric response to solar wind changes during magnetospheric substorms (as shown by AE-indices) was carefully investigated (Anh et al., 2008; Pulkkinen et al., 2006; Rypdal and Rypdal, 2010), showing how the complex feature of the Earth's magnetospheric dynamics requires the adoption of novel approaches, such as those based on fractional calculus. Indeed, several features of magnetic field fluctuations are compatible with those of a multifractional signal (Consolini et al., 2013).

Nowadays, the scenario that better describes the Earth's magnetospheric dynamics is that of a system out-of-equilibrium displaying *dynamical complexity*. In what follows, we will introduce and discuss several of the preceding concepts along with methods to unveil the complex nature of the Earth's magnetospheric dynamics by means of the analysis of time series of geomagnetic activity indices.

2 ON COMPLEXITY AND DYNAMICAL COMPLEXITY

The word *complexity* is generally associated with something that is difficult to disentangle in simple elements and/or to explain. In everyday life the word *complexity* is usually taken as a synonym of complicated and/or hard to separate. In science, the word *complexity* assumes a completely different meaning. First of all, it is not taken as a synonym of complicated, but it is associated with the emergence of features in a physical system that cannot be understood (predicted) using the traditional approach, based on *reductionism*. For example, in physics, a *complex system* is a macroscopic system that results from the sum of a multitude of elementary homogeneous/heterogeneous interacting parts and whose *phenomenological laws* describing the dynamics of the entire system cannot be directly described in terms of the elemental laws that regulate the evolution of the single entities forming the system (Holland, 2014). In a complex system the dynamics crucially depends on the details of the interaction among the different elements (Parisi, 1999), which generate multiple levels of collective structure and organization. According to Nicolis and Nicolis (2007), complexity is, indeed, a real, natural phenomenon that emerges from the laws of the nature in the systems consisting of many interacting subunits.

Thus, the term *complexity* refers to the *emergence* of nontrivial large-scale structures that dominate the system's overall dynamics and result from the multiple interactions (often nonlinear) among the elementary parts. However, *complexity* should not be confused with *collectivity*. In a complex system the total is more than the mere sum of the elements. For example, a fractal structure, such as the lung-bronchial tree, can be considered a complex structure in which the ratio between the surface and the volume of the structure is significantly larger than any simple geometrical structure.

Another crucial element of a complex system is that the large-scale structures and features can also affect the small-scale interactions, creating a feedback. This property of complex systems is called *Immergence*, and it is crucial for the understanding of global dynamics.

Consequently, in several situations, a complex system cannot be approached using the reductionist point of view, but requires development of new concepts, mental categories, and approaches that go beyond the traditional ones. For example, complex systems show features that have a universality character (Carlson and Doyle, 2002) and dynamics that might be regulated by the emergence of mesoscale coherent structures (Chang, 2015). This is, for instance, the case of magnetized space plasma systems, which, when forced outside of equilibrium, might evolve toward a dynamical state where multiscale coherent structures are formed (Chang, 1999, 2001, 2015). The interaction among these coherent structures that typically take the form of flux tubes is responsible for the large-scale dynamics of the entire plasma system. This behavior is an example of *dynamical complexity*. Indeed, *dynamical complexity* can be defined as the phenomenon, shown by a dynamical system consisting of a multitude of nonlinearly interacting multiscale coherent structures, which cannot be directly summarized from the elemental dynamical equations governing the small-scale dynamics (Chang et al., 2006). In such a case, dynamical complex systems can display a very large number of states or conformations (Frauenfelder, 2002) that regulate the dynamics in a way that it is difficult to handle. Thus, dynamical complexity is the tendency of a system to show a certain degree of spatiotemporal coherent features resulting from the competition of different basic spatial patterns/states.

Complexity in natural systems can manifest in several ways from power law statistics of event sizes and spatiotemporal patterns to scale-invariant fluctuations and self-organization near a spontaneous and/or forced critical state.

In the framework of space plasmas, as a consequence of the inherent nonlinearity of the dynamic equations, we may assist the formation of multiscale coherent structures, which can assume the shapes of convective forms, nonlinear solitary structures, pseudo-equilibrium configurations, and so forth. Examples of coherent structures are, for instance, field-aligned Alfvénic flux tubes (Chang, 1999; Bruno et al., 2001), which can be understood in terms of bundles of nonpropagating fluctuations arising from Alfvénic resonance sites (Chang, 1999). The diffusion, migration, and merging among these coherent magnetic flux tubes is responsible for the global dynamics of the entire plasma system. The interaction among these coherent structures is at the basis of the intermittent and coarse-grained dissipation that is observed in several space plasma regions from the solar-wind to the Earth's magnetospheric CPS (Bruno et al., 2001; Consolini et al., 2005).

3 COHERENCE AND INTERMITTENT FEATURES IN TIME SERIES GEOMAGNETIC INDICES

Here, some features of geomagnetic indices are presented and discussed in relation to the emergence of dynamical complexity in the Earth's magnetospheric response to solar wind changes. In particular, some features of the variability of geomagnetic indices (*AE* and *SYM-H*) are shown to provide the evidence of the occurrence of coherent temporal processes. To this end, a step-by-step analysis of time series of some geomagnetic indices will be presented. This procedure will allow the reader to become familiar with some methods of analysis and the interpretation of the results.

Before moving to the description of the information on the magnetospheric dynamics that can be inferred from the analysis of geomagnetic indices' time series, it is necessary to make a brief introduction on the meaning of the most relevant geomagnetic indices.

Geomagnetic indices have been introduced to provide information on the geomagnetic disturbance level at different latitudes. Each of these indices is representative of one of the main magnetospheric and/or ionospheric current systems and is generally computed on the basis of the variations of some component (typically the horizontal component) of the geomagnetic field measured on the ground in a set of geomagnetic observatories. The set of *AE*-indices (*AE*, *AU*, *AL*, and *AO*) (Davis and Sugiura, 1966) have been designed to be representative of the high-latitude electrojet current systems that flow in the polar ionosphere. These indices are computed from the geomagnetic field horizontal component measured on a longitudinal chain of 12 observatories, located near the auroral oval. Among the set of *AE*-indices, *AE* and *AL* indices are the most interesting, being proxies of the overall auroral electrojet current and of the westward electrojet current which is directly related to the magnetotail current systems by the field aligned current (FAC) systems. In particular, the *AE*-index is also used as a proxy of the energy deposition rate in the auroral regions (Ahn et al., 1983), and, thus, it is a very relevant quantity to investigate the overall magnetospheric dynamics with a special emphasis on the magnetotail dynamics and its impact on the high-latitude polar ionosphere during the geomagnetic substorms.

Another relevant geomagnetic index is *SYM-H*, which refers to low-latitude geomagnetic perturbations. In particular, the *SYM-H*-index can be considered as a high-resolution version (1 min) of the well-known *disturbance storm time* (*Dst*) index (Sugiura and Poros, 1971). This index was introduced to estimate the variations of the geomagnetic field due to magnetospheric currents flowing in the equatorial regions, with a special emphasis on the equatorial ring current. This index is computed by averaging the geomagnetic symmetric disturbance field as measured by the horizontal component H of the geomagnetic field at six ground-based geomagnetic observatories located near the equatorial region. A detailed description of the differences between *SYM-H* and *Dst* can be found in Wanliss and Showalter (2006).

One of the main features of the Earth's magnetospheric dynamics in response to solar wind and IMF changes is its intermittent and sporadic character, which principally manifests during geomagnetic substorms, and is very evident in the case of Auroral Electrojet indices *AE* and *AL*. Here, the intermittent and sporadic character of geomagnetic indices indicates that the Earth's magnetospheric dynamics is characterized by periods of relative stasis, interrupted/punctuated by periods of intense activity.

Fig. 1 shows the behavior of *SYM-H* and *AE*, along with the 1 min increments of the *AE* index ($\delta AE(t) = AE(t+1) - AE(t)$), in the time interval from December 1, 2015 to December 31, 2015. Data came from the NSSDC CDAweb and refers to OMNI merged 1-min data.

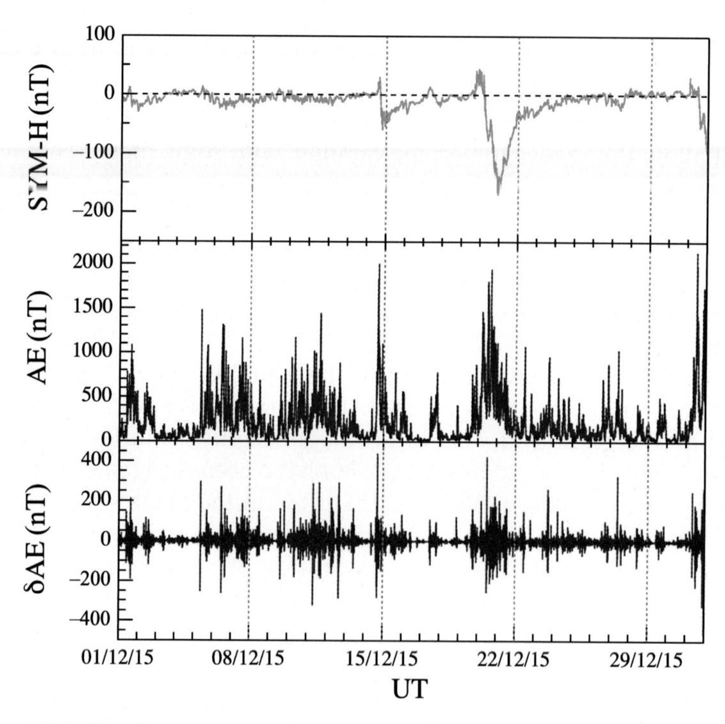

FIG. 1 A sample of *SYM-H* and *AE* indices for the time interval from December 1, 2015 to December 31, 2015. The *third panel* shows the 1-min increments of the *AE* index for the same period.

The intermittent and sporadic character of the Earth's magnetospheric dynamics is clearly evident from the high-latitude *AE*-index, which shows a sequence of activity bursts both during storm and nonstorm time. In particular, the time series of 1-min *AE*-index increments, δAE, shows intervals characterized by periods of stasis interrupted by events of crisis, corresponding to periods of enhanced geomagnetic activity. This is the fingerprint of the intermittent nature of geomagnetic activity, whose origin can be due to the loading-unloading process occurring in the Earth's magnetotail in response to solar wind energy, mass, and momentum transfer to the magnetosphere (Consolini and De Michelis, 1998, 2005). According to Akasofu (2004) a magnetospheric substorm comprises a wide variety of different processes and disturbances that spread throughout the entire Earth's magnetosphere, such as, auroral displays, which are the manifestation of the enhancement of plasma transfer toward the high-latitude polar regions. This enhancement is generally due to the increase of plasma convection and the impulsive energy relaxation events occurring in the CPS tail region.

One peculiar feature of *AE*-index activity bursts is that these take the form of coherent energy relaxation events, where the term "coherence" refers to the frequency time structure of these burst events.

This feature can be revealed by performing the wavelet transform (WT) decomposition and the *local intermittency measure* (*LIM*) analysis of the *AE*-index time series. Wavelet decomposition is, indeed, a powerful tool to investigate nonstationary time series and is capable of revealing the local features of a signal with a detail matching the scale (Kumar and Foufoula-Georgiou, 1997). Given a time series $x(t)$ and a *mother wavelet* $\psi(t)$, wavelet decomposition consists of decomposing/projecting the original time series on a continuous (or discrete) set of *daughter wavelets* $\psi((t-t_0)/a)$, obtained by dilating by a scale factor a ($a > 0$) and translating by a time shift t_0 the mother wavelet. The WT $W(a, t_0)$ is then obtained in terms of a convolution between the actual time series $x(t)$ and the daughter wavelets, that is

$$W(a, t_0) = \frac{1}{\sqrt{(a)}} \int_{-\infty}^{+\infty} \psi\left(\frac{t - t_0}{a}\right) x(t)dt \tag{1}$$

where $W(a, t_0)$ represents the WT coefficients at the scale a and point t_0 (Mallat, 1998). The knowledge of the WT coefficients allows the computation of a dynamical spectrum, that is, a scale-time distribution of the energy density, named as *scalogram* and defined as the $\mid W(a, t_0) \mid^2$. Furthermore, following Farge et al. (1990) it is possible to introduce a quantity, defined as *LIM*, capable of identifying within a time series the events that are responsible for the intermittent nature of the signal. The *LIM* is defined from the scalogram as follows,

$$LIM_{a,t_0} = \frac{\mid W(a, t_0) \mid^2}{\langle \mid W(a, t_0) \mid^2 \rangle_T} \tag{2}$$

where $\langle \mid W(a, t_0) \mid^2 \rangle_T$ is the average energy density at the scale a. Thus, those points/times, where $LIM_{a,t_0} > 1$, identify intervals of the signal $x(t)$ that have more power in respect to the average value, allowing location of those events in time and scale, which are related to the intermittent character of the geomagnetic activity.

Fig. 2 shows an example of the WT and *LIM* analysis of the *AE*-index during the famous St. Patrick's day 2015 geomagnetic storm (March 17, 2015). Here, we have used the well-known Morlet wavelet to perform the WT. The scalogram (mid-panel) clearly shows the

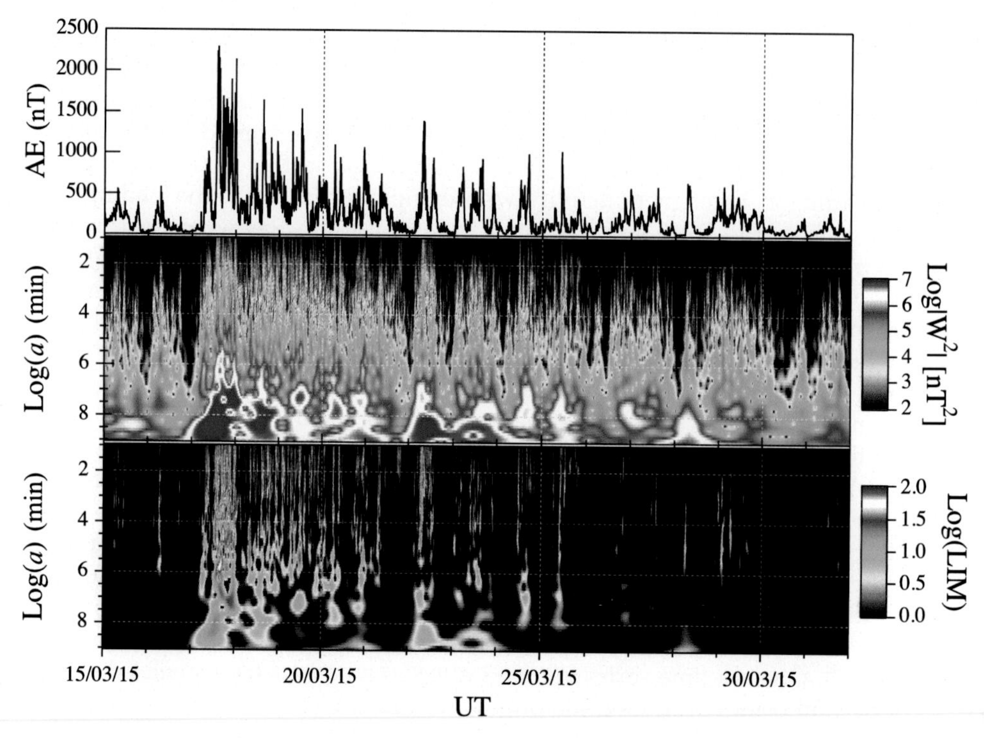

FIG. 2 A sample of wavelet and LIM analysis applied to the *AE*-index during the St. Patrick's day 2015 geomagnetic storm. *Upper panel* reports the *AE*-index time series, *mid panel* shows the wavelet scalogram, and the *lower panel* refers to the LIM analysis.

nonstationary nature of the *AE*-index and the large increase of the activity during disturbed periods. Furthermore, the *LIM* analysis (lower panel) reveals the occurrence of coherent scale-time structures associated with the activity bursts, showing how the intermittent nature of magnetospheric dynamics in response to solar wind changes is due to coherent energy relaxation events.

Another evidence of the intermittent character of the Earth's magnetospheric dynamics is given by the non-Gaussian character of the probability distribution functions (PDFs) of the small-scale increments of the geomagnetic indices, that is, the effect of sporadic and high amplitude fluctuations/increments occurring at the smallest scales. This feature is sometimes read as the evidence of the occurrence of turbulence in the internal magnetospheric dynamics in response to solar wind changes. Fig. 3 displays the PDFs of small-scale increments, $\delta x(\tau) = x(t+\tau) - x(t)$, of some geomagnetic indices (*AE*, *AL*, and *SYM-H*) in the period from January 1, 2015 to May 31, 2016 without distinguishing between quiet and disturbed periods.

The PDFs exhibit a clear departure from Gaussianity, indicating how the magnetospheric current systems are characterized by fast and sporadic intensity enhancements. We are reminded, indeed, that all the considered geomagnetic indices are proxies of the main geomagnetic current systems, the high-latitude auroral electrojets (*AE* and *AL*), and the

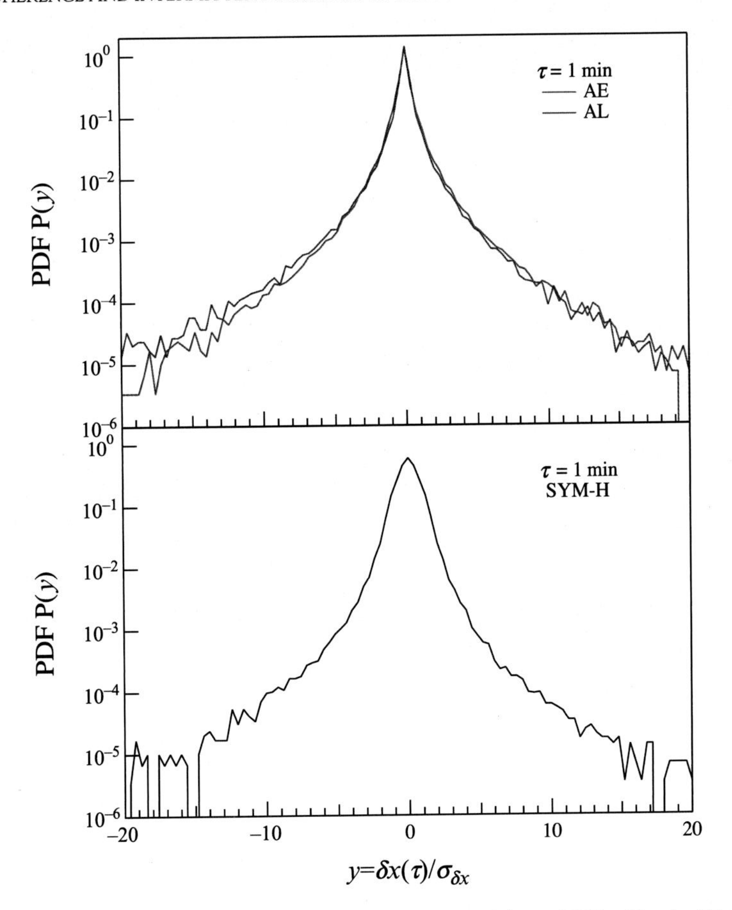

FIG. 3 The PDFs of the increments of geomagnetic indices. *Upper panel* shows PDFs of 1-min *AE* and *AL* increments, while the *lower panel* refers to 10-min increments of *SYM-H* index. Increments have been normalized to unit standard deviation ($\sigma_{\delta x}$). Note that, to evaluate the PDF of *SYM-H* increments at the time scale of 1 min, a random white noise in the interval (−0.5, 0.5) has been added to *SYM-H* time series to reduce the 1 nT discretization effects.

equatorial ring current (*SYM-H*). Although in the present analysis, we have not made any distinction between quiet and disturbed periods, the observed departure from Gaussianity seems to be a general feature of both periods (Consolini and De Michelis, 1998). Clearly, during disturbed periods, the departure from Gaussianity is expected to be more pronounced. In 1998 an analysis from Consolini and De Michelis (1998) has clearly shown that in the case of the *AE*-index, a strong departure from Gaussianity is found for time scales, τ, in the interval from 1 to 120 min, both for quiet and disturbed periods. However, the differences observed between quiet and disturbed periods suggest that the magnetospheric response is characterized by an intermittent character that depends on the geomagnetic activity level. This is, again, an evidence of the dynamical complexity of the Earth's magnetospheric response to changes of solar wind and IMF conditions.

4 SCALE-INVARIANCE AND SELF-SIMILARITY IN GEOMAGNETIC INDICES

Another interesting aspect of the Earth's magnetospheric response that can be inferred from the study of geomagnetic time series is the occurrence of *scale-invariance* and *self-similarity* over a wide interval of time scales. In the early years of the 1990s, Takalo et al. (1993) and Takalo and Timonen (1994) showed that the *AE*-index fluctuations have scaling features up to a characteristic time scale of about 113 (± 9) min. It was found that the *AE*-index is *self-affine*, with a scaling exponent $H \sim 0.5$ and the time scales in which these scaling features were observed are well in agreement with those where the *AE*-index spectral features do not follow the spectral features of the IMF (Tsurutani et al., 1990).

The study of scaling features in geomagnetic indices can be generally performed by using the structure function analysis, which is, for instance, widely used to characterize the scaling properties of velocity and magnetic field increments and/or fluctuations in fluid and MHD turbulence (Frisch, 1995). In particular, given a time series or a signal $x(t)$ defined over a time interval Ω, the qth-order generalized structure function is given by the expression:

$$S_q(\tau) = \langle | \, x(t+\tau) - x(t) \, |^q \rangle_\Omega \tag{3}$$

Here, the term *generalized* refers to the fact that Eq. (3) contains the absolute value of the increments and computation is done in respect to time. Indeed, in fluid turbulence, the structure functions of the velocity field are computed in respect to spatial differences along the flow direction and do not contain the absolute value (Frisch, 1995). The second-order generalized structure function, $S_2(\tau)$, is sometimes called *variogram* (Hergarten, 2002) and represents the mean-squared displacement after a time interval τ. Furthermore, this quantity is related to the power spectral density PSD by the well-known Wiener-Khinchin theorem (see, e.g., de Groot and Mazur, 1984).

In the case of scale-invariant signals the generalized structure functions $S_q(\tau)$ are expected to scale with τ according to a power law,

$$S_q(\tau) \sim \tau^{\zeta(q)} \tag{4}$$

where $\zeta(q)$ are the scaling exponents.

The scaling exponent $\zeta(1)$ is called *Hurst (or Hausdorff) exponent H* and provides a measure of the *self-affine* nature of the signal (Bunde and Havlin, 1994; Hergarten, 2002). A *self-affine* signal, $y = f(x)$, is an object that is scale invariant under the transformation, $x \rightarrow bx$ and $y \rightarrow ay$ so that,

$$f(bx) = af(x) \equiv b^H f(x) \tag{5}$$

Furthermore, the H exponent provides a measure of the *persistence character* of the signal fluctuations/increments and is also a measure of the non-Markovian features in correlated random walks. In detail, persistence in a signal refers to memory effects in the sign of the fluctuations. In particular, $H > 0.5$ ($H < 0.5$) indicates a *persistent* (*antipersistent*) signal, where the persistent character refers to a biased probability of observing the same sign or not ($++$ or $--$ for persistent signals, $+-$ or $-+$ for antipersistent signals) in two successive fluctuations/increments. $H = 0.5$ is generally associated with purely stochastic Brownian signals, for which the sign of two successive fluctuations/increments is equiprobable. The investigation of the dependence of the q-order generalized structure functions as a function of the scale τ is equivalent to investigate the dependence of the moments of the PDFs of the signal increments at different time scales.

The scaling features in a signal can be simple or anomalous, and this different character is associated with the behavior of the scaling exponents $\zeta(q)$ as a function of q. In particular, for a signal characterized by a linear trend of the scaling exponents with the moment order, $\zeta(q) \sim cq$, the scaling features are called simple and the knowledge of one scaling exponent is sufficient to characterize the complex feature and the scale invariance of the signal. Conversely, when the relationship between the scaling exponents $\zeta(q)$ and the moment order q is nonlinear (typically convex), the scaling features are called *anomalous*. In this case, the signal exhibits a higher degree of complexity, being no longer sufficient with the knowledge of a simple scaling exponent to characterize the complexity degree of the signal. This is analogous to the difference between *fractal* and *multifractal* features (Mandelbrot, 1989; Frisch, 1995). We remind readers that a multifractal measure is a more complex object than a simple fractal one. Indeed, a multifractal measure can be thought of as the superposition of an infinite number of intertwined subsets characterized by different scaling features of the measure over fractal subsets of different dimensions (Mandelbrot, 1989). The concept of multifractal measure is widely applied in the field of turbulence with reference to the heterogeneity of the scaling features of the dissipation field (Frisch, 1995).

Fig. 4 shows the behavior of the first-order generalized structure function, $S_1(\tau)$, for *AE*-index data in the interval from January 2015 to May 2016. As reported on some works (see, e.g., Takalo et al., 1993; Takalo and Timonen, 1994) the $S_1(\tau)$ follows a power law scaling for $\tau \in (4, 120)$ min with a scaling exponent $\zeta(1) = [0.512 \pm 0.002]$. This value of the scaling exponent

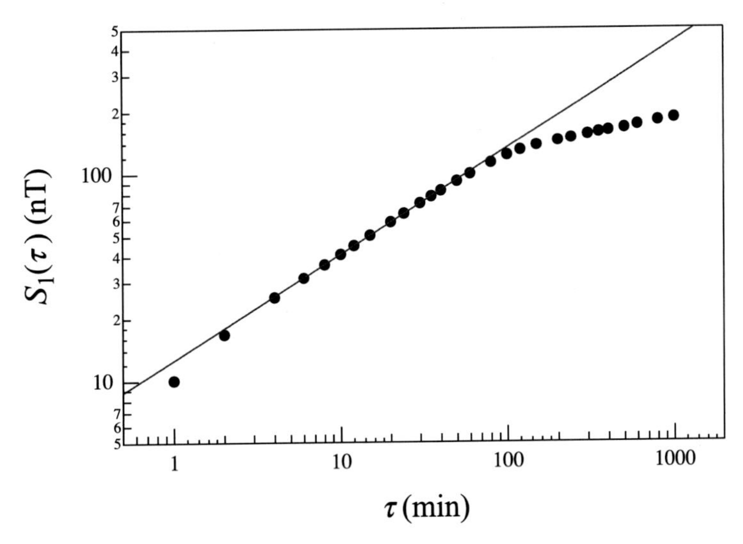

FIG. 4 The first-order generalized structure function for *AE*-index data in the time interval from January 1, 2015 to May 31, 2016. The *blue line* is a power-law fit with a scaling exponent $\zeta(1) = [0.512 \pm 0.002]$.

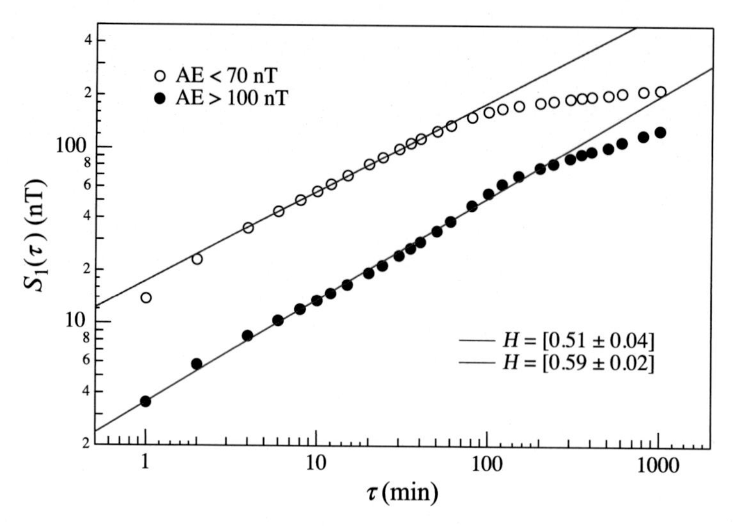

FIG. 5 The conditioned first-order generalized structure functions for $AE < 70$ nT and $AE > 100$ nT in the time interval from January 1, 2015 to May 31, 2016. *Blue* and *red lines* are power-law fits with a scaling exponent $\zeta(1) = H = [0.51 \pm 0.04]$ and $\zeta(1) = H = [0.59 \pm 0.02]$, respectively.

suggests that the *AE*-index is characterized by a small degree of persistence, being $H \equiv \zeta(1) > 0.5$. However, the scaling features of geomagnetic indices show a dependence on the geomagnetic activity level (Consolini and De Michelis, 1998; Uritsky and Pudovkin, 1998; Wanliss, 2005). Indeed, for quiet intervals, the persistent character of the *AE*-index decreases, as reported in Fig. 5, where the conditioned first-order generalized structure functions for $AE < 70$ nT and $AE > 100$ nT are shown, respectively. In particular, during disturbed periods, an increase of the persistent character of fluctuations/increments is observed. This change in the scaling features from quiet to disturbed periods has been widely observed in the literature and it is considered evidence for the occurrence of a dynamical phase transition in the magnetospheric dynamics, or better, as the increase of the time correlation/memory effect of the geomagnetic indices' increments during disturbed magnetospheric conditions (Balasis et al., 2006; Uritsky and Pudovkin, 1998; Wanliss, 2005; Wanliss et al., 2005; Wanliss and Dobias, 2007). We will return to this point in Section 6.

The emergence of the scaling feature and scale-invariance in the geomagnetic indices' time series is the evidence of an inherent multiscale dynamic, and that over a certain interval of time scales, it is not possible to identify a characteristic time scale for the magnetospheric response to solar wind and IMF changes. This issue is extremely relevant when we would like to develop systems/methods to forecast the geomagnetic activity.

As shown, the self-similar nature of geomagnetic indices is generally characterized by continuous scale invariance. However, there is also evidence of the occurrence of a *discrete scale invariance* (DSI) (Sornette, 1998), which manifests in log-periodic corrections to scaling. For example, Balasis et al. (2011) showed that the *Dst*-index time series (the *Dst*-index is a 1-h geomagnetic index that is a proxy of the geomagnetic ring current intensity), exhibit DSI, discussing the possible implications of this feature on space weather forecasting efforts.

Although geomagnetic indices show scale-invariance, the nature of their fluctuations/increments is more complex. They are characterized by anomalous scaling features and a multifractal nature. Evidence of this multifractal feature of the magnetospheric dynamics has been found by analyzing several geomagnetic indices using different techniques, such as the canonical multifractal analysis (Consolini et al., 1996), the rank ordered multifractal analysis (ROMA) (Consolini and De Michelis, 2011), and other methods (Wanliss et al., 2005). For example, Consolini et al. (1996) showed that to correctly describe the *AE*-index scaling features and their intermittent nature, it was necessary to introduce a hierarchy of dimensions (*the Renyi dimensions* D_q). As a consequence of the multifractal nature of geomagnetic index fluctuations/increments, it is reasonable to expect that the scaling exponents $\zeta(q)$ of the generalized structure functions display an anomalous scaling with the moment order.

One way to study the occurrence of anomalous scaling features in a signal is to investigate the relative scaling of the qth-order structure function from the pth-order one (see Frisch, 1995 and references therein), that is,

$$S_q(\tau) \sim \left[S_p(\tau)\right]^{\xi(p,q)} \tag{6}$$

For monofractal signals, or in the case of simple scaling features, $\xi(p,q) = q/p$, while in the case of anomalous scaling features, this returns to a convex function of q for fixed p.

Fig. 6 reports an example of the relative scaling of $S_q(\tau)$ versus $S_2(\tau)$ in the case of the *AE*-index time series from January 2015 to May 2016. The relative scaling allows evaluation of the $\xi(q,2)$ exponents as a function of the moment order q. Fig. 7 shows the trend of the relative scaling exponents $\xi(q,2)$ as a function of the moment order for the structure functions reported in Fig. 6. The nonlinear trend of $\xi(q,2)$ is the signature of the occurrence of anomalous

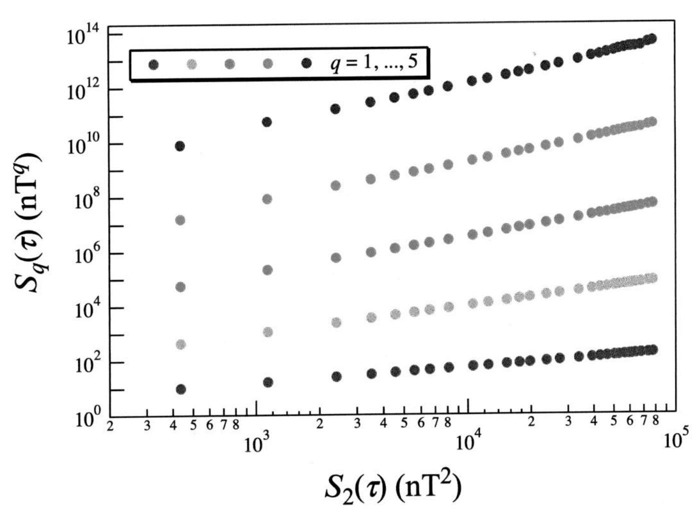

FIG. 6 The relative scaling of $S_q(\tau)$ versus $S_2(\tau)$ for $q \in [1,5]$ in the case of *AE*-index time series from January 1, 2015 to May 31, 2016.

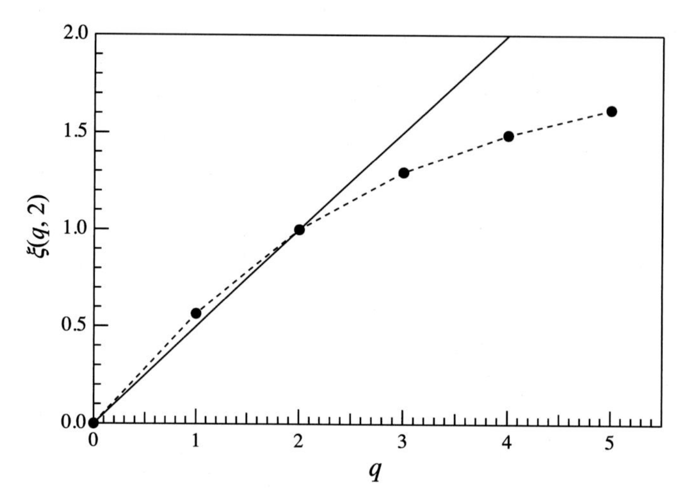

FIG. 7 The relative scaling exponents $\xi(q, 2)$ as a function of the moment order q for the structure functions reported in Fig. 7. The *continuous line* refers to the linear trend expected for simple scaling features, $\xi(q, 2) = q/2$.

scaling features for *AE*-index fluctuations/increments. As in the case of fluid turbulence (Frisch, 1995), this feature can be read as complementary evidence of the multifractal character of *AE*-index data.

The existence of anomalous scaling features in geomagnetic indices has been, however, conjectured and criticized in some papers (Chapman et al., 2005; Hnat et al., 2002, 2005; Watkins et al., 2005). The main issue concerns the fact that the multifractal and anomalous scaling features of the *AE*-index may be due to statistically poorly resolved behavior of the largest fluctuations. A possible alternative scenario for the high variability and intermittent nature of geomagnetic indices could be that of a fractional-Lèvy flight on time scales less than ~100 min (Watkins et al., 2005). However, Rypdal and Rypdal (2010) evidenced how a smoothly truncated Lèvy flight is not able to explain the complexity of geomagnetic indices (*AE*-index in particular) and how a multifractal description is more appropriate. Successively, Consolini and De Michelis (2011), using the ROMA method proposed by Chang and Wu (2008), provided evidence of the multifractal character of the *AE*-index over a wide range of time scales.

The anomalous scaling observed in the case of the *AE*-index is a general feature of the main geomagnetic indices (Wanliss et al., 2005) and can be observed also in the change of the shape of the PDFs of indices' increments as a function of the time scale (Consolini and De Michelis, 1998, 2011). Indeed, the shape of the PDFs $p(\delta x, \tau)$ of indices' increments, $\delta x(\tau) = x(t + \tau) - x(t)$, evolves with the time scale τ from a very pronounced leptokurtotic shape at short-time scales (typically of the order of few minutes) toward a quasi-exponential/Gaussian one at long-time scales (i.e., at a time scale much larger than 1000 min). The absence of a universal, nonscale dependent shape of the PDFs is more evidence of the anomalous scaling nature of geomagnetic indices' fluctuations/increments. Indeed, in the absence of anomalous scaling properties, it should be possible to observe PDFs collapsing by applying the following scale transformation,

$$\begin{cases} \delta x(\tau) \longrightarrow & \frac{\delta x(\tau)}{\sigma_{\delta x}} \\ p(\delta x, \tau) \longrightarrow & \sigma_{\delta x} p(\delta x, \tau) \end{cases} \tag{7}$$

where $\sigma_{\delta x}$ is the standard deviation of the increments $\delta x(\tau)$ at the time scale τ. This scaling procedure to get PDFs' data collapsing and to explore the existence of a scale-invariant shape for the PDFs (a *master curve*) is quite common in studying scale-invariant features of both stochastic motions and fluid turbulence (Frisch, 1995; Stauffer and Stanley, 1995; Warhaft, 2002). To clarify the link between anomalous scaling features and the lack of collapsing of the PDFs $p(\delta x, \tau)$, it is useful to note that the qth-order structure function, $S_q(\tau)$, is essentially the behavior of the qth-order moment of the PDFs $p(\delta x, \tau)$ with the scale τ. Thus, because a way to study the scaling features of fluctuations/increments of larger and larger amplitude is to increase the moment order q, the occurrence of anomalous scaling features is analogous to the fact that the scaling index acquires a dependence on the amplitude of the fluctuation/increment. This consideration is at the basis of the novel method ROMA (Chang and Wu, 2008; Chang, 2015) to estimate the occurrence of anomalous scaling and multifractal features in signals/objects.

Fig. 8 presents the evolution of the shape PDFs of the increments *AE* and *AL* indices at different timescales for the time interval from January 1, 2015 to May 31, 2016. PDFs have been rescaled according to Eq. (7). PDFs do not collapse into a single master shape, but evolve with the time scale τ. This behavior, along with the anomalous scaling features, is

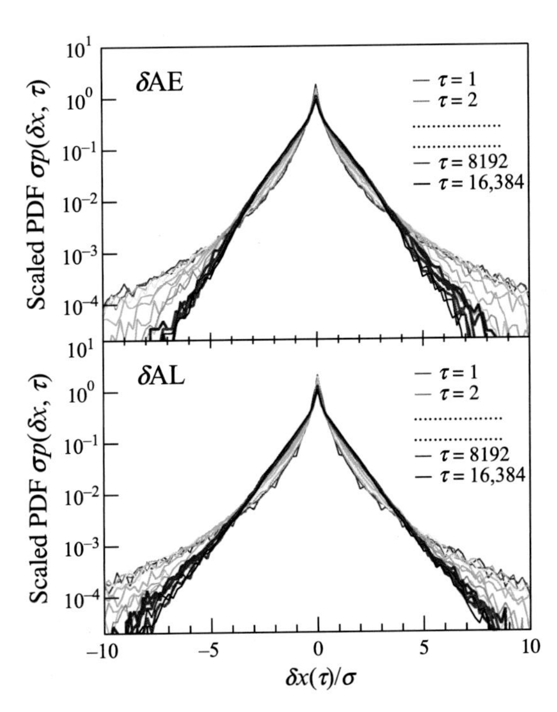

FIG. 8 Evolution of the shape of PDFs of *AE* and *AL* increments with the time scale $\tau \in [1, 16{,}384]$ min.

common to turbulent media, such as, for instance, fluid and space plasmas at MHD scales (Frisch, 1995; Sorriso-Valvo et al., 1999), and it is a manifestation of the multifractal nature and anomalous scaling features of the signal.

The occurrence of anomalous scaling features and non-Gaussian PDFs of geomagnetic indices' increments suggests that intermittent turbulence might be the physical process responsible for the Earth's magnetospheric dynamics at short-time scales. This aspect still needs to be investigated, together with the role that magnetotail turbulent processes play in the emergence of dynamical complexity in the Earth's magnetospheric dynamics in response to solar wind and IMF changes.

5 NEAR-CRITICALITY DYNAMICS

In Section 3 we have discussed the emergence of coherence in the high-latitude energy releases during geomagnetic substorms as evidenced from the coherent activity bursts in *AE*-indices (*AE* and *AL*). However, one of the central issues to understanding the origin of the magnetospheric dynamical complexity is if there is a characteristic size and/or duration of these coherent energy relaxation events.

In 1997, by investigating the statistics of the *AE*-index burst sizes during disturbed periods, Consolini (1997) evidenced how the PDF of the size of these bursts follows a power-law over many orders of magnitude. This observation, along with other results on the dynamical changes of *AE*-index scaling features during geomagnetic substorms (Uritsky and Pudovkin, 1998) and numerical simulations (Chapman et al., 1998), opened a novel scenario for the Earth's magnetospheric dynamics, suggesting that the coupled solar wind-magnetospheric system might exhibit *self-organized criticality* (SOC) (Bak et al., 1987, 1988) that can be described by avalanche models. Indeed, the presence of a power-law distribution function of *AE*-index activity burst sizes, of self-similarity and $1/f^{\alpha}$ spectral features in the time series of the geomagnetic indices indicates that the Earth's magnetospheric dynamics is scale-free and that they can be assimilated to that of an avalanching system (Chapman and Watkins, 2001; Consolini and Chang, 2001; Consolini, 2002; Klimas et al., 2000; Lui et al., 2000; Lui, 2002a, 2004).

Fig. 9 shows an example of the PDFs of the *AE*-index activity burst size and duration for different activity thresholds (AE_{thr}) in the range (70, 200) nT for the *AE*-index dataset relative to the interval January 2015 to May 2016. The definition of *AE* burst size s and duration T is done following the procedure described in Consolini (2002), that is,

$$\begin{cases} s = \int_{\Delta} (AE(t) - AE_{thr})\, dt \\ T = \int_{\Delta} \theta\left(\frac{AE(t)}{AE_{thr}} - 1\right) dt \end{cases} \tag{8}$$

where $\theta(x)$ is the Heaviside step function and the integration is done over each interval Δ where $AE(t) \geq AE_{thr}$. Both the PDFs show a quasi power law behavior over a wide range of scales (up to four decades for the size s and about three decades for the duration T). Furthermore, the scaling exponents of the two PDFs (α and β) are essentially independent on the threshold value.

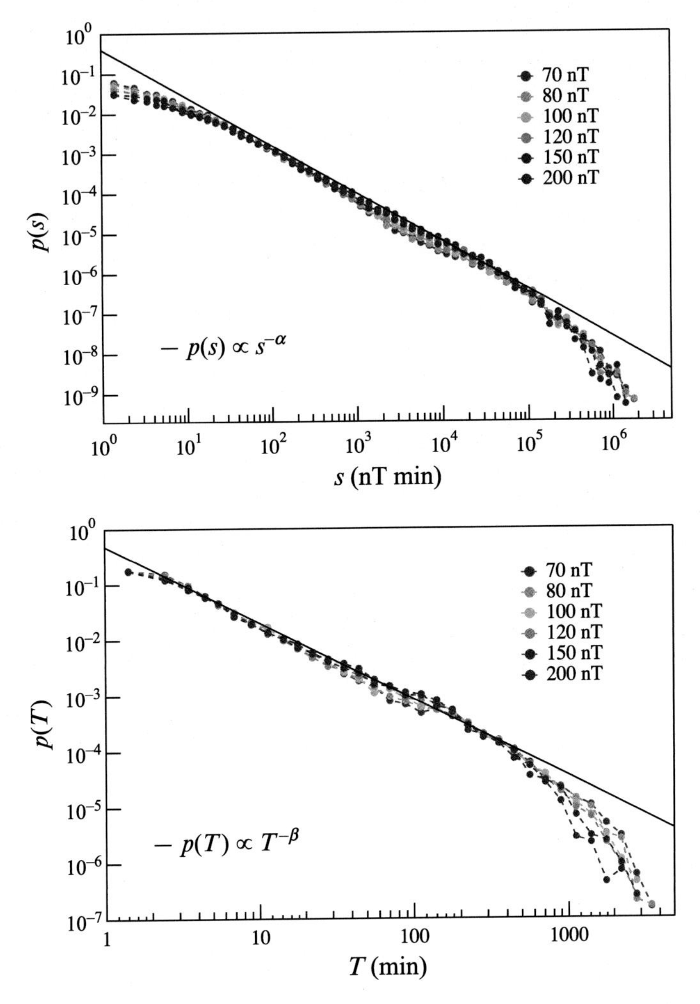

FIG. 9 A sample of the PDFs of the *AE*-index burst size s (*upper panel*) and duration T (*lower panel*) for different activity thresholds AE_{thr}. The *solid lines* are power-laws characterized by power indices $\alpha = 1.2$ and $\beta = 1.38$, respectively.

A possible origin of the observed scale invariance in the PDFs of *AE*-index activity bursts has been associated with the emergence of a complex topology of magnetic field and plasma coherent structures in the magnetotail CPS (Chang, 1992, 1999) that could evolve near a critical state/point under the continuous driving from the solar wind. It has been suggested that these coherent structures were due to the continuous forcing of the solar wind so that their dynamics (formation, migration, and mixing) could generate a complex topology that could accumulate a certain amount of *configurational free energy* to be released during geomagnetic substorms as nonideal relaxation events and/or plasma energization processes (Chang, 1999; Consolini and Chang, 2001). Because in the case of the Earth's magnetospheric dynamics, different from the SOC, it was not possible to exclude a certain degree of tuning of the near criticality configuration of the Earth's magnetospheric CPS, Chang (1992, 1999) coined the term FSOC.

The hypothesis that the observed scale-invariant features of the magnetospheric dynamics, as shown by the *AE*-index, are of an internal origin, has been opposed in some papers (see, e.g., Freeman et al., 2000). However, comparing *AE*-index dynamics with the features of solar wind input vB_{yz}, Uritsky et al. (2001) found that the scaling properties of the *AE*-index at scales less than 3.5 h were independent on the solar wind disturbances, supporting, thus, that at these time scales, the origin of scale-invariance in the Earth's magnetospheric dynamics had an internal origin. Recently, Alberti et al. (2017) quantitatively confirmed the existence of a time scale separation between the internal processes and externally driven ones in the Earth's magnetospheric dynamics during magnetospheric storms. The time scale separating internal and external processes has been found to be about 200 min, a value that agrees very well with previous findings of nearly 3.5 h and with the characteristic time scales of unloading processes (Consolini et al., 2005; Kamide and Kokubun, 1996; Tsurutani et al., 1990; Uritsky et al., 2001). However, the scale-free activity bursts at time scales of less than 200 min, although they are not directly driven by solar wind input, can still be triggered by changes of the solar wind and IMF conditions and the enhancement of the plasma convection inside the Earth's magnetosphere. In other words, the observed near-criticality dynamics can be partially tuned and triggered by changes in interplanetary conditions and the continuous flow of energy, mass, and momentum by the solar wind.

Other indications of the observed scale-invariance and near-criticality dynamics have been found by analyzing other geomagnetic indices (Wanliss and Uritsky, 2010), auroral emissions (Lui et al., 2000; Uritsky et al., 2002), and finite-size effects (Uritsky et al., 2006).

6 MULTIFRACTIONAL FEATURES AND DYNAMICAL PHASE TRANSITIONS

Another relevant aspect of the scaling features of the geomagnetic indices is the possible dependence on time and on the level of geomagnetic activity that these properties may acquire (Balasis et al., 2006; Uritsky and Pudovkin, 1998; Wanliss, 2005; Wanliss and Dobias, 2007). For example, in Section 5 we have seen that there is a dependence of the scaling features of *AE*-index data on the geomagnetic activity level. That seems to suggest that the magnetospheric dynamics as shown by the time series of geomagnetic indices, as well as ground-based geomagnetic measurements, is not only characterized by scale invariance and intermittency, but also by what is called *multifractionality* (Benassi et al., 1997; Peltier and Lèvy-Vehel, 1995; Wanliss, 2005; Consolini et al., 2013), that is, a *multifractional* character.

A *multifractional signal* or a *MultiFractional Brownian Motion* (MFBM) is a generalization of the *fractional signal* (or *fractional Brownian motion*) (Mandelbrot and Van Ness, 1968; Bunde and Havlin, 1994) where the Hurst exponent H (which characterizes the persistence character of the signal) acquires a dependence on time. Consequently, the scaling features of the signal are only valid locally on time (Benassi et al., 1997; Peltier and Lèvy-Vehel, 1995). We note that there is a substantial difference between a signal with anomalous scaling and multifractal features and a multifractional one. While the former refers to scaling features of a measure defined on

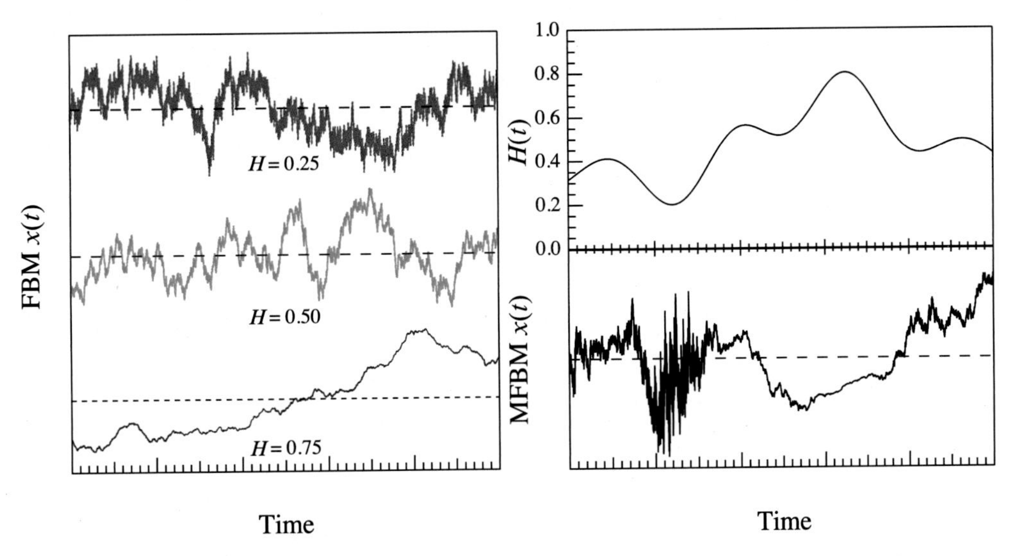

FIG. 10 A comparison between fractional Brownian signals (FBM), *left panel*, and multifractional Brownian one (MFBM), *right panel*. The FBM signals are for $H = 0.25$, 0.50, and 0.75, respectively. The *upper plot of the right panel* shows the Hurst exponent $H(t)$ as a function of time t, while the *lower panel* is the obtained MFBM signal, generated using the algorithm described in Muniandy and Lim (2001).

the signal, the latter indicates the occurrence of a dependence on time, that is, $H = f(t)$, of scaling properties. As an example of the different character of FBM in respect to MFBM, we show in Fig. 10 some samples of FBM signals for different values of the Hurst exponent and an MFBM signal obtained by means of the algorithm described in Muniandy and Lim (2001). A very notable feature of such signals is that the PDFs of increments are all Gaussian, so that the different nature of the signals is only due to the persistent character.

Fig. 11 displays the behavior of the local Hurst exponent, $H(t)$, for *SYM-H* and *AE* indices during March 2015, characterized by the famous St. Patrick's day 2015 geomagnetic storm. The computation of the local Hurst exponent has been done by applying the *detrended moving average* (DMA) technique introduced by Alessio et al. (2002) and Carbone et al. (2004). This method is based on the evaluation of a scale-dependent detrended standard deviation, $\sigma_{\mathrm{DMA}}(n)$, defined as

$$\sigma_{\mathrm{DMA}}(n) = \sqrt{\frac{1}{N_{\max} - n} \sum_{i=n}^{N_{\max}} \left[x(i) - \bar{x}(i)\right]^2} \sim n^H \tag{9}$$

where $x(i)$ is the signal, $\bar{x}(i)$ is the local average measured on a moving window of dimension n, and $N_{\max}$ is the dimension of the moving window where the scaling features are investigated. With reference to Fig. 10 the computation is done over a set of scales in the interval $n \in [2, 100]$ min (in our case $n \equiv \tau$), using a moving window of dimension $N_{\max} = 801$ min (for more details please refer to Consolini et al., 2013). A similar analysis has been done by Wanliss (2005) over an extended period in the case of the *SYM-H* index.

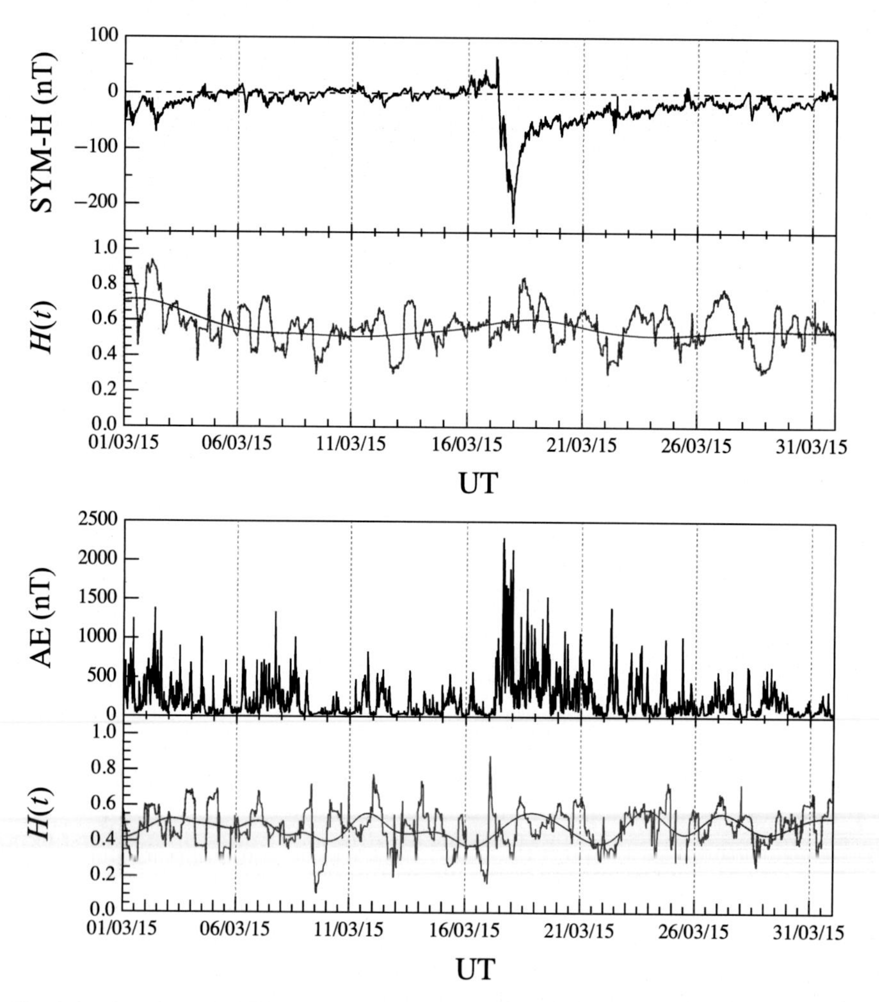

FIG. 11 The behavior of the local Hurst exponent $H(t)$ as a function of time in the period of March 2015, during which occurred the famous St. Patrick's day geomagnetic storm. The *upper/lower panel* shows the results for $SYM\text{-}H/AE$ index, respectively. *Blue lines* are the $H(t)$ long-time trend, computed by applying the Empirical Mode Decomposition filtering technique (see Wu et al., 2007 and references therein).

The behavior of the local Hurst exponent, $H(t)$, on time clearly shows that we are not in the presence of a signal, characterized by a pointwise regularity that is the same along their paths. This is evidence of the occurrence of *multifractionality* in geomagnetic time series (Wanliss, 2005). Furthermore, the local Hurst exponent tends to increase during disturbed periods. That is more evident by applying a filtering technique capable of revealing the long-time trend (see blue curves in Fig. 10). This feature can be interpreted as an increase in organization with magnetospheric activity (Wanliss, 2005), or as the evidence of near-criticality dynamics (Uritsky and Pudovkin, 1998) and has been widely observed, also in more extended studies (Balasis et al., 2006; Uritsky and Pudovkin, 1998; Wanliss, 2005).

In general, it has been observed that there is a change of the persistency features of the geomagnetic indices' fluctuations in proximity or during periods where the geomagnetic activity increases (see e.g., Balasis et al., 2006; Uritsky and Pudovkin, 1998). This point can be very interesting in the framework of space weather-related studies, being able to be used as a proxy for some dynamical changes in the Earth's magnetosphere before the occurrence of a geomagnetic storm/substorm. Analogous features have also been found by Consolini et al. (2013) investigating the scaling features of geomagnetic time records.

A possible interpretation of the observed changes of the local scaling features is that the occurrence of geomagnetic storms and substorms is accompanied by a sort of dynamical phase transition, which manifests in a change of the persistency features of geomagnetic indices' fluctuations (Balasis et al., 2006, 2008, 2009; Wanliss, 2005; Wanliss and Dobias, 2007). This scenario on the occurrence of a dynamical phase transition is also corroborated from some results and analyzes by Sitnov et al. (2000, 2001). Indeed, Sitnov and coauthors, using a different approach based on *singular spectrum analysis* of a set of input-output data, have evidenced that the magnetospheric dynamics during geomagnetic substorms could be very well described in terms of the more general framework of nonequilibrium phase transitions. In particular, the Earth's magnetospheric dynamics during geomagnetic substorms seems to be characterized by both multiscale features, which are typical of second-order phase transitions, and features typical of first-order nonequilibrium phase transitions, such as the occurrence of hysteresis (Sitnov et al., 2001).

7 SUMMARY

In the previous sections some features of geomagnetic indices' time series have been presented and discussed in connection with the emergence of dynamical complexity in the Earth's magnetospheric response to solar wind and IMF changes. These features can be summarized as follows:

1. Intermittency and coherent relaxation events (e.g., *AE*-index activity bursts) are relevant features of geomagnetic indices' time series. The relaxation events display a clear structure in the scale-time plane (wavelet and LIM analysis) and are at the origin of the intermittent nature of geomagnetic indices.
2. Geomagnetic indices' time series display scale-invariance and self-similarity over a wide range of time scales. However, these scale-invariance features are not simple as in the case of fractal signals, but show a clear multifractal (multiscaling) nature that manifests in anomalous scaling properties.
3. Activity bursts in geomagnetic indices are characterized by scale invariant PDFs for the events size s and duration T. This has been read as an evidence of the avalanching nature of the magnetospheric dynamics.
4. The scale invariance nature of the geomagnetic indices' time series displays a clear dependence on time. This feature is compatible with a quasi multifractional character of geomagnetic indices and with the occurrence of nonequilibrium phase transitions near or during geomagnetic storms and substorms.

All the preceding features have been long interpreted as a piece of evidence of near-criticality dynamics (SOC and FSOC) of the Earth's magnetosphere. However, some of the observed features are also common to turbulent systems (Frisch, 1995) or to out-of-equilibrium systems displaying a complex dynamics. Most of the observed features of geomagnetic indices can be linked to the dynamics of the Earth's magnetotail, which is the region of the Earth's magnetosphere that mostly contributes to the internal response and is responsible for the high-latitude geomagnetic activity. The CPS and the current sheet of the Earth's magnetotail are, indeed, highly dynamical plasma regions showing large and turbulent magnetic field fluctuations (Borovsky et al., 1997; Hoshino et al., 1994; Vörös et al., 2005) and rapid particle energization processes during geomagnetic disturbed periods (Angelopoulos et al., 1992, 1999; Lui et al., 1998; Lui, 2002b). Very recently, Klimas and Uritsky (2017) have shown, by 2D resistive MHD simulations of current sheet plasma dynamics, that both intermittent turbulence and avalanching scaling features, as for SOC systems, can coexist and be responsible for some of the observed features of geomagnetic indices fluctuations. Consequently, the whole variety of the magnetospheric activity features during magnetic storms and substorms cannot be described taking into consideration a simple class of models/phenomena, for example, SOC or turbulence.

A better framework for the dynamics of Earth's magnetosphere in response to the changes of interplanetary conditions could be that of a continuously perturbed stochastic system, driven outside of equilibrium near a critical point (Chang, 1992) and displaying dynamical complexity, that is, a certain degree of spatiotemporal coherent features resulting from the competition of different basic states. In such a context, the dynamical complexity emerges from the interplay of different dynamical processes and states, like turbulence, near-criticality dynamics, nonlinear external forcing, out-of-equilibrium pattern formations, topological complexity, and so forth.

ACKNOWLEDGMENTS

The author wish to thank T.S. Chang (MIT, USA), A.T.Y. Lui (APL-JHU, USA), and P. De Michelis (INGV, Italy) for the long-standing collaboration. The author also acknowledges the NSSDC OMNIWeb database (P.I. J.H. King, N. Papatashvilli) for providing data used in this work.

References

Ahn, B.H., Akasofu, S.I., Kamide, Y., 1983. The Joule heat-production rate and the particle energy injection rate as a function of the geomagnetic indices AE and AL. J. Geophys. Res. 88, 6275–6287.

Akasofu, S., 2004. Several controversial issues on substorms. Space Sci. Rev. 113, 1.

Alberti, T., Consolini, G., Lepreti, F., Laurenza, M., Vecchio, A., Carbone, V., 2017. Timescale separation in the solar wind-magnetosphere coupling during St. Patrick's day storms in 2013 and 2015. J. Geophys. Res. 122, A023175. https://doi.org/10.1002/2016JA023175.

Alessio, E., Carbone, A., Castelli, G., Frappietro, V., 2002. Second order moving average and scaling of stochastic time series. Eur. Phys. J. B 27, 197–200.

Angelopoulos, V., Baumjohann, W., Kennel, C.F., Coroniti, F.V., Kilevson, M.G., Pellat, R., Walker, R.J., Luher, H., Paschmann, G., 1992. Bursty bulk flows in the inner central plasma sheet. J. Geophys. Res. 97, 4027–4039.

Angelopoulos, V., Mukai, T., Kokubun, S., 1999. Evidence for intermittency in Earth's plasma sheet and implications for self-organized criticality. Phys. Plasmas 6, 4161–4168.

Anh, V.V., Yong, J.M., Yu, Z.G., 2008. Stochastic modeling of the auroral electrojet index. J. Geophys. Res. 113, A10215. https://doi.org/10.1029/2007JA012851.

Bak, P., Tang, C., Wiesenfeld, K., 1987. Self-organized criticality—an explanation of 1/f noise. Phys. Rev. Lett. 59, 381–384.

Bak, P., Tang, C., Wiesenfeld, K., 1988. Self-organized criticality. Phys. Rev. A 38, 364–374.

Baker, D.N., Klimas, A.J., McPherron, R.I., Büchner, J., 1990. The evolution from weak to strong geomagnetic activity: an interpretation in terms of deterministic chaos. Geophys. Res. Lett. 17, 41–44.

Balasis, G., Daglis, I.A., Kapiris, P., Mandea, M., Vassiliadis, D., Eftaxias, K., 2006. From pre-storm activity to magnetic storms: a transition described in terms of fractal dynamics. Ann. Geophys. 24, 3557–3567. https://doi.org/10.5194/angeo-24-3557-2006.

Balasis, G., Daglis, I.A., Papadimitriou, C., Kalimeri, M., Anastasiadis, A., Eftaxias, K., 2008. Dynamical complexity in Dst time series using nonextensive Tsallis entropy. Geophys. Res. Lett. 35, L14102. https://doi.org/10.1029/2008GL034743.

Balasis, G., Daglis, I.A., Papadimitriou, C., Kalimeri, M., Anastasiadis, A., Eftaxias, K., 2009. Investigating dynamical complexity in the magnetosphere using various entropy measures. J. Geophys. Res. 114, A00D06. https://doi.org/10.1029/2008JA014035.

Balasis, G., Papadimitriou, C., Daglis, I.A., Anastasiadis, A., Athanasopoulou, L., Eftaxias, K., 2011. Signature of discrete scale invariance in Dst time series. Geophys. Res. Lett. 38, L13103. https://doi.org/10.1029/2011GL048019.

Benassi, A., Jaffard, S., Roux, D., 1997. Elliptic Gaussian random processes. Rev. Mat. Iberoam. 13, 19–90.

Borovsky, J.E., Elphic, R.C., Funsten, H.O., Thomsen, M., 1997. The Earth's plasma sheet as a laboratory for flow turbulence in high-β MHD. J. Plasma Phys. 57, 1–34.

Bruno, R., et al., 2001. Identifying intermittency events in the solar wind. Planet. Space Sci. 49, 1201–1210.

Bunde, A., Havlin, S., 1994. Fractals in Science. Springer-Verlag, Berlin, Heidelberg.

Carbone, A., Castelli, G., Stanley, H.E., 2004. Time-dependent Hurst exponent in financial time series. Phys. A 344, 267–271.

Carlson, J.M., Doyle, J., 2002. Complexity and robustness. Proc. Natl Acad. Sci. USA 99, 2538–2545.

Chang, T., 1992. Low-dimensional behavior and symmetry breaking of stochastic systems near criticality: can these effects be observed in space and in the laboratory. IEEE Trans. Plasma Phys. 20, 691.

Chang, T., 1999. Self-organized criticality, multi-fractal spectra, sporadic localized reconnections and multiscale intermittent turbulence in the magnetotail. Phys. Plasmas 6, 4137.

Chang, T., 2001. Colloid-like behavior and topological phase transitions in space plasmas,: intermittent low-frequency turbulence in the auroral zone. Phys. Scripta T89, 80–83.

Chang, T., 2015. An Introduction to Space Plasma Complexity. Cambridge University Press, Cambridge.

Chang, T.S., Wu, C.C., 2008. Rank-ordered multifractal spectrum for intermittent fluctuations. Phys. Rev. E 77, 045401. https://doi.org/10.1103/PhysRevE.77.045401.

Chang, T., Tam, S.W.Y., Wu, C.C., 2006. Complexity in space plasmas: a brief review. Space Sci. Rev. 122, 281.

Chapman, S.C., Watkins, N.W., 2001. Avalanching and self-organised criticality, a paradigm for geomagnetic activity? Space Sci. Rev. 95, 293–307.

Chapman, S.C., Watkins, N.W., Dendy, R.O., Helander, P., Rowlands, G., 1998. A simple avalanche model as an analogue for magnetospheric activity. Geophys. Res. Lett. 25, 2397–2400.

Chapman, S.C., Hnat, B., Rowlands, G., Watkins, N.W., 2005. Scaling collapse and structure functions: identifying self-affinity in finite length time series. Nonlinear Process. Geophys. 12, 767–774. https://doi.org/10.5194/npg-12-767-2005.

Consolini, G., 1997. Sandpile cellular automata and the magnetospheric dynamics. In: Aiello, S., et al. (Eds.), Proceedings of VIII GIFCO Conference. Cosmic Physics in the Year 2000. SIF, Bologna, 123 pp.

Consolini, G., 2002. Self-organized criticality: a new paradigm for the magnetotail dynamics. Fractals 10, 275–283.

Consolini, G., Chang, T.S., 2001. Magnetotail field topology and criticality in geotail dynamics: relevance to substorm phenomena. Space Sci. Rev. 95, 309–321.

Consolini, G., De Michelis, P., 1998. Non-Gaussian distribution of AE-index fluctuations. Evidence for time intermittency. Geophys. Res. Lett. 25, 4087–4090.

Consolini, G., De Michelis, P., 2005. Local intermittency measure analysis of AE index: the directly driven and unloading component. Geophys. Res. Lett. 32, L05101.

Consolini, G., De Michelis, P., 2011. Rank ordering multifractal analysis of the auroral electrojet index. Nonlinear Process. Geophys. 18, 277–285.

Consolini, G., Marcucci, M.F., Candidi, M., 1996. Multifractal structure of auroral electrojet index data. Phys. Rev. Lett. 76, 4082–4085.

Consolini, G., Chang, T., Lui, A.T.Y., 2005. Complexity and topological disorder in the Earth's magnetotail dynamics. In: Sharma, A.S., Kaw, P.K. (Eds.), Nonequilibrium Phenomena in Plasmas. Springer, The Netherlands.

Consolini, G., De Michelis, P., Tozzi, R., 2008. On the Earth's magnetospheric dynamics: nonequilibrium evolution and the fluctuation theorem. J. Geophys. Res. 113, A08222. https://doi.org/10.1029/2008JA013074.

Consolini, G., De Marco, R., De Michelis, P., 2013. Intermittency and multifractional Brownian character of geomagnetic time series. Nonlinear Process. Geophys. 20, 455–466.

Davis, T., Sugiura, M.J., 1966. Auroral electrojet activity index AE and its universal time variations. J. Geophys. Res. 71, 785–791.

de Groot, S.R., Mazur, P., 1984. Non-Equilibrium Thermodynamics. Dover Publication Inc., New York, NY.

Dobias, P., Wanliss, J.A., 2009. Intermittency of storms and substorms: is it related to the critical behaviour? Ann. Geophys. 27, 2011–2018. https://doi.org/10.5194/angeo-27-2011-2009.

Farge, M., Holschneider, M., Colonna, J.F., 1990. Wavelet analysis of coherent structures in two dimensional turbulent flows. In: Moffat, H.K. (Ed.), Topological Fluid Mechanics. Cambridge University Press, New York, NY, 765 pp.

Frauenfelder, H., 2002. Proteins: paradigms of complexity. Proc. Natl Acad. Sci. USA 99, 2479–2480.

Freeman, M.P., Watkins, N.W., Riley, D.J., 2000. Evidence for a solar wind origin of the power law burst lifetime distribution of the AE indices. Geophys. Res. Lett. 27, 1087–1090.

Frisch, U., 1995. Turbulence: The Legacy of A. N. Kolmogorov. Cambridge University Press, Cambridge.

Gonzalez, W., Joselyn, J., Kamide, Y., Kroehl, H., Rostoker, G., Tsurutani, B., Vasyliunas, V., 1994. What is a geomagnetic storm? J. Geophys. Res. 99, 5771. https://doi.org/10.1029/93JA02867.

Hergarten, S., 2002. Self-Organized Criticality in Earth Systems. Springer-Verlag, Berlin, Heidelberg.

Hnat, B., Chapman, S.C., Rowlands, G., Watkins, N.W., Freeman, M.P., 2002. Scaling of solar wind and AU, AL and AE indices as seen by WIND. Geophys. Res. Lett. 29, 2078. https://doi.org/10.1029/2002GL016054.

Hnat, B., Chapman, C., Rowlands, G., 2005. Scaling and Fokker-Planck model for fluctuations in geomagnetic indices and comparison with solar wind as seen by Wind and ACE. J. Geophys. Res. 110. https://doi.org/10.1029/2004JA010824.

Holland, J.H., 2014. Complexity. A Very Short Introduction. Oxford University Press, Oxford.

Hoshino, M., Nishida, A., Yamamoto, T., Kokubun, S., 1994. Turbulent magnetic field in the distant magnetotail: bottom-up process of plasmoid formation? Geophys. Res. Lett. 21, 2935–2938.

Kamide, Y., Chian, A., 2007. Handbook of the Solar-Terrestrial Environment. Springer-Verlag, Berlin, Heidelberg.

Kamide, Y., Kokubun, S., 1996. Two-component auroral electrojet: importance for substorm studies. J. Geophys. Res. 101, 13027.

Klimas, A.J., Uritsky, V.M., 2017. Criticality and turbulence in a resistive magnetohydrodynamic current sheet. Phys. Rev. E 95, 023209.

Klimas, A.J., Vassiliadis, D.V., Baker, D.N., Roberts, D.A., 1996. The organized nonlinear dynamics of the magnetosphere. J. Geophys. Res. 101, 13089.

Klimas, A.J., Valdivia, J.A., Vassiliadis, D., Baker, D.N., Hesse, M., Takalo, J., 2000. Self-organized criticality in the substorm phenomenon and its relation to localized reconnection in the magnetospheric plasma sheet. J. Geophys. Res. 105, 18765–18780.

Kumar, P., Foufoula-Georgiou, E., 1997. Wavelet analysis for geophysical applications. Rev. Geophys. 35, 385.

Lui, A.T.Y., 2002a. Evaluation on the analogy between the dynamic magnetosphere and a forced and/or self-organized critical system. Nonlinear Process. Geophys. 9, 399–407.

Lui, A.T.Y., 2002b. Multiscale phenomena in the near-earth magnetosphere. J. Atmos. Sol. Terr. Phys. 64, 125–143.

Lui, A.T.Y., 2004. Testing the hypothesis of the Earths magnetosphere behaving like an avalanching system. Nonlinear Process. Geophys. 11, 701–707.

Lui, A.T.Y., Liou, K., Newell, P.T., Meng, C.I., Othani, S., Oginoi, T., Kokubun, S., Brittnacher, M., Parks, G.K., 1998. Plasma and magnetic flux transport associated with auroral breakups. Geophys. Res. Lett. 25, 4059–4062.

Lui, A.T.Y., Chapman, S.C., Liou, K., Newell, P.T., Meng, C.I., Brittnacher, M., Parks, G.K., 2000. Is the dynamic magnetosphere an avalanching system? Geophys. Res. Lett. 27, 911–914.
Mallat, S., 1998. A Wavelet Tour of Signal Processing. Elsevier, New York, NY.
Mandelbrot, B.B., 1989. Multifractal measures, especially for geophysicist. Pure Appl. Geophys. 131, 5–42. https://doi.org/10.1007/BF00874478.
Mandelbrot, B.B., Van Ness, J.W., 1968. Fractional Brownian motions, fractional noises and applications. Soc. Ind. Appl. Math. Rev. 10, 422–437.
Muniandy, S.V., Lim, S.C., 2001. Modeling of locally self-similar processes using multifractional Brownian motion of Riemann-Liuouville type. Phys. Rev. E 63, 046104. https://doi.org/10.1103/PhysRevE.63.046104.
Nicolis, G., Nicolis, C., 2007. Foundations of Complex Systems: Nonlinear Dynamics, Statistical Physics, Information and Prediction. World Sci. Pub. Co. Pte. Ltd, Singapore.
Parisi, G., 1999. Complex systems: a physicist's viewpoint. Phys. A 263, 557–564.
Peltier, R.F., Lèvy-Vehel, J., 1995. Multifractional Brownian motion: definition and preliminary results. INRIA Preprint No. 2645.
Pulkkinen, A., Klimas, A., Vassiliadis, D., Uritsky, V., 2006. Role of stochastic fluctuations in the magnetosphere-ionosphere system: a stochastic model for the AE index variations. J. Geophys. Res. 111, A10218. https://doi.org/10.1029/2006JA011661.
Roberts, D.A., Baker, D.N., Klimas, A.J., Bargatze, L.F., 1991. Indications of low dimensionality in magnetospheric dynamics. Geophys. Res. Lett. 18, 151–154.
Rypdal, M., Rypdal, K., 2010. Stochastic modelling of the AE index and its relation to fluctuations in Bz of the IMF on time scales shorter than substorm duration. J. Geophys. Res. 115, A11216. https://doi.org/10.1029/2010JA015463.
Sharma, A.S., Sitnov, M.I., Papadopoulos, K., 2001. Substorms as nonequilibrium transitions of the magnetosphere. J. Atmos. Sol. Terr. Phys. 63, 1399.
Sitnov, M.I., et al., 2000. Phase transition-like behavior of the magnetosphere during substorms? J. Geophys. Res. 105, 12955.
Sitnov, M.I., Sharma, A.S., Papadopoulos, K., Vassiliadis, D., 2001. Modeling substorm dynamics of the magnetosphere: from self-organization and self-organized criticality to non-equilibrium phase transitions. Phys. Rev. E 65, 16116.
Sornette, D., 1998. Discrete-scale invariance and complex dimensions. Phys. Rep. 297, 239–270.
Sorriso-Valvo, L., Carbone, V., Veltri, P., Consolini, G., Bruno, R., 1999. Intermittency in the solar wind turbulence through probability distribution functions of fluctuations. Geophys. Res. Lett. 26, 1801–1804.
Stauffer, D., Stanley, H.E., 1995. From Newton to Mandelbrot. Springer-Verlag, Berlin, Heidelberg.
Sugiura, M., Poros, D.J., 1971. Hourly Values of Equatorial Dst for Years 1957 to 1970. Goddard Space Flight Center, Greenbelt, MD.
Takalo, J., Timonen, J., 1994. Characteristic time scale of auroral electrojet data. Geophys. Res. Lett. 21, 617–620.
Takalo, J., Timonen, J., Koskinen, H., 1993. Correlation dimension and affinity of *ae* data and bicoloured noise. Geophys. Res. Lett. 20, 1527–1530.
Tsurutani, B., et al., 1990. The nonlinear response of AE to the IMF B_S. Geophys. Res. Lett. 17, 279–282.
Uritsky, V.M., Pudovkin, M.I., 1998. Low frequency $1/f$-like fluctuations of the AE-index as a possible manifestation of self-organized criticality in the magnetosphere. Ann. Geophys. 16, 1580.
Uritsky, V.M., Klimas, A.J., Vassiliadis, D., 2001. Comparative study of dynamical critical scaling in the auroral electrojet index versus solar wind fluctuations. Geophys. Res. Lett. 28, 3809–3812.
Uritsky, V.M., et al., 2002. Scale-free statistics of spatiotemporal auroral emissions as depicted by POLAR UVI images: the dynamic magnetosphere is an avalanching system. J. Geophys. Res. 107, 1426.
Uritsky, V.M., Klimas, A.J., Vassiliadis, D., 2006. Critical finite-size scaling of energy and lifetime probability distributions of auroral emissions. Geophys. Res. Lett. 33, L08102.
Vassiliadis, D.V., Sharma, A.S., Eastman, T.E., Papadopoulos, K., 1990. Low-dimensional chaos in magnetospheric activity from AE time series. Geophys. Res. Lett. 17, 1841–1844.
Vörös, Z., Baumjohan, W., Nakamura, R., Runov, A., Volwerk, M., Schwarlz, H., Balogh, A., Reme, H.R., 2005. Dissipation scales in the Earths plasma sheet estimated from Cluster measurements. Nonlinear Process. Geophys. 12, 725.
Wanliss, J.A., 2005. Fractal properties of SYM-H during quiet and active times. J. Geophys. Res. 110, A03202. https://doi.org/10.1029/2004JA010544.

Wanliss, J.A., Dobias, P., 2007. Space storms as a phase transition. J. Atmos. Sol. Terr. Phys. 69, 675–684. https://doi.org/10.1016/j.jastp.2007.01.001.
Wanliss, J.A., Showalter, K.M., 2006. High-resolution global storm index: Dst versus SYM-H. J. Geophys. Res. 111, A02202.
Wanliss, J.A., Uritsky, V.M., 2010. Understanding bursty behavior in midlatitude geomagnetic activity. J. Geophys. Res. 115, A03215. https://doi.org/10.1029/2009JA014642.
Wanliss, J.A., Ahn, V.V., Yu, Z.G., Watson, S.S., 2005. Multifractal modeling of magnetic storms via symbolic dynamics analysis. J. Geophys. Res. 110, A08214. https://doi.org/10.1029/2004JA010996.
Warhaft, Z., 2002. Passive scalar turbulent flows. Annu. Rev. Fluid Mech. 32, 203–240.
Watkins, N.W., Credgington, D., Hnat, B., Chapman, S.C., Freeman, M.P., Greenhough, J., 2005. Towards synthesis of solar wind and geomagnetic scaling exponents: a fractional Lèvy motion model. Space Sci. Rev. 121, 271–284. https://doi.org/10.1007/s11214-006-4578-2.
Wu, Z., Huang, N.E., Long, S.R., Peng, C.K., 2007. On the trend, detrending, and variability of nonlinear and nonstationary time series. Proc. Natl Acad. Sci. USA 104, 14889–14894.

CHAPTER

8

Applications of NARMAX in Space Weather

Richard Boynton, Michael Balikhin, Hua-Liang Wei, Zi-Qiang Lang

Department of Automatic Control and Systems Engineering, University of Sheffield, Sheffield, UK

CHAPTER OUTLINE

1 INTRODUCTION

Nature is rich with physical systems that are so complex that at our current level of knowledge, we are not able to deduce their mathematical models from physical principles.

Machine Learning Techniques for Space Weather
https://doi.org/10.1016/B978-0-12-811788-0.00008-1

Living organisms and human brains are examples of such systems. The complementary approach to advance our understanding of such systems is provided by systems science, which has developed advanced experimental data analysis methodologies to identify patterns in the system evolution. Methodologies developed within the field of systems science are aimed not only at development of system models that can mimic the dynamics of natural objects, but also to provide information about types of physical processes involved in the dynamics of that particular object in both the time and frequency domains. Systems science often deals with objects for which very limited information is available. In many cases, even the number of degrees of freedom are not known. What is usually known is that a particular system evolves under the influence of a particular factor. One example is the geospace, which evolves under the influence of the solar wind. The parameters of the solar wind can be measured, and they constitute the input into the system. Some parameters of the system can also be measured, and are assumed to be functions of the system state, which are referred to as the output of the system. In the case of the magnetosphere-ionosphere system, examples of the system output are geomagnetic indices, total electron content (TEC), fluxes of energetic particles in the radiation belts, and so forth. Some of the methodologies developed in systems science deal with these input-output data sets with the aim to develop the mathematical/numerical models of dynamical systems that can be used to both forecast their evolution, and deduce information about physical processes involved in the dynamics of a particular complex system. Nonlinear AutoRegressive Moving Average eXogenous (NARMAX) methodologies comprise of some of the most powerful tools of systems science. The main strength of the NARMAX is that it provides physically interpretable models.

The NARMAX philosophy, which is similar to other machine learning techniques, consists of five steps. The first step is structure detection, which involves identifying the structure of the nonlinear model. The second step is to determine the coefficients of the terms in the nonlinear model, and is called the parameter estimation step. The third step is called model validation, and is to determine if the model provides unbiased predictions. The fourth step is prediction, where the model is used to provide a forecast of the system. The final step is analysis, which investigates the underlying dynamical physical properties behind the system, which can be accomplished in both the time and frequency domain.

2 NARMAX METHODOLOGY

The NARMAX model was first introduced in 1981 to represent a wide class of nonlinear systems (Billings and Leontaritis, 1981) and can be defined as

$$\begin{aligned} y(t) = F[&y(t-1),\ldots,y(t-n_y), \\ &u_1(t-1),\ldots,u_1(t-n_{u_1}),\ldots, \\ &u_m(t-1),\ldots,u_m(t-n_{u_m}), \\ &e(t-1),\ldots,e(t-n_e)] + e(t) \end{aligned} \tag{1}$$

where the output, y, at time, t, is represented as a nonlinear function F of the past output with a maximum lag n_y, the past m inputs, u, each with a maximum lag $n_{u_1},\ldots,n_{u_m}$ and past noise, e, with a maximum lag n_e.

The nonlinear function, F, can be expanded in a number of ways (e.g., polynomial, rational, radial basis function, wavelets). The most commonly employed representation of the NARMAX model is the polynomial expansion of F to a specified degree of nonlinearity. The expansion of the polynomial will result in many monomials, for example, when the fourth-degree polynomial is considered with four inputs and the maximum lags of the output, the four inputs and the noise are set to 4, then there will be 20,475 candidate monomials for the NARMAX model. To capture the dynamics of a system, it has been shown that in almost all cases only a small number of candidate monomials are needed for the final NARMAX model (Billings, 2013). The vast majority of the monomials will have no influence on the output and many monomials will be similar to one another. The challenge was to identify the small set of significant monomials that influence the system, which is the first step and most fundamental part of the NARMAX approach: structure detection.

2.1 Forward Regression Orthogonal Least Square

An effective way to deal with the structure detection problem was presented by Billings et al. (1989) in the form of the Forward Regression Orthogonal Least Squares (FROLS) algorithm. After the expansion of F in terms of a polynomial, the NARMAX model of Eq. (1) can be represented as

$$y(t) = \sum_{i=1}^{M} p_i(t)\Theta_i + e(t) \tag{2}$$

where $p_i(t)$ is the ith monomial Θ_i is the coefficient of the ith monomial and M is the total number of monomials. In compact matrix form

$$\mathbf{y} = \mathbf{P}\Theta + \mathbf{e} \tag{3}$$

where M is the length of the data

$$\mathbf{y} = \begin{bmatrix} y(1) \\ y(2) \\ \vdots \\ y(N) \end{bmatrix}, \quad \mathbf{P} = \begin{bmatrix} p_1(1) & p_2(1) & \cdots & p_M(1) \\ p_1(2) & p_2(2) & \cdots & p_M(2) \\ \vdots & \vdots & & \vdots \\ p_1(N) & p_2(N) & \cdots & p_M(N) \end{bmatrix},$$

$$\Theta = \begin{bmatrix} \Theta_1 \\ \Theta_2 \\ \vdots \\ \Theta_M \end{bmatrix}, \quad \mathbf{e} = \begin{bmatrix} e(1) \\ e(2) \\ \vdots \\ e(N) \end{bmatrix}$$

In the first step ($k = 1$) of the FROLS algorithm, for each of the i monomials, the Error Reduction Ratio (ERR) is calculated

$$\mathrm{ERR}_1(i) = \frac{g^2(i)\mathbf{p_i^T p_i}}{\mathbf{y^T y}} \tag{4}$$

where g is called the auxiliary coefficient and is calculated by

$$g_1(i) = \frac{\mathbf{p_i^T y}}{\mathbf{p_i^T p_i}} \tag{5}$$

The ERR of a monomial is effectively the proportion of output variance that the monomial can explain. Therefore, the monomial with the highest ERR is selected as the first model term for the final model, for example, letting $h_1 = \arg\{\max\{\mathrm{ERR}_1(i)\}\}$ means that $\mathbf{p_{h_1}}$ is the first model term. The ERR for the first term of the final model $\mathrm{err}(1) = \mathrm{ERR}_1(h_1)$ and the first auxiliary coefficient set to $g(1) = g_1(h_1)$.

To select a subsequent monomial for the final model in the kth step (where $k > 1$), the contribution of monomials selected in the previous steps needs to be removed from remaining monomials and the individual increment to the output variance of each monomial in the final model needs to be separated. This can be achieved by employing the classical or modified Gram-Schmidt procedure to make the remaining candidate monomials orthogonal to all the monomials already selected for the final model. The orthogonalization of Eq. (3) yields $\mathbf{y} = \mathbf{P}(\mathbf{R}^{-1}\mathbf{R})\Theta + \mathbf{e}$, so the orthogonalized monomial matrix $\mathbf{W} = \mathbf{PR}^{-1}$ and the auxiliary coefficient vector $\mathbf{g} = \mathbf{R}\Theta$ where

$$\mathbf{R} = \begin{bmatrix} 1 & r_{12} & r_{13} & \cdots & r_{1M} \\ 0 & 1 & r_{23} & \cdots & r_{2M} \\ \vdots & \vdots & \ddots & & \vdots \\ 0 & & \cdots & 1 & r_{(M-1)M} \\ 0 & & \cdots & 0 & 1 \end{bmatrix}$$

and is called the upper right triangular matrix. Eq. (2) then becomes

$$\mathbf{y} = \mathbf{Wg} + \mathbf{e} \tag{6}$$

Therefore, for the second and subsequent steps ($k > 1$), the remaining $M - k + 1$ candidate monomials are made orthogonal. A classical Gram-Schmidt procedure could achieve this; however, this can be shown to be very sensitive to rounding errors. A simpler scheme is the three-term recurrence method, which can be adapted to Eq. (6) by

$$\mathbf{w_k^*}(i) = \mathbf{p_i} - \sum_{j=1}^{k-1} r_{jk}(i)\mathbf{w_j} \tag{7}$$

where $\mathbf{w_k^*}(n)$ is the nth orthogonalized candidate monomial vector for the kth step, $\mathbf{w_j}$ are the previous selected orthogonalized monomials, where the first column of the orthogonalized monomial matrix $\mathbf{w_1} = \mathbf{p_{h_1}}$, and

$$r_{jk}(n) = \frac{\mathbf{w_j}^T \mathbf{p_n}}{\mathbf{w_j}^T \mathbf{w_j}} \tag{8}$$

The candidate orthogonalized monomials, $\mathbf{w_k^*}(n)$, are then used to calculate the ERR of each of the monomials

$$\mathrm{ERR}_k(i) = \frac{g_k{}^2(i)\mathbf{w}_\mathbf{k}^{*T}(i)\mathbf{w}_\mathbf{k}^*(i)}{\mathbf{y}^T\mathbf{y}} \tag{9}$$

where the kth auxiliary coefficient for the ith monomial is

$$g_k(i) = \frac{\mathbf{w}_\mathbf{k}^{*T}(i)\mathbf{y}}{\mathbf{w}_\mathbf{k}^{*T}(i)\mathbf{w}_\mathbf{k}^*(i)}$$

In the second step ($k = 2$), for $i = 1, \ldots, M$ where $i \neq h_1$, the candidate orthogonalized monomial vectors $\mathbf{w}_\mathbf{2}^*(i)$, auxiliary coefficients $g_2(i)$, and ERR $\mathrm{ERR}_2(i)$ are calculated. Each of the $M - 1$ orthogonalized candidate monomial is calculated by

$$\mathbf{w}_\mathbf{2}^*(i) = \mathbf{p_i} - r_{1,2}(i)\mathbf{w_1} \tag{10}$$

where $\mathbf{w_1} = \mathbf{p_{h_1}}$ and

$$r_{1,2}(i) = \frac{\mathbf{w_1}^T\mathbf{p_i}}{\mathbf{w_1}^T\mathbf{w_1}} \tag{11}$$

This allows the calculation of the auxiliary coefficient

$$g_2(i) = \frac{\mathbf{w}_\mathbf{2}^{*T}(i)\mathbf{y}}{\mathbf{w}_\mathbf{2}^{*T}(i)\mathbf{w}_\mathbf{2}^*(i)} \tag{12}$$

which is required for the calculation of the ERR

$$\mathrm{ERR}_2(i) = \frac{g_2{}^2(i)\mathbf{w}_\mathbf{2}^{*T}(i)\mathbf{w}_\mathbf{2}^*(i)}{\mathbf{y}^T\mathbf{y}} \tag{13}$$

The monomial with the highest ERR with the index $h_2 = \arg\{\max\{\mathrm{ERR}_2(i)\}\}$ is selected as the second model term for the final model and the second orthogonalized monomial vector is selected as $\mathbf{w_2} = \mathbf{w}_\mathbf{2}^*(h_2)$.

For the kth step, as with the first and second steps, the monomial with the highest ERR is selected as the kth model term for the final model, where the index for the monomial will be $h_k = \arg\{\max\{\mathrm{ERR}_k(i)\}\}$, the kth orthogonalized monomial vector $\mathbf{w_k} = \mathbf{w}_\mathbf{k}^*(h_k)$, the ERR of the kth term of the final model $\mathrm{err}(k) = \mathrm{ERR}_k(h_k)$ and the kth auxiliary coefficient set to $g(k) = g_k(h_k)$.

The procedure of orthogonalizing the remaining candidate monomials with respect to the terms selected for the final model, then designating the orthogonalized monomial with the highest ERR for the final model, is continued until the model has the optimum number of model monomials, M_p. A number of methods to decide the optimum number of model terms have been suggested, such as the tolerance test

$$1 - \sum_{i=1}^{M_m} \mathrm{err}_i \leq \rho \tag{14}$$

where ρ is a small tolerance or the Adjustable Prediction Error Sum of Squares (Billings and Wei, 2008).

Once the model structure has been identified, the second step of the NARMAX philosophy needs to be achieved: Parameter Estimation. This is simply achieved during the FROLS algorithm. The estimates of the monomial coefficients, $\hat{\Theta}$, can then be computed backward from M_p using $\mathbf{g} = \mathbf{R}\hat{\Theta}$.

$$\hat{\Theta}_{M_\mathrm{p}} = g_{M_\mathrm{p}} \tag{15}$$

$$\hat{\Theta}_i = g_i - \sum_{n=i+1}^{M_\mathrm{p}} r_{in}\hat{\Theta}_n \tag{16}$$

The variance of NARMAX model parameters is measured and evaluated using t-test (Billings et al., 1989; Chen et al., 1989; Wei and Billings, 2008).

2.2 The Noise Model

Extended least squares (ELS) methods have been extensively studied and applied to linear system identification for a long time (Norton, 1986; Ljung and Soderstrom, 1983; Ljung, 1987). The basic idea is that the correlated noise can be reduced to a white noise sequence by introducing a noise model and combining with the process model. Ideally, using the combined model, the process-model and noise-model parameters are alternatively estimated, in an iterative manner, until the bias is reduced to zero. The idea and procedure were introduced to nonlinear system identification (Billings and Zhu, 1991; Billings and Wei, 2005).

The FROLS algorithm is able to identify the process-related submodel, the part of the NARMAX model that excludes the monomials that contain past noise. To identify the noise-related submodel, the monomials that contain the past noise, the initial residuals are calculated ($s = 0$)

$$\varepsilon_0(t) = y(t) - \hat{y}(t) = y(t) - \sum_{i=1}^{M_\mathrm{p}} g_i w_i(t) \tag{17}$$

where $\hat{y}(t)$ is the model estimated output. The FROLS algorithm is then used to identify the noise-related submodel of the residuals, terminating the procedure when the optimum number of model monomials, M_n, has been reached. The identified process and noise submodels are combined to form a NARMAX model of $(M_\mathrm{p} + M_n)$ monomials.

In the subsequent steps in determining the noise model coefficients, set $s = s + 1$. An ELS algorithm is then initiated to determine the final coefficients. For each iteration, the coefficient of each of the $(M_\mathrm{p} + M_n)$ monomials are reestimated (denoted as $\hat{\Theta}_i(s)$) using least square with $\varepsilon_{s-1}(t)$ for the noise. Then the new residuals $\varepsilon_s(t)$ are calculated recursively using

$$\varepsilon_s(t) = y(t) - \hat{y}(t) = y(t) - \sum_{i=1}^{M_\mathrm{p}+M_n} \hat{\Theta}_i(s)_i p_i(t) \tag{18}$$

This is terminated when

$$\sum_{t=1}^{N} \left|\varepsilon_s(t) - \varepsilon_{s-1}(t)\right|^2 \leq \rho \tag{19}$$

is satisfied, where ρ is a small tolerance. In practice, this is typically achieved in 5 to 10 iterations; however, a perfect convergence cannot be guaranteed (Billings and Zhu, 1991).

2.3 Model Validation

The primary objective of system identification is to find the simplest possible model that can represent the system. One of the commonly used model tests is based on the performance of the predictions (e.g., the mean square error [MSE]). MSE, as a measure, can be used as a reference, but MSE alone is far from sufficient for model validation. A model with the smallest MSE value may fit the training data well; however, this may not be the best model because the model could be biased. A biased model may fit the data set extremely well, but it is essentially a curve fit to one set of data and is not a model that captures the inherent dynamics of the system (Billings, 2013). The meanings of the terms "biased" and "unbiased" (model, prediction, etc.) are slightly different from the conventional "bias-variance" definition commonly used in the Machine Learning context (e.g., Geman et al., 1992; Friedman et al., 2009; James et al., 2013). Here, we put more emphasis on whether models are an adequate representation of the inherent dynamics hidden in the recorded data set of a system. Roughly speaking, models that do not satisfactorily or sufficiently represent the data set can be biased.

In the NARMAX method, statistical tests are used to test model validity. One of the important considerations of these tests is that the residuals of unbiased models should be uncorrelated to the system input and output variables. So an important test for model validation is the correlation tests (Billings and Voon, 1986), which are designed to test the correlation relation between the model residuals and system input and output signals. The correlation tests assume that the residuals $\varepsilon(t)$ are a zero mean white sequence, $\varepsilon(t)$ is independent of the monomials, $p_i(t)$, and all odd order moments are zero. The correlation tests check the bias of the model by confirming that the residuals $\varepsilon(t)$ are uncorrelated with all linear and nonlinear combinations of past inputs and outputs. To satisfy the correlation tests

$$\phi_{\varepsilon\varepsilon}(\tau) = \delta(\tau) \quad \forall\tau \tag{20}$$

$$\phi_{u\varepsilon}(\tau) = 0 \quad \forall\tau \tag{21}$$

$$\phi_{\varepsilon(\varepsilon u)}(\tau) = 0 \quad \tau \geq 0 \tag{22}$$

$$\phi_{(u^2)'\varepsilon}(\tau) = 0 \quad \forall\tau \tag{23}$$

$$\phi_{(u^2)'\varepsilon^2}(\tau) = 0 \quad \forall\tau \tag{24}$$

where u is the input, $(u^2)' = u^2(t) - \bar{u}$, ϕ is the cross-correlation function

$$\phi_{xy}(\tau) = \frac{\sum_{t=1}^{N-\tau}\left[(x(t)-\bar{x})(y(t+\tau)-\bar{y})\right]}{\sqrt{\sum_{t=1}^{N}\left[(x(t)-\bar{x})\right]^2}\sqrt{\sum_{t=1}^{N}\left[(y(t)-\bar{y})\right]^2}} \tag{25}$$

N is the data length and the bar donates the mean. To class the tests as satisfied, 95% confidence bands are used, which are approximately $\pm 1.96/\sqrt{N}$. The tests were first developed for single-input single-output (SISO) nonlinear systems by Billings and Voon (1986) and modified for multiinput multioutput nonlinear systems by Billings et al. (1989). A more detailed description on model validity tests can be found in Billings (2013).

2.4 Summary

The FROLS algorithm (also known as the orthogonal least squares algorithm) was designed in the late 1980s for nonlinear dynamic system model structure detection and model estimation. It remains popular for practical data-based modeling, due to the fact that the algorithm is simple and efficient and is capable of producing parsimonious linear-in-the-parameters models with good generalization performance. Similar to other prediction error based algorithms, FROLS uses the MSE as a measure to monitor the model construction process. It is known that the training MSE typically decreases as the model size increases, and the resulting model is prone to over-fitting, especially for cases where data are severely contaminated by noise. Although a number of information theoretic methods (e.g., Akaike information criterion and Bayesian information criterion) can be employed to help find an appropriate model size, these methods do not always work well for the length determination of nonlinear dynamic models.

One effective solution is using a criterion of model generalization capability directly in the structure detection procedure, rather than only using it as a measure of model complexity (Chen et al., 2004). In the past decade, the excellent work and success of least absolute shrinkage and selection operator (LASSO) and least angle regression (LAR) (Efron et al., 2004; Zou and Hastie, 2005; Friedman et al., 2009) has facilitated the application of L1 regularization to nonlinear system identification. For example, Chen et al. (2004) proposed a regularized FROLS algorithm combined with PRESS statistics; Kukreja et al. (2006), Bonin et al. (2010), and Zhao et al. (2017) proposed using L1-norm regularization for NARX/NARMAX model identification; Chen et al. (2008) introduced a regularized FROLS algorithm for basis hunting; Qin et al. (2012) suggested using weighted least squares method and L1-norm regularization for NARX model selection and estimation; Hong and Chen (2012) proposed an improved FROLS algorithm regularized with L1 and L2 norm simultaneously. The computational complexity of the basic FROLS is $O(N \times M^2)$, where the number of candidate regressors is M and the sample size (number of observations) is N. This is the same computational complexity of L1 (e.g., LASSO and LAR) (Efron et al., 2004).

3 NARMAX AND SPACE WEATHER FORECASTING

3.1 Geomagnetic Indices

The modeling of geomagnetic indices lends well to systems science and machine learning techniques, as there are long-term data on which the models can be trained. The Dst index, which has often been used as a measure of geomagnetic storms, has data from 1957. Together with solar wind data from missions such as IMP 8, WIND, and ACE, these data have allowed the development of many data-based models.

3.1.1 SISO Dst Index

The NARMAX approach was first applied to the forecast of the Dst index by Boaghe et al. (2001). The Dst index SISO NARMAX model used the solar wind velocity v multiplied by the southward interplanetary magnetic field (IMF) component B_s ($B_s = -B_z$ for $B_z < 0$ and $B_s = 0$

for $B_s > 0$) as the input. A maximum lag of 5 was used for the Dst and the noise and 15 for vB_s resulting in a NARMAX model of the form

$$\begin{aligned} Dst(t) = F[&Dst(t-1),\ldots,Dst(t-5), \\ &vB_s(t-1),\ldots,vB_s(t-15), \\ &e(t-1),\ldots,e(t-5)] + e(t) \end{aligned} \quad (26)$$

where F was chosen to be a quadratic polynomial. The model was trained on 1000 h of data from January 21, 1979 to March 4, 1979. They validated the model using the correlation tests and assessed the model forecast using the correlation coefficient (CC). The CC is defined as Eq. (25) with $\tau = 0$

$$\rho_{y\hat{y}} = \frac{\sum_{t=1}^{N}\left[(y(t)-\bar{y}(t))\left(\hat{y}(t)-\bar{\hat{y}}(t)\right)\right]}{\sqrt{\sum_{t=1}^{N}\left[(y(t)-\bar{y}(t))^2\right]\sum_{t=1}^{N}\left[\left(\hat{y}(t)-\bar{\hat{y}}(t)\right)^2\right]}} \quad (27)$$

The CC is bound between −1 and 1, where a CC close to 1 illustrates that there is a high linear dependence between the measured and estimated Dst, a value of 0 means no linear dependence, and a value close to −1 indicates an anticorrelation. The model forecast had a CC of 0.989 on a period from March 5, 1979 to June 31, 1979.

The model developed by Boynton et al. (2011a) also employed an SISO NARMAX model, but with the input $p^{1/2}v^{4/3}B_T \sin^6(\theta/2)$, where p is the solar wind dynamic pressure, B_T is the tangential IMF in the GSM y–z plane ($B_T = \sqrt{B_y^2 + B_z^2}$) and θ is the clock angle of the IMF in the GSM y−z plane. A quadratic polynomial was set as the nonlinear function F and the maximum lags of the past output and noise were 2 and for the input the maximum lag was 6. The data employed for the training of the model was for the period from 1000 UTC March 18, 2001 to 0800 UTC September 7, 2001. Fig. 1 displays the output Dst index and input coupling function, C, used as the training data for the model.

The model was validated using the correlation tests and the predictive performance of the model was assessed over a period of 11 years of data from the start of 1998 to the end of 2008, excluding the training period, using the CC, the prediction efficiency (PE), and coherency function. The CC for the model was 0.9751, indicating the that model closely follows the trends of the measured Dst index. The PE is defined as

$$E_{PE} = 1 - \frac{\sum_{t=1}^{N}\left[(y(t)-\hat{y}(t))^2\right]}{\sum_{t=1}^{N}\left[(y(t)-\bar{y}(t))^2\right]} \quad (28)$$

and gives an indication of the accuracy of the predictions. A high PE (~1) will indicate a low error and a low PE (≤0) shows a poor accuracy because the mean squared error of the model is greater than the variance of the measured output. The PE for the Dst model was 0.9518, indicating an accurate model for this 11-year period.

The coherency function is defined as

$$C_{y\hat{y}} = \frac{\left|P_{y\hat{y}}(f)\right|}{P_{yy}(f)P_{\hat{y}\hat{y}}(f)} \quad (29)$$

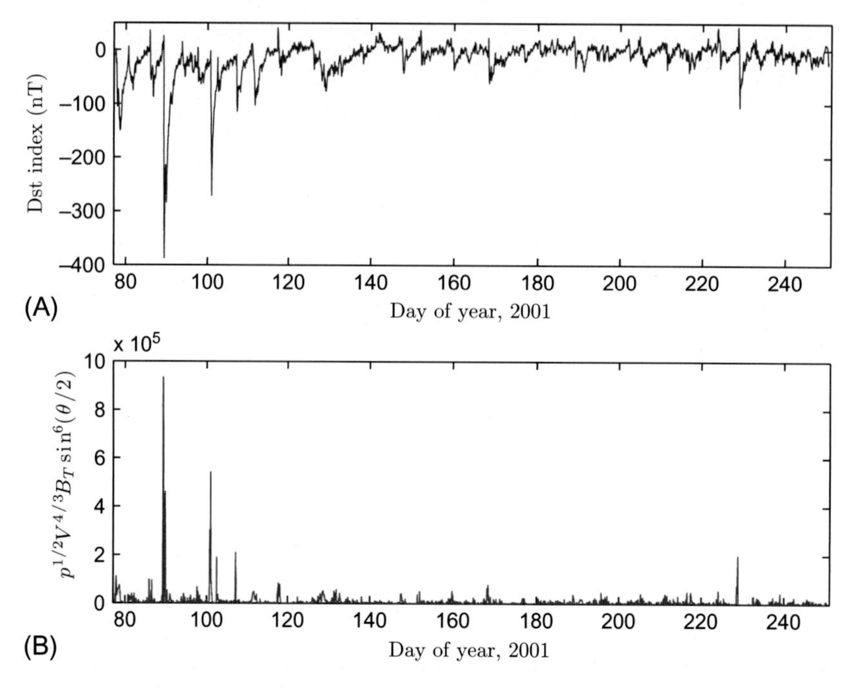

FIG. 1 Output and input data employed for the training of the NARMAX model. (A) The output of the system, the Dst index. (B) The input to the system, the coupling function C.

where P is the cross-spectral density for the subscripts at frequency f. The coherency function shows the relationship of the frequencies between the measured and estimated Dst and can therefore ascertain how well the model performs for the low-frequency events of a geomagnetic storm. A geomagnetic storm takes place on a time scale from 1 to 5 days. Therefore, a large coherency at these frequencies would imply the model is able to forecast the main phase and recovery phase of the storm. To calculate the coherency, the data were divided into 32 bins of 1024 data point intervals, over which the coherency could be averaged. Fig. 2 displays the coherency of the model. The low frequencies of a geomatic storm are around 0.01 to 0.04 h^{-1}, at which Fig. 2 shows a very high coherency for the model, suggesting these frequencies are forecast with a very high accuracy.

3.1.2 Continuous Time Dst model

Zhu et al. (2006) developed a continuous time model of the Dst index. The procedure employed the FROLS algorithm to derive an SISO NARMAX model of the Dst index, where vB_s was the input. The model was trained on two periods, on from August 16, 1998 to October 20, 1998 and the other from August 15, 1999 to October 20, 1999. After the model was validated by the correlation tests, the discrete time NARMAX model was transferred to the continuous time domain. To achieve this, the NARMAX models are first transferred to the frequency domain using the generalized frequency response functions (GFRF), which have been shown to represent nonlinear systems in the frequency domain (Billings and Tsang, 1989). The nth-order GFRF, H_n, of a system is defined as

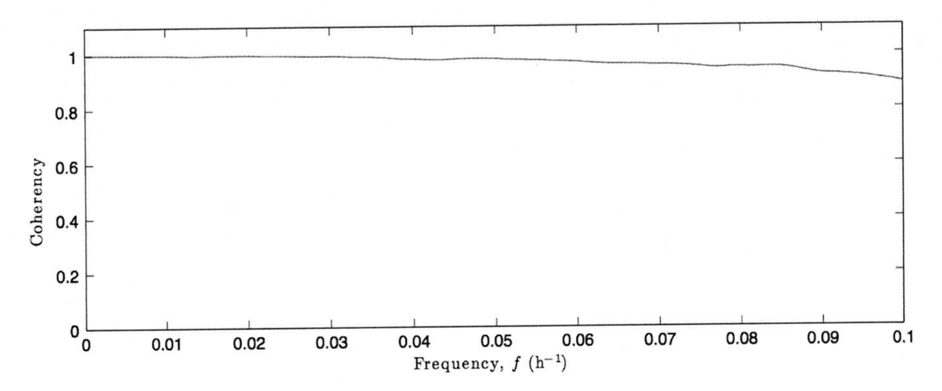

FIG. 2 Coherency between the measured and model *Dst*.

$$H_n(f_1,\ldots,f_n) = \int_{-\infty}^{\infty}\cdots\int_{-\infty}^{\infty} h(\tau_1,\ldots,\tau_n)e^{-j2\pi(f_1\tau_1+\cdots+f_n\tau_n)}d\tau_1,\ldots,d\tau_n \tag{30}$$

where $h_n(\tau_1,\ldots,\tau_n)$ is the nth Volterra Kernel. if $n = 1$, this is the linear frequency response function, while for $n > 1$, the GFRF represent the nonlinear system dynamics. The GFRF were computed from the NARMAX model using the recursive GFRF algorithm (Billings and Peyton Jones, 1990).

It is possible to represent a continuous time system by many different discrete time models, because the discrete time model can be sampled at different rates. However, the GFRF that results from these different discrete time models will all possess the same features in the frequency domain (e.g., resonances) that correspond to the dynamics of the underlying continuous time system. Therefore, the continuous time models that result from the GFRF of the discrete time models should be the same. The continuous time model can be reconstructed from the GFRF using the algorithm by Li and Billings (2001). Zhu et al. (2006) reconstructed the continuous time model as

$$20.8\frac{dDst}{dt} + Dst + 1.29vB_sDst + 34.8vB_s + 48.4\frac{dvB_s}{dt} + 24(vB_s)^2 = 0 \tag{31}$$

The forecasting ability of the continuous time model was assessed on a time period from August 15, 1999 to October 20, 1999 where the PE was 0.9974.

3.1.3 *MISO Dst*

Zhu et al. (2007) deduced an MISO Dst model using a NARMAX approach, using vB_s and the solar wind dynamic pressure p as the two inputs. The maximum lags of the output, inputs, and noise were 1, 3, and 1, respectively, resulting in a NARMAX Dst model

$$\begin{aligned} Dst(t) = F[&Dst(t-1),\\ &vB_s(t-1), vB_s(t-2), vB_s(t-3),\\ &p(t-1), p(t-2), p(t-3),\\ &e(t-1)] + e(t) \end{aligned} \tag{32}$$

where F was a quadratic polynomial.

The main aim of the study by Zhu et al. (2007) was to capture the dynamics of the sudden commencements observed in the Dst index, the positive increases in Dst that often occur before geomagnetic storms. These are thought to be magnetopause currents generated by the compression of the magnetosphere under high solar wind dynamic pressure (Burton et al., 1975).

The resultant model was trained on a period from August 15, 1999 to October 20, 1999 and validated by the correlation tests. The model was subsequently tested on four geomagnetic storms between 1998 and 2000. The 1-h ahead forecast of these storms had a high CC, averaging 0.987. They concluded that the two-input NARMAX model provided a more comprehensive model of the Dst index as the introduction of the solar wind pressure as an input allowed them to capture the dynamics of the sudden commencements in the model.

3.1.4 Kp Index

Another index that the NARMAX approach has been used to model is the Kp index. To achieve this, Ayala Solares et al. (2016) employed an MISO NARX model, with velocity v, vB_S, B_S, p, and $\sqrt{p}$ as inputs. One of the problems with forecasting a one-step-ahead estimate of the Kp index is that the Kp index has a cadence of 3 h, yet the solar wind measurements taken at the Lagrange point L1 can influence the Kp within an hour. Therefore, rather than forecast a 3-h-ahead forecast for Kp, Ayala Solares et al. (2016) computed a model to forecast 1 h ahead. As a result, the lags of the Kp employed in the model were 3 and 6 h, while the inputs had time lags of 1 and 2 h, resulting in

$$\begin{aligned} Kp(t) = F[&Kp(t-3), Kp(t-6), \\ &v(t-1), v(t-2), B_S(t-1), B_S(t-2), \\ &vB_S(t-1), vB_S(t-2), p(t-1), p(t-2), \\ &\sqrt{p}(t-1), \sqrt{p}(t-2)] + e(t) \end{aligned} \tag{33}$$

where F was a quadratic polynomial.

The models were trained on a period from January 1, 2000 to June 31, 2000 using a modified version of the FROLS algorithm by Billings and Wei (2008) and then validated by the correlation tests. The resulting model was assessed using the CC and also the PE. For a period between July 1, 2000 to December 31, 2000 the CC was 0.9156 and the PE was 0.8287.

3.2 Radiation Belt Electron Fluxes

With long-term electron flux data from the geostationary satellites, the outer radiation belts have been another attractive area in which systems science techniques were applied, to model the outer radiation belts.

3.2.1 GOES High Energy

Boynton et al. (2015) developed an MISO NARMAX model of the >800 keV and >2 MeV electron flux channels on the GEO GOES satellites. The main objective of this study was to provide an accurate online forecast for 1 day ahead. The inputs employed for this model were the solar wind velocity v, density n, z component of the GSM IMF B_z, Dst index, and the fraction of time that the solar wind remains southward within each day, τ_{B_s}. F was set to be a cubic polynomial and the maximum lags for output, inputs, and noise were set to 4.

The >800 keV electron flux model was trained on data starting on April 10, 2010 and ending on December 31, 2010. While the >2 MeV electron flux model was chosen to start on July 11, 2004 and end on October 11, 2005. Both models were validated by the correlation tests. The FROLS algorithm only selected monomials consisting of past output, v, n, and τ_{B_s}, indicating that B_z and the Dst index have a negligible role in influencing the electron fluxes at high energies.

The model's predictive performance was assessed using the PE and the CC. The predictive performance of the >0.8 MeV model was tested on a period between January 1, 2011 and June 30, 2012, which resulted in a PE of 0.700 and CC of 0.847. The >2 MeV model was tested on a period starting on April 14, 2010 and ending on June 30, 2012 and yielded a PE of 0.786 and CC of 0.894. These results indicate that both models have an error much smaller than the measured electron flux variance and have a high dependence between the measured and estimate, with the >2 MeV model having a slightly higher performance than the >0.8 MeV model. This is reflected in Fig. 3 where >0.8 MeV model seems less accurate due to the observed overshoots.

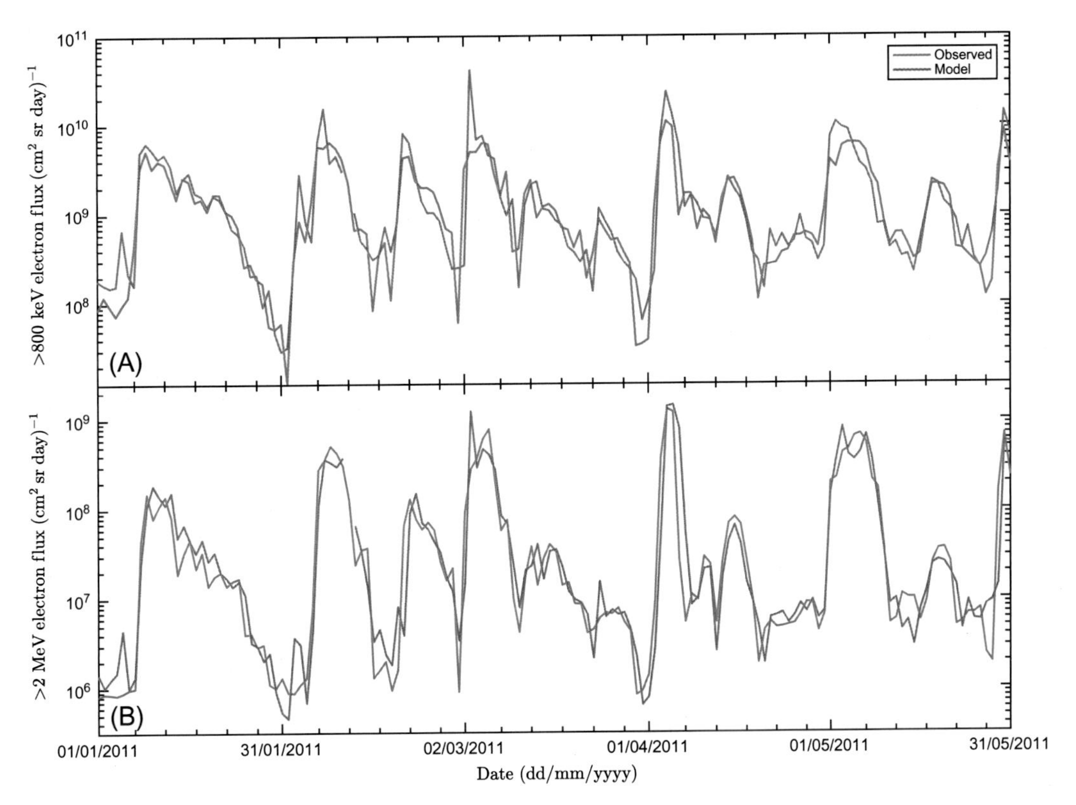

FIG. 3 Model forecast showing measured electron flux in *blue* and the model estimate in *orange* for (A) >0.8 MeV electron flux and (B) >2 MeV electron flux from January 1, 2011 to May 31 2011.

Since July 2012, these models have been implemented in real time to provide an online forecast at http://www.ssg.group.shef.ac.uk/USSW/UOSSW.html. The model, named SNB^3GEO, uses real time data provided by the National Oceanic and Atmospheric Administration (NOAA) Space Weather Prediction Center (SWPC) to provide a forecast of the electron fluxes at GEO for the following day.

3.2.2 SNB^3GEO Comparison With NOAA REFM

Balikhin et al. (2016) compared the 1 day ahead >2 MeV SNB^3GEO electron flux model with the NOAA relativistic electron flux model (REFM). The REFM has been running at NOAA SWPC since 1996. The model is based on the linear prediction filter developed by Baker et al. (1990), which also forecasts the 1 day ahead >2 electron flux at GEO using past values >2 electron flux and solar wind velocity as inputs. The REFM also uses a function of the model's residuals to improve the short-term predictions.

The data employed for the comparison only used the archived forecasts from both models for the same period of time: from March 2, 2012 to January 1, 2014. These were compared with the measured data, which was from the observations of the >2 MeV channel of the westward looking energetic proton electron and alpha detector (EPEAD) on the GOES 13 satellite (Hanser, 2011), as this data is provided in real time.

The assessment criteria used to compare the accuracy of the 1-day ahead predictions of the models were the PE, CC, and the Heidke skill score (HSS). The HSS can be used to assess models on their forecast of particular events and is defined as

$$E_{\mathrm{HSS}} = \frac{2(xw - yz)}{y^2 + z^2 + 2xw + (y+z)(x+w)} \tag{34}$$

where x is the number of successful forecasts of the event happening, w is the number of successful forecasts of the event not happening, y is the number of times the model forecasts the event but the event does not happen, and z is the number of times an event occurs but the model does not forecast the event. It represents the total number of correct forecasts divided by the total number of observations with the expected number of correct forecasts by chance subtracted.

The PE and CC were calculated for both fluxes and $\log_{10}$(fluxes). Table 1 is a reprint of Table 1 in Balikhin et al. (2016) and shows that the SNB^3GEO has a 5–10% better accuracy rate than the REFM, discounting the PE for REFM fluxes. The PE for REFM fluxes is negative, indicating the MSE is greater than the variance and is considerably more inaccurate than the forecasts by SNB^3GEO.

TABLE 1 A Comparison of the PE and CC Obtained by Comparing the Forecasts of the >2 MeV Electron Flux and $\log_{10}$(Flux) From the REFM and SNB^3GEO Models With Measurements From the GOES 13 Satellite (Balikhin et al., 2016)

Model	PE Flux	Correlation Flux	*PE* $\log_{10}$ Flux	Correlation $\log_{10}$ Flux
REFM	−1.31	0.73	0.70	0.85
SNB^3GEO	0.63	0.82	0.77	0.89

TABLE 2 Contingency Tables and Heidke Skill Scores for the REFM Predictions (Balikhin et al., 2016)

Fluence (cm^{-2} sr^{-1} day^{-1})	$>10^8$	$>10^{8.5}$	$>10^9$
REFM HSS	0.666	0.482	0.437
SNB^{3}GEO HSS	0.738	0.634	0.612

The HSS was used to assess the model forecasts during events where the fluxes are high, because producing accurate estimates of high fluxes is the most essential demand for a relativistic electron flux model. Therefore, the binary events chosen for the HSS were if the fluxes were $>10^8$, $>10^{8.5}$, and $>10^9$. Table 2 shows the HSS of the two models for the three cases. SNB^{3}GEO has a higher HSS than REFM for each of the three cases, indicating that SNB^{3}GEO is more accurate at forecasting high fluxes.

3.2.3 *GOES Low Energy*

Boynton et al. (2016) developed a set of MISO NARMAX models for five differential electron flux energies at GEO: 30–50, 50–100, 100–200, 200–350, and 350–600 keV. These energies are measured by the MAGnetospheric Electron Detector (MAGED) (Hanser, 2011) onboard the GOES 13 spacecraft. Each of the models was trained on data from March 1, 2011 to February 28, 2013 and validated using the correlation tests.

The algorithm to model the 24-h average electron fluxes used the solar wind velocity v and density n, the amount of time the IMF is southward in a 24-h period τ_{Bs}, the Dst index, and the term resulting from the coupling function proposed by Balikhin et al. (2010) and Boynton et al. (2011b), $B_T \sin^6(\theta/2)$ (where $B_T = \sqrt(B_y^2 + B_z^2)$ is the tangential IMF and $\theta = \tan^{-1}(B_y/B_z)$ is the clock angle of the IMF). There were two lags for the outputs and noise at 24 to 48 h, and 47 lags for each of the inputs from 2 to 48 h. Therefore, the NARXAX model for each of the electron flux energies J was

$$\begin{aligned} J(t) = F[&J(t-24), J(t-48), \\ &v(t-2), v(t-3), \ldots, v(t-48), \\ &n(t-2), n(t-3), \ldots, n(t-48), \\ &T_{Bs}(t-2), T_{Bs}(t-3), \ldots, T_{Bs}(t-48), \\ &Dst(t-2), Dst(t-3), \ldots, Dst(t-48), \ldots, \\ &B_T \sin^6(\theta/2)(t-2), B_T \sin^6(\theta/2)(t-3), \ldots, B_T \sin^6(\theta/2)(t-48), \\ &e(t-24), e(t-48)] + e(t) \end{aligned} \tag{35}$$

where the lags are in hours. The nonlinear function F employed a cubic polynomial.

The forecast time of each model (i.e., the amount of time the model is able to forecast into the future) depends on the minimum exogenous lag selected within the final model. For example, with a minimum exogenous lag of 6 h $J(t) = F[I(t-6), \ldots]$, using the value of the input at the present time t, it is possible to estimate the electron flux 6 h into the future, $J(t+6) = F[I(t), \ldots]$. As a result, the models have different forecast horizons.

TABLE 3 The Performance of the Five Low-Energy Electron Flux Models as Well as the Forecast Horizon

Model (keV)	Forecast Horizon (h)	PE	CC
30–50	10	0.669	0.820
50–100	12	0.692	0.835
100–200	16	0.732	0.856
200–350	24	0.716	0.849
350–600	24	0.736	0.859

The models were validated and then the forecast performance was analyzed statistically using the PE and CC on data from March 1, 2013 to February 28, 2015. Table 3 shows the PE and CC for each model along with the forecast horizon of the models. The PE for each of the models are vary between 0.669 and 0.736, with the lowest energy having the lowest PE and the highest energy having the highest PE. These values indicate the models are accurate as the MSE is well within the variance of the fluxes. The CC vary between 0.82 and 0.859, which demonstrates that the models have a strong relationship with the measured electron flux.

Fig. 4 shows the electron flux measured by GOES 13 in blue with the forecast by the model in orange, for each of the five energies in panels (A)–(E) and the Dst index in panel (F) between April 15, 2013 and May 15, 2013. A geomagnetic storm of ~ -50 nT occurs on April 24, which leads to an increase in the five energies of the electron flux. This is forecast in each of the models, with the forecast's electron flux increasing plus/minus a few hours of the actual timing of the flux increase. A subsequent geomagnetic storm of ~ -65 nT occurs on May 1, 2016. This corresponds with an enhancement in the two energies <100 keV but a dropout of electron fluxes for the three energies >100 keV, which recover back to the prestorm levels the next day. The low-energy models are able to forecast the enhancement for the two lower energies, but miss the dropout of the three higher energies. An explanation as to why the models struggle to forecast the dropouts could be due to the faster time scales of the dropouts, within a couple of hours. If the time scales of the dropouts occur more quickly than the forecast horizon, then the model will be unable to predict the dropouts.

All of these models have since been implemented in real time to forecast the electron fluxes at GEO. They have been providing forecasts at the University of Sheffield Space Weather website (www.ssg.group.shef.ac.uk/USSW2/UOSSW.html) since 2016.

3.3 Summary of NARMAX Models

Tables 4 and 5 summarize the characteristics of the NARMAX geomagnetic indices models and radiation belt models discussed in this section, respectively. The tables show the training period of the respective model, with the assessment period and the performances of the models for the given metric. The CC is the most used metric for assessing the performance of the models. However, a model with a high CC close to 1 may still miss important aspects of the system dynamics. For example, the Dst index model by Boynton et al. (2011a) had a CC of 0.9751 over the 11 years of data that the model was assessed, but when analyzing the model

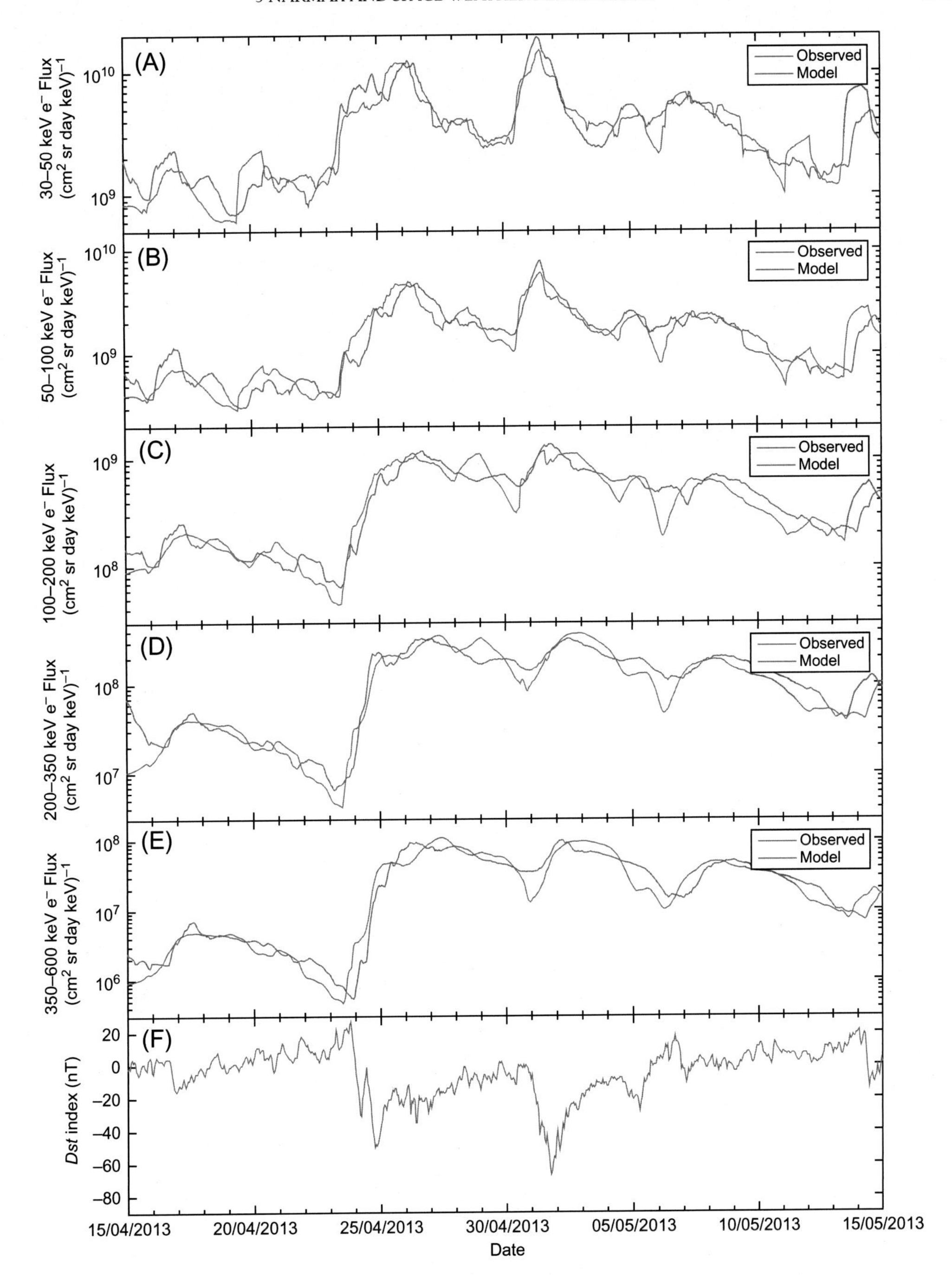

FIG. 4 The daily average electron flux measured by GOES in *blue* and the model forecast in *orange* for the period from April 15, 2013 to May 15, 2013 (Panel (A) 30–50 keV; (B) 50–100 keV; (C) 100–200 keV; (D) 200–350 keV; (E) 350–600 keV, with the Dst index in Panel (F).

TABLE 4 The Performance of the NARMAX Geomagnetic Indices Models

Model	Training Period	Assessment Period	Performance
SISO Dst (Boaghe et al., 2001)	January 21, 1979 to March 14, 1979	March 15, 1979 to June 31, 1979	CC = 0.989
SISO Dst (Boynton et al., 2011a)	March 18, 2001 to September 7, 2001	January 1, 1998 to December 31, 2008 excluding training	CC = 0.9751 PE = 0.9518
Continuous Dst (Zhu et al., 2006)	August 15, 1999 to October 20, 1999	August 15, 1999 to October 20, 1999	PE = 0.9974
MISO Dst (Zhu et al., 2007)	August 15 to October 20, 1999	4 storms between 1998 and 2000	CC = 0.987
Kp (Ayala Solares et al., 2016)	January 11 to June 31, 2000	July 1, 2000 to December 31, 2000	PE = 0.8287 CC = 0.9156

TABLE 5 The Performance of the NARMAX Radiation Belt Models

Model	Training Period	Assessment Period	Performance
>2 MeV (Boynton et al., 2015)	July 11, 2004 to October 11, 2005	April 14, 2010 to June 30, 2012	PE = 0.786 CC = 0.894
>800 keV (Boynton et al., 2015)	April 10, 2010 to December 31, 2010	January 1, 2011 to June 30, 2012	PE = 0.700 CC = 0.847
30–50 keV (Boynton et al., 2016)	March 1, 2013 to February 28, 2015	March 1, 2013 to February 28, 2015	PE = 0.669 CC = 0.820
50–100 keV (Boynton et al., 2016)	March 1, 2013 to February 28, 2015	March 1, 2013 to February 28, 2015	PE = 0.692 CC = 0.835
100–200 keV (Boynton et al., 2016)	March 1, 2013 to February 28, 2015	March 1, 2013 to February 28, 2015	PE = 0.732 CC = 0.856
200–350 keV (Boynton et al., 2016)	March 1, 2013 to February 28, 2015	March 1, 2013 to February 28, 2015	PE = 0.716 CC = 0.849
350–600 keV (Boynton et al., 2016)	March 1, 2013 to February 28, 2015	March 1, 2013 to February 28, 2015	PE = 0.736 CC = 0.859

in detail, it would miss the small increases in Dst before the main phase of the storm, known as sudden commencements. Another issue for systems such as the Dst index, which is for the majority of the signal constant with occasional negative peaks, is that the persistence model ($Dst(t) = Dst(t-1)$) will also have a high CC. As such, when assessing forecast models, it is best to use a variety of metrics and also compare with the persistence model.

The NARMAX models of the radiation belt are each for a different energy range, so no comparison can be made between them; however, many NARMAX models have been developed for the Dst index. The first Dst model, developed by Boaghe et al. (2001), with an aim more toward analyzing the model in the frequency domain, which is discussed in Section 4.3. Therefore, the predictive ability of the model was not discussed in detail and only the CC was given for a three-and-a-half month period. The continuous time model by Zhu et al. (2006) was mainly used to investigate its similarities with an analytically derived model from first principles by Burton et al. (1975). The forecasting ability was only tested on the same period on which it was trained, which is not good practice for assessing a model's performance. The MISO Dst model by Zhu et al. (2007) employed solar wind dynamic pressure as a second input in an attempt to capture the dynamics of the sudden commencements. The SISO Dst model by Boynton et al. (2011a) was to test the solar wind-magnetosphere coupling function, which is discussed in Section 4.1.

Trying to compare the forecast ability of all these models from Table 4 is difficult, as each model is assessed on a different period using different metrics. However, all models are simple polynomials and are given in their respective publications. Thus, the models can be easily implemented and can then be compared on the same period of data. The test data were chosen to be from January 1, 2011 to December 31, 2016. The PE and CC of the 1-h ahead Dst model forecasts for this period are displayed in Table 6, as well as the persistence model, Dst($t-1$). This shows that the model by Boynton et al. (2011a) had the more accurate performance over this period, followed by the MISO model by Zhu et al. (2007), then the continuous model by Zhu et al. (2006) and the model by Boaghe et al. (2001) had the lowest CC and PE. The persistence model had a higher PE and CC than the model by Boaghe et al. (2001) during this period. One of the reasons why the model by Boaghe et al. (2001) has the poorest performance could be due to the data on which the model was trained. The Boaghe et al. (2001) model was trained on solar wind data from the IMP-8 spacecraft, while the other models were trained on data from the ACE spacecraft. The data that the models were tested on in Table 6 were from the ACE spacecraft, which may be disadvantageous to the model by Boaghe et al. (2001).

Fig. 5 shows the model output of the models in Panel (A) and their errors with measured Dst in Panel (B) for a geomagnetic storm on March 9, 2012, highlighting the sudden commencement (blue-dashed line) and peak in the main phase of the storm (red-dashed line). The figure shows a small increase in the Dst index (green line), known as a sudden commencement, at 0100 UTC on March 9, 2012 (blue-dashed line), which no model manages to forecast. After this small increase, the Dst index starts to decrease in the main phase of the storm, which is initially closely matched by the Zhu et al. (2006) model (orange) and the Boynton et al. (2011a) model (purple), while the Zhu et al. (2007) model (yellow) seems to lag the measured Dst and

TABLE 6 The Performance of the NARMAX Dst Index Models Between January 1, 2011 to December 31, 2016

Model	PE	CC
SISO Dst (Boaghe et al., 2001)	0.9281	0.9648
SISO Dst (Boynton et al., 2011a)	0.9615	0.9807
Continuous Dst (Zhu et al., 2006)	0.9515	0.9766
MISO Dst (Zhu et al., 2007)	0.9544	0.9770
Persistence, Dst($t-1$)	0.9484	0.9742

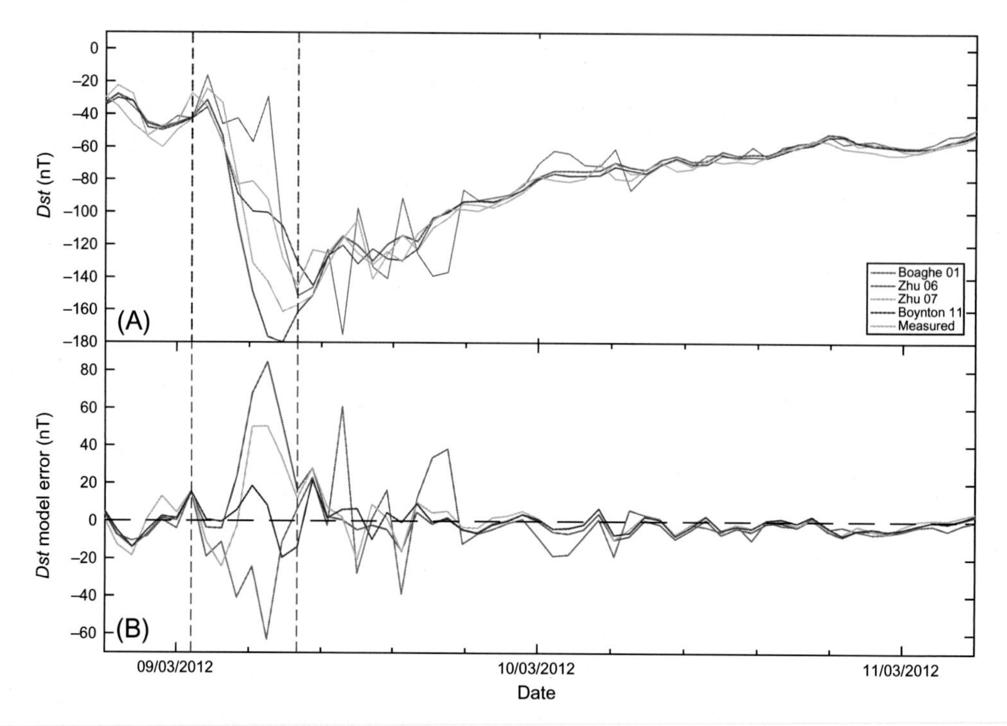

FIG. 5 Dst index model comparison on a geomagnetic storm on the March 9, 2012. Panel (A) shows the Dst index (*green*), Boaghe et al. (2001) model (*blue*), Zhu et al. (2006) model (*orange*), Zhu et al. (2006) model (*yellow*), Boynton et al. (2011b) model (*purple*). Panel (B) shows the residuals of the models. The *blue-dashed line* indicates the sudden commencement and the *red-dashed line* indicates the peak of the main phase of the storm.

the Boaghe et al. (2001) misses the start of the main phase. There is a pause in the main phase, which no model manages to forecast, and instead the Zhu et al. (2006) and the Zhu et al. (2007) models both overshoot the measured model with errors of ~80 and ~50 nT, respectively. The peak in the main phase of the storm occurs at 0800 UTC on March 9, 2012, which is forecast closest by the Boaghe et al. (2001) model; however, this model also has some of the largest errors throughout the storm. During the recovery phase of the storm, after the peak as the Dst returns to prestorm levels, the Zhu et al. (2006, 2007) and Boynton et al. (2011a) models closely match the Dst index, while the Boaghe et al. (2001) model has some significant errors. Overall, the model by Boynton et al. (2011a) has, on average, the fewest errors during this storm.

4 NARMAX AND INSIGHT INTO THE PHYSICS

One of the advantages the NARMAX approach has over other data-based modeling techniques is that the models can be physically interpretable. The NARMAX models can result in a simple polynomial, where it can be seen how the inputs change the output. This has led to some insight into the physics of space weather.

4.1 NARMAX Deduced Solar Wind-Magnetosphere Coupling Function

There have been many suggested solar wind-magnetosphere coupling functions (CF), a function based on solar wind parameters that is able to predict the dynamics of the magnetosphere. One such example is the solar wind velocity multiplied by the southward IMF, vB_s, which was proposed by Burton et al. (1975). These CF are often used as the inputs to data deduced models of the geomagnetic indices (Klimas et al., 1999). Various methodologies to obtain the CF have been attempted. Burton et al. (1975) employed the dawn-dusk electric field and empirically found a vB_s relationship. Kan and Lee (1979) deduced a CF from first principles of dayside reconnection. Newell et al. (2007) employed the CC to determine the solar wind factor that had the highest correlation with geomagnetic indices, and then assembled these factors into a CF.

Boynton et al. (2011b) implemented the NARMAX approach to automatically identify a CF. The study aimed to utilize the structure selection procedure from the FROLS algorithm to construct a nonlinear monomial from solar wind input parameters. The ERR is able to identify the most influential monomials within the NARMAX polynomial, which is comprised of all possible cross-coupled combinations of past inputs, outputs, and noise terms.

The Dst index was chosen as the output, as the Dst index is a good representation of the condition of the magnetosphere, while the inputs were different solar wind parameters. These inputs, which were comprised of the factors within other CFs, were the square root of the pressure $p^{1/2}$, the sixth root of the density $n^{1/6}$, the velocities v and $v^{4/3}$, the southward IMF B_s, and the tangential IMF with the different functions of clock angle $B_T \sin^4(\theta/2)$ and $B_T \sin^6(\theta/2)$.

The data were split into 64 datasets consisting of 1000 equally spaced sampled data points from a period between 1998 and 2008. For each of the 64 datasets, the NARMAX FROLS algorithm was run, using the same inputs, lags, and NARMAX polynomial degree. A polynomial of degree 4 was used with maximum lags of 5 for output, input, and noise. This resulted in 64 different sets of monomials with their corresponding ERR. The autoregressive monomial $Dst(t-1)$ had the highest ERR in all the 64 sets. Because the aim of the study was to determine the most influential monomial consisting of solar wind parameters, the other monomials were normalized by the ERR of the $Dst(t-1)$ in each of the 64 sets. This normalized ERR (NERR) was then averaged across the 64 sets.

Table 7 shows the results in order of the average NERR. The higher the NERR, the higher the output variance explained by the monomial. The three monomials with the highest NERR are all of a similar structure, where each consists of density, velocity, tangential IMF, and a

TABLE 7 CFs Assembled by the NARMAX FROLS Algorithm, in Order of the Average NERR

Coupling Function	NERR (%)
$p^{1/2}v^{4/3}B_T \sin^6(\theta/2)(t-1)$	5.46
$p^{1/2}v^2B_T \sin^6(\theta/2)(t-1)$	3.18
$n^{1/6}v^2B_T \sin^4(\theta/2)(t-1)$	3.15

function of the IMF clock angle (because $p = \frac{1}{2}nv^2$, where the top two functions only differ by the power of velocity).

The top coupling function in Table 7 was used as the input to develop the SISO Dst model by Boynton et al. (2011a) described in Section 3.1.1. The main reason for the development of this model was to compare other data-based models that used vB_s as the sole input. This coupling function derived by Boynton et al. (2011b), like the vB_s function, is composed of multiple physical solar wind parameters, so they are, in a sense, "multiinput," but implemented in the NARMAX procedure as a single input. The model by Boynton et al. (2011a) was shown to have the better performance between January 1, 2011 to December 31, 2016 than the other models that used vB_s in Section 3.3.

The top two monomials according to the ERR both have a $\sin^6(\theta/2)$ factor. Balikhin et al. (2010) justified this from first principles using a similar methodology to Kan and Lee (1979). Kan and Lee (1979) deduced a CF containing a factor of $\sin^4(\theta/2)$ from the geometric relationship between the electric and magnetic fields. Balikhin et al. (2010) followed the procedure described by Kan and Lee (1979), starting off by calculating the reconnection electric field. The reconnection electric field, E_r, at the dayside magnetosphere is parallel to the line x_1x_2 in Fig. 6 and can be expressed as (Sonnerup, 1974)

$$E_r = v_{MS}B_{MS}\sin\left(\frac{\theta}{2}\right) \tag{36}$$

where B_{MS} is the magnetic field in the magnetosheath, v_{MS} the inflow velocity perpendicular to the magnetic field, and the angle θ is the angle between the magnetosheath magnetic field and the Earth's magnetic field (displayed in Fig. 6). Kan and Lee (1979) reasoned that

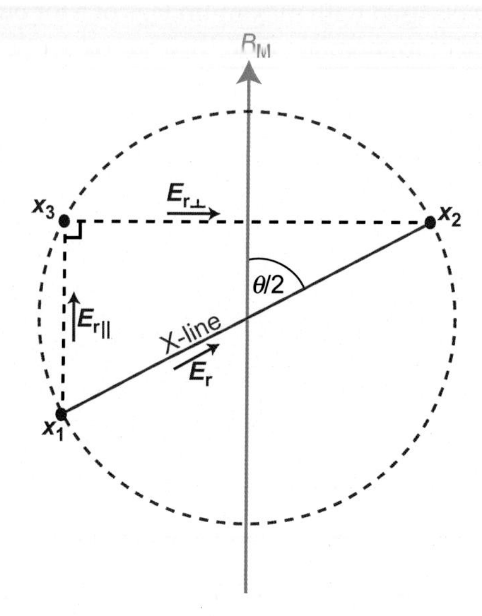

FIG. 6 The components of the reconnection electric field E_r, where the line x_1x_2 is the length of the X-line, l_0 and B_M is the Earth's magnetic field.

the reconnection field is the only component of the electric field in the magnetosheath that penetrates into the magnetosphere. Therefore, the potential difference across the polar cap, measured between x_2 and x_3, is due to the perpendicular component of the reconnection electric field

$$E_{r\perp} = E_r \sin\left(\frac{\theta}{2}\right) = v_{\rm MS} B_{\rm MS} \sin^2\left(\frac{\theta}{2}\right) \tag{37}$$

Balikhin et al. (2010) calculated the potential along the projection of the perpendicular component of the reconnection electric field

$$\Phi_M = v_{\rm MS} B_{\rm MS} \sin^2(\theta/2) l_0 \sin(\theta/2) \tag{38}$$

$$\Phi_M = v_{\rm MS} B_{\rm MS} \sin^3(\theta/2) l_0 \tag{39}$$

however; during the calculation of the potential, Kan and Lee (1979) missed a factor of $\sin(\theta/2)$ in the substitution of the integration path between x_2 and x_3. When calculating the potential of the perpendicular component, the length of integration also needs to change so that it is along the perpendicular component of the electric field. As such, l_0 also needed to be multiplied by $\sin(\theta/2)$.

Kan and Lee (1979) used the expression for the cross-polar cap potential, Φ_M, to deduce the power produced by the solar wind dynamo as $P = \Phi_M^2/R$, where R is the total equivalent resistance. Employing the same method, but using the correct expression for the cross-polar cap potential, Balikhin et al. (2010) calculated the power produced by the solar wind dynamo as

$$P = \frac{\Phi_m^2}{R} = \frac{v^2 B^2}{R} \sin^6(\theta/2) l_0 \tag{40}$$

assuming that $v_{\rm MS} B_{\rm MS} = vB$ due to the magnetic flux conservation.

This results in a theoretical explanation of the $\sin^6(\theta/2)$ factor, automatically deduced from the NARMAX FROLS algorithm. The work by Balikhin et al. (2010) shows how the results of NARMAX can help identify the analytical explanation of the physical model.

4.2 Identification of Radiation Belt Control Parameters

The influence of the solar wind over the radiation belt is one of the most complex components of the magnetosphere system. Paulikas and Blake (1979) originally observed that the radiation belt electron fluxes had a linear correlation with solar wind velocity. However, with more data, Reeves et al. (2011) showed that the high energies had a far more complex relationship with solar wind velocity. The solar wind velocity was found to have triangular distribution with GEO electron fluxes.

Boynton et al. (2013) investigated the complex coupling between the solar wind and the radiation belt's electron fluxes at GEO. This was achieved using a similar NARMAX approach to that developed in Boynton et al. (2011b), by utilizing the structure selection stage of the NARMAX FROLS algorithm to indicate which combination of solar wind parameters influence a broad energy range of the daily averaged electron fluxes in the radiation belt.

The electron flux data, for 14 energies ranging from 24.1 keV to 3.5 MeV, were obtained from the LANL spacecraft. The data used in this study covered the period from September 22, 1989 to December 31, 2009. These were used as the output data. The input data were the solar wind velocity v, density n, dynamic pressure p, and the x, y, and z components of the IMF B_x, B_y, and B_z, respectively, in GSM coordinates.

The data were split into eight datasets consisting of equally spaced sampled data. The FROLS algorithm was run on each dataset using a quadratic polynomial and was set to five time lags for each of the inputs. Because the data is daily averaged, it is possible for the averaged solar wind parameter for 1 day to causally influence the average fluxes observed on that day. Therefore, these time lags were the current day at time t (0 time lag) and the time lags for the previous 4 days, $t-1, \ldots, t-4$; therefore, the input time lags were 0–4. This resulted in eight sets of monomials, for each electron flux energy range, which were then averaged to get the average ERR for each monomial.

Table 8 displays the top three monomials in order of ERR for each of the electron flux energies. A number of interesting points can be gained from this table. The first point is that for energies <2 MeV, the solar wind velocity can explain the majority of the electron fluxes' variance at GEO. Another point is that for energies >2 MeV, the solar wind density explains most of the electron fluxes variance. A final point concerns the relationship between the lags of the velocity and energy of the electron flux, where the lags increase with the energy.

TABLE 8 Results of the Solar Wind-Electron Flux NARMAX FROLS Analysis

Energy	M1	ERR_{M1} (%)	M2	ERR_{M2} (%)	M3	ERR_{M3} (%)
24.1 keV	$v(t)$	96.928	$v^2(t)$	2.824	$n(t)$	0.082
31.7 keV	$v(t)$	96.944	$v^2(t)$	2.825	$n(t)$	0.071
41.6 keV	$v(t)$	96.968	$v^2(t)$	2.819	$n(t)$	0.057
62.5 keV	$v(t)$	97.014	$v^2(t)$	2.798	$n(t)$	0.035
90.0 keV	$v(t)$	97.062	$v^2(t)$	2.769	$nv(t)$	0.026
127.5 keV	$v(t)$	74.880	$v(t-1)$	22.252	$v^2(t)$	2.082
172.5 keV	$v(t-1)$	65.687	$v(t)$	31.563	$v^2(t-1)$	1.736
270 keV	$v(t-1)$	97.476	$v^2(t-1)$	2.339	$B_z(t-1)$	0.022
407.5 keV	$v(t-1)$	84.116	$v(t-2)$	13.726	$v^2(t-1)$	1.626
625 keV	$v(t-1)$	75.876	$v(t-2)$	22.275	$v^2(t-1)$	0.610
925 keV	$v(t-2)$	96.162	$n(t)$	0.279	$v(t-4)$	0.238
1.3 MeV	$v^2(t-2)$	76.508	$nv(t-1)$	2.211	$nv(t)$	1.900
2.0 MeV	$n(t-1)$	53.692	$nv(t-1)$	13.561	$n^2(t-1)$	5.550
2.65 MeV	$n(t-1)$	51.504	$n^2(t-1)$	15.111	$v^2(t-2)$	6.128

Notes: The top three monomials in the order of ERR for each energy of the electron flux.

4.2.1 *Solar Wind Density Relationship With Relativistic Electrons at GEO*

For the two highest energies studies by Boynton et al. (2013), the solar wind density surprisingly explains the majority of the electron flux variance. The previous day's density accounts for over 50% of the ERR. The density plays an increasing role in the electron fluxes for energies $\geq$925 keV, according to the ERR. Balikhin et al. (2011) investigated the high ERR of density on relativistic electron fluxes at GEO. One of the first puzzles was to understand why the ERR criteria, solar wind density, seemed to have the most control over the fluxes, when other studies treated the solar wind velocity as the most significant parameter. To solve this problem, they investigated the relationship between solar wind velocity and density, as the result could be explained if density had a dependence on velocity, because the velocity would influence the fluxes through the density.

Fig. 7 shows the scatter plot of the velocity against the density during the period the ERR analysis was performed over, from September 22, 1989 to December 31, 2009. From this figure, Balikhin et al. (2011) stated that there is no simple functional relationship between the density and velocity. The red points on Fig. 7 highlight the points where the 2.65 MeV electron fluxes were $>10^{0.5}$ (cm^2 s sr keV)$^{-1}$ and shows that the high fluxes of electrons mainly occur at times of low density.

Balikhin et al. (2011) explained the electron flux-velocity-density relationship through a simple series of scatter plots to those displayed in Fig. 8. This shows the dependence of the 2.65 MeV electron flux on solar wind velocity for different solar wind densities' ranges. The lowest density range of $n \leq 0.8\,\mathrm{cm}^{-3}$ shows an increase of electron flux with an increase in

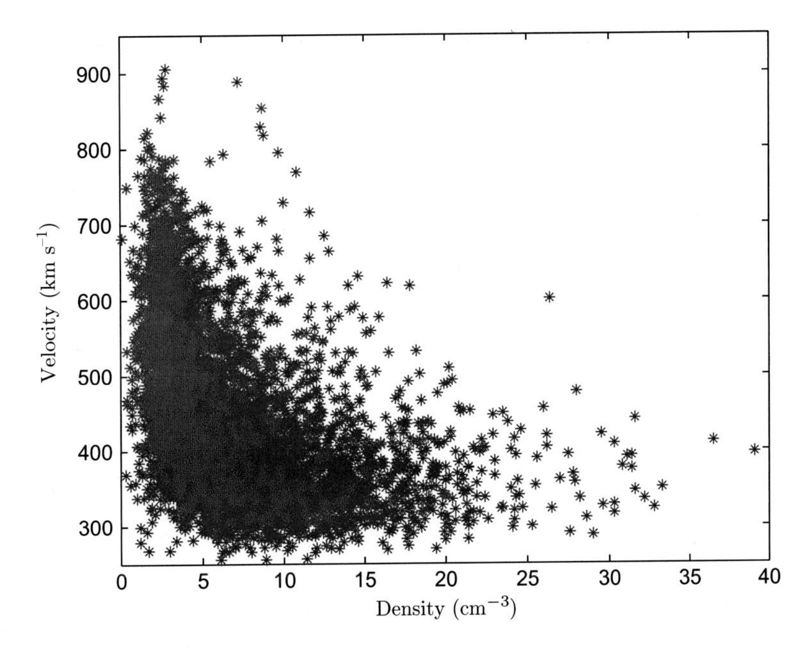

FIG. 7 Scatter plot showing the relationship between the daily averaged solar wind velocity and density, where the points in *red* indicate high fluxes greater than $10^{0.5}$ (cm^{2}ssrkeV)$^{-1}$ for *E*234.

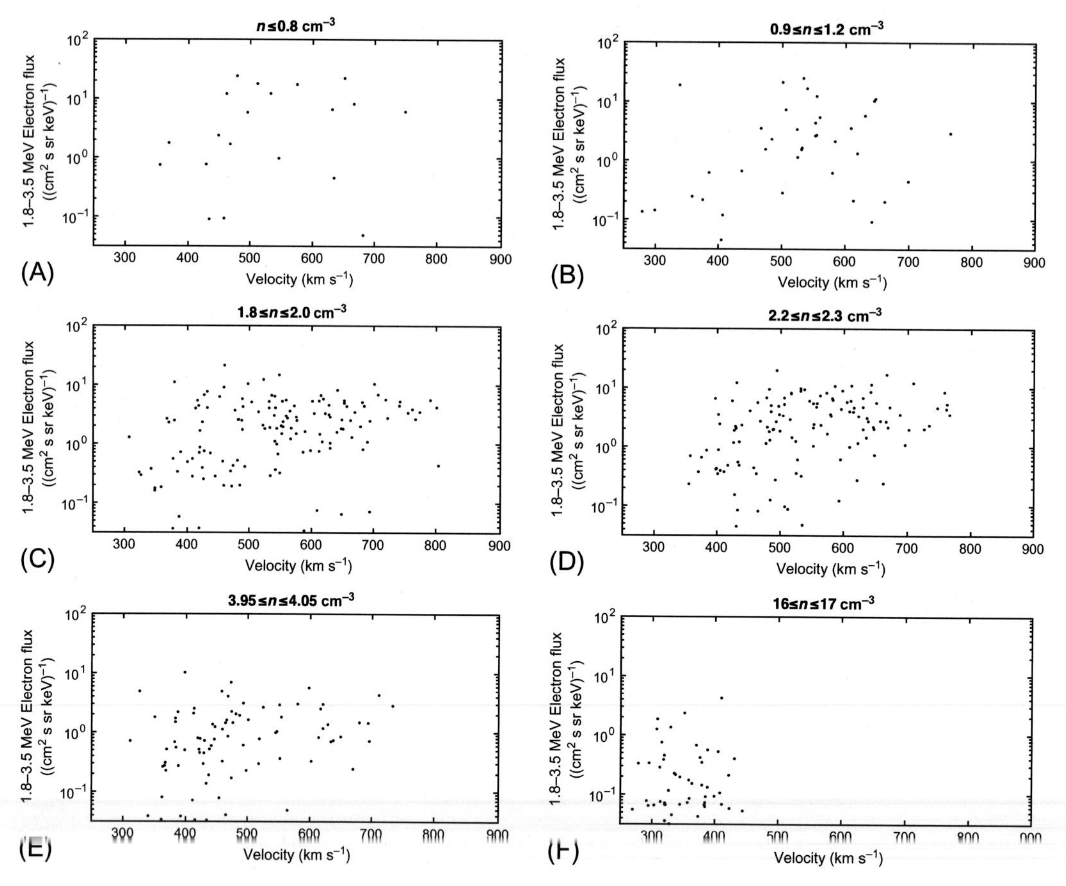

FIG. 8 The dependence of the 2.65 MeV electron flux on solar wind velocity for solar wind densities in the ranges (A) $n \leq 0.8\,\text{cm}^{-3}$, (B) $0.9 \leq n \leq 1.2\,\text{cm}^{-3}$, (C) $1.8 \leq n \leq 2.0\,\text{cm}^{-3}$, (D) $2.2 \leq n \leq 2.3\,\text{cm}^{-3}$, (E) $3.95 \leq n \leq 4.0\,\text{cm}^{-3}$, and (F) $16 \leq n \leq 17\,\text{cm}^{-3}$.

velocity until around $500\,\text{km}^{-1}$, where the fluxes saturate at approximately $10^{1.5}$ (cm^2 s sr keV)$^{-1}$ and there are no increases in the fluxes with further increases in velocity. The same pattern is observed in the next density window of $0.9 \leq n \leq 1.2\,\text{cm}^{-3}$ in Fig. 8C. The fluxes increase with velocity until a saturation at $\sim 500\,\text{km}^{-1}$ where the fluxes level off $\sim 10^{1.5}$ (cm^2 s sr keV)$^{-1}$. The subsequent figures show that this saturation point decreased in velocity and flux with density. Therefore, Balikhin et al. (2011) concluded that, for a fixed density, the values of the 2.65 MeV electron fluxes increase with velocity until they reach a saturation velocity where the maximum flux is reached and this saturation velocity and maximum flux are both inversely dependent on the density.

4.2.2 *Geostationary Local Quasilinear Diffusion vs. Radial Diffusion*

Table 8 shows that the time lags between the solar wind velocity and fluxes of energetic electrons are not constant, but have a relationship with the energy. For the five lowest energies, from 24.1 to 90.0 keV, the current day's velocity accounts for most of the ERR. For the electrons at 127.5 keV, the current day's velocity has the highest ERR, but the previous day's velocity also provides a significant contribution. For the next energy up, 172.5 keV, the lags switch and the contribution of the previous day's solar wind velocity has the higher ERR, while the current day's velocity has a smaller contribution. At 270 keV, the previous day's solar wind velocity becomes dominant. Between 407 and 625 keV, there is an increasing significance of the velocity from 2 days in the past, while the previous day's solar wind velocity has the higher ERR. For the energies of 925 keV and 1.3 MeV, the solar wind velocity from 2 days has the highest ERR.

Balikhin et al. (2012) sought to explain the velocity time lag relationship with the electron flux energies by comparing this to an estimate of the electron energy increase rate by local diffusion due to the interaction with waves. The upper limit of the wave-particle interaction electron flux increase rate was deduced by solving the energy diffusion equation (Horne et al., 2005)

$$\frac{\partial F}{\partial t} = \frac{\partial}{\partial E}\left[A(E)D\frac{\partial}{\partial E}\left[\frac{F}{A(E)}\right]\right] - \frac{F}{\tau_L} \tag{41}$$

where F is the electron distribution function, which describes how the electrons evolve in kinetic energy E over a time t, A is

$$A = (E + E_0)(E + 2E_0)^{1/2}E^{1/2}$$

D is the bounce-averaged energy diffusion coefficient, τ_L is the effective timescale for losses to the atmosphere, and E_0 is the rest energy of the electron.

An analytical solution to Eq. (41) is not possible in the general case for arbitrary dependences of A, D, and τ_L on energy. As such, Balikhin et al. (2012) considered no losses because the timescale of the losses are much larger than those of the electron flux increases, $\tau_L \to \infty$. The diffusion coefficients were shown by Horne et al. (2005) to be the constant over the energies between 100 and 1000 keV, so Balikhin et al. (2012) assumed D had no dependence on energy. However, the energy dependence on A does still not allow for a compact analytical solution. Therefore, Balikhin et al. (2012) simplified the problem by considering A for three energy ranges relative to the rest energy: (1) subrelativistic $E \ll E_0$, (2) $E \approx E_0$, which means $E - E_0 \ll E_0$, and (3) relativistic $E \gg E_0$. In each case, $A \propto E^\beta$, where β is equal to 1/2, 0, and 2 correspondingly. For the second case, $A \propto E^0$, then the solution to Eq. (41) becomes a standard diffusion equation with a known time scaling of $\sqrt{t}$. Therefore, Balikhin et al. (2011) solved this for a subrelativistic case (where $E \ll E_0$)

$$F(E,t) = KE(t + t_0)^{-5/4}\exp\left(-\frac{E^2}{4DE_0^2(t + t_0)}\right) \tag{42}$$

and the relativistic case (where $E \gg E_0$)

$$F(E,t) = KE^2(t + t_0)^{-3/2}\exp\left(-\frac{E^2}{4DE_0^2(t + t_0)}\right) \tag{43}$$

where K is a constant and t_0 is the initial time.

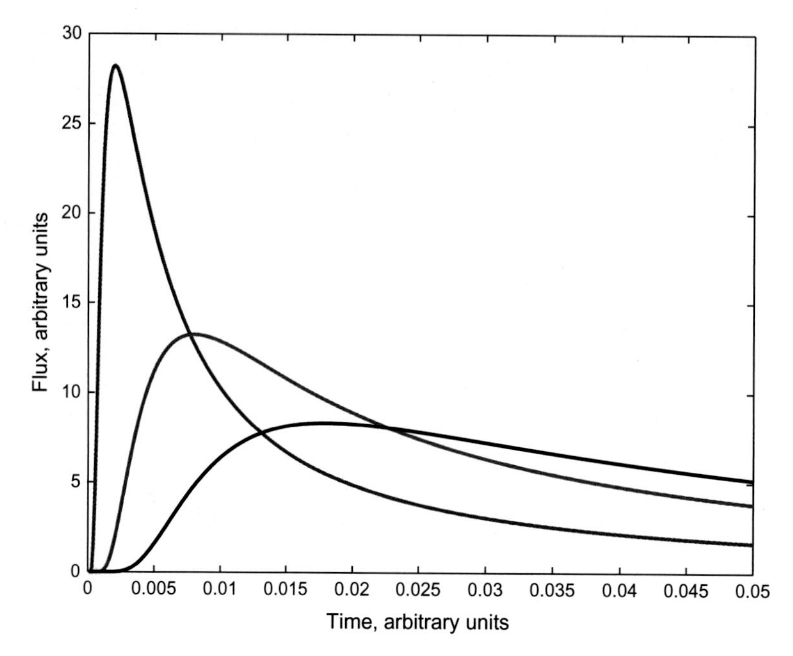

FIG. 9 The subrelativistic case, showing the evolution of fluxes in time for the normalized energies of 0.1 (*solid blue*), 0.3 (*solid red*), and 0.5 (*solid black*).

To translate the change of the distribution function into the change of fluxes J

$$F(E, \alpha_{eq}) = \frac{E + E_0}{c(E + 2E_0)^{1/2}E^{1/2}} J(E, \alpha_{eq}) \tag{44}$$

where α_{eq} is the equatorial pitch angle.

Figs. 9 and 10 show the electron fluxes for three values of normalized energy with respect to the rest energy, calculated from the solutions to the subrelativistic (Eq. 42) and relativistic (Eq. 43) cases, respectively. The normalized energies displayed in Fig. 9 for the subrelativistic case are 0.1 (solid blue), 0.3 (solid red), and 0.5 (solid black) and in Fig. 10 for the relativistic case are 3 (solid blue), 4 (solid red), and 5 (solid black). To quantify the energy time relationship, the moment in time when the flux reached 10% of maximum was chosen for a steep flux increase time. For the subrelativistic case, these times were $t_{E=0.1} = 4.60 \times 10^{-4}$, $t_{E=0.2} = 1.89 \times 10^{-3}$, and $t_{E=0.3} = 4.19 \times 10^{-3}$. Therefore, the ratio $t_{E=0.1} : t_{E=0.2} : t_{E=0.3} = 1 : 4.10 : 9.10$, which shows a time scaling that is approximately $\sqrt{t}$. For the relativistic case, $t_{E=3} = 0.386$, $t_{E=4} = 0.685$, and $t_{E=5} = 1.067$, resulting in a ratio of $t_{E=3} : t_{E=4} : t_{E=5} = 1 : 1.78 : 2.77$, which again is a time scaling of approximately $\sqrt{t}$. Both cases are similar to a diffusion equation with constant coefficients.

Balikhin et al. (2012) then compared these results with the energy-velocity time lag relationship found by Boynton et al. (2013). The relationship is plotted in Fig. 11, which displays the energy of the electron flux, against the effective time delay of the solar wind velocity on a log-log plot. The figure shows a much steeper relationship, where the time scaling is $\sim t^{1.5}$, than

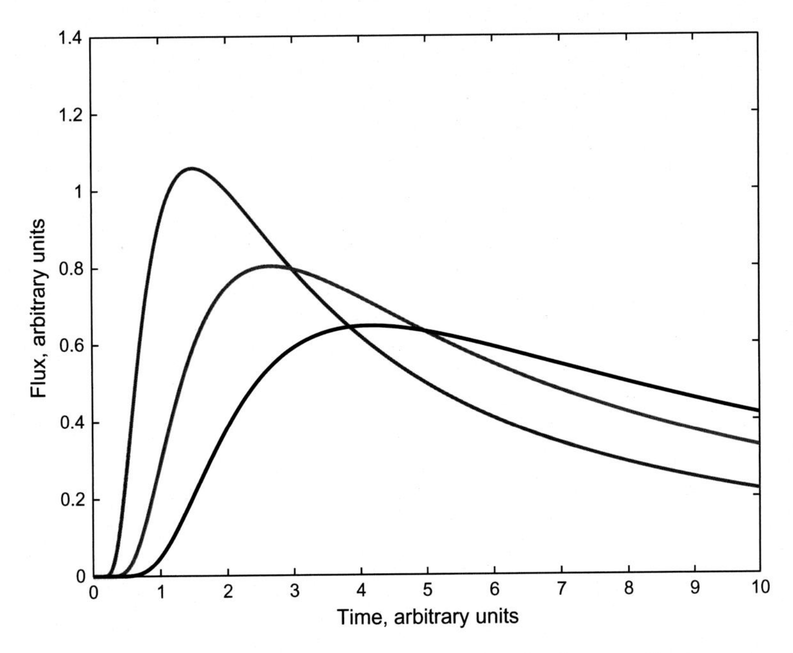

FIG. 10 The relativistic case, showing the evolution of fluxes in time for the normalized energies of 0.1 (*solid blue*), 0.3 (*solid red*), and 0.5 (*solid black*).

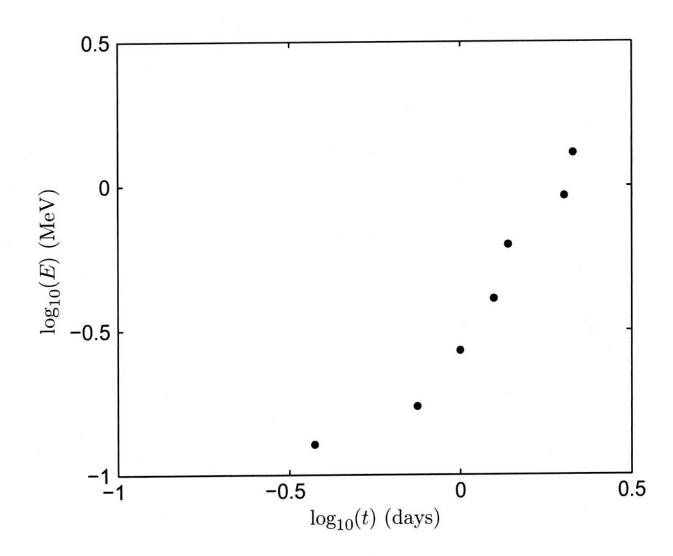

FIG. 11 The log-log plot showing the energy of the electron flux, against the effective time delay of the solar wind velocity calculated from the NARMAX FROLS results.

those of Eqs. (42), (43). In the case of Eqs. (42), (43), if the acceleration of a seed population to energies of about 270 keV takes 1 day, it should take more than 11 days to reach 925 keV, assuming a much lower energy seed population. Fig. 11 shows that 925 keV are reached much faster than this case, only 2 days. The conclusions of Balikhin et al. (2012) were that the time scaling of the energy diffusion equation is too slow to explain the increase of fluxes at GEO and that radial diffusion should also play a key role in the evolution of high-energy electron distributions at GEO.

4.3 Frequency Domain Analysis of the Dst Index

The model of the Dst index derived by Boaghe et al. (2001) was analyzed in the frequency domain to obtain insight into the nonlinearities involved in the energy storage process in the magnetosphere. The frequency domain analysis is more suitable for the analysis for physical aspects of the dynamics. The NARMAX model was used to calculate the first-, second-, and third-order GFRF (Eq. 30) using the algorithm by Billings and Peyton Jones (1990), which is able to directly map NARMAX models into the frequency domain. On examining the first-order GFRF $H_1(f_1)$, which represents the linear amplification of the spectral components, Boaghe et al. (2001) found the maximum amplification at very low frequencies $f \approx 0\,\mathrm{h}^{-1}$ and a second maximum, much lower than the first, at $f \approx 3\,\mathrm{h}^{-1}$. This resembles a linear system with two resonant frequencies.

The second-order GFRF $H_2(f_1,f_2)$ contained several peaks in the surface. The maxima in the surface of $H_2(f_1 = f_i, f_2 = f_j)$, where f_i and f_j are the coordinates of the maxima, can be physically interpreted as a nonlinear coupling between two spectral components of the input that transfers energy to the spectral component of the output at the summation of the frequencies $f_k = f_i + f_j$. Boaghe et al. (2001) found that the position of these maxima were along the line of $f_2 = -f_1$. This indicates that the coupling of the spectral components occurs at frequencies of the same magnitude, but opposite sign. The third-order GFRF $H_3(f_1,f_2,f_3)$ also exhibited maxima where the summation frequency $f_1 + f_2 + f_3 = 0$. This implies that there is a transfer of energy to the summation frequency of $f_k = 0$, which is the same frequency as the main maximum in $H_1(f_1)$. This was explained by Boaghe et al. (2001) in terms of a known well energy storage phenomena in nonlinear systems, where energy from finite frequencies is transferred to very low frequencies, for example, the Langmuir collapse (Zaharov, 1972).

Lang and Billings (1996) derived an expression of the output frequency response

$$Y(j2\pi f) = \sum_{n=1}^{D} Y_n(j2\pi f) \tag{45}$$

$$Y_n(j2\pi f) = \frac{1}{\sqrt{n}(2\pi)^{n-1}} \int_{f_1+\cdots+f_n=f} H_n(f_1,\ldots,f_n) \prod_{i=1}^{n} U(j2\pi f_i) d\sigma_{f_n} \tag{46}$$

where Y_n is the output spectrum, U is the input spectrum, n is the order, and σ_{f_n} is the area of a small element on the hyperplane $f = f_1 + \cdots + f_n$. Eqs. (45), (46) detail the relationship between the nonlinear system output frequency response, the GFRF, and the input spectrum and shows that the output frequency response is comprised of the summation the contributions of different orders. This interprets why the output energy could be stored at the summation frequency $f = 0$ when there is a ridge in the surfaces of $H_2(f_1,f_2)$ and $H_3(f_1,f_2,f_3)$ in their respective hyperplanes.

Another feature observed in the second-order GFRF, $H_2(f_f,f_2)$, was a ridge-like local maxima, which corresponded to a release of energy from very low frequencies to a finite frequency. However, this energy release feature was less prominent and could have been a result of the coarse time resolution of the data.

Balikhin et al. (2001) also examined the Dst index using GFRF, where they found the low-frequency maxima in $H_1(f_1)$ and another two local maxima corresponding to frequencies of 0.12 and $0.3\,h^{-1}$. As with Boaghe et al. (2001), $H_1(f_1)$ decreased with increasing frequency, suggesting that the linear response of the Dst decreases in the characteristic time scale of solar wind variation. This was in accordance with the analytical model by Burton et al. (1975), which introduced a special normalization in order to account for a lack of response to the short-time scale variations. However, the model by Burton et al. (1975) also assumed a linear relationship between Dst and a time delayed vB_s. If there is a time delay in magnetospheric response to solar wind, there should be an increase of H_1 with frequency, which was not found by either Balikhin et al. (2001) or Boaghe et al. (2001). Balikhin et al. (2001) also found a maxima in the second-order GFRF, $H_2(f_1,f_2)$, where the summation frequency was zero. Such nonlinear coupling can provide an alternative explanation of the energy storage instead of the time delay proposed by Burton et al. (1975).

5 DISCUSSIONS AND CONCLUSION

The NARMAX FROLS methodology has been applied to a number of space weather problems. The algorithms were first applied to the modeling of the Dst index, and then analyzing the model in the frequency domain (Boaghe et al., 2001). The subsequent studies also focused on the Dst index, developing a continuous time model (Zhu et al., 2006) and a multiinput model. The structure detection part of the FROLS algorithm was used to identify a multisolar wind parameter function for the Dst index (Boynton et al., 2011b), which was validated analytically by Balikhin et al. (2010). This coupling function was then used as an input in a Dst forecast model (Boynton et al., 2011a). In Section 3.3, these NARMAX Dst models were compared to find the model that has the best performance. To do this in a fair manner, the models were tested on the same data period from January 1, 2011 to December 31, 2016. The model by Boynton et al. (2011a) was found to have the highest PE and CC for this period, and therefore produced the more accurate forecasts.

The other area of space weather where the NARMAX methodology has been applied is the radiation belts, specifically the electron fluxes at GEO. At first, the FROLS algorithm was used to determine the solar wind parameters that have control over the GEO electron fluxes (Balikhin et al., 2012; Boynton et al., 2013), the results of which were used to verify acceleration processes in the radiation belts (Balikhin et al., 2012). Forecast models of the electron fluxes at GEO have also been developed for a wide range of energies (Boynton et al., 2015, 2016).

Open source software is available for developing NARMAX models that can be downloaded online, for example, https://github.com/rnwatanabe/FROLSIdentification (Watanabe and Kohn, 2014). This toolbox contains the FROLS algorithm and tools to develop GFRF. However, the authors would recommend anyone wanting to apply NARMAX to develop their own FROLS algorithm, as there are many variants on the basic FROLS algorithm that are easy to implement by modifying the basic FROLS algorithm.

The application of NARMAX methodologies to space weather, as well as many other machine learning approaches, is still in its infancy. Also, progress is still ongoing in the development of more advanced variants of the FROLS algorithm such as the iterative orthogonal forward regression by Guo et al. (2014) or the ultra orthogonal forward regression by Guo et al. (2016). It would be interesting to compare the models deduced by these newer algorithms to those that have already been produced by the standard FROLS algorithm using the same inputs and training data. There are many more aspects of space weather to explore using these techniques. To start with, there are a wide range of geomagnetic indices to analyze and develop forecast models. There are also completely different aspects of space weather, such as ionospheric TEC and geomagnetic induced currents, where there is an ever-increasing amount of data that can be utilized by systems science and machine learning techniques.

References

Ayala Solares, J.R., Wei, H.L., Boynton, R.J., Walker, S.N., Billings, S.A., 2016. Modeling and prediction of global magnetic disturbance in near-Earth space: a case study for Kp index using NARX models. Space Weather 14 (10), 2016SW001463. ISSN 1542-7390. https://doi.org/10.1002/2016SW001463.

Baker, D.N., McPherron, R.L., Cayton, T.E., Klebesadel, R.W., 1990. Linear prediction filter analysis of relativistic electron properties at 6.6 RE. J. Geophys. Res. 95 (A9), 15133–15140. ISSN 0148-0227. https://doi.org/10.1029/JA095iA09p15133.

Balikhin, M.A., Boaghe, O.M., Billings, S.A., Alleyne, H.S.C.K., 2001. Terrestrial magnetosphere as a nonlinear resonator. Geophys. Res. Lett. 28 (6), 1123–1126.

Balikhin, M.A., Boynton, R.J., Billings, S.A., Gedalin, M., Ganushkina, N., Coca, D., Wei, H., 2010. Data based quest for solar wind-magnetosphere coupling function. Geophys. Res. Lett. 37 (24), L24107. ISSN 0094-8276. https://doi.org/10.1029/2010GL045733.

Balikhin, M.A., Boynton, R.J., Walker, S.N., Borovsky, J.E., Billings, S.A., Wei, H.L., 2011. Using the NARMAX approach to model the evolution of energetic electrons fluxes at geostationary orbit. Geophys. Res. Lett. 38 (18), L18105. ISSN 0094-8276. https://doi.org/10.1029/2011GL048980.

Balikhin, M.A., Gedalin, M., Reeves, G.D., Boynton, R.J., Billings, S.A., 2012. Time scaling of the electron flux increase at GEO: the local energy diffusion model vs observations. J. Geophys. Res. 117 (A10), A10208. ISSN 0148-0227. https://doi.org/10.1029/2012JA018114.

Balikhin, M.A., Rodriguez, J.V., Boynton, R.J., Walker, S.N., Aryan, H., Sibeck, D.G., Billings, S.A., 2016. Comparative analysis of NOAA REFM and SNB3GEO tools for the forecast of the fluxes of high-energy electrons at GEO. Space Weather 14 (1), 22–31. ISSN 1542-7390. https://doi.org/10.1002/2015SW001303.

Billings, S.A., 2013. Nonlinear System Identification: NARMAX Methods in the Time, Frequency, and Spatio-Temporal Domains. Wiley, Chichester.

Billings, S.A., Leontaritis, I.J., 1981. Identification of nonlinear systems using parametric estimation techniques. In: Proceedings of the IEEE Conference on Control and Its Application, Warick, UK, pp. 183–187.

Billings, S.A., Peyton Jones, J.C., 1990. Mapping non-linear integro-differential equations into the frequency domain. Int. J. Control 52, 863–879.

Billings, S.A., Tsang, K.M., 1989. Spectral analysis for nonlinear systems, part i: parametric non-linear spectral analysis. Mech. Syst. Signal Proc. 3, 319–339.

Billings, S.A., Voon, W.S.F., 1986. Correlation based model validity tests for non-linear models. Int. J. Control 44, 235–244.

Billings, S.A., Wei, H.L., 2005. The wavelet-NARMAX representation: a hybrid model structure combining polynomial models with multiresolution wavelet decompositions. Int. J. Syst. Sci. 36 (3), 137–152. ISSN 0020-7721. https://doi.org/10.1080/00207720512331338120.

Billings, S.A., Wei, H.L., 2008. An adaptive orthogonal search algorithm for model subset selection and non-linear system identification. Int. J. Control. 81 (5), 714–724. ISSN 0020-7179. https://doi.org/10.1080/00207170701216311.

Billings, S.A., Zhu, Q.M., 1991. Rational model identification using an extended least-squares algorithm. Int. J. Control 54 (3), 529–546. ISSN 0020-7179. https://doi.org/10.1080/00207179108934174.

Billings, S.A., Chen, S., Korenberg, M.J., 1989. Identification of MIMO non-linear systems using a forward-regression orthogonal estimator. Int. J. Control 49 (6), 2157–2189.

Boaghe, O.M., Balikhin, M.A., Billings, S.A., Alleyne, H., 2001. Identification of nonlinear processes in the magnetospheric dynamics and forecasting of Dst index. J. Geophys. Res. 106 (A12), 30047–30066.

Bonin, M., Seghezza, V., Piroddi, L., 2010. LASSO-enhanced simulation error minimization method for NARX model selection. In: Proceedings of the 2010 American Control Conference, pp. 4522–4527.

Boynton, R.J., Balikhin, M.A., Billings, S.A., Sharma, A.S., Amariutei, O.A., 2011a. Data derived NARMAX Dst model. Ann. Geophys. 29 (6), 965–971. https://doi.org/10.5194/angeo-29-965-2011. Available from: http://www.ann-geophys.net/29/965/2011/.

Boynton, R.J., Balikhin, M.A., Billings, S.A., Wei, H.L., Ganushkina, N., 2011b. Using the NARMAX OLS-ERR algorithm to obtain the most influential coupling functions that affect the evolution of the magnetosphere. J. Geophys. Res. 116 (A5), A05218. ISSN 0148-0227. https://doi.org/10.1029/2010JA015505.

Boynton, R.J., Balikhin, M.A., Billings, S.A., Reeves, G.D., Ganushkina, N., Gedalin, M., Amariutei, O.A., Borovsky, J.E., Walker, S.N., 2013. The analysis of electron fluxes at geosynchronous orbit employing a NARMAX approach. J. Geophys. Res. Space Phys. 118 (4), 1500–1513. ISSN 2169-9402. https://doi.org/10.1002/jgra.50192.

Boynton, R.J., Balikhin, M.A., Billings, S.A., 2015. Online NARMAX model for electron fluxes at GEO. Ann. Geophys. 33 (3), 405–411. ISSN 1432-0576. Available from: http://www.ann-geophys.net/33/405/2015/.

Boynton, R.J., Balikhin, M.A., Sibeck, D.G., Walker, S.N., Billings, S.A., Ganushkina, N., 2016. Electron flux models for different energies at geostationary orbit. Space Weather 14 (10), 2016SW001506. ISSN 1542-7390. https://doi.org/10.1002/2016SW001506.

Burton, R.K., McPherron, R.L., Russull, C.T., 1975. An empirical relationship between interplanetary conditions and Dst. J. Geophys. Res. 80 (31), 4204–4214.

Chen, S., Billings, S.A., Luo, W., 1989. Orthogonal least squares methods and their application to non-linear system identification. Int. J. Control 50 (5), 1873–1896.

Chen, S., Hong, X., Harris, C.J., Sharkey, P.M., 2004. Sparse modeling using orthogonal forward regression with PRESS statistic and regularization. IEEE Trans. Syst. Man Cybern. B Cybern. 34 (2), 898–911. ISSN 1083-4419.

Chen, S., Wang, X.X., Harris, C.J., 2008. NARX-based nonlinear system identification using orthogonal least squares basis hunting. IEEE Trans. Control Syst. Technol. 16 (1), 78–84. ISSN 1063-6536.

Efron, B., Hastie, T., Johnstone, I., Tibshirani, R., 2004. Least angle regression. Ann. Statist. 32 (2), 407–499. ISSN 0090-5364. Available from: https://projecteuclid.org:443/euclid.aos/1083178935.

Friedman, J., Hastie, T., Tibshirani, R., 2009. The Elements of Statistical Learning. Springer-Verlag, New York, NY.

Geman, S., Bienenstock, E., Doursat, R., 1992. Neural networks and the bias/variance dilemma. Neural Comput. 4 (1), 1–58. ISSN 0899-7667. https://doi.org/10.1162/neco.1992.4.1.1.

Guo, Y., Guo, L.Z., Billings, S.A., Wei, H.L., 2014. An iterative orthogonal forward regression algorithm. Int. J. Syst. Sci. 46 (5), 776–789. ISSN 0020-7721. https://doi.org/10.1080/00207721.2014.981237.

Guo, Y., Guo, L.Z., Billings, S.A., Wei, H.L., 2016. Ultra-orthogonal forward regression algorithms for the identification of non-linear dynamic systems. Neurocomputing 173 (Part 3), 715–723. ISSN 0925-2312. Available from: http://www.sciencedirect.com/science/article/pii/S0925231215011741.

Hanser, F.A., 2011. EPS/HEPAD Calibration and Data Handbook. Tech. Rep. GOESN-ENG-048D. Assurance Technol. Corp., Carlisle, MA.

Hong, X., Chen, S., 2012. An elastic net orthogonal forward regression algorithm. IFAC Proc. 45 (16), 1814–1819. ISSN 1474-6670. Available from: http://www.sciencedirect.com/science/article/pii/S1474667015382203.

Horne, R.B., Thorne, R.M., Glauert, S.A., Albert, J.M., Meredith, N.P., Anderson, R.R., 2005. Timescale for radiation belt electron acceleration by whistler mode chorus waves. J. Geophys. Res. 110 (A3), A03225. ISSN 0148-0227. https://doi.org/10.1029/2004JA010811.

James, G., Witten, D., Hastie, T., Tibshirani, R., 2013. An Introduction to Statistical Learning. Springer-Verlag, New York, NY.

Kan, J.R., Lee, L.C., 1979. Energy coupling function and solar wind-magnetosphere dynamo. Geophys. Res. Lett. 6, 577–580.

Klimas, A.J., Vassiliadis, D., Baker, D.N., Valdivia, J.A., 1999. Data-derived analogues of the solar wind-magnetosphere interaction. Phys. Chem. Earth Part C 24 (1–3), 37–44. ISSN 1464-1917. Available from: http://www.sciencedirect.com/science/article/B6VPW-3W5ST23-5/2/d4bce8cce21f3a02b0c77bc1da38b330.

Kukreja, S.L., Löfberg, J., Brenner, M.J., 2006. A least absolute shrinkage and selection operator (LASSO) for nonlinear system identification. IFAC Proc. 39 (1), 814–819. ISSN 1474-6670. Available from: http://www.sciencedirect.com/science/article/pii/S1474667015353647.

Lang, Z.Q., Billings, S.A., 1996. Output frequency characteristics of nonlinear systems. Int. J. Control 64 (6), 1049–1067. ISSN 0020-7179. https://doi.org/10.1080/00207179608921674.

Li, L.M., Billings, S.A., 2001. Continuous time non-linear system identification in the frequency domain. Int. J. Control 74, 1051–1052.

Ljung, L., 1987. System Identification: Theory for the User. Prentice-Hall, Inc., Upper Saddle River, NJ.

Ljung, L., Soderstrom, T., 1983. Theory and Practice of Recursive Identification. MIT Press, Cambridge, MA.

Newell, P.T., Sotirelis, T., Liou, K., Meng, C.I., Rich, F.J., 2007. A nearly universal solar wind-magnetosphere coupling function inferred from 10 magnetospheric state variables. J. Geophys. Res. 112 (A1), A01206. ISSN 2156-2202. https://doi.org/10.1029/2006JA012015.

Norton, J.P., 1986. An Introduction to Identification. Dover Publications, Inc., New York, NY.

Paulikas, G.A., Blake, J.B., 1979. Effects of the solar wind on magnetospheric dynamics: energetic electrons at the synchronous orbit, Geophys. Monogr. Ser., vol. 21. AGU, Washington, D.C., pp. 180–202.

Qin, P., Nishii, R., Yang, Z.J., 2012. Selection of NARX models estimated using weighted least squares method via GIC-based method and l 1-norm regularization methods. Nonlinear Dyn. 70 (3), 1831–1846. ISSN 1573-269X. https://doi.org/10.1007/s11071-012-0576-y.

Reeves, G.D., Morley, S.K., Friedel, R.H.W., Henderson, M.G., Cayton, T.E., Cunningham, G., Blake, J.B., Christensen, R.A., Thomsen, D., 2011. On the relationship between relativistic electron flux and solar wind velocity: Paulikas and Blake revisited. J. Geophys. Res. 116 (A2), A02213. ISSN 0148-0227. https://doi.org/10.1029/2010JA015735.

Sonnerup, B.U.O., 1974. Magnetopause reconnection rate. J. Geophys. Res. 79 (10), 1546–1549. ISSN 2156-2202. https://doi.org/10.1029/JA079i010p01546.

Watanabe, R.N., Kohn, A.F., 2014. System identification of a motor unit pool using a realistic neuromusculoskeletal model. In: 5th IEEE RAS/EMBS International Conference on Biomedical Robotics and Biomechatronics, pp. 610–615.

Wei, H.L., Billings, S.A., 2008. Model structure selection using an integrated forward orthogonal search algorithm assisted by squared correlation and mutual information. Int. J. Model. Identif. Control. 3, 341–356.

Zaharov, V.E., 1972. Collapse of langmuir waves. Sov. Phys. 35, 908.

Zhao, W., Beach, T.H., Rezgui, Y., 2017. Efficient least angle regression for identification of linear-in-the-parameters models. Proc. R. Soc. A Math. Phys. Eng. Sci. 473 (2198). Available from: http://rspa.royalsocietypublishing.org/content/473/2198/20160775.abstract.

Zhu, D., Billings, S.A., Balikhin, M., Wing, S., Coca, D., 2006. Data derived continuous time model for the Dst dynamics. Geophys. Res. Lett. 33 (4), L04101. ISSN 0094-8276. https://doi.org/10.1029/2005GL025022.

Zhu, D., Billings, S.A., Balikhin, M.A., Wing, S., Alleyne, H., 2007. Multi-input data derived Dst model. J. Geophys. Res. 112 (A6), A06205. ISSN 0148-0227. https://doi.org/10.1029/2006JA012079.

Zou, H., Hastie, T., 2005. Regularization and variable selection via the elastic net. J. R. Stat. Soc. Ser. B Stat. Methodol. 67 (2), 301–320. ISSN 1467-9868. https://doi.org/10.1111/j.1467-9868.2005.00503.x.

CHAPTER

9

Probabilistic Forecasting of Geomagnetic Indices Using Gaussian Process Models

Mandar Chandorkar, Enrico Camporeale

Centrum Wiskunde & Informatica, Amsterdam, The Netherlands

CHAPTER OUTLINE

Machine Learning Techniques for Space Weather
https://doi.org/10.1016/B978-0-12-811788-0.00009-3

1 GEOMAGNETIC TIME SERIES AND FORECASTING

The Earth's magnetosphere is the region above the ionosphere that is defined by the extent of the Earth's magnetic field in space. It extends several tens of thousands of kilometers into space, a region that is impinged upon constantly by the solar wind. Ionized plasma ejected by the Sun couples with the Earth's magnetic field in a complex manner leading to highly nonlinear and chaotic processes that determine the state of the magnetosphere. It is quite common to reduce these complex dependencies by condensing the state-of-the-Earth's magnetosphere into a set of geomagnetic indices.

Geomagnetic indices come in various forms; they may take continuous or discrete values and may be defined with varying time resolutions. Their values are often calculated by averaging or combining a number of readings taken by instruments, usually magnetometers, around the Earth. Each geomagnetic index is a proxy for a particular kind of phenomenon. Some common indices are the *Kp*, the *Dst*, and the *AE* index.

1. *Kp*: The *Kp*-index is a discrete-valued global geomagnetic activity index and is based on 3-h measurements of the *K*-indices (Bartels and Veldkamp, 1949). The *K*-index itself is a 3-h-long quasilogarithmic local index of the geomagnetic activity, relative to a quiet day curve for the given location.
2. *AE*: The auroral electrojet index (*AE*) is designed to provide a global, quantitative measure of auroral zone magnetic activity produced by enhanced ionospheric currents flowing below and within the auroral oval (Davis and Sugiura, 1966). It is a continuous index that is calculated every hour.
3. *Dst*: A continuous hourly index that measures the magnitude of the magnetic field induced by near equatorial ring currents (Dessler and Parker, 1959). A large negative value of *Dst* ($\leq -100\,\text{nT}$) indicates occurrence of geomagnetic storm events that are of particular importance in space weather prediction due to their adverse effects on telecommunications infrastructure, navigation systems, power grids, satellites, and so forth.

Space weather forecasting systems usually use in situ measurements of solar wind parameters, taken by satellites, as well as historical data of indices to produce forecasts for various geomagnetic time series. Geomagnetic indices are often recorded in ground-based measurements. The *Space Physics Data Facility* hosted by NASA is a gateway for accessing data from a variety of sources, including those mentioned herein, and it is meant for use in research and validation. In order to deploy space weather prediction systems, real-time data of solar wind parameters can be obtained from the *ACE* and *DISCOVR* missions.

In this chapter, for the purpose of exposition, we will focus on 1-h ahead prediction of the *Dst* time series. From the data we will use time histories of *Dst*, solar wind speed V, and the z-component of the interplanetary magnetic field (IMF) B_z. Our choice is motivated by both physical and practical considerations: solar wind speed and IMF are important external drivers for particle injection into the magnetosphere, and they have extensive coverage in the *OMNI* data with relatively fewer missing records.

2 DST FORECASTING

2.1 Models and Algorithms

A number of modeling techniques have been applied for the prediction of the *Dst* index. One of the earliest forecasting techniques involves calculating the $Dst(t)$ as a solution of an *Ordinary Differential Equation* that expressed the rate of change of $Dst(t)$ as a combination of two terms: decay and injection $\frac{dDst(t)}{dt} = Q(t) - \frac{Dst(t)}{\tau}$, where $Q(t)$ relates to the particle injection from the plasma sheet into the inner magnetosphere. This method was presented first by Burton et al. (1975) and later modified and extended in works such as Wang et al. (2003), O'Brien and McPherron (2000), Ballatore and Gonzalez (2014), and others.

Important empirical geomagnetic prediction models include the *Nonlinear Auto-Regressive Moving Average with eXogenous inputs* (NARMAX) methodology (see Billings et al., 1989; Balikhin et al., 2001; Zhu et al., 2006, 2007; Boynton et al., 2011a,b, 2013) and *Artificial Neural Networks*-based models (Lundstedt et al., 2002; Wing et al., 2005; Bala et al., 2009; Pallocchia et al., 2006) for time series prediction of *Dst* and *Kp* indices from IMF data and solar wind parameters.

2.2 Probabilistic Forecasting

Although much research has been done on prediction of the *Dst* index, much less has been done on probabilistic forecasting of *Dst*. One such work described in McPherron et al. (2013) involves identification of high-speed solar wind streams using the WSA model, using predictions of high-speed streams to construct ensembles of *Dst* trajectories, that yield the quartiles of *Dst* time series. McPherron et al. (2004) use a similar methodology for probabilistic forecasting of the 3-h ahead *Ap* index.

Probabilistic forecasting is of particular importance in geophysics applications as the end users of forecasts often require confidence bounds on said forecasts. It is in this context where *Gaussian processes* become especially attractive due to their inherent probabilistic formulation and tractability of exact inference.

3 GAUSSIAN PROCESSES

Gaussian processes first appeared in machine learning research in Neal (1996), as the limiting case of Bayesian inference performed on neural networks with infinite neurons in the hidden layers. Although their inception in the machine learning community is recent, their origins can be traced back to the geo-statistics research community where they are known as *Kriging* methods (Krige, 1951). In pure mathematics, *Gaussian processes* have been studied extensively, and their existence was first proven by Kolmogorov's extension theorem (Tao, 2011). The reader is referred to Rasmussen and Williams (2005) for an in-depth treatment of Gaussian processes in machine learning. In space weather forecasting literature, *Gaussian process regression* has been used for 1-h ahead prediction of the *Dst* index (Chandorkar et al., 2017).

Without going into too many details, we give a quick recap of the formulation and exact inference in GPR models.

3.1 Gaussian Process Regression: Formulation

Our aim is to infer an unknown function $f(\mathbf{x})$ from its noise corrupted measurements $(\mathbf{x}_i, y_i)$ where $y_i = f(\mathbf{x}_i) + \epsilon$ and $\epsilon \sim \mathcal{N}(0, \sigma^2)$ is independent and identically distributed Gaussian noise.

A *Gaussian process* model represents the finite dimensional probability distribution of $f(\mathbf{x}_i)$ by a multivariate Gaussian having a particular structure for its mean and covariance as shown in Eqs. (2), (3).

$$\mathbf{f} = \begin{pmatrix} f(\mathbf{x}_1) \\ f(\mathbf{x}_2) \\ \vdots \\ f(\mathbf{x}_N) \end{pmatrix} \tag{1}$$

$$\mathbf{f}|\mathbf{x}_1, \ldots, \mathbf{x}_N \sim \mathcal{N}(\mathbf{m}, \mathbf{K}) \tag{2}$$

$$\mathbb{P}(\mathbf{f} \mid \mathbf{x}_1, \ldots, \mathbf{x}_N) = \frac{1}{(2\pi)^{n/2} det(\mathbf{K})^{1/2}} \exp\left(-\frac{1}{2}(\mathbf{f} - \mathbf{m})^T \mathbf{K}^{-1} (\mathbf{f} - \mathbf{m})\right) \tag{3}$$

In order to uniquely define the distribution of $\mathbf{f}$, it is necessary to specify the mean $\mathbf{m}$ and the covariance matrix $\mathbf{K}$. For this probability density to be valid, there are further requirements imposed on $\mathbf{K}$:

1. Symmetry: $\mathbf{K}_{ij} = \mathbf{K}_{ji}\ \forall i, j \in 1, \ldots, N$
2. Positive semidefiniteness: $\mathbf{z}^T \mathbf{K} \mathbf{z} \geq 0\ \forall \mathbf{z} \in \mathbb{R}^N$

In *Gaussian processes* the individual elements of $\mathbf{x}$ and $\mathbf{K}$ are specified in the form of functions shown as follows.

$$\mu_i = \mathbb{E}[f(\mathbf{x}_i)] := m(\mathbf{x}_i) \tag{4}$$

$$\Lambda_{ij} = \mathbb{E}[(f(\mathbf{x}_i) - \mu_i)(f(\mathbf{x}_j) - \mu_j)] := K(\mathbf{x}_i, \mathbf{x}_j) \tag{5}$$

In the machine learning community, $m(.)$ and $K(.,.)$ are known as the *mean function* and *covariance function* or *kernel function* of the process, respectively. Giving a closed form expression for $m(.)$ and $K(.,.)$ uniquely specifies a particular *Gaussian process*, and so a GP model is often expressed in the following notation.

$$f(\mathbf{x}) \sim \mathcal{GP}(m(\mathbf{x}), K(\mathbf{x}, \mathbf{x}')) \tag{6}$$

3.2 Gaussian Process Regression: Inference

In order to generate predictions $f(\mathbf{x}_i^*)$ for a set of test points $\mathbf{x}_i^*$: $\forall i \in 1, \ldots, M$. Using the multivariate Gaussian distribution in Eq. (3) we can construct the joint distribution of $f(\mathbf{x})$ over the training and test points.

$$\mathbf{f}_* = \begin{pmatrix} f(\mathbf{x}_1^*) \\ f(\mathbf{x}_2^*) \\ \vdots \\ f(\mathbf{x}_\mathbf{M}^*) \end{pmatrix}_{M\times 1} \tag{7}$$

$$\left(\begin{array}{c} \mathbf{y} \\ \mathbf{f}_* \end{array} \right) \Bigg| \; \mathbf{X}, \mathbf{X}_* \sim \mathcal{N}\left(\mathbf{0}, \begin{bmatrix} \mathbf{K} + \sigma^2\mathbf{I} & \mathbf{K}_* \\ \mathbf{K}_*^T & \mathbf{K}_{**} \end{bmatrix}\right) \tag{8}$$

Probabilistic predictions f_* can be generated by constructing the conditional distribution $\mathbf{f}_*|\mathbf{X}, \mathbf{y}, \mathbf{X}_*$ which is also a multivariate Gaussian as shown in Eq. (9).

$$\mathbf{f}_*|\mathbf{X}, \mathbf{y}, \mathbf{X}_* \sim \mathcal{N}(\bar{\mathbf{f}}_*, \Sigma_*) \tag{9}$$

$$\bar{\mathbf{f}}_* = \mathbf{K}_*^T[\mathbf{K} + \sigma^2\mathbf{I}]^{-1}\mathbf{y} \tag{10}$$

$$\Sigma_* = \mathbf{K}_{**} - \mathbf{K}_*^T \left(\mathbf{K} + \sigma^2\mathbf{I}\right)^{-1} \mathbf{K}_* \tag{11}$$

4 ONE-HOUR AHEAD *DST* PREDICTION

4.1 Data Source: OMNI

The *OMNI* data set is a collation of in situ measurements taken by orbiting satellites and ground-based measurements of geomagnetic indices. The data itself is represented as time averaged versions of the various quantities and is available at two frequencies, by hour and by minute. The quantities recorded can be broadly classified into two groups.

1. Plasma parameters and magnetic fields: Measurements of the solar wind, particularly the solar wind speed, flow angles, pressure, proton temperature, proton density, and field measurements such as strength of electric and magnetic fields as well as their components in various geomagnetic coordinate systems.
2. Indices: *Kp*, *Ae*, *Ap*, *Dst*, sunspot number.

To get a more in-depth understanding about the *OMNI* data, interested readers are encouraged to visit the NASA omni web service (https://omniweb.gsfc.nasa.gov).

4.2 Gaussian Process *Dst* Model

In Eqs. (12)–(14) we outline a *Gaussian process* formulation for *OSA* prediction of *Dst*. A vector of features $\mathbf{x}_{t-1}$ is used as input to an unknown function $f(\mathbf{x}_{t-1})$.

The features $\mathbf{x}_{t-1}$ can be any collection of quantities in the hourly resolution OMNI data set. Generally $\mathbf{x}_{t-1}$ are time histories of *Dst* and other important variables such as plasma pressure $p(t)$, solar wind speed $V(t)$, z component of the IMF $B_z(t)$.

$$Dst(t) = f(\mathbf{x}_{t-1}) + \epsilon \tag{12}$$

$$\epsilon \sim \mathcal{N}(0, \sigma^2) \tag{13}$$

$$f(x_t) \sim \mathcal{GP}(m(\mathbf{x}_t), K_{osa}(\mathbf{x}_t, \mathbf{x}_s)) \tag{14}$$

We consider two choices for the input features $\mathbf{x}_{t-1}$ leading to two variants of *Gaussian process* regression for the *Dst* time series prediction.

4.3 Gaussian Process Auto-Regressive (GP-AR)

The simplest auto-regressive models for *OSA* prediction of *Dst* are those that use only the history of *Dst* to construct input features for model training. The input features $\mathbf{x}_{t-1}$ at each time step are the history of $Dst(t)$ until a time lag of p hours.

$$\mathbf{x}_{t-1} = \big(Dst(t-1), \ldots, Dst(t-p+1)\big)$$

4.4 GP-AR With eXogenous Inputs (GP-ARX)

Auto-regressive models can be augmented by including exogenous quantities in the inputs $\mathbf{x}_{t-1}$ at each time step, in order to improve predictive accuracy. *Dst* gives a measure of ring currents, which are modulated by plasma sheet particle injections into the inner magnetosphere during substorms. Studies have shown that the substorm occurrence rate increases with solar wind velocity (high-speed streams) (Kissinger et al., 2011; Newell et al., 2016). Prolonged southward IMF z-component (B_z) is needed for substorms to occur (McPherron et al., 1986). An increase in the solar wind electric field, $V_{sw}B_z$, can increase the dawn-dusk electric field in the magnetotail, which in turn determines the amount of plasma sheet particles that move into the inner magnetosphere (Friedel et al., 2001).

Although there are also other solar wind inputs, such as dynamic pressure p and proton density ρ, it is known that the dynamic pressure is a function of the solar wind velocity and the particle densities. Further, there are gaps in the OMNI data for these quantities, making them less attractive as compared with the solar wind velocity and the IMF data.

Therefore, our exogenous parameters consist of solar wind velocity V_{sw} and IMF B_z. We choose distinct time lags p, p_v, and p_b for *Dst*, V, and B_z, respectively.

$$\begin{aligned}\mathbf{x}_{t-1} = (&Dst(t-1), \ldots, Dst(t-p+1),\\ &V_{sw}(t-1), \ldots, V_{sw}(t-p_v+1),\\ &B_z(t-1), \ldots, B_z(t-p_b+1))\end{aligned}$$

5 ONE-HOUR AHEAD *DST* PREDICTION: MODEL DESIGN

5.1 Choice of Mean Function

Mean functions in GPR models encode trends in the data; they are the baseline predictions the model falls back to in case the training and test data have little correlation as predicted by the kernel function. If there is no prior knowledge about the function to be approximated, Rasmussen and Williams (2005) state that it is perfectly reasonable to choose $m(\mathbf{x} = 0)$ as the mean function, as long as the target values are normalized. In the case of the *Dst* time series, it is known that the so-called *persistence model* $\hat{Dst}(t) = Dst(t-1)$ performs quite well in the context of OSA prediction. We therefore choose the *persistence model* as the mean function in our OSA Dst models.

5.2 Choice of Kernel

For the success of a *Gaussian process* model, an appropriate choice of kernel function is paramount. The symmetry and positive semidefiniteness of *Gaussian process* kernels implies that they represent innerproducts between some basis function representation of the data. The interested reader is referred to Berlinet and Thomas-Agnan (2004), Scholkopf and Smola (2001), and Hofmann et al. (2008) for a thorough treatment of kernel functions and the rich theory behind them. Some common kernel functions used in machine learning are listed in Table 1.

The quantities l in the *Radial basis function kernel*, and b and d in the polynomial kernel are known as *hyperparameters*. Hyperparameters give flexibility to a particular kernel structure, for example $d = 1, 2, 3, \ldots$ in the polynomial kernel represents linear, quadratic, cubic, and higher-order polynomials, respectively. The process of assigning values to the *hyperparameters* is crucial in the model building process, and is known as *model selection*.

In this text, we construct GPR models with a combination of the *maximum likelihood perception* kernel and *Student t* kernel, as shown in Eq. (15). The *maximum likelihood perception* kernel is the *Gaussian process* equivalent of a single hidden layer feed-forward neural network model as demonstrated in Neal (1996).

$$K_{osa}(\mathbf{x}, \mathbf{y}) = K_{mlp}(\mathbf{x}, \mathbf{y}) + K_{st}(\mathbf{x}, \mathbf{y}) \tag{15}$$

$$K_{mlp}(\mathbf{x}, \mathbf{y}) = \sin^{-1}\left(\frac{w\mathbf{x}^{\intercal}\mathbf{y} + b}{\sqrt{w\mathbf{x}^{\intercal}\mathbf{x} + b + 1}\sqrt{w\mathbf{y}^{\intercal}\mathbf{y} + b + 1}}\right) \tag{16}$$

$$K_{st}(\mathbf{x}, \mathbf{y}) = \frac{1}{1 + \|\mathbf{x} - \mathbf{y}\|_2^d} \tag{17}$$

5.3 Model Selection: Hyperparameters

Given a GPR model with a kernel function K_θ, the problem of model selection consists of finding appropriate values for the kernel hyperparameters $\theta = (\theta_1, \theta_2, \ldots, \theta_i)$. In order to

TABLE 1 Popular Kernel Functions Used in GPR Models

Name	**Expression**	**Hyperparameters**
Radial basis function	$\frac{1}{2}\exp(-\|\mathbf{x} - \mathbf{y}\|^2/l^2)$	$l \in \mathbb{R}$
Polynomial	$(\mathbf{x}^{\intercal}\mathbf{y} + b)^d$	$b \in \mathbb{R}, d \in \mathbb{N}$
Laplacian	$\exp(-\|\mathbf{x} - \mathbf{y}\|_1/\theta)$	$\theta \in \mathbb{R}^+$
Student t	$1/(1 + \|\mathbf{x} - \mathbf{y}\|_2^d)$	$d \in \mathbb{R}^+$
Maximum likelihood perception	$\sin^{-1}\left(\frac{w\mathbf{x}^{\intercal}\mathbf{y}+b}{\sqrt{w\mathbf{x}^{\intercal}\mathbf{x}+b+1}\sqrt{w\mathbf{y}^{\intercal}\mathbf{y}+b+1}}\right)$	$w, b \in \mathbb{R}^+$

assign a value to θ, we must define an objective function ($Q(\theta)$) that represents our confidence that the GP model built from a particular value of θ is the best performing model. Because GP models are constructed from assumptions about the conditional probability distribution of the data $p(\mathbf{y}|\mathbf{X})$, it is natural to use the negative log-likelihood of the training data as a model selection criterion.

$$\begin{aligned} Q(\theta) &= -\log\, p(\mathbf{y}|\mathbf{X}, \theta) \\ &= \frac{1}{2}\mathbf{y}^{\mathsf{T}}(\mathbf{K}_\theta + \sigma^2\mathbf{I})^{-1}\mathbf{y} + \frac{1}{2}|\mathbf{K}_\theta + \sigma^2\mathbf{I}| + \frac{N}{2}\log(2\pi) \\ \mathbf{K}_\theta &= [K_\theta(\mathbf{x}_i, \mathbf{x}_j)]_{N\times N} \end{aligned}$$

Because a GP model expresses prior assumptions about the finite dimensional distribution of the quantity of interest, the expression for $Q(\theta)$ follows directly from the expression of the probability density function of a multivariate Gaussian.

The model selection problem can now be expressed as the minimization problem shown as follows.

$$\theta^* = \arg\min_{\theta}\; Q(\theta)$$

Rasmussen and Williams (2005) note that there are ready interpretations to the three terms that add up to give $Q(.)$. The first term $\frac{1}{2}\mathbf{y}^{\mathsf{T}}(\mathbf{K}_\theta + \sigma^2\mathbf{I})^{-1}\mathbf{y}$ quantifies the fit of the model with respect to the training data, the second term $\frac{1}{2}|\mathbf{K}_\theta + \sigma^2\mathbf{I}|$ is a complexity penalty, in nonparametric models such as *Gaussian processes*, the model complexity grows with the number of training data points N. The third term $\frac{N}{2}log(2\pi)$ is a normalization term.

The objective function $Q(\theta)$ can have multiple local minima, and evaluating the value of $Q(.)$ at any given θ requires inversion of the matrix $\mathbf{K}_\theta + \sigma^2\mathbf{I}$, which has a time complexity $O(N^3)$ as noted herein. In the interest of saving computational cost, one should not use an exhaustive search through the domain of the hyperparameters to inform one's choice for θ.

The model selection problem has historically received less attention due to a combination of difficulty and numerous design decisions such as choice of hyperparameter optimization techniques, which makes model selection as much of an art as a science.

Choosing a suitable heuristic or algorithm for model selection for GP models is quite often a balancing exercise between the computational limitations of the modeler and the need to explore the hyperparameter space in a manner that yields a satisfactory predictive model. Some of the techniques used for model selection in the context of GPR include the following.

5.3.1 Grid Search

This is the simplest procedure that may be applied for performing model selection in GPR models. This routine constructs a grid of values for θ as the Cartesian product of one-dimensional grids for each θ_i, evaluate $Q(.)$ at each such grid point and choose the configuration that yields a minimum value of $Q(.)$. The advantage of this technique is its simplicity and control over the computational cost of training GP models for practical problems. The computational cost of this procedure scales with the total number of points on the grid; this is controlled by the scales and number of grid points for each hyperparameter θ_i.

5.3.2 Coupled Simulated Annealing

Introduced in Xavier-De-Souza et al. (2010), *Coupled simulated annealing* (CSA) is a family of optimization techniques that are a generalization of *simulated annealing* (SA) (Kirkpatrick et al., 1983). The CSA optimization technique can be understood as a set of parallel SA processes that are coupled by their acceptance probabilities.

From an implementation point of view, CSA follows the same procedure as *grid search*, but after evaluation of the objective $Q(.)$ on the grid, each grid point is iteratively mutated in a random walk fashion. This mutation is accepted or rejected according to the new value of $Q(.)$ as well as its value on the other grid points. This procedure is iterated until some stop criterion is reached.

To illustrate how CSA updates are made and accepted, consider a population or ensemble of hyperparameter values x_i that can be considered to be initialized in a grid-based fashion. For each configuration x_i, a probing solution $y_i = x_i + \epsilon$ is generated where ϵ is drawn from some random variable. The acceptance probability of the solution is calculated as follows.

$$\mathbb{P}(x_i \to y_i|x_i) = \frac{\exp(-E(y_i)/T)}{\exp(-E(y_i)/T) + \gamma}$$

$$\gamma = \sum_j \exp\left(\frac{-E(x_j)}{T}\right)$$

The quantity γ defines the *coupling* between the states x_i of the ensemble, when the number of configurations in the ensemble x_i is one, this rule reduces to the classical *Metropolis-Hastings* acceptance rule of the SA algorithm. Up to four variants of the CSA algorithm have been defined with different acceptance rules and temperature annealing schedules; the reader is advised to refer to Xavier-De-Souza et al. (2010) for an in-depth treatment of the CSA optimization class.

5.3.3 Maximum Likelihood

This technique, as described in Rasmussen and Williams (2005), is a form of *gradient descent*. It involves starting with an initial guess for θ and iteratively improving it by calculating the gradient of $Q(.)$ with respect to θ. The gradient of $Q(.)$ with respect to each hyperparameter θ_i can be calculated analytically, as follows.

$$\frac{\partial}{\partial \theta_i} \log p(\mathbf{y}|\mathbf{X}, \theta) = \frac{1}{2}\mathbf{y}^\intercal K^{-1} \frac{\partial K}{\partial \theta_i} K^{-1}\mathbf{y} - \frac{1}{2} tr\left(K^{-1}\frac{\partial K}{\partial \theta_i}\right)$$

$$= \frac{1}{2} tr\left((\alpha\alpha^\intercal - K^{-1})\frac{\partial K}{\partial \theta_i}\right) \quad \text{where } \alpha = K^{-1}\mathbf{y}$$

Although gradient-based *maximum likelihood* looks very similar to the *gradient descent* used to learn parametric regression models such as *feed forward neural networks*, in the GPR model selection context it introduces an extra computational cost of calculating the gradient of $Q(\theta)$, with respect to each θ_i in every iteration. Applying this method can sometimes lead to overfitting of the GPR model to the training data (Rasmussen and Williams, 2005).

5.4 Model Selection: Auto-Regressive Order

Apart from having a practical method for choosing values of kernel hyperparameters, we must also address the issue of choosing the time delays p and p_v, p_b for the GP-AR and GP-ARX models. One might argue that we group these quantities with the kernel hyperparameters in the model selection procedure, in the context of the GP-AR and GP-ARX models there is one reason why this might not be beneficial.

The value of kernel hyperparameters θ^* that maximize the marginal likelihood $Q(.)$ are conditional on the chosen model orders p and p, p_v, p_b for GP-AR and GP-ARX, respectively. Because the model orders determine the size of the input space of the predictive models, they influence the values of characteristic length scales and other kernel hyperparameters that yield a good fit.

Not only do the model orders influence the value of kernel hyperparameters θ^* that maximize model fit, but increasing values of p, p_v, p_b also make it more difficult for the model selection algorithms to search the hyperparameter space, due to the curse of dimensionality.

An empirical way to choose the values of p, p_v, p_b for the GP-AR and GP-ARX models is by choosing an independent validation data set that is used to evaluate the GP models obtained from the model selection procedures of the previous section. It is then possible for the modeler to choose these model orders in a way to balance performance and dimensionality of input features.

6 GP-AR AND GP-ARX: WORKFLOW SUMMARY

Section 5 brings together all the practical issues with respect to building of GP-AR and GP-ARX models for predicting *Dst*, they can be condensed to form the following steps.

1. *Data Preprocessing*: Three separate data sets for model training and hyperparameter selection, model order selection, and model evaluation must be created. Care must be taken that the size of the training sets does not delay the processes of kernel hyperparameter selection. The training, validation, and test data sets created from this procedure must all contain nonoverlapping geomagnetic storm events.
   We selected OMNI data sections 00:00 January 3, 2010 to 23:00 January 23, 2010 and 20:00 August 5, 2011 to 22:00 August 6, 2011 for training the GP-AR and GP-ARX models. The first training data section consists of ambient fluctuations of *Dst* while the second contains a geomagnetic storm. A set of 24 storm events listed in Table 2 is kept aside as a validation data set for selecting model orders p, p_v, p_b while a list of 63 storm events listed in Table 3 is used for final model evaluation.
2. *Choice of Model Components*: Kernel, mean function, model selection procedure.
   We choose a kernel K_{osa} of the form given in Eq. (15). In order to find appropriate values of the hyperparameters of K_{osa}, we apply *grid search*, CSA, and *maximum likelihood* methods. We fix the parameters d and σ^2 of K_{st} and model noise to values 0.01 and 0.2, respectively; the remaining parameters w and b are kept free to be calculated by model selection. Table 4 summarizes the settings used to run each model selection procedure.
3. *Choice of Performance Metrics*: To compare the performance of GP and GP-ARX models of varying auto-regressive orders, their predictive performance during the validation (and

TABLE 2 Storm Events Used for Model Selection of GP-AR and GP-ARX

Event ID	Start Date	Start Hour	End Date	End Hour	Min. Dst
1	March 26, 1995	0500	March 26, 1995	2300	−107
2	April 7, 1995	1300	April 9, 1995	0900	−149
3	September 27, 1995	0100	September 28, 1995	0400	−108
4	October 18, 1995	1300	October 19, 1995	1400	−127
5	October 22, 1996	2200	October 23, 1996	1100	−105
6	April 21, 1997	1000	April 22, 1997	0900	−107
7	May 15, 1997	0300	May 16, 1997	0000	−115
8	October 10, 1997	1800	October 11, 1997	1900	−130
9	November 7, 1997	0000	November 7, 1997	1800	−110
10	November 22, 1997	2100	November 24, 1997	0400	−108
11	June 12, 2005	1700	June 13, 2005	1900	−106
12	August 31, 2005	1200	September 1, 2005	1200	−122
13	December 14, 2006	2100	December 16, 2006	0300	−162
14	September 26, 2011	1400	September 27, 2011	1200	−101
15	October 24, 2011	2000	October 25, 2011	1400	−132
16	March 8, 2012	1200	March 10, 2012	1600	−131
17	April 23, 2012	1100	April 24, 2012	1300	−108
18	July 15, 2012	0100	July 16, 2012	2300	−127
19	September 30, 2012	1300	October 1, 2012	1800	−119
20	October 8, 2012	0200	October 9, 2012	1700	−105
21	November 13, 2012	1800	November 14, 2012	1800	−108
22	March 17, 2013	0700	March 18, 2013	1000	−132
23	May 31, 2013	1800	June 1, 2013	2000	−119
24	February 18, 2014	1500	February 19, 2014	1600	−112

later, final evaluation) phase must be calculated using some clearly defined metrics. We use the following as performance metrics for our modeling experiments. The reader is advised to refer to Wilks (2011) for background on the various performance metrics used in meteorological and space sciences.

a. The mean absolute error.

$$MAE = \sum_{t=1}^{n} \left| (Dst(t) - \hat{D}st(t)) \right| \Big/ n \quad (18)$$

TABLE 3 Storm Events Used to Evaluate GP-AR and GP-ARX Models

Event ID	Start Date	Start Time	End Date	End Time	Min. Dst
1	February 17, 1998	1200	February 18, 1998	1000	−100
2	March 10, 1998	1100	March 11, 1998	1800	−116
3	May 4, 1998	0200	May 5, 1998	0200	−205
4	August 26, 1998	1000	August 29, 1998	0700	−155
5	September 25, 1998	0100	September 26, 1998	0000	−207
6	October 19, 1998	0500	October 20, 1998	0800	−112
7	November 9, 1998	0300	November 10, 1998	1600	−142
8	November 13, 1998	0000	November 15, 1998	0400	−131
9	January 13, 1999	1600	January 14, 1999	2000	−112
10	February 18, 1999	0300	February 19, 1999	2100	−123
11	September 22, 1999	2000	September 23, 1999	2300	−173
12	October 22, 1999	0000	October 23, 1999	1400	−237
13	February 12, 2000	0500	February 13, 2000	1500	−133
14	April 6, 2000	1700	April 8, 2000	0900	−288
15	May 24, 2000	0100	May 25, 2000	2000	−147
16	August 10, 2000	2000	August 11, 2000	1800	−106
17	August 12, 2000	0200	August 13, 2000	1700	−235
18	October 13, 2000	0200	October 14, 2000	2300	−107
19	October 28, 2000	2000	October 29, 2000	2000	−127
20	November 6, 2000	1300	November 7, 2000	1800	−159
21	November 28, 2000	1800	November 29, 2000	2300	−119
22	March 19, 2001	1500	March 21, 2001	2300	−149
23	March 31, 2001	0400	April 1, 2001	2100	−387
24	April 11, 2001	1600	April 13, 2001	0700	−271
25	April 18, 2001	0100	April 18, 2001	1300	−114
26	April 22, 2001	0200	April 23, 2001	1500	−102
27	August 17, 2001	1600	August 18, 2001	1600	−105
28	September 30, 2001	2300	October 2, 2001	0000	−148
29	October 21, 2001	1700	October 24, 2001	1100	−187
30	October 28, 2001	0300	October 29, 2001	2200	−157
31	March 23, 2002	1400	March 25, 2002	0500	−100

TABLE 3 Storm Events Used to Evaluate GP-AR and GP-ARX Models—cont'd

Event ID	Start Date	Start Time	End Date	End Time	Min. Dst
32	April 17, 2002	1100	April 19, 2002	0200	−127
33	April 19, 2002	0900	April 21, 2002	0600	−149
34	May 11, 2002	1000	May 12, 2002	1600	−110
35	May 23, 2002	1200	May 24, 2002	2300	−109
36	August 1, 2002	2300	August 2, 2002	0900	−102
37	September 4, 2002	0100	September 5, 2002	0000	−109
38	September 7, 2002	1400	September 8, 2002	2000	−181
39	October 1, 2002	0600	October 3, 2002	0800	−176
40	October 3, 2002	1000	October 4, 2002	1800	−146
41	November 20, 2002	1600	November 22, 2002	0600	−128
42	May 29, 2003	2000	May 30, 2003	1000	−144
43	June 17, 2003	1900	June 19, 2003	0300	−141
44	July 11, 2003	1500	July 12, 2003	1600	−105
45	August 17, 2003	1800	August 19, 2003	1100	−148
46	November 20, 2003	1200	November 22, 2003	0000	−422
47	January 22, 2004	0300	January 24, 2004	0000	−149
48	February 11, 2004	1000	February 12, 2004	0000	−105
49	April 3, 2004	1400	April 4, 2004	0800	−112
50	July 22, 2004	2000	July 23, 2004	2000	−101
51	July 24, 2004	2100	July 26, 2004	1700	−148
52	July 26, 2004	2200	July 30, 2004	0500	−197
53	August 30, 2004	0500	August 31, 2004	2100	−126
54	November 7, 2004	2100	November 8, 2004	2100	−373
55	November 9, 2004	1100	November 11, 2004	0900	−289
56	November 11, 2004	2200	November 13, 2004	1300	−109
57	January 21, 2005	1800	January 23, 2005	0500	−105
58	May 7, 2005	2000	May 9, 2005	1000	−127
59	May 29, 2005	2200	May 31, 2005	0800	−138
60	June 12, 2005	1700	June 13, 2005	1900	−106
61	August 31, 2005	1200	September 1, 2005	1200	−131
62	April 13, 2006	2000	April 14, 2006	2300	−111
63	December 14, 2006	2100	December 16, 2006	0300	−147

TABLE 4 Settings of Model Selection Procedures

Procedure	Grid Size	Step	Max Iterations
Grid search	10	0.2	NA
Coupled simulated annealing	4	0.2	30
Maximum likelihood	NA	0.2	150

b. The root mean square error.

$$RMSE = \sqrt{\sum_{t=1}^{n}(Dst(t) - \hat{D}st(t))^2/n} \tag{19}$$

c. Correlation coefficient between the predicted and actual value of *Dst*.

$$CC = Cov(Dst, \hat{D}st)/\sqrt{Var(Dst)Var(\hat{D}st)} \tag{20}$$

4. *Model Order Selection*: In order to find appropriate values for the auto-regressive time lags p, p_v, p_b for *Dst*, solar wind speed and IMF B_z, respectively, create models up to a certain maximum time lag; that is, $3 \leq p \leq 12$ for GP-AR and $3 \leq p + p_v + p_b \leq 12$ for GP-ARX. For each such model, perform the model selection routine to select values of the kernel hyperparameters and evaluate the resultant model on the validation set. Choose the value of p, p_v, p_b, which achieves the predictive performance on the validation set (i.e., minimizes the *mean absolute error*).
The reader should remember that range of time lags to search ($3 < p < 12$) is a balance between computational effort and predictive performance. This choice is application-specific and can be better informed by exploratory analysis of the geomagnetic time series.

5. *Final Evaluation*: After selecting GP-AR and GP-ARX models with particular values of auto-regressive orders and kernel hyperparameters in the model selection phase, the best performing GP-AR and GP-ARX models are then evaluated on the test data set in order to get an unbiased picture of their predictive performance. In our case we use the already chosen test set outlined in Table 3 for final model evaluation. The GP-based OSA prediction models are compared against the naive *Persistence* model, which serves as a performance baseline for the OSA prediction of *Dst*.

7 PRACTICAL ISSUES: SOFTWARE

Scientific computing, whether it be in the form of simulation, data analysis, or modeling is brought to fruition through well-thought choices and usage of software packages. *Gaussian processes* have been implemented in a number of software distributions, we discuss two of them here.

1. *PlasmaML*: An open source project https://gitlab.com/MD-CWI-NL/PlasmaML maintained at the CWI Amsterdam, it is a software toolbox for machine learning applications in space and plasma physics. It is comprised of a number of modules; we use codes from the *omni* module for the experiments and results shown in Section 8. It leverages the DynaML machine learning toolkit https://github.com/transcendent-ai-labs/DynaML, which has a well-developed library for building GP and kernel-based models, and also includes built-in support for a number of kernel functions, as well as the *grid search*, *maximum likelihood*, and CSA model selection routines applied in this chapter. *PlasmaML* is written in the *Scala* programming language, making it distributable as a *Java Runtime Environment jar* file.
2. *GPML* (Rasmussen and Nickisch, 2016): A *MATLAB*-based toolbox for building GP models that has support for a variety of features. The software includes a number of kernel functions supported off the shelf, approximate as well as exact inference techniques, and support for non-Gaussian likelihood functions.

8 EXPERIMENTS AND RESULTS

After running the workflow steps as outlined in Section 6, we are in a position to analyze the performance of the GP-AR and GP-ARX family of models on the task of OSA *Dst* prediction.

8.1 Model Selection and Validation Performance

The model validation workflow described in Section 6 produces data containing model validation scores for GP-AR and GP-ARX models with different values of autoregressive orders p_t. In order to better understand the overall trend, we group the performance scores by unique values of $p_t = p + p_v + p_b$ and analyze the summary statistics with increasing p_t.

Figs. 1 and 2 show how the mean absolute error and coefficient of correlation as calculated on the validation set storm events of Table 2 vary with increasing model order for GP-AR and GP-ARX. The results are represented as box and whisker plots, in which a rectangle is drawn to represent the first and third quartiles, with a horizontal line inside to indicate the median value, outlying points are shown as dots, while the whiskers indicate the smallest and largest nonoutliers. In both cases, the predictive performance first improves, and then stagnates or worsens with increasing model order.

From the preceding charts, we can choose the model order that yields the best *mean absolute error*, for GP-AR it is $p_t = 6$ while for GP-ARX it is $p_t = 11$. Further examination of the validation results shows that in the scheme $p_t = 11$, choosing $p = 7, p_v = 1, p_b = 3$ gives superior results.

8.2 Comparison of Hyperparameter Selection Algorithms

Figs. 3 and 4 break down the results for GP-ARX by the model selection routine used. Apart from the general trend observed in Figs. 1 and 2, we also observe that *grid search* and CSA give superior performance as compared with gradient-based *maximum likelihood*.

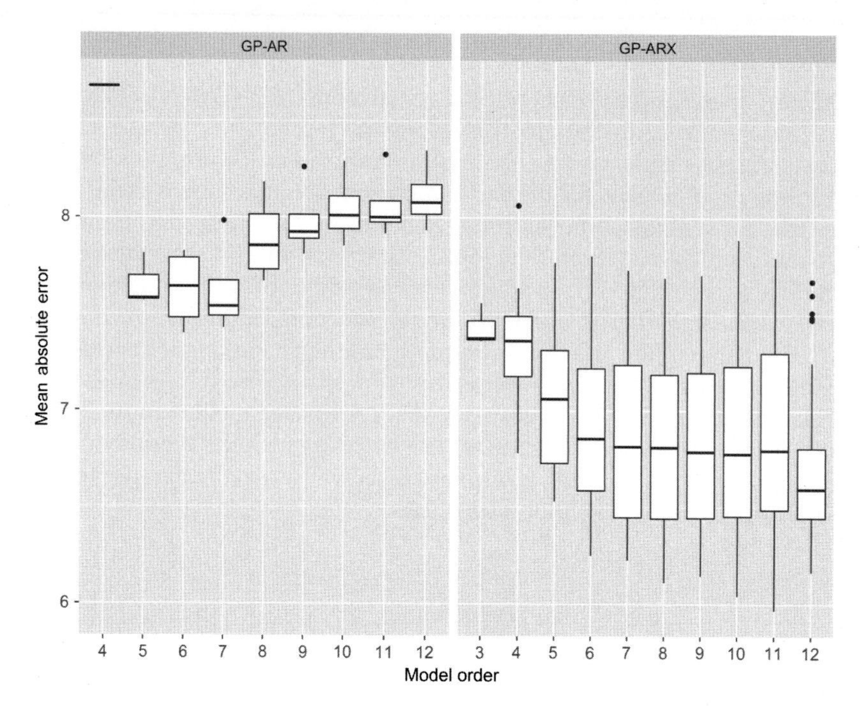

FIG. 1 Mean absolute error on validation set storms versus model order for GP-AR and GP-ARX. *Rectangle borders* represent the first and third quartiles, with a *horizontal line* inside to indicate the median value, outlying points are shown as *dots* and *whiskers* indicate the smallest and largest nonoutliers.

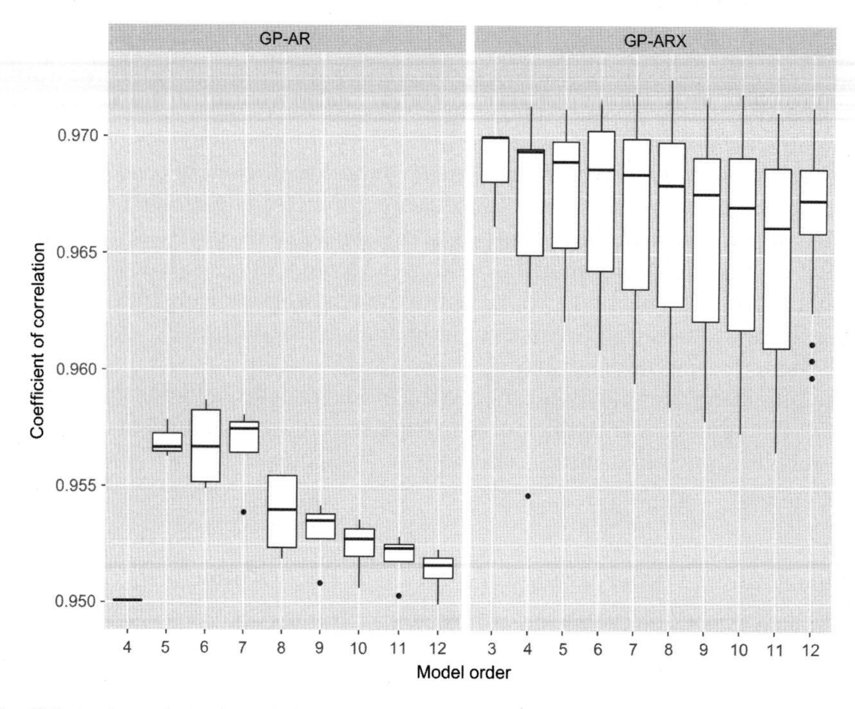

FIG. 2 Coefficient of correlation on validation set storms versus model order for GP-AR and GP-ARX. *Rectangle borders* represent the first and third quartiles, with a *horizontal line* inside to indicate the median value, outlying points are shown as *dots* and *whiskers* indicate the smallest and largest nonoutliers.

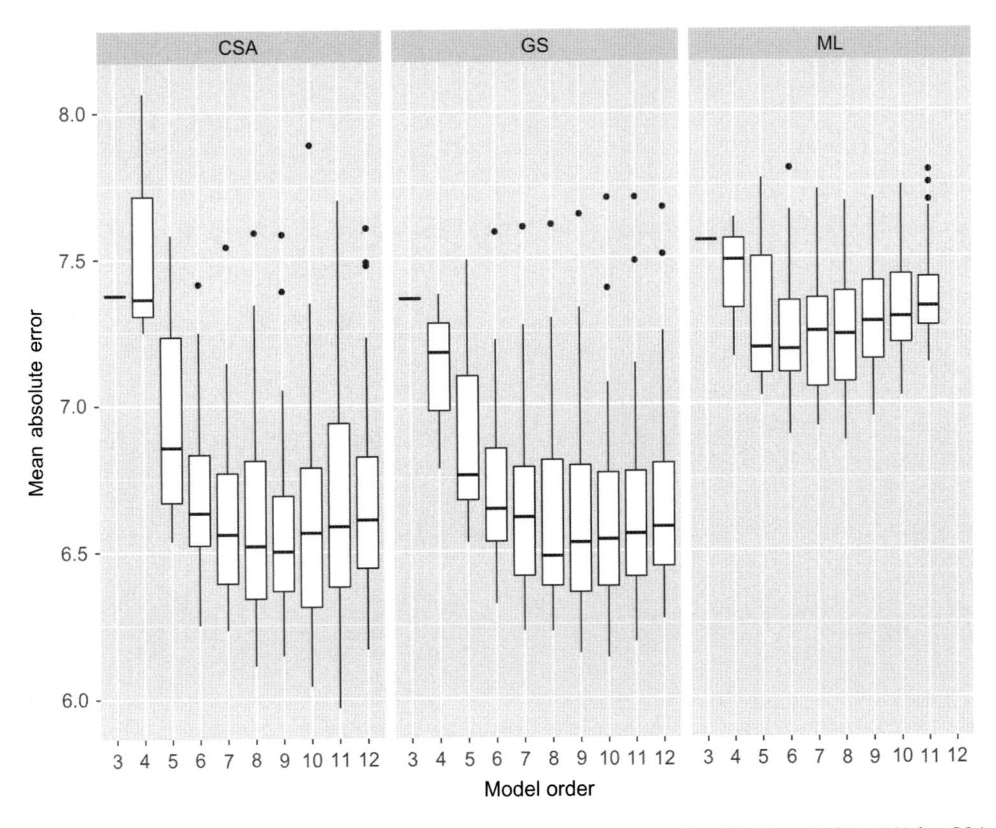

FIG. 3 Mean absolute error on validation set storms versus model order for GP-AR and GP-ARX for *CSA*, *GS*, and *ML* model selection routines. *Rectangle borders* represent the first and third quartiles, with a *horizontal line* inside to indicate the median value, outlying points are shown as *dots* and *whiskers* indicate the smallest and largest nonoutliers.

8.3 Final Evaluation

After choosing the best performing GP-AR and GP-ARX models, we calculate their performance on the test set of Table 3. The results of these model evaluations are summarized in Table 5, the GP-AR and GP-ARX models improve upon the performance of the *persistence model*.

8.4 Sample Predictions With Error Bars

Figs. 5–7 show OSA predictions of the GP-ARX model with $\pm\sigma$ error bars for three storm events in the time period between 1998 and 2003. The GP-ARX model gives accurate predictions along with plausible error bars around its mean predictions.

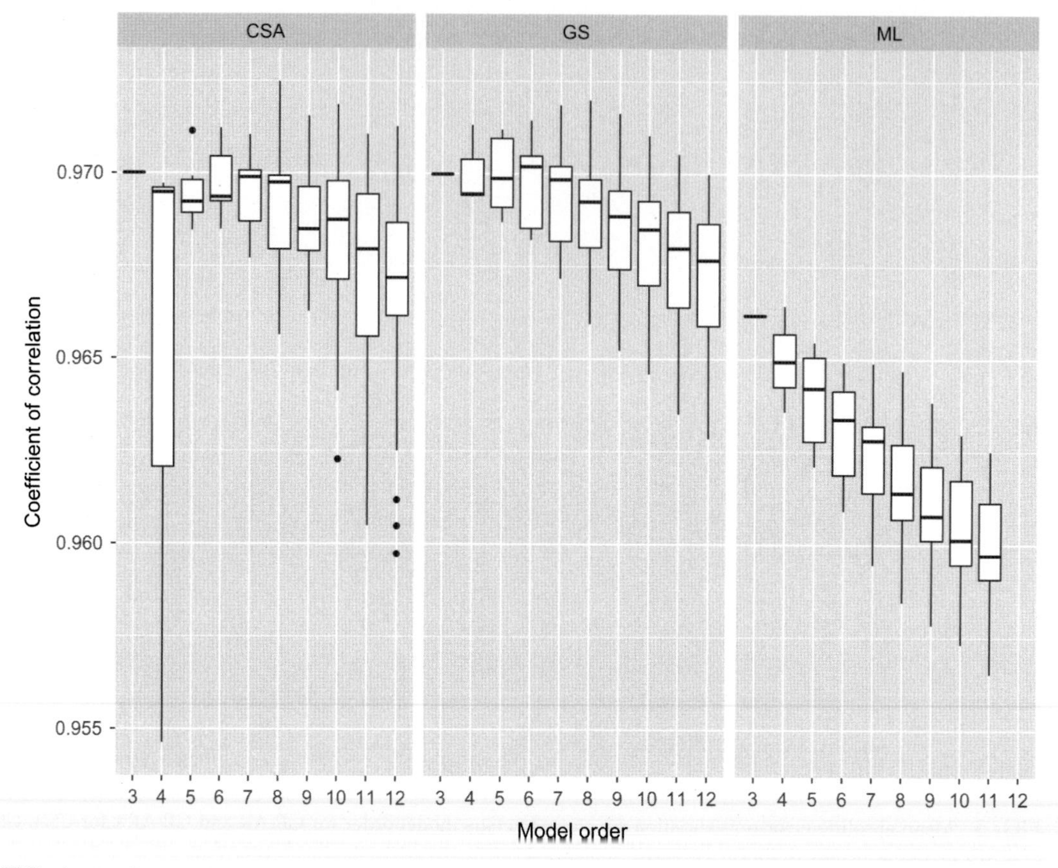

FIG. 4 Coefficient of correlation on validation set storms versus model order for GP-AR and GP-ARX for *CSA*, *GS*, and *ML* model selection routines. *Rectangle borders* represent the first and third quartiles, with a *horizontal line* inside to indicate the median value, outlying points are shown as *dots* and *whiskers* indicate the smallest and largest nonoutliers.

TABLE 5 Evaluation Results for Models on Storm Events Listed in Table 3

Model	Mean Absolute Error	Root Mean Square Error	Coefficient of Correlation
GP-ARX	7.252	11.93	0.972
GP-AR	8.37	14.04	0.963
Persistence	9.182	14.94	0.957

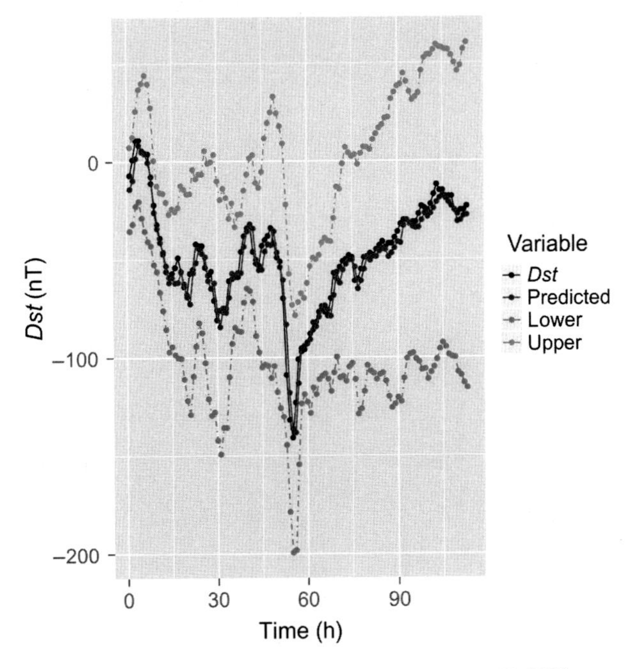

FIG. 5 OSA predictions with $\pm\sigma$ error bars for event: June 17, 2003 to June 19, 2003.

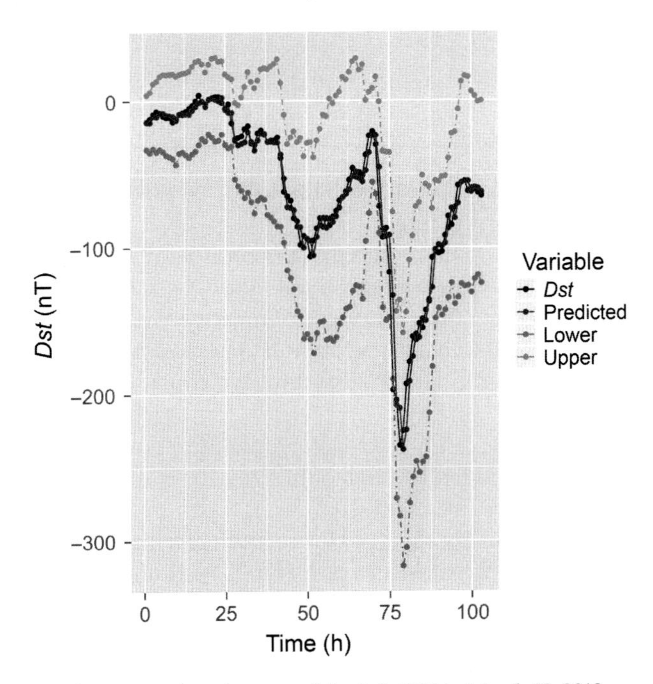

FIG. 6 OSA predictions with $\pm\sigma$ error bars for event: March 8, 2012 to March 10, 2012.

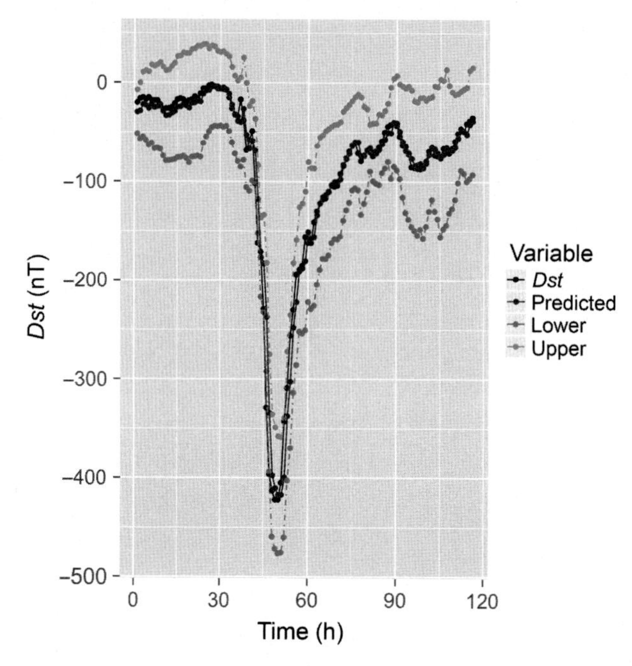

FIG. 7 OSA predictions with $\pm\sigma$ error bars for event: November 20, 2003 to November 22, 2003.

9 CONCLUSION

In this chapter, we introduced the reader to the *Gaussian process* prediction methodology in the context of space weather forecasting.

In order to successfully apply a Gaussian process model for a prediction problem, there are a number of design choices that the modeler must take. These include choice of model trend, kernel, and hyperparameter selection, among others; we gave a detailed description of each step of the process, from the data, to an operational Gaussian process forecast model.

To elucidate each step of the modeling workflow, we considered as an example, 1-h ahead prediction of the *Dst* index.

We demonstrated that in the 1-h ahead Dst prediction context, GP models yield reliable probabilistic predictions with relatively small data sizes: the training set contains 243 instances, while the validation set contains 782 instances.

To understand how well the models perform, their performance is compared with the trivial 1-h ahead baseline; the *persistence model*. The GP-AR and GP-ARX both improved on the performance of the *persistence model*.

Gaussian processes present the data-driven space weather practitioner with a tractable and expressive method for generating probabilistic forecasts of geomagnetic time series. Leveraging the strengths of the Bayesian approach, they are able to learn robust predictors from data.

References

Bala, R., Reiff, P.H., Landivar, J.E., 2009. Real-time prediction of magnetospheric activity using the Boyle Index. Space Weather 7 (4), S04003. ISSN 1542-7390. https://doi.org/10.1029/2008SW000407.

Balikhin, M.A., Boaghe, O.M., Billings, S.A., Alleyne, H.S.C.K., 2001. Terrestrial magnetosphere as a nonlinear resonator. Geophys. Res. Lett. 28 (6), 1123–1126. ISSN 1944-8007. https://doi.org/10.1029/2000GL000112.

Ballatore, P., Gonzalez, W.D., 2014. On the estimates of the ring current injection and decay. Earth Planets Space 55 (7), 427–435. ISSN 1880-5981. https://doi.org/10.1186/BF03351776.

Bartels, J., Veldkamp, J., 1949. International data on magnetic disturbances, second quarter, 1949. J. Geophys. Res. 54 (4), 399–400. ISSN 2156-2202. https://doi.org/10.1029/JZ054i004p00399.

Berlinet, A., Thomas-Agnan, C., 2004. Reproducing Kernel Hilbert Spaces in Probability and Statistics. Springer US, New York, NY. ISBN 978-1-4613-4792-7, 355 pp. https://doi.org/10.1007/978-1-4419-9096-9. Available from: http://www.springer.com/us/book/9781402076794.

Billings, S.A., Chen, S., Korenberg, M.J., 1989. Identification of MIMO non-linear systems using a forward-regression orthogonal estimator. Int. J. Control 49 (6), 2157–2189. https://doi.org/10.1080/00207178908559767.

Boynton, R.J., Balikhin, M.A., Billings, S.A., Sharma, A.S., Amariutei, O.A., 2011a. Data derived NARMAX Dst model. Ann. Geophys. 29 (6), 965–971. https://doi.org/10.5194/angeo-29-965-2011. Available from: http://www.ann-geophys.net/29/965/2011/.

Boynton, R.J., Balikhin, M.A., Billings, S.A., Wei, H.L., Ganushkina, N., 2011b. Using the NARMAX OLS-ERR algorithm to obtain the most influential coupling functions that affect the evolution of the magnetosphere. J. Geophys. Res. Space Phys. 116 (A5), A05218. ISSN 2156-2202. https://doi.org/10.1029/2010JA015505.

Boynton, R.J., Balikhin, M.A., Billings, S.A., Reeves, G.D., Ganushkina, N., Gedalin, M., Amariutei, O.A., Borovsky, J.E., Walker, S.N., 2013. The analysis of electron fluxes at geosynchronous orbit employing a NARMAX approach. J. Geophys. Res. Space Phys. 118 (4), 1500–1513. ISSN 2169-9402. https://doi.org/10.1002/jgra.50192.

Burton, R.K., McPherron, R.L., Russell, C.T., 1975. An empirical relationship between interplanetary conditions and Dst. J. Geophys. Res. 80 (31), 4204–4214. ISSN 2156-2202. https://doi.org/10.1029/JA080i031p04204.

Chandorkar, M., Camporeale, E., Wing, S., 2017. Probabilistic forecasting of the disturbance storm time index: an autoregressive Gaussian process approach. Space Weather. ISSN 1542-7390. https://doi.org/10.1002/2017SW001627.

Davis, T.N., Sugiura, M., 1966. Auroral electrojet activity index AE and its universal time variations. J. Geophys. Res. 71 (3), 785–801. ISSN 2156-2202. https://doi.org/10.1029/JZ071i003p00785.

Dessler, A.J., Parker, E.N., 1959. Hydromagnetic theory of geomagnetic storms. J. Geophys. Res. 64 (12), 2239–2252. ISSN 2156-2202. https://doi.org/10.1029/JZ064i012p02239.

Friedel, R.H.W., Korth, H., Henderson, M.G., Thomsen, M.F., Scudder, J.D., 2001. Plasma sheet access to the inner magnetosphere. J. Geophys. Res. Space Phys. 106 (A4), 5845–5858. ISSN 2156-2202. https://doi.org/10.1029/2000JA003011.

Hofmann, T., Schölkopf, B., Smola, A.J., 2008. Kernel methods in machine learning. Ann. Statist. 36 (3), 1171–1220. https://doi.org/10.1214/009053607000000677.

Kirkpatrick, S., Gelatt, C.D., Vecchi, M.P., 1983. Optimization by simulated annealing. Science 220 (4598), 671–680. ISSN 0036-8075. https://doi.org/10.1126/science.220.4598.671. Available from: http://science.sciencemag.org/content/220/4598/671.

Kissinger, J., McPherron, R.L., Hsu, T.S., Angelopoulos, V., 2011. Steady magnetospheric convection and stream interfaces: relationship over a solar cycle. J. Geophys. Res. Space Phys. 116 (A5), A00I19. ISSN 2156-2202. https://doi.org/10.1029/2010JA015763.

Krige, D.G., 1951. A Statistical Approach to Some Mine Valuation and Allied Problems on the Witwatersrand. Available from: https://books.google.co.in/books?id=M6jASgAACAAJ.

Lundstedt, H., Gleisner, H., Wintoft, P., 2002. Operational forecasts of the geomagnetic Dst index. Geophys. Res. Lett. 29 (24), 34-1–34-4. ISSN 1944-8007. https://doi.org/10.1029/2002GL016151.

McPherron, R.L., Terasawa, T., Nishida, A., 1986. Solar wind triggering of substorm expansion onset. J. Geomagn. Geoelectr. 38 (11), 1089–1108. https://doi.org/10.5636/jgg.38.1089.

McPherron, R.L., Siscoe, G., Arge, N., 2004. Probabilistic forecasting of the 3-h Ap index. IEEE Trans. Plasma Sci. 32 (4), 1425–1438.

McPherron, R.L., Siscoe, G., Crooker, N.U., Arge, N., 2013. Probabilistic forecasting of the Dst index. American Geophysical Union. pp. 203–210. ISBN 9781118666098, https://doi.org/10.1029/155GM22.

Neal, R.M., 1996. Bayesian Learning for Neural Networks. Springer-Verlag New York, Inc., Secaucus, NJ. ISBN 0387947248.

Newell, P.T., Liou, K., Gjerloev, J.W., Sotirelis, T., Wing, S., Mitchell, E.J., 2016. Substorm probabilities are best predicted from solar wind speed. J. Atmos. Sol. Terr. Phys. 146, 28–37. ISSN 1364-6826. https://doi.org/10.1016/j.jastp.2016.04.019. Available from: http://www.sciencedirect.com/science/article/pii/S1364682616301195.

O'Brien, T.P., McPherron, R.L., 2000. An empirical phase space analysis of ring current dynamics: solar wind control of injection and decay. J. Geophys. Res. Space Phys. 105 (A4), 7707–7719. ISSN 2156-2202. https://doi.org/10.1029/1998JA000437.

Pallocchia, G., Amata, E., Consolini, G., Marcucci, M.F., Bertello, I., 2006. Geomagnetic Dst index forecast based on IMF data only. Ann. Geophys. 24 (3), 989–999. Available from: https://hal.archives-ouvertes.fr/hal-00318011.

Rasmussen, C.E., Nickisch, H., 2016. GPML Gaussian processes for machine learning toolbox. Available from: http://mloss.org/software/view/263/.

Rasmussen, C.E., Williams, C.K.I., 2005. Gaussian Processes for Machine Learning (Adaptive Computation and Machine Learning). The MIT Press, Cambridge, MA. ISBN 026218253X.

Scholkopf, B., Smola, A.J., 2001. Learning With Kernels: Support Vector Machines, Regularization, Optimization, and Beyond. MIT Press, Cambridge, MA. ISBN 0262194759.

Tao, T., 2011. An Introduction to Measure Theory, Graduate Studies in Mathematics. American Mathematical Society. ISBN 9780821869192. Available from: https://books.google.nl/books?id=HoGDAwAAQBAJ.

Wang, C.B., Chao, J.K., Lin, C.H., 2003. Influence of the solar wind dynamic pressure on the decay and injection of the ring current. J. Geophys. Res. Space Phys. 108 (A9), 1341. ISSN 2156-2202. https://doi.org/10.1029/2003JA009851.

Wilks, D.S., 2011. Statistical Methods in the Atmospheric Sciences, vol. 100. Academic Press, San Diego, CA.

Wing, S., Johnson, J.R., Jen, J., Meng, C.I., Sibeck, D.G., Bechtold, K., Freeman, J., Costello, K., Balikhin, M., Takahashi, K., 2005. Kp forecast models. J. Geophys. Res. Space Phys. 110 (A4), A04203. ISSN 2156-2202. https://doi.org/10.1029/2004JA010500.

Xavier-De-Souza, S., Suykens, J.A.K., Vandewalle, J., Bolle, D., 2010. Coupled simulated annealing. IEEE Trans. Syst. Man Cybern. B Cybern. 40 (2), 320–335. ISSN 10834419. https://doi.org/10.1109/TSMCB.2009.2020435.

Zhu, D., Billings, S.A., Balikhin, M., Wing, S., Coca, D., 2006. Data derived continuous time model for the Dst dynamics. Geophys. Res. Lett. 33 (4), L04101. ISSN 1944-8007. https://doi.org/10.1029/2005GL025022.

Zhu, D., Billings, S.A., Balikhin, M.A., Wing, S., Alleyne, H., 2007. Multi-input data derived Dst model. J. Geophys. Res. Space Phys. 112 (A6), A06205. ISSN 2156-2202. https://doi.org/10.1029/2006JA012079.

CHAPTER

10

Prediction of MeV Electron Fluxes and Forecast Verification

Adam Kellerman

UCLA, Los Angeles, CA, United States

CHAPTER OUTLINE

1 RELATIVISTIC ELECTRONS IN EARTH'S OUTER RADIATION BELT

Three primary forms of radiation may adversely affect spacecraft systems in Earth's outer magnetosphere: solar flare particles (primarily protons with energy, E, 10–200 MeV), magnetospheric substorm particles (energetic electrons and ions $\sim E <$ a few hundred keV), and very-high-energy (relativistic, or $\sim E > 1$ MeV) electrons (Baker et al., 1987). This chapter is focused on the MeV electron population, what effect it has on space hardware, where these particles come from, and what controls their dynamics. Later in the chapter, numerical and statistical methods are introduced, which aid in understanding how well a given radiation-belt forecast model may characterize, and forecast, the state of the Earth's radiation belts.

https://doi.org/10.1016/B978-0-12-811788-0.00010-X

Baker et al. (1987) detailed a long-lasting increase in fluxes, at the maximum of which the GOES-5 spacecraft experienced a filament failure in an encoder lamp. This damage was detrimental to the spacecraft's purpose in providing weather observations, thus demonstrating that high-energy electron fluxes may damage space hardware and have an immediate impact on our society. The process of energy deposition into spacecraft materials through impact from a relativistic electron is known as *deep-dielectric charging*, and was first postulated by Meulenberg (1976). This phenomenon may have led to several recorded failures of spacecraft electrical systems during geomagnetic storms, such as the well-known attitude control system failure on the Galaxy 4 spacecraft in 1998 (Baker et al., 1998).

The likelihood of a spacecraft encountering sufficient flux for deep-dielectric charging was discussed by Vampola (1987). An excellent example of the unexpected behavior resulting from deep-dielectric charging occurred during Voyager 1's pass of Jupiter on March 5, 1979. The spacecraft was to execute a close approach to the planet, while conducting a photographic sequence. As the spacecraft was traversing Jupiter's radiation belts, a series of power-on resets (POR) occurred. The cold plasma density was too high to suggest surface charging as a culprit, and a significant increase in >10 MeV electrons was observed at the same time as the POR's. It was determined that one of the cables carrying the POR signal was exposed to the exterior radiation environment, and thus laboratory tests on an equivalent cable were conducted. The results suggested that spurious pulses in the exposed cable, as a result of embedded charge accumulation, caused the POR's. The embedded charge resulted due to the large fluxes of high-energy electrons, and thus deep-dielectric charging was responsible for the POR's. Vampola (1987) recommended (a) to not expose sensitive elements to the environment (in that case, cables), and (b) if a component needs to be exposed, then shield it, or design any connected system so that it will not respond to a signal that may be caused by deep-dielectric charging.

The increased awareness of potential radiation hazards to operational spacecraft as a result of deep-dielectric charging motivated further research into the source, loss, transport, and acceleration mechanisms of relativistic electrons. In this chapter, the focus is entirely on the Earth's inner magnetosphere. In the remainder of this section, the knowledge that has led to current radiation belt model development is briefly introduced. In Section 2, different numerical approaches used for forecasting are covered, and the concept of data assimilation is introduced. In Section 3, forecast verification is described, along with several metrics that may be used for radiation belt forecast verification. Toward the end of Section 3, an example application of forecast verification for forecasting the dynamics of Earth's radiation belts is presented.

1.1 Source, Loss, Transport, and Acceleration, Variation

Many natural systems are defined by a large number of variables, and as researchers, we seek to understand how those variables are related to each other, and to the state and evolution of the system. The first step in understanding the system is usually to observe it, visually, or through one or more measurement techniques. Through observation, a theoretical and/or empirical model of the system may be developed. Given the large number of variables in most natural systems, accurate modeling and forecasting of such systems may appear, at first glance, to be a tremendously difficult, if not impossible task. In many cases though, a system can be well described with the use of fewer, though key, variables, and by introducing the notion of a *phase space*. A phase space is simply a geometrical representation of the hypothetically possible states of a given system. In the case of radiation belt particles, trapped within Earth's magnetic field, the kinematical state of a particle may be described using the three coordinates of position, and the three components of canonical momentum (Schulz and Lanzerotti, 1974). These six variables will define the position of the particle in a six-dimensional phase space.

In a planetary dipole magnetic field, relativistic charged particles undergo periodic motions as a result of a force described by the Lorentz equation

$$\mathbf{F} = q\mathbf{E} + q(\mathbf{v} \times \mathbf{B}) \tag{1}$$

where q is the particle charge, $\mathbf{E}$ is the vector electric field, $\mathbf{v}$ is the particle velocity, and $\mathbf{B}$ is the vector magnetic field. Note that this is a simplified, and nonrelativistic version of the momentum equation, where the force due to gravity and collision term have been excluded. The equation is separated in this manner as the first, $q\mathbf{E}$, term will be ignored for the rest of this chapter. The term is ignored because the motion of sufficiently energetic particles is dominated by the second term.

There are three periodic motions that define a radiation belt particle:

(1) Gyration around a field line, due to the force perpendicular to the direction of $\mathbf{v}$ and $\mathbf{B}$. In a uniform magnetic field, a charged particle will experience the same perpendicular force, toward the center of gyration, at all points on its circular gyration path. The center of that gyration, the *guiding center*, will remain fixed in space.

(2) Force-free motion along the magnetic field near the magnetic equator, until the particle approaches a magnetic pole, and the Lorentz force is no longer strictly perpendicular to the local $\mathbf{B}$. In this situation, a component known as the *mirror force* will act against the direction of motion, increasing as the particle approaches the north or south pole. If the mirror force overcomes the particle's poleward momentum, then it will *mirror* and move back toward the equator. If no other significant force acts on the particle, and an equally strong magnetic field is encountered at the opposite pole, the particle will continue to mirror or *bounce* between the mirror points.

(3) Drift motion: If the magnetic field is nonuniform, and increases in a given direction, then the force exerted on a particle will be greater on that half of the gyration orbit, maximizing at the point of strongest magnetic field; a stronger magnetic field has the effect of decreasing the gyration radius. The result is a drift of the guiding center perpendicular to the direction of increasing magnetic field, dependent on the charge of the particle (Eq. 1).

In the case of a dipole field, similar to the field observed in the near-Earth space, the field strength increases radially inward along the magnetic equator. The field strength then must decrease slightly along straight-line trajectories in both directions perpendicular to the radial direction. The effect is that a charged particle will always drift along a path that is tangent to the direction of increasing field strength. In the case of a dipole field, the guiding center of that particle will trace out a perfect circle around the Earth. If the particle is also bouncing, then it will experience a different magnetic field during its bounce-drift orbit, and hence a bouncing particle may trace out a different path than an equatorial particle, depending on the magnetic field configuration.

At a few Re or more, the Earth's magnetic field may not be approximated correctly by a dipole, and the computation of a drift path can only be accomplished numerically. In a realistic field, a bouncing particle usually traces out a significantly different path than an equatorial particle, which may lead to different dynamics for the two populations. For more information and a good introduction into single-particle motion, see, for example, Walt (2005).

If the trapped particles are of sufficiently high energy, then their gyration, bounce, and drift motions may occur on a time scale faster than most large-scale changes in the global magnetic field. In this case, these motions may be phase-averaged, and described in terms of three adiabatic invariants of motion, namely, μ, J, and ϕ (Roederer, 1970). The derivation of the invariants will not be discussed here, as this is covered in several textbooks in the field; however, each invariant will be introduced briefly, so that the reader may follow the reasoning behind using these as coordinates for a state space to conduct radiation belt research. The first invariant given by $\mu = p_\perp^2/(2m_0B)$, where $p_\perp$ is the particle momentum, perpendicular to **B**, and m is the particle mass. The first invariant is equal to the magnetic moment of a particle and says that the magnetic flux enclosed within a current loop, defined by the particle gyration orbit, will remain constant so long as no change to the magnetic field occurs on a time scale close to that of the gyroperiod. The second invariant $J = \oint p_\parallel ds$, is the integral of the parallel momentum along the field line on which the particle is bouncing. One may combine the second invariant with the first, to create a new invariant, $K = J/(2\sqrt{2m\mu})$ which is still invariant under an external force acting perpendicular to **B**.

The third invariant relates to the drift shell of a particle and is most easily understood if it is expressed as the magnetic flux encompassed by a drift orbit of a particle. If one traces the field lines intersecting the guiding center of the drifting particle to the northern or southern hemisphere polar cap, and integrates the magnetic flux, this value should remain constant under global field changes that are much slower than the drift period. The third invariant is expressed as $\phi = \int_{ni} \mathbf{B} \cdot ds$. In radiation belt physics, it is customary to use the *Roederer L* value instead of ϕ, which is more commonly known as L^*. This transformation adds a "spatial" element to the coordinate system of electron phase space density (PSD), and is expressed as $L^* = 2\pi B_E R_E^2/\phi$. L^* can be thought of as the distance along the magnetic equator to the field line on which the particle is located, should the nondipole field be relaxed very slowly (adiabatically) to a dipole field. From this point on, L and L^* will be used interchangeably.

The number of particles in the Earth's radiation belts is quite large, and so it is customary to work with the density of particles in phase space, the PSD. Similar to the single-particle motion example, the density of particles can be described by the three adiabatic invariants. In this formulation, the evolution of PSD may then be described by the modified Fokker-Planck equation for radiation belt particles (e.g., Roederer, 1970; Schulz and Lanzerotti, 1974). This equation is a diffusion equation, and describes the time-rate of change of the electron PSD, by incorporating diffusion coefficients that act to redistribute the density of electrons. It is the interaction of magnetospheric waves, driven by processes external or internal to the Earth's magnetosphere, that occur on the time scale of a gyration, bounce, or drift period that cause the *violation* of one or more adiabatic invariants, and the associated change in PSD. Implicit in this description is that zero diffusion describes no change in PSD and only adiabatic motion occurs in real space.

Since the discovery of the Van Allen Radiation belts, there has been much discussion on the processes responsible for the source, loss, transport, and acceleration of electrons to relativistic energies. The two primary physical mechanisms identified as responsible for electron acceleration are (1) ultra low-frequency (ULF) wave interaction with particles leading to inward diffusion in L, more commonly known as inward *radial diffusion* (Kellogg, 1959; Fälthammar, 1965)—radial diffusion conserves the first and second adiabatic invariants, while the third invariant is violated—and (2), diffusion to higher energies, or *local acceleration* by very low-frequency (VLF) electromagnetic waves (e.g., Summers et al., 1998; Horne and Thorne, 1998; Omura et al., 2007).

In the first case, as electrons are transported inward, the gyroperiod and bounce period is much faster than the change in the ambient magnetic field, and so as the particles encounter a stronger magnetic field, the perpendicular velocity increases, and they are accelerated to higher energies. This is known as betatron acceleration. In the second case, VLF waves can interact on the time scale of a gyration or bounce period, and hence act to change the direction of motion of an electron; a process known as *scattering*. The angle that an electron's velocity vector makes with $\mathbf{B}$ at the magnetic equator is known as the electron's pitch angle. If this changes due to a wave-particle interaction, then the process is known as pitch-angle scattering. Electrons of eV to a few 10s of keV energy may be pitch-angle scattered into the atmosphere, causing the Aurora Borealis and Aurora Australis. Electrons in the 100s of keV range, also known as *seed electrons*, may diffuse to a region of lower PSD, and because there are fewer high-energy (MeV) electrons, the VLF waves can act to accelerate the 100s of keV electrons, locally, to higher energies.

Meredith et al. (2001) recognized that there existed a dependence of VLF wave amplitudes on a process known as a substorm, where a large-scale reconfiguration of the tail magnetosphere occurs, leading to fast transport of ions and electrons from the distant magnetotail to geosynchronous orbit (GEO). It was later shown that the observed seed-electron fluxes (100s of keV) were higher during *acceleration events*, where enhanced relativistic electron fluxes were observed (Meredith et al., 2003). Hence, a link between MeV electrons and a sufficiently large source population was established. Hwang et al. (2004) presented events that occurred during geomagnetic storms. A geomagnetic storm is historically defined by the enhanced numbers of ions and electrons drifting around the Earth, just outside of 4 Re, leading to a current ring that intensifies, and later decays. This current is known as the *ring current*, and it causes a decrease in the magnitude of the Earth's magnetic field, as measured at the Earth's surface by magnetometers. The Dst index is, hence, a measure of geomagnetic storm intensity, and has hourly resolution. Hwang et al. (2004) looked for events accompanied by electron acceleration. One event was stronger in terms of the Dst index; however, it did not produce enhanced relativistic electron fluxes. A larger seed-electron population was observed in the other event, VLF wave activity was enhanced, and a relativistic electron enhancement was observed. These studies provided initial evidence that, although other ingredients were available for electron acceleration, an enhanced seed particle population is critical for accelerating a large number of particles to relativistic energies.

When the dynamic pressure of the solar wind is high, the Earth's magnetosphere is compressed, resulting in an overall change in the global magnetic field strength. The magnetopause moves in, electrons will conserve the third adiabatic invariant by moving outward, and may find themselves on open drift paths. These electrons may intersect the magnetopause, and possibly be lost to interplanetary space. The largest L^* at which a drifting particle may still complete an entire orbit around the Earth without intersecting the magnetopause is known as the last closed drift shell (LCDS). Shprits et al. (2006) showed in a modeling study that inclusion of a variable outer boundary in L is necessary to reproduce the dynamics of the radiation belts, which was later borne out in a combined observational and modeling study (Turner et al., 2012), where significant loss to the magnetopause coincided with no observable atmospheric precipitation.

In the current chapter, the following material focuses on observations and modeling of the radiation-belt environment around and inside of GEO. The effects of the electric field are ignored, as they are considered negligible for radiation belt electrons (e.g., Schulz and Lanzerotti, 1974). Several models have been developed to simulate radiation belt dynamics, and one such model, the versatile electron radiation belt (VERB) code (Subbotin and Shprits, 2009), will be employed to aid in discussing forecast verification.

2 NUMERICAL TECHNIQUES IN RADIATION BELT FORECASTING

Several models have been developed to accomplish the task of relativistic electron flux prediction at GEO, utilizing two popular techniques: linear prediction filters (e.g., Nagai, 1988; Baker and McPherron, 1990; Vassiliadis et al., 2002) and neural networks (e.g., Koons and Gorney, 1991; Fukata et al., 2002; Ling et al., 2010). While physics-based models allow the study of relative contributions of various acceleration and loss mechanisms (e.g., Albert et al., 2009; Subbotin and Shprits, 2009; Su et al., 2011; Subbotin et al., 2011; Kim et al., 2012; Horne et al., 2013), statistical approaches may reveal new processes or dynamics that may inform physical understanding and model development.

The long-term dataset of observations at GEO has attracted the majority of past forecasting studies, which have strived to forecast the daily averaged flux or daily integrated fluence of geosynchronous electrons. A notable exception to the majority was O'Brien and McPherron (2003), who presented a unique study that used hourly, rather than daily, data to forecast the electron radiation environment. The linear prediction filter approach to forecasting MeV electrons was used to construct the relativistic electron forecast model (REFM), which has been implemented at the National Oceanic and Atmospheric Administration (NOAA) (Baker and McPherron, 1990).

Similar in direction to the linear prediction filter approach, Simms et al. (2014) considered which specific factors contribute to the largest flux increases by applying multiple regression analysis, partial correlation analysis, and partial regression to determine the driving parameters of 1 MeV daily electron flux at geostationary orbit. Many authors have shown that multiple parameters contribute to the variation of flux at GEO. Many of those parameters are, in fact, correlated with each other. However, Simms et al. (2014) was able to show which parameters contributed the most to the flux variation by removing the partial correlation residuals.

A backward elimination technique was employed, whereby the variable resulting in the least significant contribution was eliminated. It was found that the remaining important variables were seed electrons, change in density, ground- and space-based ULF, and solar wind velocity. Simms et al. (2014) also found that the contribution of ground and space-based ULF do not differ greatly, with a correlation of 0.85, generally. Forward or backward elimination techniques may be used to determine the most significant forecast variables to include in a given model, though one needs to carefully define appropriate stopping rules or stopping criteria if utilizing such techniques (e.g., Wilks, 2011).

Another way to obtain information that may guide model development and our understanding of a system is to employ data-assimilative methods. An example of an autoregressive model is that of a Markov process, and will serve as an introduction to the data-assimilative forecast model that is introduced in the next section.

$$x_{t+1} - \mu = \phi(x_t - \mu) + \sigma_{t+1} \tag{2}$$

where x_t defines the current state of the system, μ is the mean of the time series, ϕ is the autoregressive parameter, and σ_{t+1} is the residual, assumed to be randomly selected from a Gaussian distribution. One can see that information is propagated from time t to $t+1$ through a random estimate of the error, and a linear slope ϕ, which will be positive for a time series exhibiting persistence. In this case, one does not need to use multiple values of the state prior to time t in order to forecast the future state x_{t+1}, as in the case of the linear prediction filter. Here the prior information is contained in the state x_t. However, one must estimate the error based on a basic linear model, and the error determined from the Gaussian distribution, constructed from the distribution mean and variance. In dynamic systems, this error may have a complex dependence on the past or current state of the system, making it particularly difficult to specify correctly. A more sophisticated way of estimating the current and future state of a system is to use a data assimilation technique that allows one to forecast the model error variance directly.

Many data assimilation methods have been applied successfully in radiation belt physics (e.g., Kondrashov et al., 2007; Ni et al., 2009; Shprits et al., 2012). The Kalman filter has been applied to radial transport in a 1D application (Kondrashov et al., 2007; Shprits et al., 2007; Ni et al., 2009; Daae et al., 2011; Schiller et al., 2012), while more recently several studies have employed an ensemble formulation (Evensen, 2003) in 1D (Koller et al., 2007), and 3D diffusive models (Reeves et al., 2012; Godinez and Koller, 2012; Bourdarie and Maget, 2012). However, this formulation requires multiple model runs and can also present a problem of filter divergence (e.g., Evensen, 2003; Godinez and Koller, 2012).

The computational requirements for DA become very large for high-end models in 3D state space, and especially with multiple spacecraft that provide many observations to be assimilated. It motivates development of alternative assimilation methods of reduced complexity. In particular, a so-called split-operator Kalman filter approach has been developed to assimilate radiation belt electrons using CRRES observations with the 3D VERB code (Shprits et al., 2013b). It led to the discovery of a four-zone structure and identification of local acceleration events during a historical geomagnetic super storm (Kellerman et al., 2014), and the development of the first operational data-assimilative physics-based radiation belt forecast model (http://rbm.epss.ucla.edu/realtime-forecast).

The advantage of using a Kalman filter is the aforementioned ability to forecast not just the state of the system, as in the simple Markov process in Eq. (2), but also the error variance. For more information on the Kalman filter see Kalman (1960), and for a good introduction to data assimilation and Kalman filtering, see, for example, Kalnay (2002).

The aspects of forecast performance and forecast verification will be introduced in the next section, and utilized to provide an example of quantitative forecast verification for radiation belt electrons.

3 RELATIVISTIC ELECTRON FORECASTING AND VERIFICATION

3.1 Forecast Verification

Forecast verification is a process largely developed by the atmospheric sciences, in relation to terrestrial weather forecasting. Forecast verification is an important procedure, and serves (1) to determine how accurate a particular model forecast is, and how the accuracy is changing over time, (2) to identify where the model is wrong, so that the forecast may be improved over time, and (3) to provide a robust framework so that one forecast system may be compared with another, and thus allow one to quantify why one model performs better than another.

It has become common practice to employ what is known as *scalar aspects*, or attributes to provide information on forecast performance. Some of these scalar aspects are forecast accuracy, bias, calibration, and resolution. Through experimentation, it has been revealed that these scalar aspects provide quite useful information to help with forecast verification, though they inevitably must discard some of the information available. For example, the bias of a model forecast may be represented by a single median value of all differences between model forecasts and observations. There may be thousands of forecast and observational data points that were utilized to come up with that one scalar number. Inevitably, in utilizing scalar aspects, one must discard some of the information available, in order to reduce the dimension of the forecast problem. For additional reading on frameworks for forecast verification, see Murphy and Winkler (1987) or Wilks (2011).

While some aspects have been used extensively in the field of space science, such as forecast accuracy, and relative accuracy (forecast skill), others have received less attention, such as forecast calibration, resolution, discrimination, and sharpness. Commonly applied scalar accuracy measures are the mean absolute error (MAE)

$$\mathrm{MAE} = \frac{1}{n}\sum_{k=1}^{n}|x_k - y_k| \tag{3}$$

and the mean squared error (MSE),

$$\mathrm{MSE} = \frac{1}{n}\sum_{k=1}^{n}(x_k - y_k)^2 \tag{4}$$

where x_k is the kth forecast value, y_k is the kth observation, and n is the total number of predictions and observations to be considered. Both of these metrics provide one number to quantify overall forecast accuracy, and may be combined with a baseline model, such as persistence, in order to develop a skill score. The square root of the MSE, root-mean-squared error (RMSE), is also a useful measure of forecast accuracy, as it preserves the dimension of the observations and forecast, and thus can be used to quantify accuracy directly. In addition to the MAE and MSE, the mean error (ME) is a useful method for determining the overall bias in a given forecast, and is defined as

$$\begin{aligned}\mathrm{ME} &= \frac{1}{n}\sum_{k=1}^{n}(x_k - y_k)\\ &= \bar{x} - \bar{y}\end{aligned} \tag{5}$$

where $\bar{x}$ is the mean value of x.

As an example, Kellerman et al. (2013) employed a probabilistic approach, recurrence, and persistence, with a dynamic weighting function to forecast daily geosynchronous electron fluxes. The technique was described in Kellerman and Shprits (2012). Though not exclusively applicable to this study, this example can be used to demonstrate different aspects of forecast performance. The driving parameters of the geosynchronous radiation-belt electron empirical prediction (GREEP) model were solar wind density and velocity—two parameters used widely to characterize relativistic electron dynamics at GEO (e.g., Paulikas and Blake, 1979; Baker and McPherron, 1990; Takahashi and Ukhorskiy, 2008; Lyatsky and Khazanov, 2008a,b; Tan et al., 2010; Balikhin et al., 2011). The MSE of the GREEP model and that of the average observations were combined to construct the prediction efficiency (PE) of the model, which characterizes the performance of the model with respect to that of the arithmetic mean of the data. The PE is thus characterized as a skill score.

$$\begin{aligned} PE &= 1 - \frac{MSE_{\text{model}}}{MSE_{\text{mean}}} \\ &= 1 - \frac{\sum_{i=1}^{N}(m_i - p_i)^2}{\sum_{i=1}^{N}(m_i - \bar{m})^2} \end{aligned} \tag{6}$$

The forecast performance was measured by way of "forecast skill" and "forecast score," both of which are actually skill scores, as they refer to *relative accuracy* of the forecast with respect to a reference model. In order to keep nomenclature under control, we will define the "forecast score" as skill score 1 (SS1), and the forecast skill as skill score 2 (SS2). Both skill scores are introduced as follows, along with an explanation of when to use them.

The SS1 is defined as the ratio of the PE of a given model to that of a baseline model, say, simple persistence or recurrence. In the case of persistence we have

$$\begin{aligned} SS1 &= \frac{PE_{\text{Model}}}{PE_{\text{Persist}}} \\ &= \frac{\sum_{i=1}^{N}(m_i - \bar{m})^2 - \sum_{i=1}^{N}(m_i - p_i)^2}{\sum_{i=1}^{N}(m_i - \bar{m})^2 - \sum_{i=1}^{N}(m_i - m_{i-1})^2} \end{aligned} \tag{7}$$

This skill score includes the sample data set mean in the numerator and denominator. The SS2 compares the model forecast with a standard forecast. In the case where the standard forecast is persistence we have

$$\begin{aligned} SS2 &= \frac{PE_{\text{Model}} - PE_{\text{Persist}}}{1 - PE_{\text{Persist}}} \\ &= \frac{\sum_{i=1}^{N}(m_i - m_{i-1})^2 - \sum_{i=1}^{N}(m_i - p_i)^2}{\sum_{i=1}^{N}(m_i - m_{i-1})^2} \end{aligned} \tag{8}$$

This skill score is independent of the sample set mean used to calculate the prediction efficiency, and only depends on the difference of the sum of variances between the prediction model and the baseline model, normalized to the baseline model variance. The SS2 may then be related to SS1 by:

$$SS2 = \frac{PE_{\text{Persist}}(SS1 - 1)}{1 - PE_{\text{Persist}}} \tag{9}$$

For the same SS1, a higher PE for persistence will result in a higher value for SS2 (Kellerman et al., 2013). This is particularly important when considering the baseline dataset, as SS1 includes the variance of the baseline model with respect to a mean value of that dataset in the numerator and denominator. Therefore, use of SS1 to compare two studies using different baseline data is okay, provided that the reference data variance is determined in the same way (i.e., as a 1-year mean of the data). The use of SS2 is only meaningful if the same dataset is used for each study, as the final result is normalized to the baseline model, which includes those data. The result is that for smoother, less-noisy datasets there are smaller baseline model variances, and so a small improvement in model performance over persistence performance would be accentuated.

In the following section, these skill scores are applied to radiation belt modeling and forecasting. Both scores are applied, so that one may compare results with the same, or another, dataset in the future. The data-assimilative VERB code is the model, and it is used to conduct several forecasts over a ~2-year period. Persistence of electron PSD is used as a reference model. In order to determine the quality of the forecast, several existing, and some new, techniques are employed. The hope is that these techniques will become useful to the radiation-belt electron forecasting community. The techniques may also prove useful for standard model validation, or cross-model validation. This should be the beginning of the discussion on how to construct a forecast framework that we may utilize together as a community in order to understand the quality of our radiation belt models, and how that quality is improving over time.

3.2 Relativistic Electron Forecasting

Radiation belt forecasting has historically been restricted to one set of observations, such as a 1D scatter plot of relativistic electrons at GEO versus time, usually for electrons of a particular energy, with electron flux or fluence often the desired forecast variable. In this section, a quantitative measure of forecast performance is introduced, which utilizes electron PSD, and analyzes the forecast performance conditional on the value of the first, second, and third adiabatic invariants. The innovation here is that the electron flux is implicitly forecast for many locations and energies simultaneously, though the forecast flux, or fluence, is not presented here.

The data-assimilative VERB code model utilized in this section applies the diffusion coefficients described in Subbotin and Shprits (2009), and the data-assimilation framework described in Shprits et al. (2013b) and Kellerman et al. (2014). The modeled PSD from the VERB code (Subbotin and Shprits, 2009) is combined with electron PSD derived from the Van Allen Probes Magnetic Electron Ion Spectrometer (MagEIS) (Blake et al., 2013) and Relativistic Electron Proton Telescope (REPT) (Baker et al., 2013a) instrument measurements, and Geostationary Operational Environmental Satellites (GOES) MAGnetospheric Electron Detector (MAGED) and Energetic Proton, Electron, and Alpha Detector (EPEAD) instrument measurements. The data assimilation is performed through the 1D split-operator Kalman filter (Shprits et al., 2013b; Kellerman et al., 2014). The intercalibration of the observations from different spacecraft was performed using a PSD matching algorithm, similar to that described in Kellerman et al. (2014).

The data assimilative model includes the LCDS as a function of Roederer L^* and K. The LCDS may change when the global field is reconfigured, which occurs when the driving parameters of a given field model change. In this application, the LCDS is determined as follows: (1) find the last magnetic field line at 12 MLT where only one B-field minimum is observed; (2) trace a drift shell conserving I and the mirror point B-field (Roederer, 1970). If an open field line is encountered, then take a step closer to the Earth and try again, alternatively; if a full drift shell is completed, then we are on the LCDS; (3) for each matching field line around the Earth, the flux in the polar cap is integrated to determine L^*. Because the last field line with two minimums may not actually be the LCDS, an adjustment factor of 0.5 is added to that number. It should be noted that this estimate of the LCDS is only a very approximate measure and is introduced here to include physics associated with magnetopause loss to the forecast model. The 0.5 L correction is based largely on trial and error over the past 2 years of the forecast model operation at UCLA. A more thorough analysis of LCDS determination, including code to deal with multiple B-field minimums for long-term datasets, is a topic of current study. For the purposes of the work presented here, this simple model will suffice.

An example of a radiation belt forecast is shown below, where data are assimilated for a 2-day period prior to 0 UT on March 16, 2013. After that time only the planetary-K (Kp) index and the computed LCDS are used to simulate the radiation belt response over a 4-day period. The Kp index is constructed from 13 geomagnetic observatories located at mid latitudes, and is a proxy for global geomagnetic activity over the previous 3 h. At the writing of this chapter, the current forecast model uses the NOAA Space Weather Prediction Center (SWPC) 3-day forecast Kp and the Tsyganenko (1989) model to determine the LCDS and to drive the VERB code. The use of Kp and the LCDS provides a similar set of input parameters to the real-time operational code.

The code outer boundary in L is defined as a Neumann boundary ($df/dL = 0$), the inner boundary is set to 0, the upper energy boundary PSD is set to 0, and the lower energy boundary PSD is defined by the steady-state solution to the radial diffusion equation, dependent on the current value of Kp, and utilizing a statistical spectrum (e.g., Shprits et al., 2009) at the lowest value in energy at the outer L boundary. The choice of this lower-boundary condition in energy reflects the need for a seed population of electrons, as described in Section 1.

In Fig. 1A, the electron PSD at $\mu = 200\,\text{MeV/G}$ and $K = 0.11\,\text{G}^{1/2}R_E$ is displayed in an L versus time format. The vertical-dashed line indicates the UT of the forecast. Although data are displayed after this time, they are not used in any way to guide the forecast, and are shown only for visual comparison. Fig. 1B illustrates the result after assimilation of the measured PSD and the VERB-modeled PSD (reanalysis) prior to 0 UT on the 16th, while the model-only forecast is displayed after this time, as indicated at the bottom of the panel. Fig. 1C and D is similar to the preceding, except for $\mu = 700\,\text{MeV/G}$. Fig. 1E displays the Kp during this period.

Visual inspection is an important first step in forecast verification. Often the human eye can detect many aspects of forecast performance that are difficult to quantify together numerically. Recall that the use of scalar aspects to characterize forecast performance leads to loss of information. The human eye can help to determine which information may be best to include or discard, and guide the quantitative investigation. However, visual inspection is highly skewed toward individual, subjective biases of interpretation, and thus it must be used with

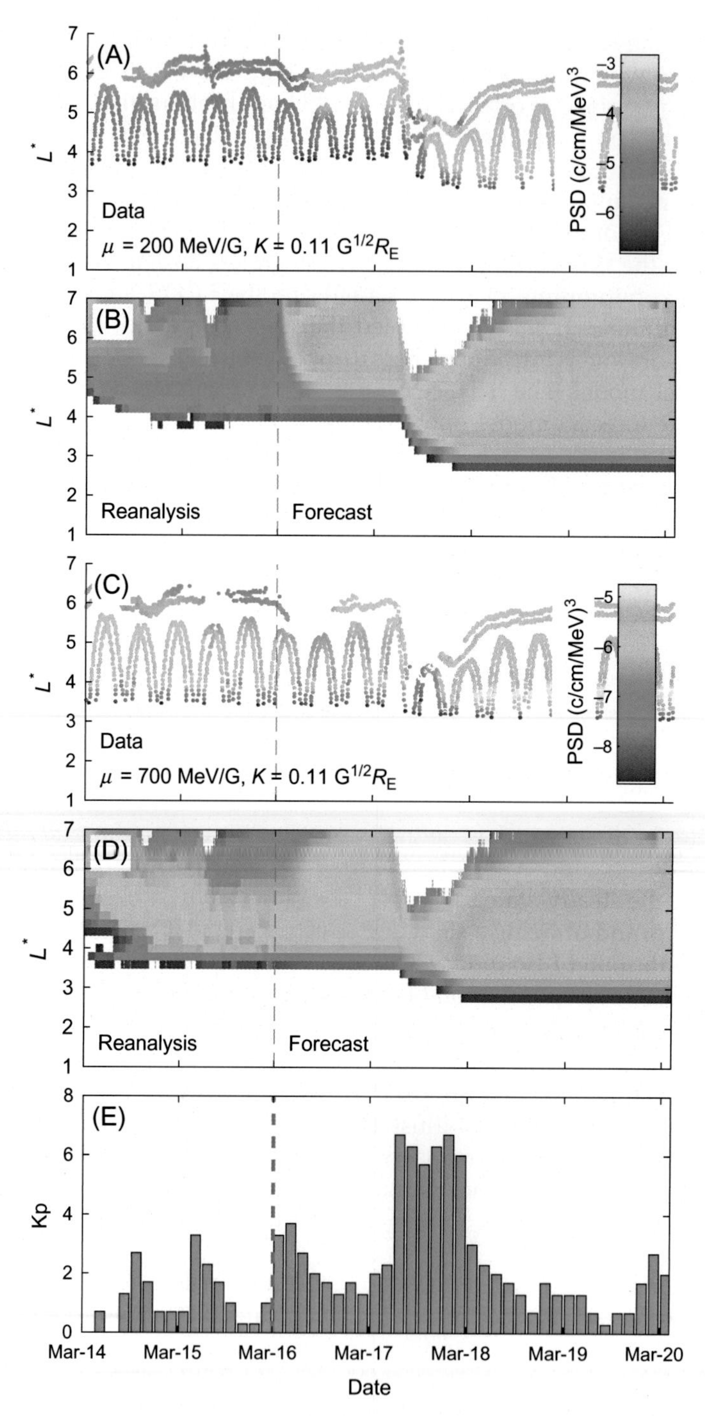

FIG. 1 Electron PSD L^* versus time profiles for fixed values of the first and second adiabatic invariants in the Tsyganenko and Sitnov (2005) model. (A) Observed PSD at $\mu = 200\,\text{MeV/G}$ and $K = 0.11\,G^{1/2}R_E$ from GOES and Van Allen Probe spacecraft; (B) reanalysis and forecast PSD at the same μ and K values; (C) Observations at $\mu = 700\,\text{MeV/G}$ and the same K; (D) reanalysis and forecast PSD at the same μ and K values; (E) Kp index. In each panel, the *vertical-dashed line* indicates the forecast UT, after which no observed data are used for the forecast.

caution as a formal verification procedure. It is best used as a "time-to-time" check to ensure that the data, model, and applied quantitative methods are appropriate and meaningful.

In Fig. 1, one may visually observe that the reanalysis provides a good measure of the *current* state of the system up to the time of the forecast, effectively removing any model bias and hysteresis effects leading up to the forecast time. The general dynamics appear to be well forecast—spatiotemporal occurrence of depletion and enhancement of PSD at both values of μ—though it is difficult to understand objectively how well the forecast model is performing from visual inspection alone.

As a first example in quantitative analysis of model performance, the model *accuracy* and *skill* are determined for this time period. The results of this analysis are displayed in Table 1. Computing the RMSE for $\mu = 200\,\text{MeV/G}$ for the reanalysis and forecast gives the two scalar values of 6.87×10^{-7} and 2.38×10^{-4}, respectively. Given that the order of the observations and model are $\sim 1 \times 10^{-3}$ during and after the storm, the former number indicates that the reanalysis agrees very well with observations, while the latter number indicates that the forecast accuracy, though not as accurate as the reanalysis, is high. In the case of $\mu = 700\,\text{MeV/G}$, the RMSE for the reanalysis and forecast are 4.59×10^{-7} and 1.93×10^{-6}, respectively. Given that the order of the observations and model are of order $\sim 1 \times 10^{-5}$ during and after the storm, once again, the former number indicates that the reanalysis agrees very well with observations, while the latter number indicates that the forecast is again accurate.

Electron PSD may vary over several orders of magnitude, both in L and time; however, the RMSE is sensitive to high values of electron PSD. In terms of forecasting for operations, sustained high levels of electron PSD and hence electron flux are of primary interest, and so this metric is appropriate for this particular application. Because the reanalysis was employed to remove any bias prior to the forecast time, a measure of model bias may be obtained by computing the ME of the model forecast. The ME of the forecast was in both cases positive, though only a small percentage of the observed enhancement. This indicates that the model is biased positive during this forecast period.

It should be noted that, since a dropout and enhancement occurred during this time, one should separately consider enhancement and dropout events in order to correctly characterize the model bias. Another important point is that the forecast is likely to diverge away from the observations with time, and so one or more fixed time intervals should be analyzed in order to reduce the effect of hysteresis on the forecast evaluation metric.

Only the model accuracy and bias have been considered thus far, and together they do not provide a quantitative measure of model performance as only one storm was considered. The

TABLE 1 Summary of Example Forecast Accuracy and Bias

μ MeV/G	**Period**	**Count**	**ME**	**RMSE**
200	Reanalysis	272	−1.78e−07	6.87e−07
200	Forecast	1021	1.80e−04	2.38e−04
700	Reanalysis	274	−1.13e−07	4.59e−07
700	Forecast	856	1.58e−06	1.93e−06

Note: ME and RMSE are in units of PSD $(c/cm/MeV)^3$.

electron PSD reanalysis agrees well with the observations, which is not surprising, as they are included in the data assimilation. Examples of testing removal of one or more data sources on the final reanalysis were explored previously, leading to a similar final result (Kellerman et al., 2014). The real purpose of applying a data assimilative approach is to provide an output dataset that is a more accurate measure of the actual electron PSD as compared with observations or model estimates alone. In the following, the reanalysis is used to quantify the model performance over many locations and energies, over a longer time period, and in a continuous manner that reduces any hysteresis effects.

The analysis was conducted as follows: The model time-step was fixed at 1 h. At 00 UT and 12 UT each day, a forecast was conducted, 1 day into the future. To set up the forecast, the reanalysis began 1.5 days prior to the current value of UT. On the first iteration, this required 36 data-assimilative model steps up to the forecast time, thereafter only 12 steps were required to reach the current value of UT. Those 12 steps covered any new data points available for assimilation. The forecast PSD for 1 day ahead (24 steps) provides the forecast variable in each case. The data assimilation is used as the truth, or observation variable at a later forecast step once available spacecraft data have been assimilated for the previously forecast time. The reference model is the persistence of electron PSD at fixed adiabatic invariants.

The use of persistence PSD helps to quantify the model performance in terms of its ability to predict nonadiabatic changes. The time step for the forecast was set to 1 h in order to balance computation time with model accuracy. The sensitivity of the model forecast to time-step and model resolution is left to future work, although this does constitute an important aspect of model forecast validation.

The PE, SS1, and SS2 were introduced earlier under the umbrella of forecast skill. These are computed for 1-day forecasts of electron PSD, with a changing forecast UT at 12-h intervals, and for various values of μ and L^*, while considering only a fixed value of $K = 0.1\,\mathrm{G}^{1/2}\mathrm{R}_E$. The purpose of this analysis is to identify model performance conditional on given fixed (binned) values of the three adiabatic invariants. If only one value was considered, it would discard too much information, whereas by separating the analysis by adiabatic invariant values, it is possible to simultaneously analyze performance over a range of locations and energies. In forecast verification, this type of analysis is known as *reliability*. The mean value of PSD is statistically higher, by several orders of magnitude, for lower μ at the same fixed K, and typically higher at large radial distances. Hence, an additional normalization factor is introduced here to allow comparison across different L^* and μ values. Normalized mean error (NME) and normalized root-mean-squared error (NRMSE) are defined as follows

$$\begin{aligned} \mathrm{NME} &= \frac{1}{n}\sum_{k=1}^{n}(x_k - y_k)/\bar{y} \\ &= (\bar{x} - \bar{y})/\bar{x} \end{aligned} \tag{10}$$

and,

$$\mathrm{NRMSE} = \left(\frac{1}{n}\sum_{k=1}^{n}(x_k - y_k).^2\right)^{1/2}\bar{y}^{-1} \tag{11}$$

where x, y, k, and n are as described earlier.

With this formulation, a measure of exact accuracy has been replaced with a measure of relative accuracy, which allows one to quantify areas where the model is performing better or worse. Fig. 2 illustrates this quantitative analysis. Each panel displays one aspect of the analysis in an L^* versus μ format. The number of points contributing to the diagram is shown

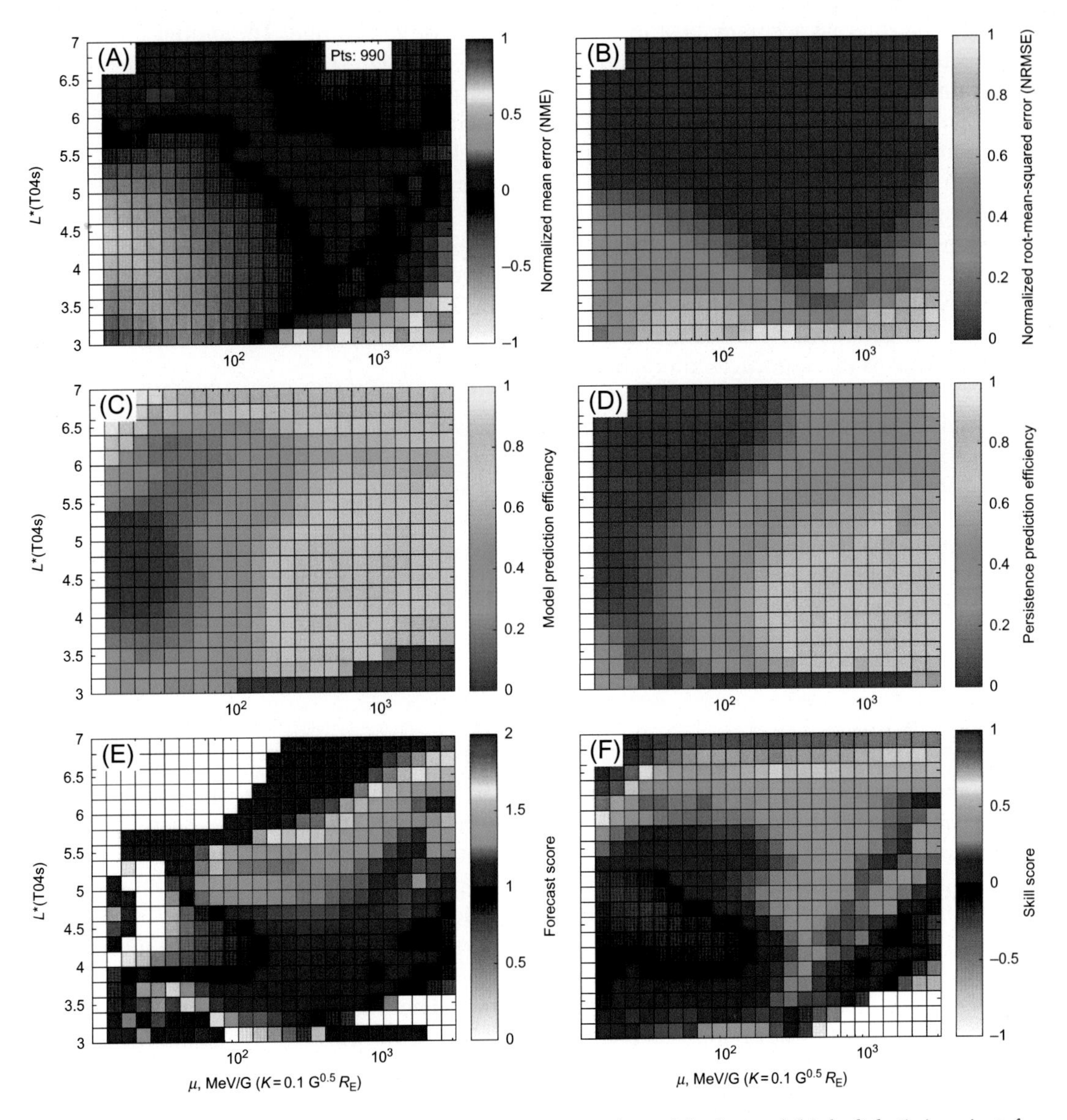

FIG. 2 Quantitative analysis of 1-day forecast PSD at various values of the first and third adiabatic invariant, for fixed $K = 0.1\ G^{1/2}\ R_E$, for all Kp, and at hourly resolution. (A) Normalized (by the mean PSD) mean error of the forecast; (B) normalized (by mean PSD) root-mean-squared error; (C) prediction efficiency of the model; (D) prediction efficiency of persistence (PSD during a 1-h interval today will be the same tomorrow during the same 1-h interval); (E) model skill score 1; and (F) model skill score 2.

in the top right of Fig. 2A. Fig. 2A–F illustrates NME, NRMSE, PE_{model}, $PE_{persistence}$, SS1, and SS2, respectively. Fig. 2A shows that the forecast at $\mu < 100$ MeV/G is positively biased at high L and negatively biased at low L. The opposite is true for ~100–1000 MeV/G, while at very low L there is large positive bias as μ increases.

In Fig. 2B, the NRMSE indicates that most of the significant error is concentrated at lower L, with the forecast accuracy best in the ~200–1000 MeV/G range. The model PE (Fig. 2C) is much higher than the electron PSD persistence PE (Fig. 2D) at lower μ and higher L, indicating that persistence plays a larger role for electron PSD at low L, and for $\mu \geq 100$ MeV/G, increasing with increasing μ. This agrees well with a recent report that ultra-relativistic electrons generally behave quite differently than relativistic electrons (Shprits et al., 2013a), and are largely unaffected by hiss waves in the plasmasphere, which may result in the stable trapping of that population for a long period of time (Baker et al., 2013b; Shprits et al., 2013a). The forecast skill is demonstrated by the SS1 and SS2 in Fig. 2E and F, respectively. Numbers greater than 1 are desirable for the SS1, while numbers greater than 0 are desired for the SS2. The model is most skillful in forecasting electron PSD at higher L and at μ values in the 100s of MeV/G range. These electrons are most susceptible to magnetopause loss, and nonadiabatic changes due to wave-particle interactions.

In the preceding example, the model was not skillful at very low and very high values of μ, especially at lower L. The poor model performance was likely due to insufficient characterization of the source population, and insufficient ultra-relativistic electron physics, respectively. This quantitative analysis technique, although only utilizing a few scalar aspects, has provided information that can directly aid in improving the model. The inclusion of an improved source population near the lower boundary in energy, and introduction of ultra-relativistic electron physics, such as electromagnetic ion cyclotron (EMIC) waves (e.g., Usanova et al., 2014; Shprits et al., 2016), may improve the forecast quality.

4 SUMMARY

In this chapter, a brief introduction to relativistic electron forecasting has been presented. The problem was briefly introduced in Section 1, numerical techniques for forecasting and data assimilation were covered in Section 2, and forecast verification was discussed in Section 3. Although several forecast models have been developed, and some have been operational for nearly three decades, there exists no community-wide framework for radiation-belt electron forecast verification, especially with respect to more sophisticated physics-based models. The first real-time operational data-assimilative radiation belt framework was utilized here to develop an initial procedure for forecast verification. The example forecast analysis demonstrated the power of applying scalar aspects to aid in forecast verification, guiding future research directions, and direct model development.

Implementing a set of standards for radiation-belt forecasting will aid in tracking model quality over time in a logical and meaningful way. Several areas for improvement were identified in the analysis presented here. The two main areas in need of improvement are better characterization of the electron source population, and the inclusion of physics associated with the ultra-relativistic electron population. Other forecast variables that may help to improve the model in the short term include more accurate indices, used for parameterizing current and new magnetic field models, improved diffusion coefficients for the VERB code, improved 1–3 day forecast boundary conditions for electron PSD, and a robust estimation of the LCDS for radiation belt particles as a function of L^* and K.

Although the presented analysis indicates how forecasting and quantitative analysis can help to inform model development, and basic research direction, true forecast verification and analysis should involve the real-time forecast model, as in that situation, one has no knowledge of the future state of the system. Analysis of the performance of the real-time forecast model is currently under investigation.

References

Albert, J.M., Meredith, N.P., Horne, R.B., 2009. Three-dimensional diffusion simulation of outer radiation belt electrons during the 9 October 1990 magnetic storm. J. Geophys. Res. 114, A09214. https://doi.org/10.1029/2009JA014336.

Baker, D.N., McPherron, R.L., 1990. Extreme energetic particle decreases near geostationary orbit: a manifestation of current diversion within the inner plasma sheet. J. Geophys. Res. 95, 6591–6599.

Baker, D., Belian, R., Higbie, P., Klebasadel, R., Blake, J., 1987. Deep dielectric charging effects due to high-energy electrons in Earth's outer magnetosphere. J. Electrostat. 20 (1), 3–19. https://doi.org/10.1016/0304-3886(87)90082 - 9.

Baker, D.N., Allen, J.H., Kanekal, S.G., Reeves, G.D., 1998. Disturbed space environment may have been related to pager satellite failure. EOS Trans. 79, 477. https://doi.org/10.1029/98EO00359.

Baker, D.N., Kanekal, S.G., Hoxie, V.C., Batiste, S., Bolton, M., Li, X., Elkington, S.R., Monk, S., Reukauf, R., Steg, S., Westfall, J., Belting, C., Bolton, B., Braun, D., Cervelli, B., Hubbell, K., Kien, M., Knappmiller, S., Wade, S., Lamprecht, B., Stevens, K., Wallace, J., Yehle, A., Spence, H.E., Friedel, R., 2013a. The relativistic electron-proton telescope (REPT) instrument on board the radiation belt storm probes (RBSP) spacecraft: characterization of Earth's radiation belt high-energy particle populations. Space Sci. Rev. 179. https://doi.org/10.1007/s11214-012-9950-9.

Baker, D.N., Kanekal, S.G., Hoxie, V.C., Henderson, M.G., Li, X., Spence, H.E., Elkington, S.R., Friedel, R.H.W., Goldstein, J., Hudson, M.K., Reeves, G.D., Thorne, R.M., Kletzing, C.A., Claudepierre, S.G., 2013b. A long-lived relativistic electron storage ring embedded in Earth's outer Van Allen belt. Science 340, 186–190. https://doi.org/10.1126/science.1233518.

Balikhin, M.A., Boynton, R.J., Walker, S.N., Borovsky, J.E., Billings, S.A., Wei, H.L., 2011. Using the NARMAX approach to model the evolution of energetic electrons fluxes at geostationary orbit. Geophys. Res. Lett. 38, L18105. https://doi.org/10.1029/2011GL048980.

Blake, J.B., Carranza, P.A., Claudepierre, S.G., Clemmons, J.H., Crain, W.R., Dotan, Y., Fennell, J.F., Fuentes, F.H., Galvan, R.M., George, J.S., Henderson, M.G., Lalic, M., Lin, A.Y., Looper, M.D., Mabry, D.J., Mazur, J.E., McCarthy, B., Nguyen, C.Q., O'Brien, T.P., Perez, M.A., Redding, M.T., Roeder, J.L., Salvaggio, D.J., Sorensen, G.A., Spence, H.E., Yi, S., Zakrzewski, M.P., 2013. The magnetic electron ion spectrometer (MagEIS) instruments aboard the radiation belt storm probes (RBSP) spacecraft. Space Sci. Rev. 179. https://doi.org/10.1007/s11214-013-9991-8.

Bourdarie, S.A., Maget, V.F., 2012. Electron radiation belt data assimilation with an ensemble Kalman filter relying on the Salammbô code. Ann. Geophys. 30, 929–943.

Daae, M., Shprits, Y.Y., Ni, B., Koller, J., Kondrashov, D., Chen, Y., 2011. Reanalysis of radiation belt electron phase space density using various boundary conditions and loss models. Adv. Space Res. 48. https://doi.org/10.1016/j.asr.2011.07.001.

Evensen, G., 2003. The ensemble Kalman filter: theoretical formulation and practical implementation. Ocean Dyn. 53. https://doi.org/10.1007/s10236-003-0036-9.

Fälthammar, C.G., 1965. Effects of time-dependent electric fields on geomagnetically trapped radiation. J. Geophys. Res. 70, 2503–2516.

Fukata, M., Taguchi, S., Okuzawa, T., Obara, T., 2002. Neural network prediction of relativistic electrons at geosynchronous orbit during the storm recovery phase: effects of recurring substorms. Ann. Geophys. 20, 947–951. https://doi.org/10.5194/angeo-20-947-2002.

Godinez, H.C., Koller, J., 2012. Localized adaptive inflation in ensemble data assimilation for a radiation belt model. Space Weather 10, S08001. https://doi.org/10.1029/2012SW000767.

Horne, R.B., Thorne, R.M., 1998. Potential waves for relativistic electron scattering and stochastic acceleration during magnetic storms. Geophys. Res. Lett. 25, 3011–3014.

Horne, R.B., Glauert, S.A., Meredith, N.P., Boscher, D., Maget, V., Heynderickx, D., Pitchford, D., 2013. Space weather impacts on satellites and forecasting the Earth's electron radiation belts with SPACECAST. Space Weather 11, 169–186. https://doi.org/10.1002/swe.20023.

Hwang, J., Min, K.W., Lee, E., Lee, C., Lee, D.Y., 2004. A case study to determine the relationship of relativistic electron events to substorm injections and ULF power. Geophys. Res. Lett. 31 (23). ISSN 1944-8007. https://doi.org/10.1029/2004GL021544.

Kalman, R.E., 1960. A new approach to linear filtering and prediction problems. Trans. ASME J. Basic Eng. 82 (Series D), 35–45.

Kalnay, E., 2002. Atmospheric Modeling, Data Assimilation and Predictability. Cambridge University Press, Cambridge, UK, 364 pp. ISBN ISBN 0521791790.

Kellerman, A.C., Shprits, Y.Y., 2012. On the influence of solar wind conditions on the outer-electron radiation belt. J. Geophys. Res. 117, A05217. https://doi.org/10.1029/2011JA017253.

Kellerman, A.C., Shprits, Y.Y., Turner, D.L., 2013. A geosynchronous radiation-belt electron empirical prediction (GREEP) model. Space Weather 11 (8), 463–475. https://doi.org/10.1002/swe.20074.

Kellerman, A.C., Shprits, Y.Y., Kondrashov, D., Subbotin, D., Makarevich, R. A. Donovan, E., Nagai, T., 2014. Three-dimensional data assimilation and reanalysis of radiation belt electrons: observations of a four-zone structure using five spacecraft and the VERB code. J. Geophys. Res. https://doi.org/10.1002/2014JA020171.

Kellogg, P.J., 1959. Van Allen radiation of solar origin. Nature 183, 1295–1297.

Kim, K.C., Shprits, Y., Subbotin, D., Ni, B., 2012. Relativistic radiation belt electron responses to GEM magnetic storms: comparison of CRRES observations with 3-D VERB simulations. J. Geophys. Res. 117, A08221. https://doi.org/10.1029/2011JA017460.

Koller, J., Chen, Y., Reeves, G.D., Friedel, R.H.W., Cayton, T.E., Vrugt, J.A., 2007. Identifying the radiation belt source region by data assimilation. J. Geophys. Res. 112, A06244. https://doi.org/10.1029/2006JA012196.

Kondrashov, D., Shprits, Y., Ghil, M., Thorne, R., 2007. A Kalman filter technique to estimate relativistic electron lifetimes in the outer radiation belt. J. Geophys. Res. 112, A10227. https://doi.org/10.1029/2007JA012583.

Koons, H.C., Gorney, D.J., 1991. A neural network model of the relativistic electron flux at geosynchronous orbit. J. Geophys. Res. 96, 5549–5556. https://doi.org/10.1029/90JA02380.

Ling, A.G., Ginet, G.P., Hilmer, R.V., Perry, K.L., 2010. A neural network-based geosynchronous relativistic electron flux forecasting model. Space Weather 8, S09003. https://doi.org/10.1029/2010SW000576.

Lyatsky, W., Khazanov, G., 2008a. A predictive model for relativistic electrons at geostationary orbit. Geophys. Res. Lett. 35, L15108. https://doi.org/10.1029/2008GL034688.

Lyatsky, W., Khazanov, G.V., 2008b. Effect of geomagnetic disturbances and solar wind density on relativistic electrons at geostationary orbit. J. Geophys. Res. 113, A08224. https://doi.org/10.1029/2008JA013048.

Meredith, N.P., Horne, R.B., Anderson, R.R., 2001. Substorm dependence of chorus amplitudes: implications for the acceleration of electrons to relativistic energies. J. Geophys. Res. 106, 13165–13178.

Meredith, N.P., Horne, R.B., Thorne, R.M., Anderson, R.R., 2003. Favored regions for chorus-driven electron acceleration to relativistic energies in the Earth's outer radiation belt. Geophys. Res. Lett. 30 (16), 1871. ISSN 1944-8007. https://doi.org/10.1029/2003GL017698.

Meulenberg, Jr., A., 1976. Evidence for a new discharge mechanism for dielectrics in a plasma. In: Studies in Condensed Matter Physics, pp. 237–246, http://adsabs.harvard.edu/abs/1976scmp.symp.237M.

Murphy, A.H., Winkler, R.L., 1987. A general framework for forecast verification. Mon. Weather Rev. 115 (7), 1330–1338. https://doi.org/10.1175/1520-0493(1987)115<1330:AGFFFV>2.0.CO;2.

Nagai, T., 1988. "Space Weather Forecast"—prediction of relativistic electron intensity at synchronous orbit. Geophys. Res. Lett. 15, 425–428.

Ni, B., Shprits, Y., Nagai, T., Thorne, R., Chen, Y., Kondrashov, D., Kim, H.J., 2009. Reanalyses of the radiation belt electron phase space density using nearly equatorial CRRES and polar-orbiting Akebono satellite observations. J. Geophys. Res. 114, A05208. https://doi.org/10.1029/2008JA013933.

O'Brien, T.P., McPherron, R.L., 2003. An empirical dynamic equation for energetic electrons at geosynchronous orbit. J. Geophys. Res. Space Phys. 108 (A3), 2156–2202. https://doi.org/10.1029/2002JA009324.

Omura, Y., Furuya, N., Summers, D., 2007. Relativistic turning acceleration of resonant electrons by coherent whistler mode waves in a dipole magnetic field. J. Geophys. Res. 112, A06236. https://doi.org/10.1029/2006JA012243.

Paulikas, G.A., Blake, J.B., 1979. Effects of the solar wind on magnetospheric dynamics: energetic electrons at the synchronous orbit. In: Quantitative Modeling of Magnetospheric Processes, vol. 21. American Geophysical Union, Washington, D.C., pp. 180–202.

Reeves, G.D., Chen, Y., Cunningham, G.S., Friedel, R.W.H., Henderson, M.G., Jordanova, V.K., Koller, J., Morley, S.K., Thomsen, M.F., Zaharia, S., 2012. Dynamic radiation environment assimilation model: DREAM. Space Weather 10, 03006. https://doi.org/10.1029/2011SW000729.

Roederer, J.G., 1970. Dynamics of Geomagnetically Trapped Radiation. Springer Verlag, New York, NY.

Schiller, Q., Li, X., Koller, J., Godinez, H., Turner, D.L., 2012. A parametric study of the source rate for outer radiation belt electrons using a Kalman filter. J. Geophys. Res. 117, A09211. https://doi.org/10.1029/2012JA017779.

Schulz, M., Lanzerotti, L.J., 1974. Particle Diffusion in the Radiation Belts. Springer, Berlin.

Shprits, Y.Y., Thorne, R.M., Friedel, R., Reeves, G.D., Fennell, J., Baker, D.N., Kanekal, S.G., 2006. Outward radial diffusion driven by losses at magnetopause. J. Geophys. Res. 111, A11214. https://doi.org/10.1029/2006JA011657.

Shprits, Y., Kondrashov, D., Chen, Y., Thorne, R., Ghil, M., Friedel, R., Reeves, G., 2007. Reanalysis of relativistic radiation belt electron fluxes using CRRES satellite data, a radial diffusion model, and a Kalman filter. J. Geophys. Res. 112, A12216. https://doi.org/10.1029/2007JA012579.

Shprits, Y.Y., Subbotin, D., Ni, B., 2009. Evolution of electron fluxes in the outer radiation belt computed with the VERB code. J. Geophys. Res. 114, A11209. https://doi.org/10.1029/2008JA013784.

Shprits, Y., Daae, M., Ni, B., 2012. Statistical analysis of phase space density buildups and dropouts. J. Geophys. Res. 117, A01219. https://doi.org/10.1029/2011JA016939.

Shprits, Y.Y., Subbotin, D., Drozdov, A., Usanova, M.E., Kellerman, A., Orlova, K., Baker, D.N., Turner, D.L., Kim, K.C., 2013a. Unusual stable trapping of the ultrarelativistic electrons in the Van Allen radiation belts. Nat. Phys. 9, 699–703. https://doi.org/10.1038/nphys2760.

Shprits, Y.Y., Kellerman, A., Kondrashov, D., Subbotin, D., 2013b. Application of a new data operator-splitting data assimilation technique to the 3-D VERB diffusion code and CRRES measurements. Geophys. Res. Lett. 40, 4998–5002. https://doi.org/10.1002/grl.50969.

Shprits, Y.Y., Drozdov, A.Y., Spasojevic, M., Kellerman, A.C., Usanova, M.E., Engebretson, M.J., Agapitov, O.V., Zhelavskaya, I.S., Raita, T.J., Spence, H.E., Baker, D.N., Zhu, H., Aseev, N.A., 2016. Wave-induced loss of ultra-relativistic electrons in the Van Allen radiation belts. Nat. Commun. 7, 12883. https://doi.org/10.1038/ncomms12883.

Simms, L.E., Pilipenko, V., Engebretson, M.J., Reeves, G.D., Smith, A.J., Clilverd, M., 2014. Prediction of relativistic electron flux at geostationary orbit following storms: multiple regression analysis. J. Geophys. Res. Space Phys. 119. https://doi.org/10.1002/2014JA019955.

Su, Z., Xiao, F., Zheng, H., Wang, S., 2011. CRRES observation and STEERB simulation of the 9 October 1990 electron radiation belt dropout event. Geophys. Res. Lett. 38, L06106. https://doi.org/10.1029/2011GL046873.

Subbotin, D.A., Shprits, Y.Y., 2009. Three-dimensional modeling of the radiation belts using the Versatile Electron Radiation Belt (VERB) code. Space Weather 7, S10001. https://doi.org/10.1029/2008SW000452.

Subbotin, D.A., Shprits, Y.Y., Gkioulidou, M., Lyons, L.R., Ni, B., Merkin, V.G., Toffoletto, F.R., Thorne, R.M., Horne, R.B., Hudson, M.K., 2011. Simulation of the acceleration of relativistic electrons in the inner magnetosphere using RCM-VERB coupled codes. J. Geophys. Res. 116, A08211. https://doi.org/10.1029/2010JA016350.

Summers, D., Thorne, R.M., Xiao, F., 1998. Relativistic theory of wave-particle resonant diffusion with application to electron acceleration in the magnetosphere. J. Geophys. Res. 103, 20487–20500.

Takahashi, K., Ukhorskiy, A.Y., 2008. Timing analysis of the relationship between solar wind parameters and geosynchronous Pc5 amplitude. J. Geophys. Res. 113, A12204. https://doi.org/10.1029/2008JA013327.

Tan, L.C., Shao, X., Sharma, A.S., Fung, S.F., 2010. Relativistic electron acceleration by compressional-mode ULF waves: evidence from correlated Cluster, Los Alamos National Laboratory spacecraft, and ground-based magnetometer measurements. J. Geophys. Res. 116, A07226. https://doi.org/10.1029/2010JA016226.

Tsyganenko, N.A., 1989. A magnetospheric magnetic field model with a warped tail current sheet. Planet. Space Sci. 37, 5–20.

Tsyganenko, N.A., Sitnov, M.I., 2005. Modeling the dynamics of the inner magnetosphere during strong geomagnetic storms. J. Geophys. Res. 110, A03208. https://doi.org/10.1029/2004JA010798.

Turner, D.L., Shprits, Y., Hartinger, M., Angelopoulos, V., 2012. Explaining sudden losses of outer radiation belt electrons during geomagnetic storms. Nat. Phys. 8, 208–212. https://doi.org/10.1038/nphys2185.

Usanova, M.E., Drozdov, A., Orlova, K., Mann, I.R., Shprits, Y., Robertson, M.T., Turner, D.L., Milling, D.K., Kale, A., Baker, D.N., Thaller, S.A., Reeves, G.D., Spence, H.E., Kletzing, C., Wygant, J., 2014. Effect of EMIC waves on relativistic and ultrarelativistic electron populations: ground-based and Van Allen Probes observations. Geophys. Res. Lett. 41, 1375–1381. https://doi.org/10.1002/2013GL059024.

Vampola, A.L., 1987. Thick dielectric charging on high-altitude spacecraft. J. Electrostat. 120, 21–30.

Vassiliadis, D., Klimas, A.J., Kanekal, S.G., Baker, D.N., Weigel, R.S., 2002. Long-term-average, solar cycle, and seasonal response of magnetospheric energetic electrons to the solar wind speed. J. Geophys. Res. 107, 1383. https://doi.org/10.1029/2001JA000506.

Walt, M., 2005. Introduction to Geomagnetically Trapped Radiation. Cambridge University Press, Cambridge.

Wilks, D., 2011. Statistical Methods in the Atmospheric Sciences. Elsevier, New York, NY.

CHAPTER

11

Artificial Neural Networks for Determining Magnetospheric Conditions

Jacob Bortnik[*], *Xiangning Chu*[*], *Qianli Ma*[*,†], *Wen Li*[†], *Xiaojia Zhang*[*], *Richard M. Thorne*[*], *Vassilis Angelopoulos*[‡], *Richard E. Denton*[§], *Craig A. Kletzing*[¶], *George B. Hospodarsky*[¶], *Harlan E. Spence*[|], *Geoffrey D. Reeves*[**], *Shrikanth G. Kanekal*[††], *Daniel N. Baker*[‡‡]

[*]University of California, Los Angeles, Los Angeles, CA, United States [†]Boston University, Boston, MA, United States [‡]Institute of Geophysics and Planetary Physics/Earth, Los Angeles, CA, United States [§]Dartmouth College, Hanover, NH, United States [¶]University of Iowa, Iowa City, IA, United States [|]University of New Hampshire, Durham, NH, United States [**]Space Science and Applications Group, Los Alamos, NM, United States [††]NASA Goddard Space Flight Center, Greenbelt, MD, United States [‡‡]University of Colorado Boulder, Boulder, CO, United States

CHAPTER OUTLINE

Machine Learning Techniques for Space Weather
https://doi.org/10.1016/B978-0-12-811788-0.00011-1

1 INTRODUCTION

The ability to specify and predict the state of the Earth's inner magnetospheric environment has been a long-standing goal of the space sciences, as reflected in numerous community "roadmap" documents, such as the recent National Research Council's decadal survey (National Research Council, 2013) or the National Space Weather Action Plan (National Science and Technology Council, 2015). There are a variety of components that comprise the magnetospheric environment, such as the ionosphere, large-scale electric and magnetic fields, plasma waves, electrons, and different species of ions ranging from cool (<1 eV) to ultrarelativistic (>5 MeV) energies. The behavior of each of these components is difficult to specify in general, because they can be driven by both external factors, such as the solar wind, and internal factors, including various instabilities, and furthermore, can couple to each other in complex ways, producing behavior that is often unexpected.

The traditional method employed by the scientific community for the purpose of specification and prediction is to delve deeply into the underlying physics of the system, understand the physical processes involved, and (hopefully) produce a set of equations that might permit modeling the spatiotemporal evolution of the system. There are indeed various researchers within the space physics community who are pursuing this goal and are developing first-principles physics-based models to simulate aspects of the inner magnetospheric environment (e.g., Wolf et al., 1991; Toffoletto et al., 2003, 2004; Jordanova et al., 2008, 2010; Wang et al., 2004; Wiltberger et al., 2004; Goodrich et al., 2004; Pembroke et al., 2014; Glauert et al., 2014; Ma et al., 2015; Shprits et al., 2015; Su et al., 2010; Tu et al., 2013). The major advantage of physics-based models is that they employ fundamental physical principles and thus should (hopefully) be useful in extrapolating to situations that have not been previously observed. The disadvantage is that these models are typically complex, computationally intensive, and inevitably cannot include all the physical processes in the simulation (either because the model becomes too computationally expensive, or because—as often happens—the physical processes that control a particular system have not been fully understood or identified yet), which can produce erroneous results.

Taking the complete opposite approach to the problem of specification and prediction, the state of the inner magnetosphere can be specified simply by plotting a particular quantity (typically collected by a satellite or other observing platform over several years) in a plane or a volume, often broadly parameterized by some controlling parameter. For example, Meredith et al. (2012) plots the intensity of whistler-mode chorus waves in the inner magnetosphere, parameterized by three ranges of the geomagnetic auroral electrojet (AE) index representing quiet, moderate, and active conditions, based on data collected by the Dynamics Explorer (DE1) satellite, Combined Release and Radiation Effects Satellite (CRRES), Cluster 1, Double Star TC1, and Time History of Events and Macroscale Interactions during Substorms (THEMIS) satellites. This approach gives a realistic, albeit static, and statistically averaged product. It is helpful in providing a general sense of the spatial distribution and intensity ranges of the quantity being plotted, but cannot capture the variation of the quantity on a case-by-case basis, nor can it capture the spatiotemporally changing conditions in the magnetosphere. To continue with the preceding example, the chorus wave distributions produced by Meredith et al. (2012) can be used to compute diffusion coefficients that describe the energetic electron scattering and acceleration, and these coefficients can then be further used to model the evolution of radiation belt electrons for a given storm (e.g., Horne et al., 2003). However, because the chorus wave distributions are averaged, the resulting radiation belt dynamics will also just produce the average storm response.

A few specialized data-based models have recently been developed by performing function-fitting to the data under various geomagnetic activity levels. For example, Orlova et al. (2014) derived a model of plasmaspheric hiss intensity in two local time sectors, and calculated weighting coefficients using linear stepwise regression for a quadratic polynomial with variables in *L*-shell, latitude, and Kp. An extension to this approach was employed by Kim et al. (2015), who fit up to seventh-order polynomials at every 2-h magnetic local time (MLT) interval to chorus wave intensity, in two latitude ranges, and driven by either solar wind conditions or geomagnetic indices. Although these functional fits make the models more flexible and dynamic, the functional forms used for fitting are only applicable to the parameter at hand, and a new approach will inevitably need to be taken if a different parameter is to be predicted.

Clearly, both physics-based models and averaged statistical distributions have their advantages and limitations in specifying and predicting the inner magnetospheric environment. An ideal model might consist of an optimal combination of the two approaches, so as to use the simplicity and realism of the data-based statistical distribution maps, together with the ability to vary in both time and space in response to geomagnetic driving conditions, similar to physics-based models (but ideally without the heavy computational cost of these models). This, essentially, is the main topic of this chapter.

The modeling approach described in this chapter relies on the use of the artificial neural network (ANN), and in particular, on "deep" neural networks (i.e., ANNs having more than one hidden layer). This approach is designed to circumvent the difficulty of having to design a new functional fitting form for every new quantity being modeled, and instead come up with a universal modeling technique that can specify any given quantity in the inner magnetosphere (based on satellite data) without necessarily knowing the functional form beforehand. The model will typically be driven by a time-history of one or more geomagnetic indices (and/or solar wind parameters), with the idea that the ANN will itself pick out the most important elements of the input data, and recombine the input in the first layer into a new feature set that is more optimal for describing the dynamics in the model. In Section 2 we provide a brief review of ANNs. We describe our methodology using the example of electron number density in Section 3, and show more advanced applications in Section 4. We conclude with a summary and discussion of how such modeling might be employed usefully in space weather prediction in Section 5.

2 A BRIEF REVIEW OF ANNS

The ANN traces its roots back to the analytical model developed by McCulloch and Pitts (1943), intended to mathematically simulate the firing of a biological neuron when a certain chemical threshold had been exceeded in the cell membrane. The McCullogh and Pitts (MP) neuron is a very simple model: given an input vector of length N, $\mathbf{x} = [x_1, x_2, \ldots, x_N]$, and a similar-length weighting vector $\mathbf{W} = [w_1, w_2, \ldots, w_N]$, the dot product $h = \mathbf{W} \cdot \boldsymbol{x}$ is taken,

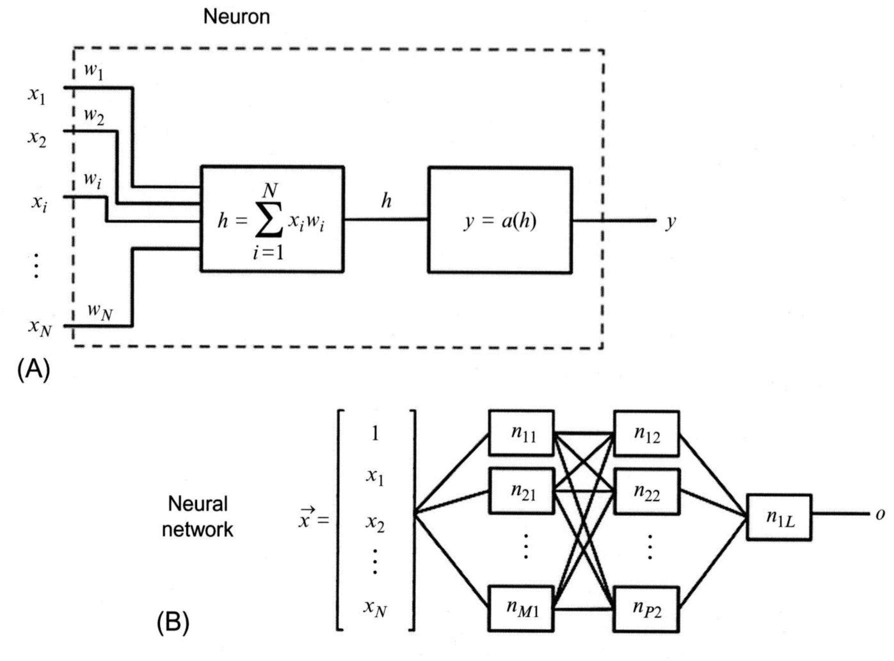

FIG. 1 (A) Schematic illustration of a single neuron with input vector x, weighting vector w, having an activation function $a(h)$ and output y. (B) A simple feedforward neural network consisting of the input layer x, two hidden layers with M and P neurons in each layer, respectively, and one neuron in the output layer, giving the output y.

and if the output exceeds a certain threshold T, the MP neuron "fires", that is, gives an output value of 1, and otherwise would output the value of 0 (e.g., Marsland, 2009, p. 14). The MP neuron is illustrated graphically in Fig. 1A, where the function $a(h)$ is known as the "activation function" and is chosen to be the Heaviside step function, that is,

$$y = a(h) = \begin{cases} 1 & \text{if} \quad h \geq T \\ 0 & \text{if} \quad h < T \end{cases} \tag{1}$$

The input vector $\boldsymbol{x}$ may be readily fed into any number of individual neurons that each produce their own individual outputs based on their specific chosen weights, such that instead of a single output value y, a vector of outputs results, $\boldsymbol{y}^{(1)} = [y_{11}, y_{21}, \ldots, y_{M1}]$, assuming there are M individual neurons. This vector of outputs $\boldsymbol{y}^{(1)}$ can then be treated as a new input vector, and itself be used as an input into a new set (or so-called "layer") of neurons, producing the next output vector $\boldsymbol{y}^{(2)} = [y_{12}, y_{22}, \ldots, y_{P2}]$, assuming there are P individual neurons in the second layer. An arbitrary number of layers may be added, each successively treating the output of the previous layer as its input, and any number of individual neurons may be placed in each layer. The set of connected neurons, including the layers and number of neurons in each individual layer (as well as the selected activation function and connectivity of the neurons to each other, which we shall not deal with here) is collectively known as the "neural network" and is illustrated graphically in Fig. 1B, where each of the boxes corresponds to a single neuron as shown in Fig. 1A. In Fig. 1B, the input vector x is often referred to as the "input layer," the middle layers 1 and 2 (consisting of M and P neurons, respectively) are known as the "hidden layers," and the final layer L is known as the "output" layer. Here, the output of every neuron becomes the input of every neuron in the next layer, and information flows from left to right without any loops or feedback, so this architecture is known as a fully connected feedforward neural network. The more layers and neurons per layer are added, the more complexity and capability is introduced to the neural network, such that it is able to model increasingly complex functions. However, the total number of free parameters (i.e., the weights of all the neurons) also grows significantly with each added layer, so the total number of neurons should be chosen judiciously.

The process of selecting the proper values for all the weights in the neural network is called "training" and typically involves an iterative numerical process. A large number of examples with known inputs and outputs are selected (called the "training set"), and for each input example, the neural net produces an output y using some arbitrarily chosen set of weights, which is then compared against a known output known as the "target" t, the error between y and t is calculated and used to adjust the weights of the neural network in the next training iteration. This process is repeated over the whole training set for many iterations until the neural network produces the "correct" output given a known input, that is, the output is within some acceptable error range of the target. As a simple example, consider a neural network designed to recognize digits in a 16×16-pixel black and white image. The input vector x then consists of the image pixels all arranged as a single column vector of length $N = 256$, and the output y gives a positive "flag" value (e.g., $y = 1$) when the appropriate digit is recognized, say when the image corresponds to the digit "9," or a negative "flag" value (e.g., $y = 0$) otherwise. Because x has length $N = 256$, each weighting vector w in layer 1 will have a similar length, and if there are $M = 10$ neurons in layer 1, the total number of free parameters will be $256 \times 10 = 2560$. If another layer is introduced, there will be 10 weights per neuron in layer 2, so that if layer 2 has $P = 20$ neurons, there will be another $\sim 10 \times 20 = 200$ free parameters in the neural network. In this way, the total number of free parameters in the neural network can quickly rise into the 100s to 1000s (sometimes millions, if larger images are considered), so that when the neural network is trained, the number of examples in the training set must be much larger than the number of free parameters, by at least a factor of 10 and preferably much more, that is, there should be >30,000 images presented to the neural network, which are each labeled "0" or "1" corresponding to whether that image has the digit "9" displayed in it. The weights of the neural net are then iteratively adjusted until the training set images are correctly classified, and then, in principle, the network should be able to classify out-of-sample data (i.e., images it has not seen before), and determine whether they display the digit "9" or not.

The difficulty arises in "training" the model, that is, algorithmically determining the optimal set of weighting coefficients **W** that would produce the "right" output. The first neural network to perform an algorithmic training consisted of a set of interconnected MP neurons in what was known as the "Mark 1 Perceptron," a custom-built machine designed to perform image recognition on an image composed of an array of 400 photocells (similar to the preceding example) (Rosenblatt, 1957). Although the perceptron algorithm was the first of its kind, and produced accurate results most of the time, it nevertheless suffered from convergence and instability problems for certain types of classification problems, primarily having to do with the sharp discontinuity inherent in the Heaviside step function.

It took almost three more decades until a general algorithm was developed by Rumelhart et al. (1986) to train neural networks. This was known as the "backpropagation algorithm," because of its unique ability to propagate back the errors between the neural network predicted values, and target values, and use that information to adjust the internal weights. The network used the logistic function $f(y) = \left(1 + e^{-y}\right)^{-1}$ as its activation function, which had the same behavior as the Heaviside step function at large negative and positive values of y, that is, tending to 0 and 1, respectively, but having the advantage of being smooth and differentiable near $y \sim 0$. The backpropagation algorithm is generally credited for the explosive growth of neural networks, a trend that continues to this day (barring a slow period in the early 1990s when computational speed and memory were insufficient for the problems of the day, but has more than caught up in the subsequent two decades).

In closing this brief review, we note that the modern-day ANN can be thought of as a general approach to modeling complex, nonlinear functions of high-dimensional vectors, and in fact has been shown to be a universal approximator to any smooth function, given that a sufficient number of neurons is included (Hornik et al., 1989; Cybenko, 1989). ANNs with two or more hidden layers are nowadays known as "deep" neural nets, and have been shown to be more efficient than "shallow" nets (Bengio, 2009), and (more importantly) that deep nets constitute better predictors because they are able to reorganize the set of input parameters into a better, or more efficient feature set than the inputs that are initially presented to the ANN (Hinton et al., 2006). The input vector presented to the ANN could take on a variety of forms, for example, (i) pixels in a static image, as in our preceding simple image classification example; (ii) a time-series of stock-market daily closing prices, where the objective of the ANN is to predict the next day's closing price; (iii) spatial values taken over some region in space where the objective of the ANN is to predict the value of a function at a different region in space; (iv) parameters describing a person, where the objective of the ANN is to make decisions about health status, insurability, and credit-worthiness; or (v) any combination of these. In this chapter, our aim is to predict a spatiotemporal distribution of values, and the problem outline, together with methodology, is described next.

3 METHODOLOGY AND APPLICATION

The general problem we aim to address can be stated as follows: given a sequence of measurements of some quantity Q_i, at time t_i and location r_i, where $i = 1, \ldots, N$ and N is a "large" number, specify the distribution of Q at every point $\mathbf{r}$ in the model domain at time t, and let that spatial distribution evolve as a function of time. In order to illustrate our technique, we present the application where the quantity Q_i is chosen to be the plasma density N_e, inferred from the spacecraft potential of three THEMIS probes (Angelopoulos, 2008), A, D, and E, using the method described in Li et al. (2010). That data set extends from June 1, 2008 to October 31, 2014 and is reduced to 5-min averages, which results in $\sim 10^6$ data points after some minimal data cleaning has been performed, that is, removal of bad data points, for example, when the probes are in the Earth's shadow.

In order to organize the data into a dynamic model, we must choose one or more time-series of input parameters to regress the data against, that will act as a suitable predictor and contain enough information embedded within it so as to explain the various behaviors of the data. As an example, we may use the preceding 5 h of the SYM-H index (also at a 5-min cadence, giving 60 points) as the only predictor, because this index is simple, readily available through the OMNI database (ftp://spdf.gsfc.nasa.gov/pub/data/omni/high_res_omni/), and has been used by previous researchers in driving more complex plasmasphere-ionosphere models (e.g., Huba and Sazykin, 2014), that is, it should contain sufficient information to act as a predictor for the cold plasma density. Our selection of SYM-H is not unique, and we could have just as easily used other geomagnetic indices such as AE or Kp (suitably subsampled to a 5-min cadence), or solar wind parameters. However, because the different geomagnetic indices are strongly correlated to each other (Borovsky, 2014), and already reflect the driving conditions in the solar wind (e.g., see the literature on coupling functions, summarized in Newell et al. (2007) and McPherron et al. (2015)), the time-series of the SYM-H index contains the necessary information as a predictor to include the storm effects on the plasmasphere. In addition, solar wind data tends to contain many data-gaps that would result in losing many of our data points, whereas SYM-H is available for long periods of time with no data gaps.

The resulting design matrix X, and target matrix T can be written as:

$$\mathbf{X} = \begin{bmatrix} \text{SYM-H}_{i-60} & \cdots & \text{SYM-H}_i & L_i & \cos\phi_i & \sin\phi_i \\ & & \vdots & & & \\ \text{SYM-H}_{N-60} & \cdots & \text{SYM-H}_N & L_N & \cos\phi_N & \sin\phi_N \end{bmatrix}, \quad T = \begin{bmatrix} Q_i \\ \vdots \\ Q_N \end{bmatrix} \tag{2}$$

Where the angle ϕ_i represents the MLT angle (MLT/24×2π), and L_i the L-shell of the satellite at time t_i, together with the corresponding electron density Q_i observed by the spacecraft at that time, and $i = 1 \ldots N$ such that X and T both have ~10^6 rows, and 63 columns. We use sin and cos functions to ensure continuity across the midnight boundary. For the purposes of our illustration, we only consider the electron density in the magnetic equatorial plane, but the model could be readily extended to 3D (as shown in the following section) by including in the design matrix X a column that contains the latitude λ_i of the satellite at time t_i.

The problem now becomes how to create a model that predicts T given the design matrix X. This problem could be solved with a variety of techniques, and essentially all machine-learning techniques that deal with regression are designed to solve a problem of this type (e.g., Marsland, 2009). The simplest approach is to create a linear model by creating a set of weights $\mathbf{A} = [a_1 \ldots a_m]^T$ such that $\mathbf{XA} = \mathbf{T}$, and obtain the ordinary least squares solution, giving: $\mathbf{A} = (\mathbf{X}^T\mathbf{X})^{-1}\mathbf{X}^T\mathbf{T}$, but there is no guarantee that the response of our target quantity (electron density) is necessarily linear, and in fact, will most likely be a nonlinear function of the inputs.

The technique we present in this chapter is based on the ANN as described in the previous section, and is a very powerful and general approach to modeling complex, nonlinear functions. The architecture of the ANN used by Bortnik et al. (2016), for example, contains an input layer with 63 features (as shown in the previous section) two hidden layers with 10 and 20 neurons in the first and second layers, respectively, each having an identical sigmoid activation function, and an output layer. We use a two-hidden layer ANN because this is the shallowest "deep" network, that is, deep in the sense that it has more than one hidden layer. The number of neurons in each hidden layer was chosen somewhat arbitrarily and no attempt has been made at either optimizing performance or producing the most parsimonious network possible.

In general, the first layer reconstructs the input parameters into its own optimal set of features, and by choosing 10 neurons in this layer we force a data compression, that is, a reduction in the dimensionality of the system from 63 to 10. This is intuitively plausible because the 63 input parameters are not all independent, and in particular, the 60-point, 5-h time history of SYM-H can be reduced significantly by extracting only the relevant information; for example, one might look for features such as the extent and intensity of dips in SYM-H, their duration, and so on. For our example of density prediction, one might guess at the most important features and present those directly to the ANN as inputs, but in the general case, we would not necessarily know which aspects of the input time-series contain the most relevant information, and it is best to let the ANN construct its own optimal set of features in the first few layers. The second layer of our ANN contains 20 neurons, and in this layer, the optimal features are recombined to capture the physical behavior of our system. In principle, the "deep" ANN should not normally need more than two hidden layers, because this architecture should be able to approximate any smooth functional mapping (see illustration in Marsland, 2009, Section 3.3.3).

In order to train the neural network, the data is divided into three groups: a training set containing 70% of the samples, a validation set containing 15% of the samples, and a test set containing the remaining 15% of the samples. The training set of points is used to train the ANN together with the time series of SYM-H for each data point (Bortnik et al., 2016), where the scaled conjugate algorithm was used to perform the training (Marsland, 2009). The training of the ANN continues and the RMS error between predicted and observed values is calculated at each time step for all three sets of points. There are a variety of ways to judge neural net "goodness," but the approach we take here is to focus on the generalizability of the network, so we continue training until the RMS error of the validation set of points stops improving for several consecutive time steps (here we arbitrarily choose six times steps). Note that the validation set of points is not involved in the training per se, and is thus independent (except that it is involved in stopping of the training step). The test set of points is truly an independent set of samples, and acts as a check on the generalizability of the ANN and its performance. In general, it is important to note how the data is divided among the training, validation, and test sets because this could affect the overall performance of the model. The data could be randomly assigned to one of the three sets on a sample-by-sample basis, it could be divided into larger chunks (e.g., days or weeks) that are then randomly divided among the sets, or the entire dataset could be divided into three nonoverlapping, temporally contiguous chunks to avoid any sort of cross-correlation among data samples. In the work of Bortnik et al. (2016), the data was divided into contiguous chunks, and in Chu et al. (2017a,b) it was randomly assigned. In practice, when training the ANN models, we typically use all three data division methods, and various hybrids of them to train several models and thus get a more accurate assessment of the true performance of our models. To date, we found that the model performance is similar, regardless of the data division method, and so the results shown in this chapter use random-sample division for simplicity and clarity, but in general we caution practitioners to try a variety of data division methods when training their ANN models to get a more accurate estimate of performance.

3.1 The DEN2D Model

The basic ANN model described in the previous section, and presented by Bortnik et al. (2016), considered the plasmaspheric dynamics that were predominantly related to storm effects, as measured by the SYM-H index. The general range of plasmaspheric dynamics is more complex than expected; however, because it also interacts with both the ionosphere and the magnetosphere. The bottomside of the plasmasphere is usually in diffusive equilibrium with the upper ionosphere; while the topside of the plasmasphere, or the plasmapause, is controlled by the strength and variability of magnetospheric convection. To study the contributions arising from various processes, the ANN model of the 2D plasma density on the equatorial plane (DEN2D) was significantly extended by including a larger number of geomagnetic indices representing those processes (Chu et al., 2017b). Therefore, the DEN2D model includes not only the storm effects as measured by the SYM-H index, but also contributions from the ionospheric effects as measured by the F10.7 (solar EUV) index and the magnetospheric effects as measured by the AL index (i.e., the lower envelope of the AE magnetic field perturbations). See Chu et al. (2017b) for detailed description of the DEN2D model. We note that here we are trying to perform extreme optimization of our general model, by selecting appropriate inputs specific to plasma density, in order to capture the last few percent of variance, not captured by the simple model using only SYM-H as input.

To illustrate its range of capabilities, the DEN2D model was applied to a moderate storm event that occurred on February 4, 2011 as shown in Fig. 2. A sudden increase in the solar wind dynamic pressure indicates the arrival of an interplanetary shock around 03 UT. The storm then starts around 17 UT as indicated by the SYM-H index, which reached a minimum of −66 nT and then slowly recovered over the next few days. A comparison between the observed (black) and modeled (blue) electron density profiles for the three THEMIS probes A, D, and E are shown in the bottom three panels. Overall, good agreement is found across the entire time interval, demonstrating that the DEN2D model was able to capture the essential behavior of the plasmaspheric spatiotemporal dynamics. If these plasmaspheric dynamics are indeed correctly captured and reproduced by the internal weights of the ANN, then the plasmaspheric electron number density should behave correctly across the entire spatial domain, and evolve realistically as a function of time, which is indeed the case. To demonstrate this capability, we have reconstructed the equatorial distribution of the electron density during the same storm event, where a few snapshots of the global evolution are shown in Fig. 2g–l. The plasmasphere is initially large and extended during the quiet period before the storm starts. As geomagnetic activity intensifies in the storm main phase, the plasmasphere gets eroded particularly in the dawn and nightside sectors, causing the plasmapause to move closer to the Earth. The plasmasphere becomes extended in the afternoon region instead, and a plasmaspheric plume is seen to develop around the minimum of the SYM-H index. Later during the recovery phase, the plasmaspheric plume disappears. The plasmasphere starts to recover due to the ionospheric outflow as the plasmapause gradually moves to higher *L*-shells. In summary, the DEN2D model successfully captures the dynamic evolution of the plasmasphere during different phases of the storm (at least, to the level of our current understanding); namely, the quiet time plasmasphere, the plasmaspheric erosion and refilling, as well as the formation of a plume.

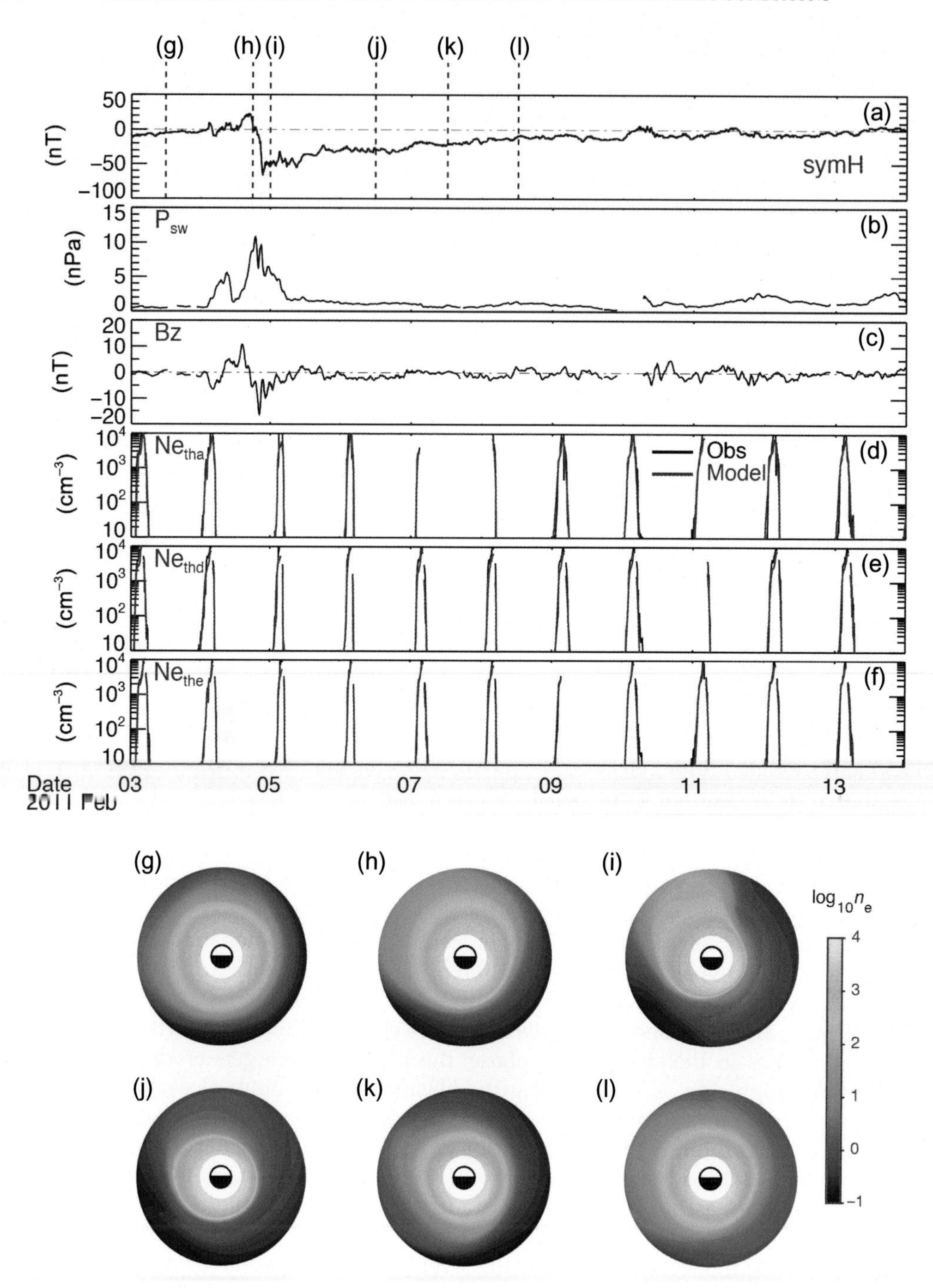

FIG. 2 A reconstruction of the equatorial electron number density produced by the DEN2D model, during the course of the February 4, 2011 geomagnetic storm. Panels (a–c) SYM-H index, solar wind dynamic pressure, and B_Z; (d–f) electron density observed (*black line*) and modeled (*blue line*) by THEMIS probes A, D, and E whose apogee is at dawn during this period. Panels (g–l): 2D dynamic equatorial density maps produced by DEN2D at the six snapshot times indicated above panel (a).

The performance of the DEN2D model can be represented by a number of correlation coefficients (r) and root of mean squared errors (RMSE). Fig. 3 shows the cross-correlation coefficients between the modeled (x-axis) and observed (y-axis) densities for the whole, training, validation, and test datasets, and color denotes the number of observation-model pairs in each bin. Most of the observation-model pairs are centered around the diagonal line ($y = x$), indicating that most observations can be accurately modeled. The Pearson correlation coefficients of all datasets are greater and close to 0.95, which means $r^2 = \sim 90.8\%$ of the observed variability is captured. Because the test dataset is not used in the training process, it can be used to evaluate the predictive capability of the DEN2D model on out-of-sample observations. The RMSE on the test dataset of the electron density ($\log_{10} n_e$) is 0.388, which can be translated to a factor of $10^{0.388} = 2.44$. This suggests that the DEN2D model can predict out-of-sample observations with an error around a factor of 2, which is very close to the factor of ~2 error inherent in the method of obtaining the electron density from the spacecraft potential (Li et al., 2010). This error is much smaller than the typical density variation during a

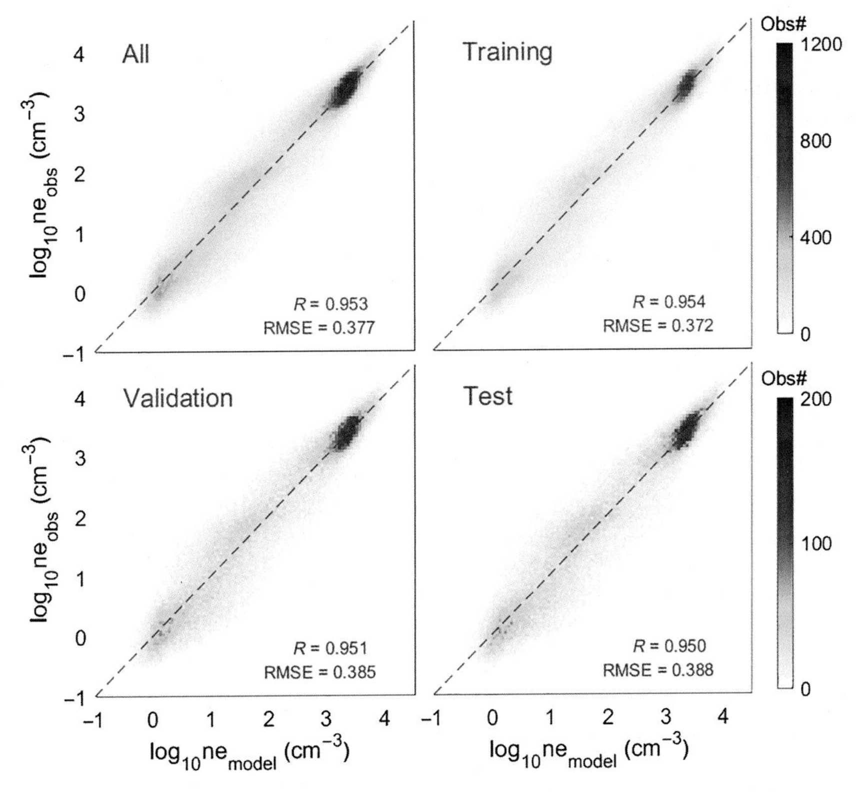

FIG. 3 The distribution of errors (in $\log_{10}$ space) for the DEN2D model along the satellite track is shown, categorized by the various classes of data (training, validation, and test samples). The DEN2D predicted densities are plotted on the x-axis and observed values on the y-axis. Zero error, that is, "perfect" fit is represented by the *orange diagonal line.*

geomagnetic storm (e.g., Fig. 2g–l) by a few orders of magnitude. As such, any further increases in the prediction efficiency of the ANN become less significant, because the model is constrained by the error in observation.

4 ADVANCED APPLICATIONS

We began our discussion in Section 3 by stating the general problem of reconstructing a sparsely measured quantity Q over an extended time (and space), into a snapshot of the quantity over an extended space, at only a single snapshot in time, and illustrated the use of the ANN technique with a 2D electron number density. We now return to the general problem of reproducing any quantity with the same technique and briefly touch on a range of applications. First, we show the simple extension of the plasmaspheric density DEN2D model into 3D, and then cover chorus and hiss waves, before delving into the important problem of radiation belt flux prediction.

4.1 The DEN3D Model

The DEN2D model can be readily extended to study the plasmaspheric dynamics in three dimensions (Chu et al., 2017a) by including the magnetic latitude of the satellite as one of the input arguments in the design matrix X, and extending the satellite dataset so that it contains a sufficiently large number of off-equatorial samples. The 3D dynamic electron density (DEN3D) model uses electron density identified using the upper hybrid resonance frequency from the wave instruments onboard four satellites, including the equatorial satellites ISEE (Gurnett et al., 1978) and CRRES (Anderson et al., 1992) and polar-orbiting satellites POLAR (Gurnett et al., 1995) and IMAGE (Reinisch et al., 2000). The architecture and training processes of the DEN3D model are described in detail in Chu et al. (2017a), and use 5-min averages (similar to the DEN2D model) that result in 217,500 data points that collectively cover all L-shells up to $L \sim 11$, MLAT between -50 and 50 degrees and all MLTs. The DEN3D model can predict the variations of the 3D plasma density dataset with a correlation coefficient of 0.954 on the test dataset and predict out-of-sample observations with errors around a factor of 2.

The DEN3D model has been applied to the same storm event that occurred on February 4, 2011 to study the variations of the electron density along magnetic field lines. We note that the data set spans 1977 to 2005, so the 2011 event is completely out of sample. As an example, Fig. 4 shows the electron density profiles in the noon-midnight meridional plane at different times, indicated by the vertical dashed lines. During the quiet times (panel a), the plasmasphere was large and extended. At the start of the main phase (panel b), the plasma density on the nightside becomes depleted, and this depletion progresses as the plasmapause contracts toward the Earth (panel c). On the other hand, the plasma density increases on the dayside and the plasmapause moves to higher L shells due to the formation of the plasmaspheric plume (panel c). The plume then disappears during the recovery phase and the electron density starts to increase due to the ionospheric outflow, causing the plasmapause to slowly move outward (panels d–f). In summary, the DEN3D model successfully reproduced various well-known dynamic features in three dimensions, such as plasmaspheric erosion and recovery, as well as the plume formation during the same storm event.

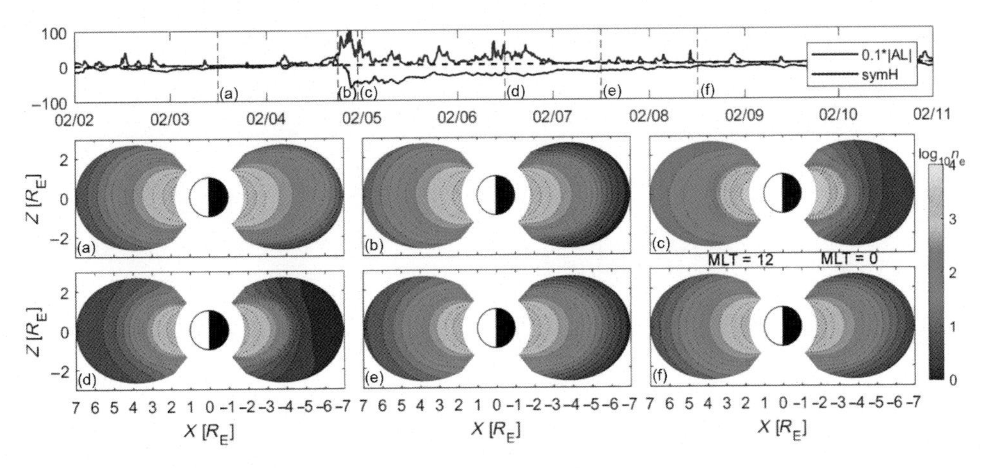

FIG. 4 Illustration of the electron number density in the noon-midnight meridional plane produced by the DEN3D model, at six instances during the course of the February 4, 2011 geomagnetic storm. (a) Quiet time, (b) initial phase, (c) main phase, (d)–(f) early to late recovery. *Vertical-dashed lines* in the top panel correspond to the snapshot times of the lower panels, and the common *colorbar* on the right shows the $\log_{10}(n_e)$.

4.2 The Chorus and Hiss Wave Models

We now demonstrate the application of the ANN technique to the inner magnetospheric wave environment, and in particular, to the reconstruction of two wave types that are known to play key roles in driving the dynamics of radiation belt particles, namely whistler-mode chorus waves and plasmaspheric hiss (Horne et al., 2003; Thorne, 2010, and references therein).

The design and target matrices are constructed in the same way shown in Section 3, but retain the latitudinal location of the satellite so as to produce a 3D model. In the case of chorus, the lower band wave magnetic field intensity was integrated between 0.1–0.5 f_{ce} (where f_{ce} is the equatorially mapped electron gyrofrequency), and reduced to 5-min average intensity values. Data was used from two missions: (1) the EMFISIS instrument (Kletzing et al., 2013) on board both Van Allen Probe-A and Probe-B (Mauk et al., 2013) in the period October 1, 2012 to July 1, 2014, and (2) the Search-Coil-Magnetometer instrument (Le Contel et al., 2008; Roux et al., 2008) in the FFF mode aboard the THEMIS mission collected in the period May 1, 2010 to July 1, 2014, from probes A, D, and E. The resulting dataset had ~372,000 values that sampled the region from ~600 km altitude to $L = 10$ (we artificially removed data samples at $L > 10$ because our focus was on the inner magnetosphere at $L < 10$, but the vast majority of THEMIS samples were taken at or near apogee, $L > 10$). Because chorus tends to be associated more closely with substorms than with storms, we chose to use the AE index as our predictor (instead of the usual SYM-H), taking the previous 5-h interval sampled at a 5-min cadence.

The results of our trained ANN chorus model are shown in Fig. 5, for the period of March 10, 2012, which was a geomagnetically active period. Fig. 5A shows a sequence of substorms occurring around our snapshot time of 02:15, and panels B–D show the distribution of chorus waves as a function of L and MLT at three values of magnetic latitude, $\lambda = 0$ degree (equatorial), $\lambda = 15$ degrees (mid-latitude), and $\lambda = 30$ degrees (high latitude). Consistent with previous studies, the equatorial distribution of chorus waves tends to remain in the

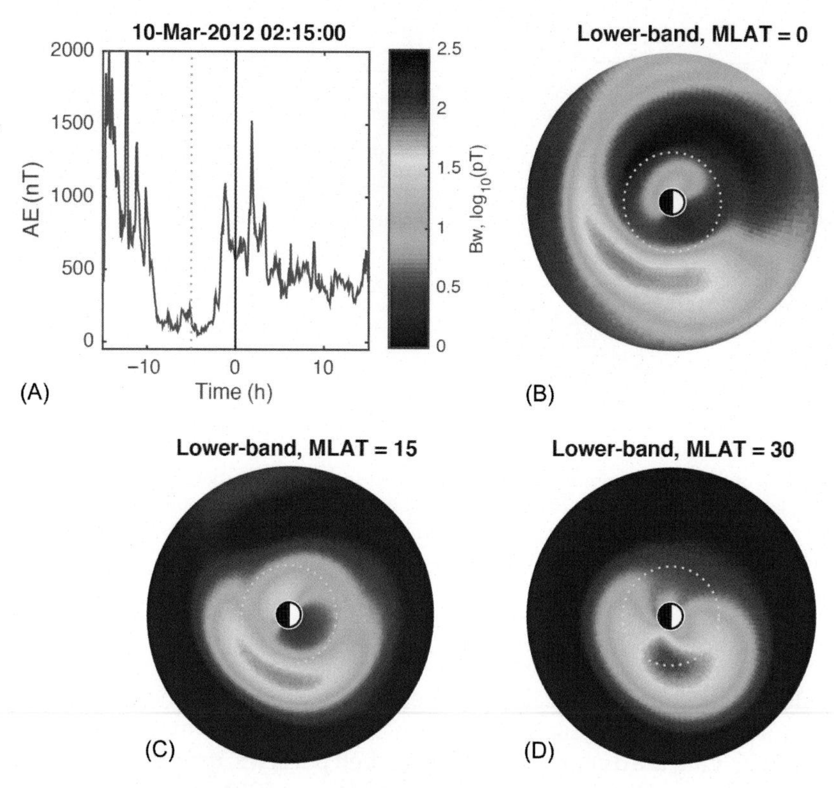

FIG. 5 The reconstructed distribution of lower-band whistler-mode chorus waves during a series of substorms that occurred in March 2012 as indicated by the AE index in panel (A). Here the reconstruction is shown as a function of L-shell and MLT for three different magnetic latitudes: (B) equatorial MLAT = 0 degree, (C) MLAT = 15 degrees, (D) MLAT = 30 degrees. The common *colorbar* shows wave intensity in $\log_{10}$ (Bw [pT]), and the *white dotted circle* acts as a reference point at $L = 4$, as a typical plasmapause location.

low-density region outside of the plasmasphere, and has a maximum power at the dawn sector, with the power maximum increasing as a function of MLT with increasing latitude and moving to lower L-shells (e.g., Bortnik et al., 2007; Li et al., 2009; Meredith et al., 2003, 2013). The dynamic simulation of this event (not shown) shows that the wave power tends to increase in intensity shortly after an increase in the AE index, and move into lower L-shells, consistent with the geomagnetic control of chorus. We have constructed a similar chorus model for upper-band chorus waves, which tends to follow the main equatorial power of the lower-band chorus waves, but does not propagate very high in latitude, remaining confined to $\lambda \sim 10$ to 15 degrees (not shown). We also note that the wave power at low L-shells in panel B on the duskside of the Earth is most likely VLF transmitter power that has leaked into the chorus frequency band, but we chose to retain it in our model to illustrate the ability of the ANN model to capture wave power from a variety of sources.

Using a very similar approach, but data taken from only the Van Allen Probes A and B satellites over the period October 1, 2012 to October 1, 2014, we construct a model of the plasmaspheric hiss intensity in the inner magnetosphere, using the trailing 10-h values of the SYM-H index (which is the appropriate timescale for plamasheet electrons to drift from the nightside to the dayside and excite/amplify hiss), sampled at a 5-min cadence. The resulting dataset consists of ~290,000 samples, covering the region from ~600 km altitude, to $L \sim 6$ at all MLT. Fig. 6 shows the results of our ANN model, with plasmaspheric hiss reconstructed over the same storm period as Fig. 5 (March 10, 2012), in the recovery phase of an intense geomagnetic storm (minimum Dst ~ -150 nT). Consistent with previous studies, hiss is seen to be confined to the plasmasphere and maximize its intensity in the dayside, with higher intensities encountered at higher latitudes (e.g., Li et al., 2015; Meredith et al., 2013). Even though no optimization has been performed on these initial chorus and hiss models, and the number of data samples is relatively low, the cross correlation coefficients are nevertheless reasonable, ~0.73 and ~0.62 for chorus and hiss, respectively, and the error is a factor of ~2 (RMSE ~ 0.3 in both cases).

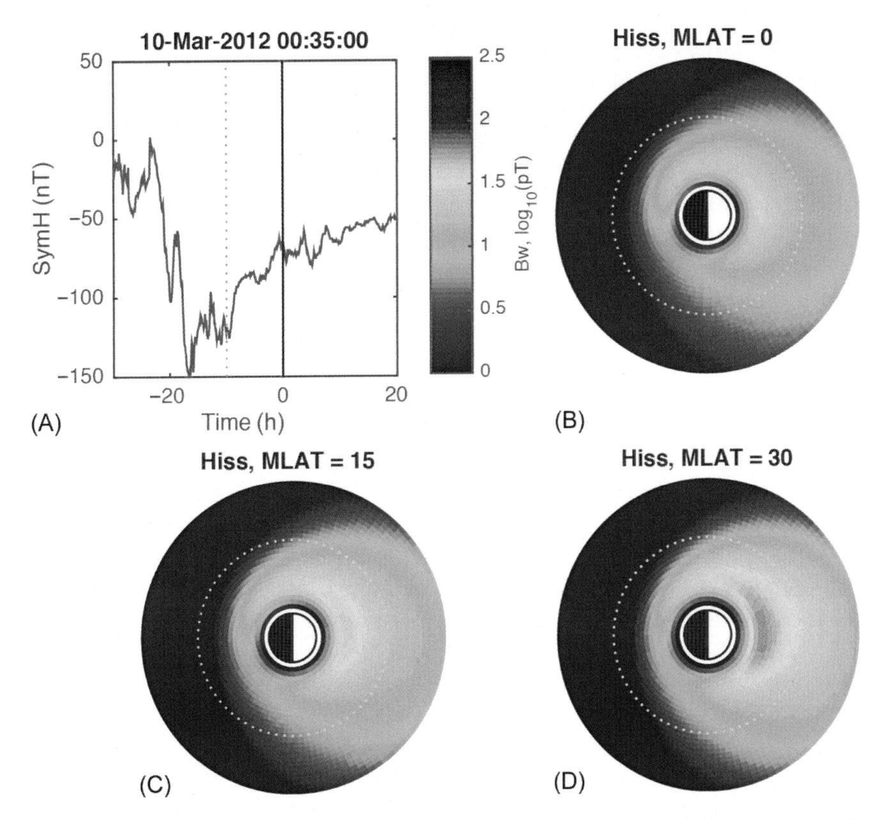

FIG. 6 The reconstructed distribution of plasmaspheric hiss waves during the same storm period as Fig. 5, but here indicated by the SYM-H index in panel (A). The reconstruction is shown as a function of *L*-shell and MLT for three different magnetic latitudes: (B) equatorial MLAT = 0 degree, (C) MLAT = 15 degrees, (D) MLAT = 30 degrees. The common *colorbar* shows wave intensity in $\log_{10}$(Bw [pT]) and the *white dotted circle* acts as a reference point at $L = 4$, as a typical plasmapause location.

It should be noted, of course, that both Figs. 5 and 6 are single snapshots of a sequence of snapshots that create a movie at a 5-min cadence of the dynamic spatiotemporal evolution of chorus and hiss waves. These types of movies contain a great deal of information and can be used as tools to study the dynamics of the chorus/hiss waves themselves, or to use the intensities of the waves for other applications, such as the prediction of radiation belt fluxes, which we discuss next.

4.3 Radiation Belt Flux Modeling

The Earth's electron radiation belts consist of highly energetic electrons (~MeV) that are trapped in the Earth's geomagnetic field, and consist of two zones: an inner zone ($L < 2.5$) and an outer zone ($L > 3$), separated by a slot region (Lyons and Thorne, 1973). The outer zone, in particular, tends to be highly variable (Thorne, 2010) and presents a space weather hazard to various spacecraft systems, due to internal charging, single event upsets, and total radiation dose (e.g., National Research Council, 2013; Cho et al., 2015; Baker and Lanzerotti, 2016 and the many references therein). Prediction and specification of radiation belt flux intensities has been a long-standing goal of space weather researchers, and here we discuss some potential strategies to achieve this goal.

We begin by applying the ANN technique directly to the prediction of high-energy fluxes measured aboard the Van Allen Probes A and B satellites, using the Relativistic Electron Proton Telescope (REPT) (Baker et al., 2012). Here, data is used in the interval October 1, 2012 to November 1, 2014, that covers eight energy channels between $E = 1.8$ MeV and $E = 7.7$ MeV, at $L < 6$. As a regressor, we use the SYM-H index sampled at a 5-min cadence over the preceding 10-h interval, which results in a total of ~188,000 individual data samples. We note that 10-h is our estimate of the right timescale for the ANN model based on the drift period of plasmasheet electrons from the nightside to the dayside, but it may also be that ultra-relativistic electrons require a much longer time period for acceleration. The typical cross-correlation coefficients between model and data fall in the range ~0.73–0.84, generally becoming progressively lower with increasing energy.

An example of our ANN model is shown in Fig. 7, where the ultra-relativistic radiation belt fluxes were reconstructed in the recovery phase of the same geomagnetic storm interval (March 10, 2012) studied in Figs. 5 and 6. There are a few points to note: while the 1.8 MeV fluxes show somewhat realistic inner and outer zones, and vary in intensity according to the SYM-H index, the behavior becomes progressively less realistic as the electron energy is increased. At E ~5–6 MeV (panels G–H), the ANN model highlights individual satellite tracks where high-intensity fluxes were encountered, which is the "correct" behavior for the ANN, given that it has no knowledge of basic physics controlling energetic electron drifts. This behavior can be further understood by examining, for example, Fig. 1 of Baker et al. (2014), which shows that in the entire interval September 1, 2012 to May 1, 2014, there was essentially only 1 intensification event in the 8.8 MeV channel, and just a small number in the 5–6 MeV range. In essence, the ultra-relativistic electron fluxes present a "data starved" environment, which implies that technique such as our ANN model should not be used directly.

Instead, we suggest an approach that employs the strengths of the ANN model in a "data-rich" environment, and projects that information with the aid of a physics-based model into

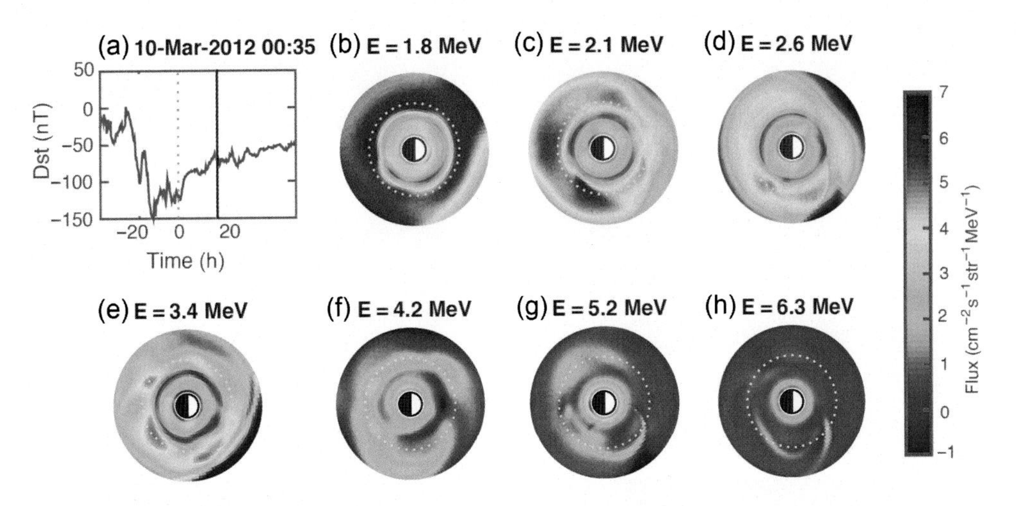

FIG. 7 A reconstruction of equatorial ultrarelativistic fluxes, based on REPT observations taken by the Van Allen Probes over the period October 1, 2012 to January 1, 2015. The same storm period in March 2012 is used, as in Figs. 5 and 6. Here the regression is based on 10-trailing hours of the SYM-H index (a), and includes energies in the ranges 1.8–6.3 MeV, as shown in panels (b–h). The common *colorbar* in $\log_{10}$(flux) is given on the right.

the data starved environment. An example of this approach is illustrated in Fig. 8, where we used a Fokker-Planck diffusion model (Ma et al., 2015), driven by our ANN-based chorus and hiss models (shown in Figs. 5 and 6) and total electron density model, which were used to calculate time-dependent, bounce- and drift-averaged diffusion coefficients. The effects of radial diffusion were incorporated using an empirical fitting of D_{LL} to Kp and L-shell by Ozeke et al. (2014).

We apply this technique to a modest geomagnetic storm (minimum Dst ~ -55 nT) during March 1–4, 2013 in Fig. 8. Reeves et al. (2016) reported the energy-dependent electron decay and the resultant radiation belt structure during the same period in detail. The energetic electron acceleration and decay during this event were later simulated by Ma et al. (2016) using a physics-based model with inputs from the available measurements. Fig. 8 shows that even though only a few physical processes were included into our model in a fairly crude way, the results are quite realistic, giving approximately the correct timescale for energization of the energetic electrons at all energies, the correct range of L-shells where energization and the (roughly) correct maximum flux intensities and electron energies that were energized (e.g., 2.6 MeV fluxes were enhanced but 6.3 MeV fluxes were not). The electron flux evolution is mostly reproduced for several hundred keV energies with small errors, especially at the L shells with important outer belt populations ($L \geq 4$). The 2.6 MeV electron fluxes are overestimated in the simulation, but the possible presence of EMIC waves may cause further loss of the 2.6 MeV electrons. The energy-dependent acceleration and decay features and the potential role of EMIC waves are consistent with the simulation results using the plasma wave inputs from available satellite measurements (Ma et al., 2016). Thus, we believe that the ANN modeling technique can be beneficial both as a model in itself, and as an input model to physics-based models, as shown in Fig. 8.

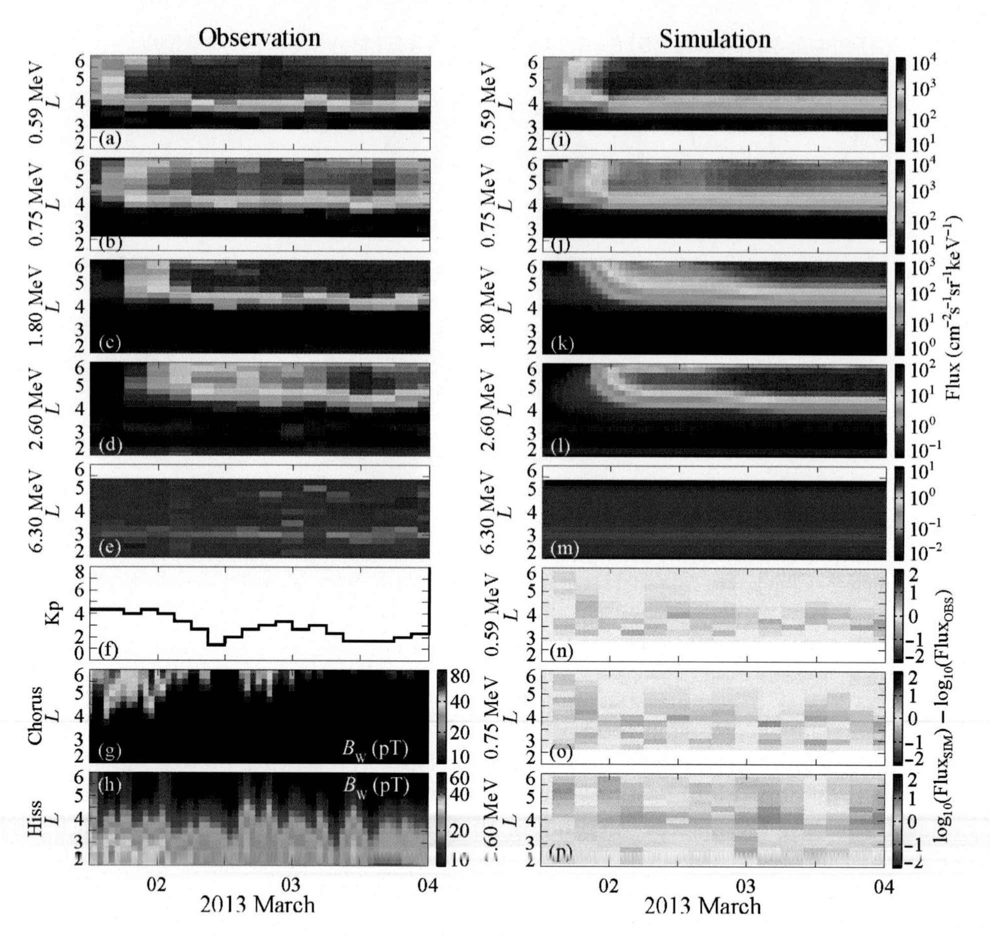

FIG. 8 A comparison of observed and simulated radiation belt fluxes during the March 1–4, 2013 geomagnetic storm. The panels represent: electron fluxes in the range 0.59–6.3 MeV from (a–e) the Van Allen Probes observations based on MagEIS and REPT data and (i–m) the Fokker-Planck simulation results; (f) Kp index; MLT-averaged (0–24 h) (g) chorus and (h) hiss wave intensities predicted by the ANN models; (n–p) the logarithmical difference of electron fluxes between simulation and observation results.

5 SUMMARY AND DISCUSSION

The basic problem we set out to address in this chapter was to find a method to reconstruct the global, time-varying distribution of some physical quantity Q, that has been sparsely sampled at various locations within the magnetosphere, and at different times. The method we used was based on the ANN, which was briefly reviewed in Section 2, and then illustrated with the use of the electron number density dataset, where we developed a 2D (equatorial) model of the inner magnetosphere based on THEMIS data alone, and using only a 5-h SYM-H index as the input. Surprisingly, even this simple model was able to capture the essential dynamical behavior of the plasmasphere and plasmatrough, and capture >90% of the variance, even for out-of-sample data.

We then showed more advanced applications of the ANN technique, first extending our simple 2D model to a full 3D model (DEN3D), then illustrating its application to the inner magnetospheric wave environment by modeling whistler-mode chorus and plasmaspheric hiss waves. Finally, we broached the subject of radiation belt flux prediction/specification and showed that a direct application of the ANN method was, in fact, inappropriate for ultra-relativistic fluxes because this is a data starved environment. This difficulty could be overcome by employing a physics-based Fokker-Planck diffusion model, driven by the ANN-model of chorus and hiss waves, and in such a way we transfer information from a data-rich (wave) environment to the data-starved (ultrarelativistic electron flux) environment.

The wave and particle data examples shown in Section 4 are illustrative and represent only the tip of the proverbial iceberg in terms of the range of applications. The ANN technique discussed in this chapter is quite general and it is conceivable that in the near future, many of the archived data from current and previous instruments and missions could be similarly "learned." For example, there are multiterabytes of data that are currently stored in data archives such as the Space Physics Data Facility, National Space Science Data Center, Coordinated Data Analysis Web, and others, which remain largely unused because they have been long-since decommissioned and thus the measurement instruments cannot be used for any of the current geomagnetic events being studied. But what if a "virtual" version of those instrument could be developed, based on the ANN technique, which would reproduce what the given instrument or dataset would have typically produced under a specific set of geomagnetic driving conditions? Although each virtually recreated instrument would be imperfect individually, having a large collection of such virtual instruments would open the door to a range of exploration that could hardly be imagined at the present time, yet it is readily achievable with the present data, computation resources, and modeling techniques. Such a virtual space environment could be the logical endpoint of the ANN technique discussed herein.

A final point that should be made is that statistical predictive techniques such as the ANN model are not, and could never be, a substitute for physical modeling/understanding as suggested in recent reports (Anderson, 2008). Models such as ANNs perform well in regions that are densely sampled, but generally do not perform well when making extrapolations (e.g., Fig. 7). Physical models, on the other hand, should perform well in extrapolating from their typical operating domains, assuming that all the physical processes have been correctly represented. In the authors' opinion, a particularly fruitful use of machine learning techniques is in the so-called "insight discovery" that is, to discover any potentially important physical processes that may not have been included in the physics-based models, so that the physics-based modeling could be used to predict and specify in situations that have not been previously encountered.

ACKNOWLEDGMENTS

The authors would like to gratefully acknowledge NASA grants NNX14AN85G and AFOSR grant FA9550-15-1-0158. Li would like to acknowledge the NASA grants NNX17AD15G and NNX17AG07G. We acknowledge the Van Allen probes data from the MagEIS and REPT instruments obtained from http://www.rbsp-ect.lanl.gov/data_pub/, and the World Data Center for Geomagnetism, Kyoto for providing Kp index (http://wdc.kugi.kyoto-u.ac.jp/kp/index.html). RD acknowledges NSF grant AGS 1105790.

References

Anderson, R.R., Gurnett, D.A., Odem, D.L., 1992. CRRES plasma wave experiment. J. Spacecr. Rockets 29 (4), 570–573. https://doi.org/10.2514/3.25501.

Angelopoulos, V., 2008. The THEMIS mission. Space Sci. Rev. 141 (1), 5–34. https://doi.org/10.1007/s11214-008-9336-1.

Baker, D.N., Lanzerotti, L.J., 2016. Resource letter SW1: space weather. Am. J. Phys. 84, 166. https://doi.org/10.1119/1.4938403.

Baker, D.N., et al., 2012. The Relativistic Electron-Proton Telescope (REPT) instrument on board the Radiation Belt Storm Probes (RBSP) spacecraft: characterization of Earth's radiation belt high-energy particle populations. Space Sci. Rev. https://doi.org/10.1007/s11214-012-9950-9.

Baker, D.N., et al., 2014. An impenetrable barrier to ultrarelativistic electrons in the Van Allen radiation belts. Nature 515, 531. https://doi.org/10.1038/nature13956.

Bengio, Y., 2009. Learning deep architectures for AI. Foundations Trends Mach. Learn. 2 (1), 1–127.

Borovsky, J.E., 2014. Canonical correlation analysis of the combined solar wind and geomagnetic index data sets. J. Geophys. Res. Space Phys. 119, 5364–5381. https://doi.org/10.1002/2013JA019607.

Bortnik, J., Thorne, R.M., Meredith, N.P., 2007. Modeling the propagation characteristics of chorus using CRRES suprathermal electron fluxes. J. Geophys. Res. 112, A08204. https://doi.org/10.1029/2006JA012237.

Bortnik, J., Li, W., Thorne, R.M., Angelopoulos, V., 2016. A unified approach to inner magnetospheric state prediction. J. Geophys. Res. Space Phys. 121, 2423–2430. https://doi.org/10.1002/2015JA021733.

Cho, J., Lee, D.Y., Kim, J.H., Shin, D.K., Kim, K.C., Turner, D., 2015. New model fit functions of the plasmapause location determined using THEMIS observations during the ascending phase of Solar Cycle 24. J. Geophys. Res. Space Phys. 120. https://doi.org/10.1002/2015JA021030.

Chu, X., et al., 2017a. A neural network model of three-dimensional dynamic electron density in the inner magnetosphere. J. Geophys. Res. Space Phys. 122, 9183–9197. https://doi.org/10.1002/2017JA024464.

Chu, X.N., Bortnik, J., Li, W., Ma, Q., Angelopoulos, V., Thorne, R.M., 2017b. Erosion and refilling of the plasmasphere during a geomagnetic storm modeled by a neural network. J. Geophys. Res. Space Phys. 122, 7118–7129. https://doi.org/10.1002/2017JA023948.

Cybenko, G., 1989. Approximation by superposition of a sigmoidal function. Math. Control Signals Syst. 2, 303–314.

Glauert, S.A., Horne, R.B., Meredith, N.P., 2014. Three-dimensional electron radiation belt simulations using the BAS Radiation Belt Model with new diffusion models for chorus, plasmaspheric hiss, and lightning-generated whistlers. J. Geophys. Res. Space Phys. 119, 268–289. https://doi.org/10.1002/2013JA019281.

Goodrich, C.C., Sussman, A.L., Lyon, J.G., Shay, M., Cassak, P., 2004. The CISM code coupling strategy. J. Atmos. Sol.-Terr. Phys. 66, 1469.

Gurnett, D.A., Scarf, F.L., Fredricks, R.W., Smith, E.J., 1978. The ISEE-1 and ISEE-2 plasma wave investigation. IEEE Trans. Geosci. Electron. 16(3), 225–230. https://doi.org/10.1109/TGE.1978.294552.

Gurnett, D.A., et al., 1995. The Polar plasma wave instrument. Space Sci. Rev. 71 (1), 597–622, https://doi.org/10.1007/bf00751343.

Hinton, G.E., Osindero, S., Teh, Y.W., 2006. A Fast Learning Algorithm for Deep Belief Nets. Neural Computation 18 (7):1527–1554.

Horne, R.B., Glauert, S.A., Thorne, R.M., 2003. Resonant diffusion of radiation belt electrons by whistler-mode chorus. Geophys. Res. Lett. 30 (9), 1493. https://doi.org/10.1029/2003GL016963.

Hornik, K., Stinchcombe, M., White, H., 1989. Multilayer feedforward networks are universal approximator. Neural Netw. 2, 359–3661.

Huang, X., Reinisch, B.W., Song, P., Green, J.L., Gallagher, D.L., 2004. Developing an empirical density model of the plasmasphere using IMAGE/RPI observations. Adv. Space Res. 33 (6), 829–832. https://doi.org/10.1016/j.asr.2003.07.007.

Huba, J.D., Sazykin, S., 2014. Storm time ionosphere and plasmasphere structuring: SAMI3-RCM simulation of the 31 March 2001 geomagnetic storm. Geophys. Res. Lett. 41, 8208–8214. https://doi.org/10.1002/2014GL062110.

Jordanova, V.K., Albert, J., Miyoshi, Y., 2008. Relativistic electron precipitation by EMIC waves from self-consistent global simulations. J. Geophys. Res. 113, A00A10. https://doi.org/10.1029/2008JA013239.

Jordanova, V.K., Zaharia, S., Welling, D.T., 2010. Comparative study of ring current development using empirical, dipolar, and self-consistent magnetic field simulations. J. Geophys. Res. 115, A00J11. https://doi.org/10.1029/2010JA015671.

Kim, J.H., Lee, D.Y., Cho, J.H., Shin, D.K., Kim, K.C., Li, W., Kim, T.K., 2015. A prediction model for the global distribution of whistler chorus wave amplitude developed separately for two latitudinal zones. J. Geophys. Res. Space Phys. 120, 2819–2837. https://doi.org/10.1002/2014JA020900.

Kletzing, C.A., et al., 2013. The Electric and Magnetic Field Instrument Suite and Integrated Science (EMFISIS) on RBSP. Space Sci. Rev. 179, 127–181. https://doi.org/10.1007/s11214-013-9993-6.

Le Contel, O., et al., 2008. First results of the THEMIS search coil magnetometers. Space Sci. Rev. 141 (1–4), 509–534. https://doi.org/10.1007/s11214-008-9371-y.

Li, W., Thorne, R.M., Angelopoulos, V., Bortnik, J., Cully, C.M., Ni, B., Le Contel, O., Roux, A., Auster, U., Magnes, W., 2009. Global distribution of whistler-mode chorus waves observed on the THEMIS space-craft. Geophys. Res. Lett. 36, L09104. https://doi.org/10.1029/2009GL037595.

Li, W., Thorne, R.M., Bortnik, J., Nishimura, Y., Angelopoulos, V., Chen, L., McFadden, J.P., Bonnell, J.W., 2010. Global distributions of suprathermal electrons observed on THEMIS and potential mechanisms for access into the plasmasphere. J. Geophys. Res. 115, A00J10. https://doi.org/10.1029/2010JA015687.

Li, W., Ma, Q., Thorne, R.M., Bortnik, J., Kletzing, C.A., Kurth, W.S., Hospodarsky, G.B., Nishimura, Y., 2015. Statistical properties of plasmaspheric hiss derived from Van Allen Probes data and their effects on radiation belt electron dynamics. J. Geophys. Res. Space Phys. 120, 3393–3405. https://doi.org/10.1002/2015JA021048.

Lyons, L.R., Thorne, R.M., 1973. Equilibrium structure of radiation belt electrons. J. Geophys. Res. 78, 2142–2149.

Ma, Q., et al., 2015. Modeling inward diffusion and slow decay of energetic electrons in the Earth's outer radiation belt. Geophys. Res. Lett. 42, 987–995. https://doi.org/10.1002/2014GL062977.

Ma, Q., et al., 2016. Simulation of energy-dependent electron diffusion processes in the Earth's outer radiation belt. J. Geophys. Res. Space Phys. 121, 4217–4231. https://doi.org/10.1002/2016JA022507.

Marsland, S., 2009. Machine Learning, an Algorithmic Perspective, Machine Learning & Pattern Recognition Series. Chapman & Hall/CRC, Taylor & Francis Group, Boca Raton, FL.

Mauk, B.H., Fox, N.J., Kanekal, S.G., Kessel, R.L., Sibeck, D.G., Ukhorskiy, A., 2013. Science objectives and rationale for the radiation belt storm probes mission. Space Sci. Rev. 179 (1), 3–27. https://doi.org/10.1007/s11214-012-9908-y.

McCulloch, W.S., Pitts, W., 1943. A logical calculus of ideas imminent in nervous activity. Bull. Math. Biophys. 5, 115–133.

McPherron, R.L., Hsu, T.S., Chu, X., 2015. An optimum solar wind coupling function for the AL index. J. Geophys. Res. Space Phys. 120, 2494–2515. https://doi.org/10.1002/2014JA020619.

Meredith, N.P., Horne, R.B., Thorne, R.M., Anderson, R.R., 2003. Favored regions for chorus-driven electron acceleration to relativistic energies in the Earth's outer radiation belt. Geophys. Res. Lett. 30 (16), 1871. https://doi.org/10.1029/2003GL017698.

Meredith, N.P., Horne, R.B., Sicard-Piet, A., Boscher, D., Yearby, K.H., Li, W., Thorne, R.M., 2012. Global model of lower band and upper band chorus from multiple satellite observations. J. Geophys. Res. 117, A10225. https://doi.org/10.1029/2012JA017978.

Meredith, N.P., Horne, R.B., Bortnik, J., Thorne, R.M., Chen, L., Li, W., Sicard-Piet, A., 2013. Global statistical evidence for chorus as the embryonic source of plasmaspheric hiss. Geophys. Res. Lett. 40. https://doi.org/10.1002/grl.50593.

National Research Council, 2013. Solar and space physics: a science for a technological society. In: 2013–2022 Decadal Survey in Solar and Space Physics.

National Science and Technology Council, 2015. Executive Office of the President (EOP). Available from: https://www.whitehouse.gov/sites/default/files/microsites/ostp/final_nationalspaceweatheractionplan_20151028.pdf.

Newell, P.T., Sotirelis, T., Liou, K., Meng, C.I., Rich, F.J., 2007. A nearly universal solar wind-magnetosphere coupling function inferred from 10 magnetospheric state variables. J. Geophys. Res. 112, A01206. https://doi.org/10.1029/2006JA012015.

Orlova, K., Spasojevic, M., Shprits, Y., 2014. Activity-dependent global model of electron loss inside the plasmasphere. Geophys. Res. Lett. 41, 3744–3751. https://doi.org/10.1002/2014GL060100.

Ozeke, L.G., Mann, I.R., Murphy, K.R., Jonathan Rae, I., Milling, D.K., 2014. Analytic expressions for ULF wave radiation belt radial diffusion coefficients. J. Geophys. Res. Space Phys. 119, 1587–1605. https://doi.org/10.1002/2013JA019204.

Pembroke, A., et al., 2014. Initial results from a dynamic coupled magnetosphere-ionosphere-ring current model. J. Geophys. Res. 117 (A2). https://doi.org/10.1029/2011JA016979.

Reeves, G.D., Friedel, R.H.W., Larsen, B.A., Skoug, R.M., Funsten, H.O., Claudepierre, S.G., Fennell, J.F., Turner, D.L., Denton, M.H., Spence, H.E., et al., 2016. Energy-dependent dynamics of keV to MeV electrons in the inner zone, outer zone, and slot regions. J. Geophys. Res. Space Phys. 121, 397–412. https://doi.org/10.1002/2015JA021569.

Reinisch, B. W., et al., 2000. The radio plasma imager investigation on the IMAGE spacecraft. Space Sci. Rev. 91 (1–2), 319–359. https://doi.org/10.1023/A:1005252602159.

Rosenblatt, F., 1957. The perceptron—a perceiving and recognizing automaton. Report 85-460-1. Cornell Aeronautical Laboratory.

Roux, A., Le Contel, O., Coillot, C., Bouabdellah, A., de la Porte, B., Alison, D., Ruocco, S., Vassal, M.C., 2008. The search coil magnetometer for THEMIS. Space Sci. Rev. 141 (1–4), 265–275. https://doi.org/10.1007/ s11214-008-9455-8.

Rumelhart, D.E., Hinton, G.E., Williams, R.J., 1986. Learning representations by back-propagating errors. Nature 323 (6088), 533–536. https://doi.org/10.1038/323533a0.

Shprits, Y.Y., Kellerman, A., Drozdov, A., Spense, H., Reeves, G., Baker, D., 2015. Combined convective and diffusive simulations: VERB-4D comparison with March 17, 2013 Van Allen Probes observations. Geophys. Res. Lett. 42. https://doi.org/10.1002/2015GL065230.

Su, Z., Xiao, F., Zheng, H., Wang, S., 2010. STEERB: a three-dimensional code for storm-time evolution of electron radiation belt. J. Geophys. Res. 115, A09208. https://doi.org/10.1029/2009JA015210.

Thorne, R.M., 2010. Radiation belt dynamics: the importance of wave-particle interactions. Geophys. Res. Lett. 37, L22107. https://doi.org/10.1029/2010GL044990.

Toffoletto, F.R., Sazykin, S., Spiro, R.W., Wolf, R.A., 2003. Modeling the inner magnetosphere using the rice convection model (review). Space Sci. Rev. 108, 175–196. WISER special issue.

Toffoletto, F.R., Sazykin, S., Spiro, R.W., Wolf, R.A., Lyon, J.G., 2004. RCM meets LFM: initial results of one-way coupling. J. Atmos. Sol.-Terr. Phys. 66, 1361.

Tu, W., Cunningham, G.S., Chen, Y., Henderson, M.G., Camporeale, E., Reeves, G.D., 2013. Modeling radiation belt electron dynamics during GEM challenge intervals with the DREAM3D diffusion model. J. Geophys. Res. Space Phys. 118, 6197–6211. https://doi.org/10.1002/jgra.50560.

Wang, W., Wiltberger, M., Burns, A.G., Solomon, S.C., Killeen, T.L., Maruyama, N., Lyon, J.G., 2004. Initial results from the coupled magnetosphere-ionosphere-thermosphere model: thermosphere-ionosphere responses. J. Atmos. Sol.-Terr. Phys. 66, 1425.

Wiltberger, M., Wang, W., Burns, A.G., Solomon, S.C., Lyon, J.G., Goodrich, C.C., 2004. Initial results from the coupled magnetosphere ionosphere thermosphere model: magnetospheric and ionospheric responses. J. Atmos. Sol.-Terr. Phys. 66, 1411–1424.

Wolf, R.A., Spiro, R.W., Rich, F.J., 1991. Extension of the rice convection model into the high-latitude ionosphere. J. Atm. Terrest. Phys. 53, 817–829.

CHAPTER

12

Reconstruction of Plasma Electron Density From Satellite Measurements Via Artificial Neural Networks

Irina S. Zhelavskaya[*,†], *Yuri Y. Shprits*[*,†,‡], *Maria Spasojevic*[§]

[*]Helmholtz Centre Potsdam, GFZ German Research Centre for Geosciences, Potsdam, Germany [†]University of Potsdam, Potsdam, Germany [‡]University of California Los Angeles, Los Angeles, CA, United States [§]Stanford University, Stanford, CA, United States

CHAPTER OUTLINE

Machine Learning Techniques for Space Weather
https://doi.org/10.1016/B978-0-12-811788-0.00012-3

1 OVERVIEW

Plasma electron density is a crucial parameter in space physics simulations and modeling and is important for predicting and preventing hazardous effects of space weather, such as satellite damage or even complete breakdown due to enhanced solar wind activity. Measuring plasma density accurately, however, has always been a challenge. One of the most accurate methods of measuring plasma electron density is to derive it from the satellite measurements of the upper hybrid resonance frequency. The upper hybrid resonance frequency is often associated with the most pronounced resonance band in dynamic spectrograms, which display electric power spectral density measured by a satellite as a function of frequency and time. In previous missions, upper hybrid resonance bands were manually identified in the dynamic spectrograms, although such manual determination is a very tedious and time-consuming process. Moreover, as new satellites for scientific exploration are being launched and more data become available, manually identifying upper hybrid resonance frequency becomes unfeasible. Research has been done in the past to automate the process of upper hybrid resonance band extraction, but the developed algorithms still require significant manual intervention and correction.

In this chapter, we present an alternative approach for automated electron density determination based on artificial neural networks. The method employs feedforward neural networks (FNNs) to derive the upper hybrid resonance bands from the dynamic spectrograms, and hence electron density, in an automated fashion. Neural networks are a powerful tool for finding the multivariate nonlinear mapping from input (in this case, dynamic spectrograms and other geophysical parameters) to output parameters (the upper hybrid frequency). Neural networks inherently require a training data set, that is, a data set for which both inputs and outputs are known. We use electric and magnetic field measurements produced by the two satellites of NASA's Van Allen Probes mission, currently the gold standard of measurements in space weather research, as input, and a large data set of upper hybrid frequency measurements produced by another recently developed semiautomated technique as output in our training data set.

The chapter is organized as follows. First, we provide the necessary background on the space weather aspects of this study related to plasma density and its importance in space physics research and delve deeper into the motivation behind this application. Then, we give a brief overview of FNNs, the type of artificial neural networks used for this application, with a focus on the importance of model validation. Next, we describe the algorithm implementation in detail and demonstrate the results. We also discuss how the developed plasma electron density data set can be used to develop a global empirical plasma density model that does not depend on satellite measurements, also using neural networks.

1.1 Space Weather-Related Aspects and Motivation

1.1.1 Plasma Density and the Plasmasphere

Plasma electron density is a parameter characterizing a number of particles in a unit volume in space (measured in cm^{-3}) (Cohen, 2007). The electron density is a fundamental parameter of plasma. The focus of this study is the density of cold particles in the near-Earth space environment. These cold particles (of temperature ~1 eV) are trapped by the closed magnetic field lines of the Earth, forming a bubble-shaped region around the Earth. This region is called the plasmasphere (Lemaire and Gringauz, 1998). The plasmasphere is a relatively dense region of plasma compared with other regions in space (density of $10-10^4$ cm^{-3}). The plasmasphere extends from the topside ionosphere (~1000 km above the ground) out to a boundary called the plasmapause that ranges from 2 to 7 Earth radii, R_E (1 R_E = 6371 km), depending on geomagnetic conditions (Gringauz, 1963; Carpenter, 1963; Grebowsky, 1970). The region outside the plasmapause is called the plasma trough, and it is a low-density region.

The plasmasphere is very dynamic, and its shape and size strongly depend on solar and geomagnetic conditions (O'Brien and Moldwin, 2003; Chappell et al., 1970b). Two mechanisms, sunward convection and corotation with the Earth, determine the configuration of the plasmasphere (Darrouzet et al., 2009; Singh et al., 2011). The corotation regime dominates during quiet geomagnetic times; plasma material trapped inside the closed magnetic field lines corotates with the Earth (Carpenter, 1966). Meanwhile, the plasmasphere is refilled with ions from the topside ionosphere and expands up to ~4–7 R_E (Goldstein et al., 2003; Singh and Horwitz, 1992; Krall et al., 2008); its shape is roughly circular with a bulge on the dusk side (Nishida, 1966). In contrast, during periods of high geomagnetic activity, the sunward magnetospheric convection starts to dominate and erodes the plasmasphere: the closed magnetic field lines at the dayside magnetopause boundary are torn apart and the plasmaspheric material is carried sunward. Due to that, the outer layers of the plasmasphere are eroded and plasmapause contracts (Carpenter, 1970; Chappell et al., 1970a; Goldstein et al., 2003). The stronger the disturbance, the more the plasmapause contracts (down to 2 R_E during severe geomagnetic storms).

Plasma density of the plasmasphere is a critical parameter in a number of important space weather applications such as GPS navigation (e.g., Mazzella, 2009; Xiong et al., 2016) and analysis of spacecraft anomalies due to spacecraft charging (e.g., Reeves et al., 2013). Plasma density is also a critical input parameter for quantifying wave-particle interactions necessary for modeling the formation and decay of Earth's radiation belt, a donut-shaped region around the Earth, hazardous for satellites electronics and crew in space (e.g., Spasojević et al., 2004; Thorne et al., 2013; Orlova et al., 2016; Shprits et al., 2016).

Plasma electron density can be measured on satellites using several methods. They include measuring the density directly with particle counters (e.g., Geiger and Müller, 1928), determining it using the spectral properties of waves (e.g., Trotignon et al., 2003) or deriving it from the spacecraft potential (e.g., Escoubet et al., 1997). A number of empirical models of plasma electron density have been developed using electron density measurements from previous missions.

The most widely used empirical models in recent years are those developed in the studies of Carpenter and Anderson (1992), Gallagher et al. (2000), and Sheeley et al. (2001). The model of Carpenter and Anderson (1992) is based on electron density measurements derived from radio measurements made with the sweep frequency receiver onboard the International Sun-Earth Explorer (ISEE-1) spacecraft and ground-based whistler measurements. This model presents the mean electron density values for different L after several days of refilling, which means that it is applicable only for quiet geomagnetic activity. The model is valid for L shells from 2.25 to 8 R_E and local times between 0 and 15 MLT (here, L can be roughly considered as the distance from the center of the Earth, and MLT stands for magnetic local time and can be considered as an angular distance around the Earth from the local midnight). The Global Core Plasma Model by Gallagher et al. (2000) combines several previously developed models (including Carpenter and Anderson, 1992; Gallagher et al., 2000), using transition equations in order to obtain a more comprehensive description of the plasma in the inner magnetosphere. The plasmasphere and plasma trough density models of Sheeley et al. (2001) present statistical density averages based on density measurements obtained using the Combined Release and Radiation Effects Satellite (CRRES) swept frequency receiver. The models are valid for L shells between 3 and 7 and all local times. The Sheeley et al. (2001) study provides the mean and the standard deviation of density in the plasmasphere and the trough to represent depleted and saturated density levels for different L (and MLT for the trough).

Despite the extensive use of these empirical density models in space physics simulations, they do not provide reliable electron density estimates during extreme events, such as geomagnetic storms, because they are parameterized only by static geomagnetic parameters such as L and MLT. The described models do not include the dynamic dependence of plasma density on geomagnetic and solar conditions, and plasma electron density is known to be highly variable during elevated geomagnetic activity (Park and Carpenter, 1970; Moldwin et al., 1995). Therefore, collecting reliable electron density measurements during varying geomagnetic conditions is still of continuing interest.

1.1.2 Determining the Electron Density From Upper-Hybrid Band Resonance Frequency

One of the most reliable techniques to measure the electron density is to use the upper hybrid resonance frequency to derive it (Mosier et al., 1973). This method will be further employed in this study.

The upper hybrid resonance frequency is a combination of the electron plasma frequency and the electron cyclotron frequency:

$$f_{\text{uhr}} = \sqrt{\left(f_{\text{ce}}^2 + f_{\text{pe}}^2\right)} \tag{1}$$

The electron plasma frequency and the electron cyclotron frequency are given as:

$$f_{\text{pe}} = \frac{1}{2\pi}\sqrt{\frac{q_{\text{e}}^2 n_{\text{e}}}{m_{\text{e}}\epsilon_0}}, \quad f_{\text{ce}} = \frac{|q_{\text{e}}|\,B}{2\pi m_{\text{e}}} \tag{2}$$

where B is the magnetic field strength, n_{e} is electron density, q_{e} is the charge of electron, ϵ_0 is the permittivity of free space, and m_{e} is the mass of an electron. Using these formulas, electron density can be easily derived. The upper boundary of the upper hybrid emission is generally the most pronounced feature in spacecraft plasma wave data. Mosier et al. (1973) found that the upper hybrid resonance can often be observed visually as the brightest emission band in the dynamic spectrograms, which display spectral properties of the electric field (see Fig. 1 for more details). In this study, we use the plasma wave data measured on the Van Allen Probes satellites.

Van Allen Probes is a dual-spacecraft NASA mission launched in August 2012, and its scientific objective is to explore the dynamic evolution of the Van Allen radiation belts (Mauk et al., 2013). The satellites have a highly elliptical orbit in a near-equatorial plane (inclination

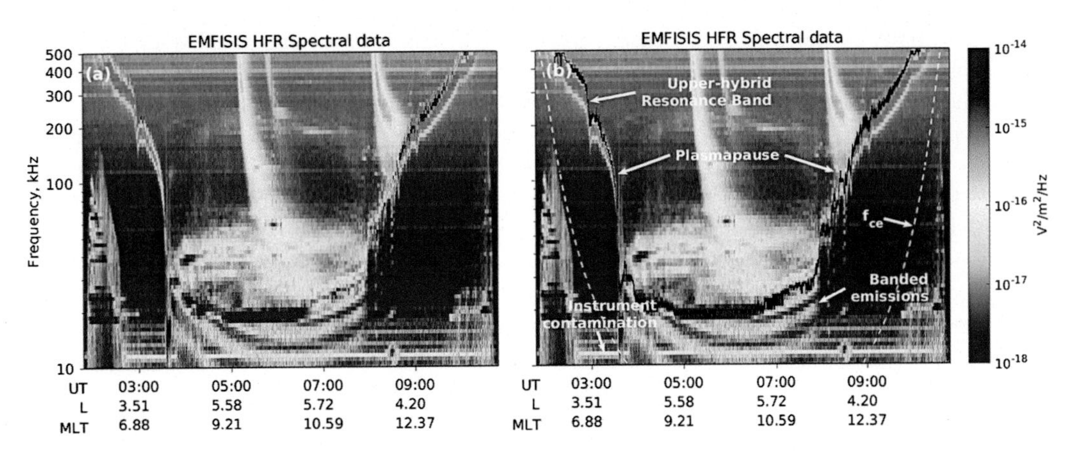

FIG. 1 An example of the EMFISIS HFR spectral data for one orbit pass (A) with various parameters and phenomena denoted (B). Upper hybrid frequency is shown with the *black curve*; electron cyclotron frequency, f_{ce}, is shown with the *white-dashed curve*.

10 degrees) with the apogee 30,414 km and perigee 618 km. A number of science instruments are deployed on the satellites. One of them is the Electric and Magnetic Field Instrument Suite and Integrated Science (EMFISIS) (Kletzing et al., 2013). The instrument performs routine measurements of the electric field in the frequency range of 10–487 kHz, thus providing the capability to determine the upper hybrid resonance band, and hence to accurately estimate the electron density. An example of measurements from the High Frequency Receiver (HFR) of the EMFISIS instrument on board Van Allen Probe A for one orbital pass No. 1612 for April 25, 2014, is illustrated in Fig. 1. Fig. 1A shows the power spectral density of the electric field as a function of frequency and time, where color indicates the spectral density, as noted in the color bar. Such a plot is referred to as a dynamic spectrogram. Dynamic spectrograms are a tool to explore spectral properties of waves and resonances in plasma. Fig. 1B shows the same spectrogram but with the upper hybrid resonance frequency indicated with black. The upper hybrid resonance frequency is often associated with the most pronounced band in dynamic spectrograms, as found by Mosier et al. (1973). The background magnetic field strength, B, is measured by the magnetometer on board the spacecraft, and therefore, f_{ce} can be determined directly; it is indicated on the spectrogram as the dashed curve.

Several processes and phenomena that might, at times, impose challenges on the upper hybrid resonance identification are denoted with white in Fig. 1B. The plasmapause, the outer border of the plasmasphere, is usually identified by a sharp density gradient and can be monitored via the sharp drop in the upper hybrid frequency. The electron density (hence, the upper hybrid frequency) might have a complex structure in the vicinity of the plasmapause, and the density gradient is not always smooth, but may sometimes have jumps. Accurately identifying those sudden changes in density may be challenging for some automated routines. Another phenomenon is the banded emissions, also referred to as "$(n+\frac{1}{2})f_{ce}$" emissions. These emissions are often observed between the harmonics of electron cyclotron frequency, f_{ce}, but not necessarily exactly in the middle between them (LaBelle et al., 1999), in the plasma trough (the low-density region). In some cases, emissions at f_{uhr} are not observed with the banded emissions (Benson et al., 2001); this brings uncertainty in the process of identification of the upper hybrid resonance and presents challenges for making definite determination of f_{uhr} without performing an appropriate visual spectral interpretation. Instrument contaminations can also present challenges for automated routines of f_{uhr} identification. They are usually observed as horizontal lines of roughly identical spectral density and might have intersections with the upper hybrid resonance.

Previously, the upper hybrid resonance band has been manually derived from dynamic spectrograms (e.g., LeDocq et al., 1994) and there have been several semiautomated techniques developed. Research on the development of semiautomated routines began in the ISEE-1 era (Trotignon et al., 1986) and still continues (Trotignon et al., 2010; Denton et al., 2012; Kurth et al., 2015). Determination of the electron density in Trotignon et al. (2010) was conducted using the active and passive wave spectra measured with the Waves of High frequency and Sounder for Probing of Electron density by Relaxation (WHISPER) instrument onboard the Cluster mission. The electron density in the work of Denton et al. (2012) was derived from the passive radio wave observations obtained from the Radio Plasma Imager (RPI) instrument onboard the IMAGE (Imager for Magnetopause-to-Aurora Global Exploration) spacecraft. The Automated Upper hybrid Resonance detection Algorithm (AURA) developed by Kurth et al. (2015) is based on the Van Allen Probes' EMFISIS HFR data and is a semiautomated algorithm to derive the upper hybrid resonance band in dynamic spectrograms. The algorithm searches a peak in the spectrum for every time step while assuming that each successive spectrum contains a peak associated with f_{uhr} near the previously determined peak. An operator then visually inspects the dynamic spectrogram for each orbital pass and corrects the resulting f_{uhr} profile identified by AURA where it is necessary. AURA significantly facilitates the processing of the HFR spectral data, but still requires manual intervention.

In this chapter, we present an alternative algorithm for automated determination of f_{uhr}, hence electron density, from the satellite measurements using artificial neural networks, which is also described in Zhelavskaya et al. (2016). Neural networks are one of the most commonly used tools for a broad range of nonlinear approximations and mappings. In this application, we use FNNs. FNNs are very efficient in solving nonlinear multivariate regression problems. The neural networks are "tuned" to a specific problem during the training using the training data set. The training data set is a data set for which both inputs (in our case, satellite measurements and geophysical parameters) and outputs (f_{uhr}) are known. We use electric (HFR spectra) and magnetic field (f_{ce}) measurements from Van Allen Probes as inputs in our training set, and the database of f_{uhr} measurements developed using AURA (courtesy of W. Kurth and the EMFISIS team) as output. The training data set covers 1091 orbital passes and thus it represents a significant set of example data. After the neural network passes the training, validation, and testing stages, it can be used in practice and be applied to a data set for which the f_{uhr} is not known. The neural network is applied to a database of 3750 orbital passes and its output is then used to produce a database of electron number density. The performance of the resulting neural network model is assessed by comparing the derived density to the density obtained in Kurth et al. (2015). The resulting density distribution is also analyzed and compared with the empirical density models of the plasmasphere and trough by Sheeley et al. (2001).

1.2 Brief Background on Neural Networks

This part provides a brief overview of artificial neural networks and discussion on important aspects of training and validation. The notions introduced here might be useful for further understanding the application described in this chapter. For more details on neural networks, the reader should refer to the works cited herein.

1.2.1 Basic Concepts Related to Neural Networks

Artificial neural networks are a family of models effective at solving problems of function approximation, pattern recognition, classification, and clustering. Artificial neural networks were inspired by biological neural networks (in the brain) and are an attempt to mimic them in a very simplified manner (e.g., McCulloch and Pitts, 1943; Hebb et al., 1949; Marr et al., 1976).

Neural networks are composed of multiple simple computational blocks called artificial neurons. An artificial neuron has a body in which computations are performed, and a number of input channels and one output channel, similar to a real biological neuron. Simply put, a neuron receives an input signal and then computes an output on it. Fig. 2A shows the construction of an artificial neuron with N inputs. Every input has a weight associated with it; the larger the weight, the more impact the corresponding input channel has on the output. A neuron also has a bias, which, for convenience, can be considered as an additional input to the neuron, x_0, that is equal to 1 and has the weight identical to the value of the bias, $w_0 = b$. Additionally, a neuron has a transfer or an activation function that defines the type of neuron. The activation function can be arbitrary; the most commonly used functions are sigmoid, hyperbolic tangent, binary, and linear. After the signal is applied to the neuron, it first computes the sum of inputs multiplied by their weights and then applies the transfer function to the resulting sum.

An artificial neuron is one of the first computational models developed in the research area of artificial neural networks (McCulloch and Pitts, 1943; Rosenblatt, 1957). A single neuron can be used to solve a limited number of problems, such as linear regression and classification of two linearly separable subsets. Two toy examples related to these problems are shown

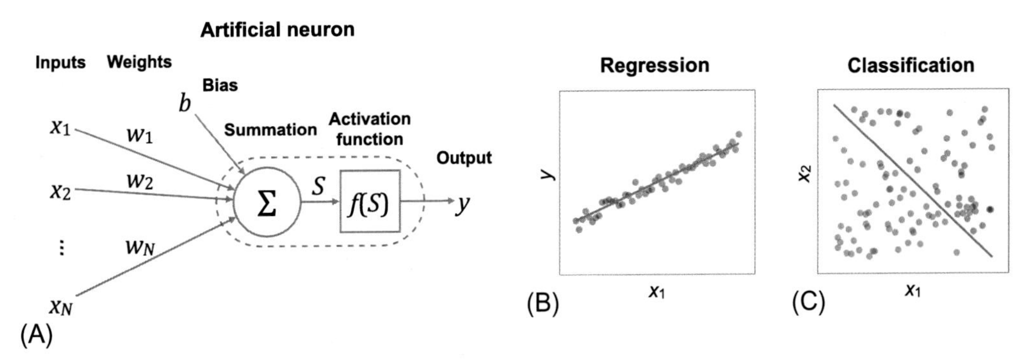

FIG. 2 (A) A scheme of an artificial neuron, a building block of a neural network. Artificial neurons can be used to solve linear problems. Simple examples of such problems for regression and classification are shown in panels (B) and (C) respectively.

in Fig. 2B and C. Fig. 2B illustrates a simple case of a linear regression problem with one dependent and one independent variables. The given data are plotted with red markers, where independent variable x_1 is plotted on the x-axis versus dependent variable y on the y-axis. The regression task consists of finding a mapping from x_1 to y. The gray line shows a linear fit to the data found by a single neuron. The neuron has a single input in this case. Fig. 2C shows a simple case of the classification problem. Here, the given data consists of two classes shown with red and blue markers correspondingly. There are two independent variables, x_1 and x_2. They are plotted on x- and y-axes correspondingly, and the single dependent variable y is represented by color (numerically, it can be represented as 0 and 1, or -1 and 1, for example). The classification task is to identify the class of each data point, given x_1 and x_2. This task can be solved by building a separation border between two classes. The gray line shows a separator found by a neuron. The neuron has two inputs in this case. These scenarios are intentionally oversimplified for demonstration purposes and can be expanded to more dimensions (hence, more inputs to the neuron). However, a single neuron cannot be used to solve more complex, nonlinear problems. Fortunately, such problems can be solved by neural networks, which are composed of multiple neurons. The main concept of artificial neural networks is that an output signal from one neuron can be used as an input to other neurons.

The way neurons are connected into a network defines the topology, or architecture, of a neural network. In this work, we use an FNN architecture. An FNN is one of the most basic and widely used neural network architectures and is effective at solving multivariate nonlinear regression and classification problems. FNNs have displayed the state-of-the-art performance in a number of applications (e.g., Salakhutdinov and Hinton, 2009; Krizhevsky and Hinton, 2011; Glorot et al., 2011; Mohamed et al., 2012). The topology of an FNN is shown in Fig. 3A. Neurons in an FNN are arranged in layers. Three types of layers exist: input, output, and hidden layers. The input layer is composed of inputs to the network and no computations are performed in this layer. Next, follow hidden layers that are composed of any number of neurons arranged in parallel. The network can have several hidden layers. The neurons of the same layer are not connected to each other, but connected to the neurons of the preceding and the subsequent layers; an output of one hidden layer serves as an input to the following layer. The output layer, and hence the network output, is formed by a weighted summation of the outputs of the last hidden layer. Neurons of one layer have the same activation function. Different layers can have different activation functions. Formally, an FNN with L hidden layers can be defined as a superposition of L activation functions $f_1, \dots, f_L$.

The neural network with at least one hidden layer can solve nonlinear regression or classification problems (Cybenko, 1989). Simple illustrations related to regression and classification are shown in Fig. 3B and C. The x- and y-axes are the same as in Fig. 2B and C, correspondingly. The neural network can fit the nonlinear function to given data (in the case of regression, Fig. 3B) and determine a nonlinear separator between classes (in the case of classification, Fig. 3C). The given examples are idealized and can be expanded to more dimensions.

The preceding explanation assumes that weights and biases of the neural network are known. In practice, the weights and biases are not given; however, they can be determined using a training data set, that is, a set of data for which inputs and outputs are known. Determining weights and biases is usually referred to as training and reduces to an optimization problem of minimizing a given cost function. The cost function is defined based on the type of application or problem we are attempting to solve. Specifically for neural networks, a number of backpropagation algorithms are used to solve the optimization problem

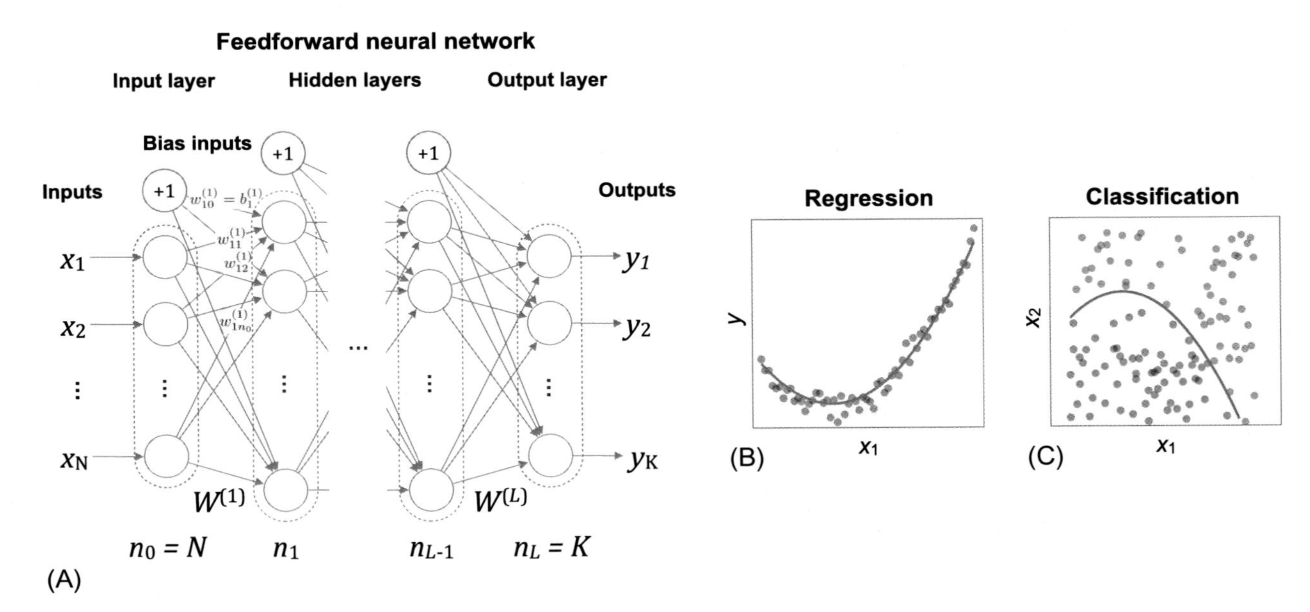

FIG. 3 (A) A scheme of a feedforward neural network. *Circles* denote artificial neurons. Feedforward neural networks arrange neurons in a layered configuration and can be used to solve nonlinear problems. (B, C) Simple examples of such problems for regression and classification.

of determining the weights (Williams and Hinton, 1986). In this study, we employ the scaled conjugate gradient (SCG) backpropagation algorithm (Møller, 1993). This algorithm is based on conjugate gradient (CG) methods, a class of optimization techniques. It works faster than most of the other algorithms for neural networks of large sizes and has relatively modest computer memory requirements (Mathworks.com, 2015).

1.2.2 Neural Network Design Flow

The neural network design flow consists of three main stages: training, validation, and testing. Consequently, the set of data for which the output is known is split into three parts: training, validation, and test data sets, in a ratio defined by a user.

The goal of the training stage is to find the weights of the neural network, which minimize the difference between the output of the training data set and the output of the neural network. An optimization algorithm is run in order to determine such weights. Before performing the training procedure, the internal parameters of the neural network, such as the number of hidden layers, the number of neurons in each hidden layer, and activation functions in each layer have to be selected. Typically, multiple neural networks with different internal parameters are trained, and in the validation stage, the model having the minimal validation error is selected for further use.

The main objective of the validation stage is to check the neural network ability to reconstruct the relation between inputs and outputs on the data it has not seen yet, that is, data not used for training. This is usually called the generalization ability of a neural network. In this stage, we measure the performance of every neural network obtained during the training stage on the validation data set and select the model with the best performance (by comparing the output of the models to the known output of the validation set). The importance of the validation stage is described in the next section.

Finally, in the testing stage, the performance of the optimal model determined in the validation stage is assessed on the test set. The calculated error of the neural network is then treated as the resulting error of the model. After the neural network is tested and the obtained results are satisfactory, it can be used on the data for which the output is not known.

1.2.3 Importance of Validation

Although neural networks are a powerful tool for building accurate multivariate nonlinear approximations, they are also inherently easy to overfit. Overfitting occurs when the model becomes too complex for the given task and therefore becomes capable of fitting the training data excellently but does not produce a reliable output on the unseen data. The more hidden layers and neurons the model has, the more complex it is. The ultimate goal of building a practical model is to find the optimal internal network parameters. The optimal internal network parameters are the parameters producing a model that generalizes well and avoids overfitting the training data, at times at the cost of a slightly reduced accuracy of the model. The optimal model has similar errors on both training and unseen data, which are acceptable for the given application.

To demonstrate this, let us consider three possible scenarios that might occur in practice: overfitting, underfitting, and desirable model performance. Overfitting occurs when the model is too complex for the given data: the model simply "memorizes" the training data points and therefore cannot produce meaningful results on the unseen data. Underfitting is directly opposite to overfitting: the model is too simple and cannot produce a reliable fit to the

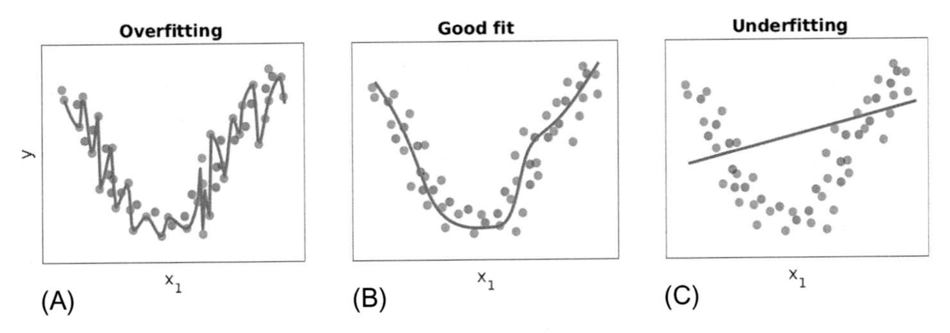

FIG. 4 Toy illustrations of three different scenarios that can occur after the training: (A) overfitting, (B) ideal case, and (C) underfitting. *Red points* represent the data to be fitted with a function. *Gray curves* show the functions fitted to the data in these three different scenarios.

given data; its errors on the training and validation data are similar, but at the same time are very high. Fig. 4 shows toy illustrations corresponding to these scenarios for the case of simple nonlinear regression problem. Here, similar to the illustrations in Fig. 2B and Fig. 3B, the given data has one independent variable, x_1 (one input) and one dependent variable, y (one output), to be modeled. The variables x_1 and y are plotted on the x- and y-axes correspondingly. The blue markers denote the training data points, which are used to find the mapping from x_1 to y; the red markers denote the validation data points that are not shown to the network during the training. The gray curve in each plot shows the resulting fit to the training data produced by the corresponding neural network. In the overfitting scenario (Fig. 4A), the model is too complex and produces a fit that passes through every point of the training data set. Therefore, the model has a poor performance on the validation set and is unacceptable for use in practice. In contrast, in the case of underfitting (Fig. 4C), the model is too simple and fits a straight line to the parabolic shape of the given data. The errors of the model, both on the training and validation sets, are unacceptably high, which is also not desirable in practice. Ideally (Fig. 4B), the model has an optimal complexity for the given data and learns the general dependency of the input, x_1, on the output, y, although it might have some error due to the presence of noise in data.

In practice, the number of inputs to the model can be much larger, reaching hundreds or even thousands of input parameters for some applications, as well as the size of the training set (reaching hundreds of thousands or millions of data points). Visualization of the results in a manner shown in Fig. 4 becomes impossible, and the quantitative assessment of the results is the only way of evaluating the model performance. Such quantitative assessment is performed in the validation stage by measuring the error of the model on the training and validation sets and comparing them. If the training error is small while the validation error is large, then the model is likely to overfit the data. The contrary, when both errors are too large, although they are similar, might be a sign of underfitting. Ideally, the training and validation errors should be similar and acceptable for the given application. The same applies to classification.

2 IMPLEMENTATION OF THE ALGORITHM

In this section, we describe the details of the application of FNNs to the plasma wave data. The inputs to the network include electric field spectral properties obtained from the satellite

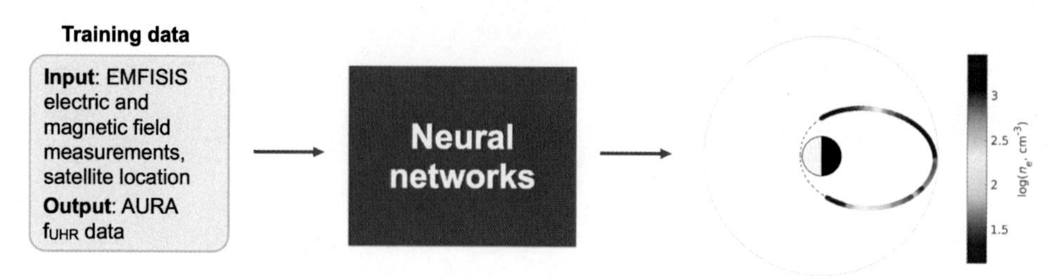

FIG. 5 Schematic presentation of the NURD algorithm. We use EMFISIS electric and magnetic field measurements as input and f_{uhr} obtained with AURA as output to the neural network. The NURD algorithm is then able to reconstruct the upper hybrid frequency, and hence density, along the satellite orbit. An example of such a reconstruction for one orbit pass is shown in the *rightmost figure*.

measurements, location of the satellite, and the parameter characterizing geomagnetic activity (*Kp* index). The network has a single output, the upper hybrid frequency, from which plasma density can be easily derived (see formulae 1, 2). We use electric field measurements from Van Allen Probes for the input and database of f_{uhr} developed by Kurth et al. (2015) as the output of the training data set. The resulting neural network is capable of reconstructing the density from the satellite measurements, hence, along the satellite orbit. The main concept of the algorithm and an example output for one orbital pass are shown in Fig. 5. Further here, we describe the details of input and output data, neural network architecture, and validation. The developed algorithm is further referred to as the Neural-network-based Upper hybrid Resonance Determination (NURD) algorithm.

2.1 Training Data Set

2.1.1 Input Data

The inputs to the neural network are listed in Table 1. A detailed explanation and rationale behind each input variable are given below.

The first 82 inputs to the model include the electric field measurements from the EMFISIS HFR instrument, specifically, the electric field power spectral density produced in the survey mode of the instrument. The instrument produces measurements in the frequency range from 10 to 487 kHz binned into 82 logarithmically spaced bins with a 6-s temporal resolution. The frequency spacing provides approximately 5% spectral resolution, $\frac{\Delta f}{f}$, which defines the resulting density resolution, $\frac{\Delta n}{n}$. As previously discussed, n_e is proportional to f_{pe}^2, which results in 10% density resolution. Furthermore, the upper frequency limit (487 kHz) restricts the maximum density that is possible to derive to $\sim$2900 cm^{-3}, and the lower frequency limit (10 kHz) restricts the minimum density to $\sim$1 cm^{-3}. The logarithm of the power spectral density of the electric field for the 82 frequency bins is used as input to the model.

The next input to the model is the logarithm of electron cyclotron frequency, f_{ce}. The electron cyclotron frequency can be directly derived from the background magnetic field, $|B|$: $f_{ce} = 28|B|$, where $|B|$ is measured in nanotesla, and f_{ce} is measured in hertz. We use measurements of the background magnetic field, B, from the EMFISIS fluxgate magnetometer.

TABLE 1 Inputs to the Neural Network

No.	Input Name	Description	Units
1–82	$\log_{10}$ spectrum	Decimal logarithm of the spectrum	$\log_{10}$ V^2/m^2/Hz
83	$\log_{10} f_{ce}$	Decimal logarithm of electron cyclotron frequency	$\log_{10}$ Hz
84	L	Magnetic field line	Earth radii
85	MLT	Magnetic local time	Hours (0–24)
86	Kp index	Geomagnetic index	Unitless (0–9)
87	f_{binmax}	Frequency bin with the largest power spectral density of the electric field from the HFR spectrum	Unitless (1, 2, …, 82)

Because plasma density is known to vary spatially, we also use the spacecraft coordinates as input to the model (84 and 85 in the table). The Van Allen probes have low inclination orbits in the near-equatorial plane; therefore, we consider two spatial coordinates corresponding to the position of the satellite in the equatorial plane. The L parameter reflects roughly the radial distance from the center of the Earth and is measured in units of Earth radii, R_E. More precisely, L denotes the distance at which the magnetic field line where the measurement was made crosses the equator. MLT stands for magnetic local time and represents the angular distance from the magnetic local midnight measured in decimal hours (0–24).

Plasma density dynamics also depends on the level of geomagnetic activity. Therefore, a parameter characterizing the global level of geomagnetic activity, the Kp index, is used as input to the model. The Kp index (originating from "planetarische Kennziffer," German for planetary index) is derived from measurements of magnetic field on the ground observatories located in different parts of the world and is an indicator of global geomagnetic disturbances. It has a 3-h cadence and ranges from 0 (lowest geomagnetic activity) to 9 (severe geomagnetic storms) in discrete steps.

The last input, f_{binmax}, corresponds to the frequency bin with the largest spectral density in the HFR spectrum. We use this parameter as input because it can serve as a rough initial approximation of f_{uhr}. This assumption might introduce errors to the neural network because, as shown in Benson et al. (2004), the upper hybrid frequency f_{uhr} is generally associated not with the maximum emission peak of the upper hybrid band, but with its upper frequency edge. However, the analysis in Benson et al. (2004) was performed using active and passive observations measured by the IMAGE RPI, which had a higher frequency resolution compared with EMFISIS HFR and therefore allowed for very accurate f_{uhr} determinations. The coarse frequency resolution of the EMFISIS HFR does not allow resolving the upper frequency edge precisely.

2.1.2 Output Data

The neural network has a single output: the logarithm of the upper hybrid frequency, f_{uhr}. The dataset of upper hybrid frequencies derived with the AURA algorithm (Kurth et al., 2015) is used in the training. AURA is a semiautomated routine for determination of the f_{uhr} band in the dynamic spectrograms. AURA uses a restricted search approach for finding the maximum value of the spectrum at a specific time point. In this algorithm, the density derived from the spacecraft potential (using EFW instrument) is used to weight the probability of a possible maximum peak occurrence in the spectrum and guide the search procedure. The weighting parameters are optimized using a bootstrapping method so that the determined spectral peaks are more likely to correspond to f_{uhr}. After AURA is run for one orbit, the operator visually detects where AURA has failed to identify the correct UHR band and manually corrects it. The produced f_{uhr} measurements have a high quality because the resulting f_{uhr} have been visually checked for each orbit. The accuracy of the AURA algorithm is $\approx$10%.

The UHR frequency dataset derived by AURA and used for training was obtained from the EMFISIS instrument website. The available UHR dataset derived by AURA at the moment of developing the algorithm consists of 1091 orbital passes and covers the period from October 2012 to March 2015. With a 6-s cadence and 9-h orbital period, the dataset comprises ~5,900,000 measurements. However, not all measurements can be used for training, because the UHR frequency may be greater than the upper limit of the HFR frequency range near the perigee. We exclude the out-of-frequency range measurements from the training dataset before performing the following procedure. We consider the spectrograms for each orbital pass separately. First, the left- and rightmost edges of the upper hybrid resonance line are identified by determining the bins with the highest values of the spectral density in the uppermost frequency spectral bin (at 487 kHz, near the perigee). We then apply the neural network only to the measurements within these boundaries. After the out-of-frequency range part of the data is excluded, the total number of measurements available for training, validation, and testing reaches 4,027,610 measurements. An 87 × 4,027,610 matrix of input variables from Table 1 (usually referred to as design or feature matrix) and a 1 × 4,027,610 vector of output values (measurements of f_{uhr}) are constructed using this data set.

2.2 Neural Network Architecture

We use an FNN with a single hidden layer shown in Fig. 6. The network has 87 input neurons as defined by the number of input variables (dimensions in the design matrix). The number of neurons in the hidden layer is initially not known and is determined during the validation procedure, as described further. For this application, 80 neurons in the hidden layer were found to be optimal in the validation stage. The network has only one output neuron because we are to model only one variable, f_{uhr}.

First, before being fed to the neural network, the input data is normalized to the common range (here, [0; 1]). Because the range of some variables could be much larger than others, large value inputs can dominate the input effect compared with small value inputs and influence the accuracy of the neural network (Li et al., 2000). Normalization ensures that inputs have identical initial importance to the neural network. Normalization is performed on the training set independently of validation and test data sets; the normalization parameters calculated on the training set are stored and applied afterward to the validation and test data sets (and not recalculated on these sets).

The input layer does not perform any computations; its function is to distribute the incoming signal to the neurons of the following hidden layer. The neurons of the hidden layer have the hyperbolic tangent activation function, which transforms the input to the range of [−1; 1]. The output of the hidden layer is then the following vector:

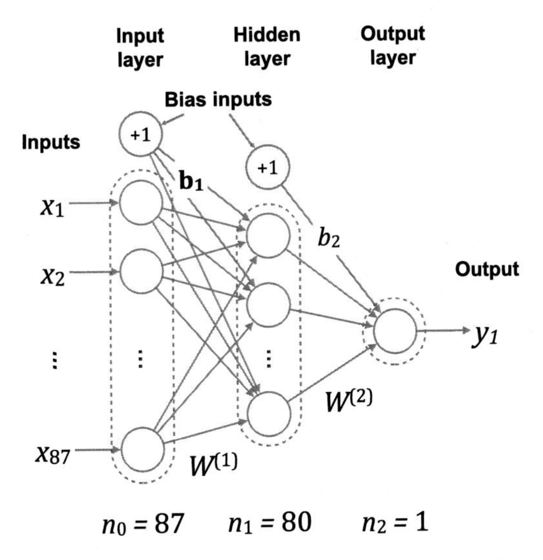

FIG. 6 The optimal architecture of the FNN determined in the validation stage. The neural network has 87 inputs, 1 output, and 80 neurons in the hidden layer. The weights of the neural network can be considered as 80×87 $W^{(1)}$ and 1×80 $W^{(2)}$ matrices; the biases can be represented as an 80×1 vector $\mathbf{b}_1$ for the hidden layer and as a scalar b_2 for the output layer. *(Adapted from Zhelavskaya, I., Spasojevic, M., Shprits, Y., Kurth, W., 2016. Automated determination of electron density from electric field measurements on the Van Allen Probes spacecraft. J. Geophys. Res. Space Phys. 121, 4611–4625. https://doi.org/10.1002/2015JA022132.)*

$$\mathbf{a}_1 = \tanh(\mathbf{b}_1 + W^{(1)}\mathbf{x}) \tag{3}$$

where $\mathbf{x}$ is an input vector (87×1, because there are 87 inputs), $\mathbf{b}_1$ is the vector of biases (80×1, because there are 80 hidden neurons), $W^{(1)}$ is the matrix of weights connecting the input to the hidden layer (dimensionality: 80×87). The transfer function of the output neuron is linear. The output of the neural network is then given as:

$$a_2 = b_2 + W^{(2)}\mathbf{a}_1 \tag{4}$$

where $\mathbf{a}_1$ is the output vector of the hidden layer, b_2 is a bias (a scalar), and $W^{(2)}$ is the matrix of weights connecting the hidden to the output layer (dimensionality: 1×80).

2.3 Steps of the Design Flow

The design matrix comprising 4,027,610 measurements is randomly divided into three parts: training, validation, and test sets, in a 34:33:33% ratio, respectively. A large subset of the available data is allocated to the validation and testing. Such division allows for a more reliable test of the generalization ability of the network. The validation set is used to determine the optimal number of neurons in the hidden layer of the network. The test set is used to estimate the accuracy of the resulting neural network model. All three sets are kept separate during the respective stages, that is, the validation and test sets are not used in training (and the test set is not used in training or in validation). It is worth noting that although in this work the division of data into training, validation, and test sets is performed in a random fashion, the more warranted way to perform division for the time series is to split data sequentially. The sequential split ensures independence of all three subsets, while random splitting might yield optimistic evaluations on the validation and test sets for the events lying outside the time period covered by the dataset. The network resulting from training performed on the random split, however, would still perform well for the reconstruction of the past events.

In order to determine the optimal number of neurons in the hidden layer, we train five neural networks with a different number of hidden neurons ranging from 40 to 120. The networks are trained on the training set of data using a conjugate gradient backpropagation algorithm. After the networks are trained, their performance is measured on the validation set. We use the mean absolute percentage error (MAPE) to assess the performance of the networks:

$$\mathrm{MAPE} = \frac{1}{M}\sum_{i=1}^{M} \frac{\left|f_i^{\mathrm{AURA}} - f_i^{\mathrm{NURD}}\right|}{f_i^{\mathrm{AURA}}} \times 100\% \tag{5}$$

where f_i^{NURD} are f_{uhr} values predicted using the constructed neural network, f_i^{AURA} and f_{uhr} measurements provided by the team of EMFISIS (we refer to them as to ground truth values), and M is the size of the validation set.

Fig. 7 displays the MAPE of the resulting neural networks plotted versus the number of neurons in their hidden layer. The blue solid line shows the MAPE on the training set, the red-dashed line—on the validation set. The MAPE on the training set decreases as the number of neurons in the hidden layer increases. This happens due to the following. As the number of hidden neurons grows, the complexity of the network increases (it has more free parameters), and it becomes capable of fitting the training data better. On the other hand, as the complexity

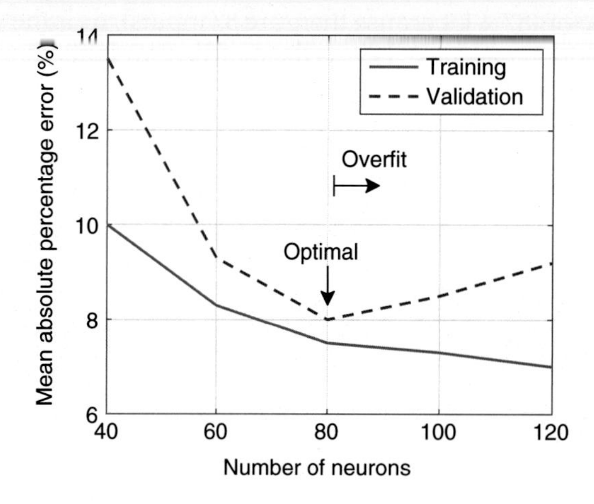

FIG. 7 Mean absolute percentage error (MAPE) as a function of the number of hidden neurons of the neural network. As the complexity of the network increases, MAPE decreases on the training set. However, for the validation set, error stops decreasing after a certain point, indicating that the network has been overfit to the training data. Here, a network with 80 neurons in the hidden layer is optimal. *(Adapted from Zhelavskaya, I., Spasojevic, M., Shprits, Y., Kurth, W., 2016. Automated determination of electron density from electric field measurements on the Van Allen Probes spacecraft. J. Geophys. Res. Space Phys. 121, 4611–4625. https://doi.org/10.1002/2015JA022132.)*

of the network increases, the MAPE on the validation set decreases until a certain point, and then it starts to increase. This is a sign of overfitting, which means that after that point, the model does not generalize well and does not produce a reliable output on the unseen data. In our case, we found that the optimal neural network contains 80 neurons in the hidden layer.

After the optimal number of hidden neurons is determined, the network with 80 neurons is retrained on the training and validation sets combined, and its performance is further examined on the test set. The MAPE of the resulting model on the test set is ~8%.

2.4 Postprocessing Step

After all the preceding steps are completed, the resulting model can be used in practice and applied to data. The output of the system, f_{uhr}, must be binned to 82 logarithmically spaced frequency bins because we are to derive the UHR band in the dynamic spectrograms with such spacing. However, the output of the neural network model is a real number by definition and is not bounded to those bins. Therefore, the raw output of the neural network must be processed and binned to the specific frequency bins to be used in practice. The implemented postprocessing procedure is described as follows.

In the first step of the postprocessing procedure, the f_{uhr} value reconstructed by the neural network is rounded off to the closest EMFISIS HFR frequency bin. Next, we consider the determined bin and the two adjacent bins in the HFR spectrum (above and below). If the identified bin is the first or the last bin, then only one adjacent bin is considered (bin 2 or 81, correspondingly). The resulting f_{uhr} is then determined as the frequency corresponding to the bin containing the maximum power spectral density (of the three considered bins).

Fig. 8 shows an example of the neural network output before ("raw" output of the neural network, blue curve) and after (black curve) the postprocessing procedure. It can be seen

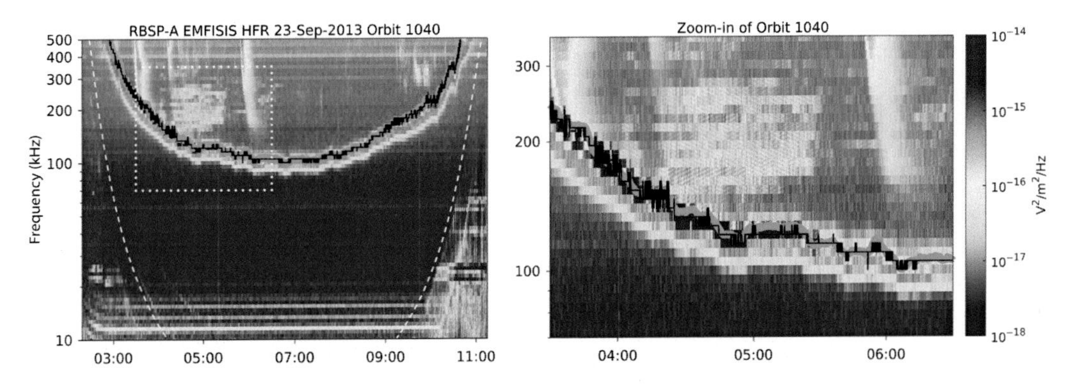

FIG. 8 An example of the postprocessing procedure of the neural network output. Postprocessing assigns the output of the neural network to the most adjacent frequency bin and removes potential noise in the neural network output. The *left panel* shows the spectrogram for orbit pass no. 1040, and the *right panel* shows the zoomed-in part of the spectrogram (from 03:30 until 06:30 UT). The *blue curve* shows the raw neural network output, and the *black curve* shows the uhr after the postprocessing step. *(Adapted from Zhelavskaya, I., Spasojevic, M., Shprits, Y., Kurth, W., 2016. Automated determination of electron density from electric field measurements on the Van Allen Probes spacecraft. J. Geophys. Res. Space Phys. 121, 4611–4625. https://doi.org/10.1002/2015JA022132.)*

from the figure that the "raw" output of the neural network can slightly fluctuate from the true uhr band. The postprocessing procedure decreases these fluctuations. The derived upper hybrid line can still toggle between adjacent bins in the spectrum (see Fig. 10 for the example), consequently introducing uncertainty in the density determination. Moreover, errors caused by using the frequency associated with the maximum emission intensity as f_{uhr} instead of the upper boundary of the upper hybrid band (Benson et al., 2004; Beghin et al., 1989) also should be taken into account. As mentioned before, the coarser frequency resolution of the EMFISIS HFR instrument in comparison with the IMAGE RPI might not allow capturing this upper edge accurately. Thus, the introduced error can be estimated by evaluating the error in density when the derived f_{uhr} differs by one frequency bin from its real value. The following formula for this error was obtained after some derivations:

$$\frac{\Delta n}{n} \approx 2\frac{\Delta f}{f}\sqrt{\left(\frac{f_{pe}}{f_{ce}}\right)^{-2} + 1} \tag{6}$$

where $\frac{\Delta f}{f}$ is the frequency resolution of 1 bin that is ≈5% as noted before. The expression under the square root is always less than 2 but greater than 1 because the $\frac{f_{pe}}{f_{ce}}$ ratio is always larger than 1. This makes the total expression always greater than ≈10% and less than ≈14%. Therefore, the error introduced due to switching between two adjacent bins is between ~10% and 14%. It is worth noting, however, that this error should be taken into account in the trough most of the time because determination of the upper hybrid line for the plasmasphere is relatively straightforward.

Finally, after the f_{uhr} determined by the neural network is binned, the electron densities are derived using expressions (1), (2).

3 RESULTS

The NURD algorithm was applied to the Van Allen Probes measurements covering the period from October 1, 2012 to July 1, 2016 and a database of 33,830,887 electron density measurements (for both probes) was produced. The output of the NURD algorithm is compared with the density derived using AURA, and several examples of such comparisons for individual satellite passes are shown as follows. Furthermore, we perform analysis of the obtained electron density database and compare the resulting density distribution with the empirical trough and plasmasphere models by Sheeley et al. (2001).

3.1 Comparison With AURA and NURD Performance

In order to compare the output of the NURD algorithm with the output provided by AURA and properly evaluate the NURD's performance, we should first introduce a classification of dynamic spectrograms into types as was done in Kurth et al. (2015). Kurth et al. (2015) classify dynamic spectrograms for each individual satellite pass (or orbit) into three types of complexity, A, B, and C, in terms of AURA performance:

(1) Type A are the spectrograms in which less than 25% of the f_{uhr} points required manual correction. These are the spectrograms in which f_{uhr} is fairly clear and easy to identify. Type A constitutes 70% of the spectrograms processed by AURA.

(2) Type B are the spectrograms in which 25% to 50% of data points must be corrected manually and comprises 20% of the spectrograms. These spectrograms are more difficult to process, as f_{uhr} is not always possible to identify unambiguously.

(3) Type C are the spectrograms containing interpretational difficulties in finding the upper hybrid band. These spectrograms might be very contaminated. This type constitutes 10% of the spectrograms.

We perform a quantitative comparison between the NURD output and AURA's f_{uhr} by computing the MAPE for each orbit type as defined in Eq. (5). It is worth noting that such evaluation does not demonstrate the true error of either NURD or AURA because the ground truth densities are ambiguous, especially for type C. Thus, it is more suitable to call this evaluation measure "average percentage divergence" rather than "mean absolute percentage error." Fig. 9 shows the average percentage divergence calculated for different orbit types on the test data set. The difference between AURA and NURD is not significant for orbits of type A, because the upper hybrid resonance bands are clear and easy to identify for this type of orbit. For orbits of type B, the difference is slightly higher (~5%), and the difference for orbits of type C is ~14%. Such deviation between the results might be caused by the ambiguity of the upper hybrid frequency determination during the geomagnetically active times and at times by the contaminated signal. During such periods, the plasma density in the trough can be very low and there might be strong electron cyclotron harmonic emissions present, making it challenging to unambiguously identify the f_{uhr} profile both for the neural network and for AURA.

Fig. 10 shows the typical examples for each type of orbit, where the f_{uhr} profiles obtained using AURA (red curve) and the resulting f_{uhr} profiles obtained with the NURD algorithm

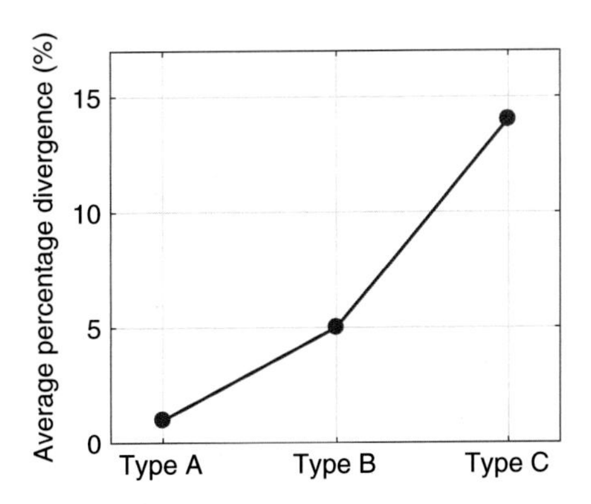

FIG. 9 The average percentage divergence of the electron plasma density determined by the NURD algorithm from the electron density determined by the AURA algorithm calculated on the test set as a function of orbit types. *(Adapted from Zhelavskaya, I., Spasojevic, M., Shprits, Y., Kurth, W., 2016. Automated determination of electron density from electric field measurements on the Van Allen Probes spacecraft. J. Geophys. Res. Space Phys. 121, 4611–4625. https://doi.org/10.1002/2015JA022132.)*

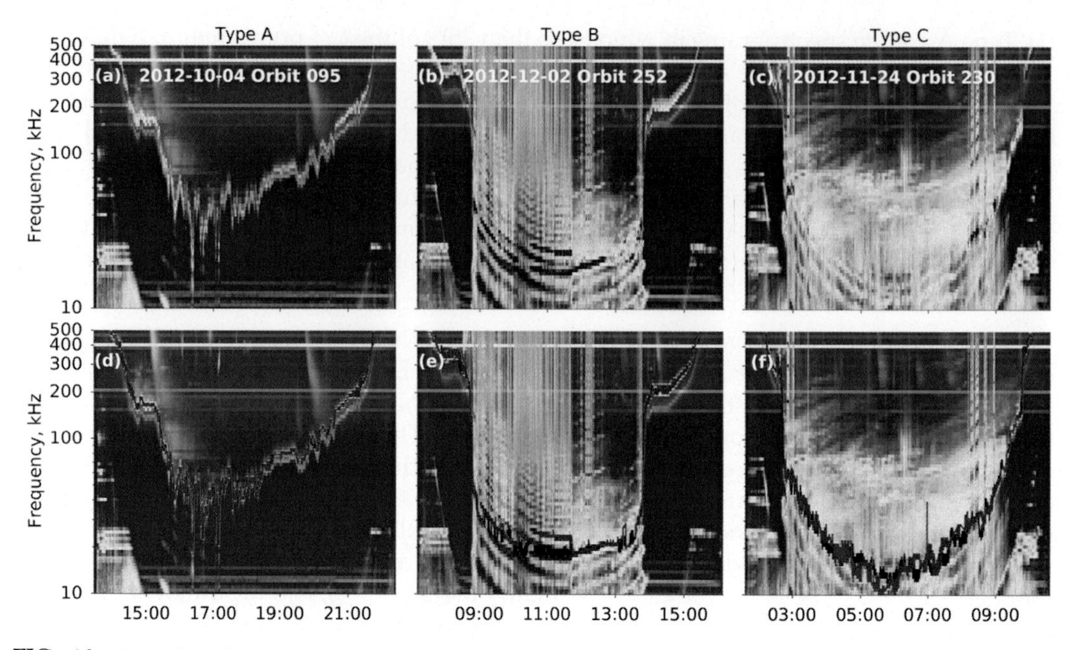

FIG. 10 Examples of spectrograms of each type (the *top panel*) and the upper hybrid frequency identified by AURA, indicated by the *red curve* and by the NURD algorithm indicated by the *black curve* (the *bottom panel*). *(Adapted from Zhelavskaya, I., Spasojevic, M., Shprits, Y., Kurth, W., 2016. Automated determination of electron density from electric field measurements on the Van Allen Probes spacecraft. J. Geophys. Res. Space Phys. 121, 4611–4625. https://doi.org/10.1002/2015JA022132.)*

are indicated (black curve). The two f_{uhr} profiles produced by AURA and NURD for orbit of type A almost overlap. For more complex cases (types B and C), the upper hybrid band identified by NURD nearly overlaps with AURA's f_{uhr} in the plasmasphere (high-density region), but differs in the trough (low-density region). Such performance can be anticipated because determining the density in the plasmasphere is relatively unambiguous. However, this is not the case for the trough, where the upper hybrid line cannot always be observed. Thus, for such complicated types of orbits, the largest contribution to the average divergence can be attributed to differences in the trough region between two algorithms. In such cases, it may be hard to definitely conclude which of the algorithms performed better. For example, in Fig. 10b and e (type B), NURD's f_{uhr} toggles between adjacent bins more than AURA's f_{uhr}. However, NURD's f_{uhr} tends to follow one resonance band that seems to be selected by it as the band corresponding to the intense emission extending beyond the banded emissions, while AURA's f_{uhr} might switch from band to band. It is very difficult to determine the correct diagnosis visually, and sometimes, both algorithms might be wrong. Most of such cases refer to parts of orbits of type C, where the electron density in the trough is extremely low and strong electron cyclotron harmonic emissions are observed. Fig. 10c and f illustrates the case, when NURD and AURA may both provide wrong f_{uhr} determination, and the f_{uhr} band can be higher than the algorithms predicted (at 07:30 near the relatively strong emission observed near 40 kHz). More comprehensive analysis involving spectral interpretation of the emissions is needed to determine the actual value of f_{uhr}.

3.2 Comparison With Empirical Model of Sheeley et al. (2001)

Initial analysis of the obtained density database comprising 33,830,887 measurements is presented as follows. Here, we compare the derived plasma density distribution to the empirical density models of the trough and the plasmasphere developed by Sheeley et al. (2001) as functions of L and MLT. For the sake of comparison, we separate density values into plasmasphere- and trough-like data by applying the criteria used in Sheeley et al. (2001) for the threshold density (in cm^{-3}):

$$n_b = 10\left(\frac{6.6}{L}\right)^4 \tag{7}$$

Density values at or above n_b for the given L shell are considered plasmasphere-like; values below n_b are considered trough-like.

Fig. 11 shows the two-dimensional plots of normalized occurrence of the measurements as a function of plasma electron density and L (the top row) and as a function of electron density and MLT (the bottom row). The data set is divided into 26 bins in the logarithm of electron density and into 16 bins in L/14 bins in MLT. For the MLT plots, the data is limited to the range of $3 < L < 5$. The value in a particular bin is calculated as the number of measurements corresponding to n_e and L/MLT in that particular range divided by the number of measurements in that range of L/MLT. Hence, the color scale shows the normalized occurrence in different L/MLT bins. The minimum number of measurements in a bin required to calculate the normalized occurrence is 10; the mean number of measurements per bin is 77,278 for L and 29,220 for MLT plots.

Fig. 11a and d shows the normalized occurrence for all density measurements, Fig. 11b and e—for the plasmasphere-like density, and Fig. 11c and f—for the trough-like density, versus L and MLT, respectively. The intense red color indicates the regions where density measurements tend to be clustered; the blue color indicates that measurements are sparse in those regions. The black-dashed line shows the threshold density, n_b. For the plots in the bottom row (vs. MLT), the threshold density was calculated using the median of L for all data (both plasmasphere- and trough-like), which is approximately $L \approx 4.2$. The uppermost black curve corresponds to the plasmasphere density model of Sheeley et al. (2001), and the lowermost black curve corresponds the trough density model for MLT = 0. Again, for the plots in the bottom row, the plasmasphere and the trough density values were calculated using the median of L for measurements in the plasmasphere ($L \approx 4.1$) and the median of L for measurements in the trough ($L \approx 4.6$) correspondingly. The upper and lower white-dotted curves show the mean of the logarithm of electron density obtained using NURD for plasmasphere and trough, correspondingly.

Fig. 11a shows that between $L = 2$ and 3.5, density measurements cluster around a narrow range of densities in the plasmasphere. At higher L, however, we see a bimodal structure with a distinct separation between plasmasphere and trough measurements. When we examine the plasmasphere and trough density regions separately (Fig. 11b, c, e, and f)

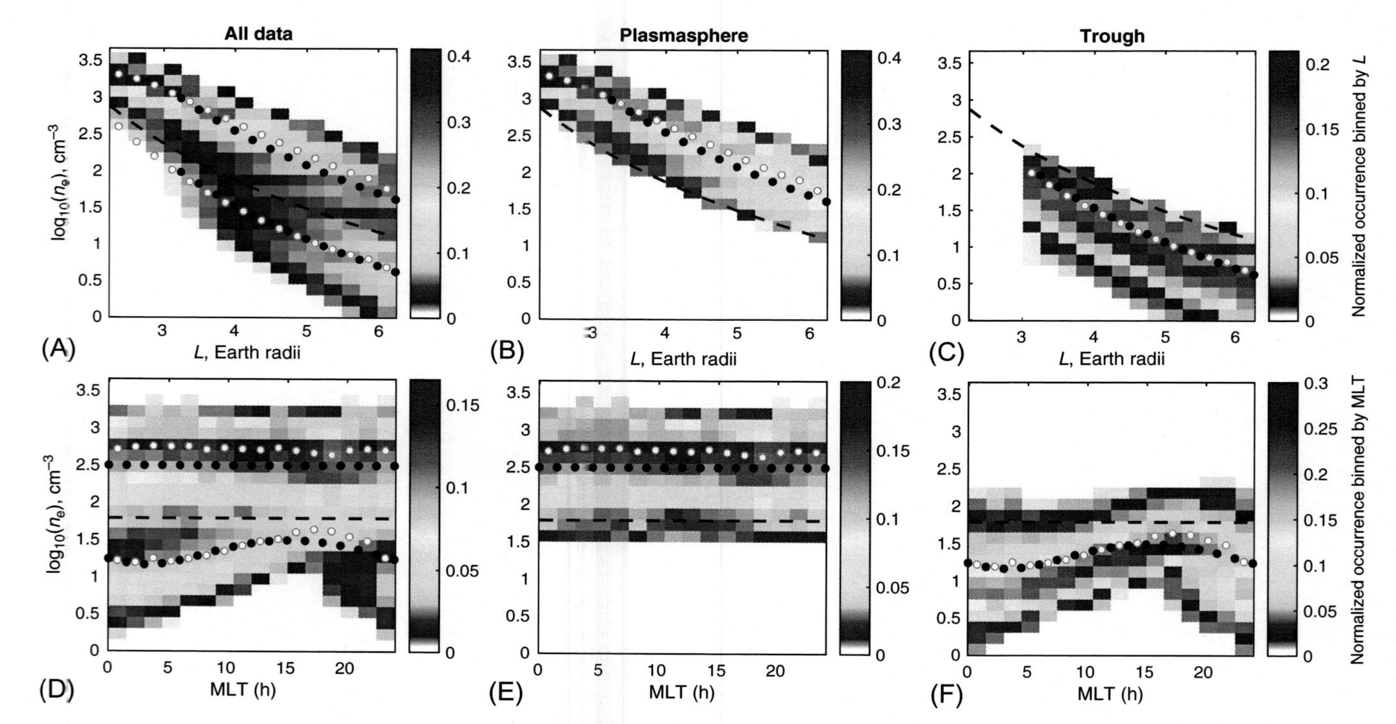

FIG. 11 The occurrence of density measurements normalized by the number of data points in different L (the *top row*) or MLT (the *bottom row*) bins as a function of electron density on the logarithmic scale and L (top) or MLT (*bottom*). The *dashed line* indicates the separation between the trough-like and plasmasphere-like data as $n_b = 10\left(\frac{6.6}{L}\right)^4$ (the same as used in Sheeley et al., 2001). The *black-dotted curves* indicate the plasmasphere and trough density model by Sheeley et al. (2001). The *white-dotted curves* indicate the mean of the log(ne) determined with the NURD algorithm for the plasmasphere and the trough, correspondingly. *(Adapted from Zhelavskaya, I., Spasojevic, M., Shprits, Y., Kurth, W., 2016. Automated determination of electron density from electric field measurements on the Van Allen Probes spacecraft. J. Geophys. Res. Space Phys. 121, 4611–4625. https://doi.org/10.1002/2015JA022132.)*

and compare those distributions to the Sheeley et al. (2001) model, we find good agreement with both the plasmasphere and the trough models. However, we also find that the peak of occurrence distributions of measurements produced by Van Allen Probes is slightly higher than the average density value measured by the CRRES satellite in the plasmasphere. The mean density value is shifted by approximately 280 cm^{-3} ($\approx$0.14 on a logarithmic scale) on average. For the trough (Fig. 11c), the peak of occurrence densities distributions is slightly lower than the empirical model. The shift in the occurrence peak is approximately 5.32 cm^{-3} on average ($\approx$0.05 on a log scale). Such a shift may be caused by the fact that the CRRES mission operated during the peak phase of the solar cycle, while Van Allen Probes were launched at the increasing phase.

4 DISCUSSION AND FUTURE DIRECTIONS

Our results have shown that neural networks can be successfully applied to plasma wave data and produce reliable and accurate density determinations. Such performance of the developed algorithm was reached due to the proficient quality of the AURA density set used for training and to the proper neural network design process. However, it is worth acknowledging that although the electron density data set obtained using AURA and employed for training is currently the most reliable source of the electron plasma density measurements for the Van Allen Probes mission, there are still cases of uncertainties in density determination that cannot be resolved, even by the manual inspection. Such uncertainties may be a source of errors in the training set, especially in the low-density trough region. As a result, the errors in the training set may influence the neural network performance. Due to that the accuracy of both NURD and AURA may be lower in the trough. Therefore, while the use of density values marked "questionable" in the electron density data set produced by NURD is safe in statistical studies, the exact values with this flag should be used with caution.

At the moment, NURD is tuned specifically to the Van Allen Probes data, but using these data, we have shown that neural networks, and potentially other machine-learning algorithms, are applicable and produce good results on such types of data. The created framework can be used to retrain the neural network model using other datasets, for which the input parameters can be adjusted if necessary.

Although FNNs were employed as a modeling technique in this work, it is worth-mentioning alternative methods that are also effective at nonlinear multivariate regression problems. These methods include recurrent neural networks (RNNs) (e.g., Hochreiter and Schmidhuber, 1997; Goller and Kuchler, 1996), nonlinear least squares (e.g., Teunissen, 1990; Moré, 1978), regression trees (e.g., Breiman et al., 1984), and so forth, which could be also potentially used for determining $f_{\rm uhr}$ from plasma wave measurements. These methods naturally have their own advantages and disadvantages, and a method that is more suitable for the data and problem at hand should be preferably selected. In the study described in this chapter, the FNNs were chosen over the methods mentioned herein for the following reasons. Unlike nonlinear least squares, FNNs do not require specification of the form (model) of the nonlinear function to be modeled, which is not known in the problem at hand. They are also less susceptible to noise in data in comparison with regression trees, if trained properly. FNNs were chosen over RNNs in this study with a goal to first employ a simpler network architecture without taking into account any temporal dependencies. In the future, however, the ability of RNNs to preserve the information about the past can be used to explore the influence of temporal dependencies on the NURD's performance.

The neural network-based methodology described in this chapter can also be employed in other space weather applications. A particular extension of this work is the development of the global plasmasphere density model driven by the solar and geomagnetic activity and therefore not bound by the satellite measurements. In this case, a neural network model will have the time history of solar wind and geomagnetic parameters and the satellite location as input, and a single output, plasma density. Such a model would be extremely useful in space physics, specifically radiation belt modeling, and also for the applications discussed in the overview section. As also mentioned in the overview, the existing empirical models tend to be oversimplified and are parameterized by location. The model driven by the solar wind and geomagnetic parameters would be a significant advancement in the field.

5 CONCLUSIONS

In this chapter, we present our algorithm for the automated determination of plasma electron density from electric and magnetic field measurements made by Van Allen Probe spacecraft, the Neural-network-based Upper hybrid Resonance Determination algorithm (NURD). The algorithm uses an FNN to derive upper hybrid frequency profiles from the dynamic spectrograms and magnetic field measurements obtained from the EMFISIS instrumentation suite. Plasma density is then derived from the inferred upper hybrid frequency profiles. The plasma density database developed using another semiautomated routine AURA (Kurth et al., 2015) is used to train the network. The NURD algorithm is valid for L-shells $1.25 < L < 6.2$ and all local time sectors.

The developed algorithm was applied to the available database of electric and magnetic field measurements from October 2012 until July 2016, comprising 3750 orbital passes of the Van Allen Probes satellites. Comparison with AURA showed that electron densities obtained using the proposed method are in good agreement with the densities inferred with AURA. To demonstrate this, we adopted the classification of the dynamic spectrograms into three levels of difficulty, as introduced in Kurth et al. (2015): type A (upper hybrid frequency is relatively straightforward to identify, 70% of all orbits); type B (interpretation from an expert is needed, 20% of orbits); and type C (concealed signal, 10% of orbits). The mean average percentage divergence between the density values produced by the NURD and AURA was $\sim$1% for type A, $\sim$5% for type B, and $\sim$14% for type C. The overall error of the derived density is $\sim$14%. However, the error may be larger in cases of high uncertainty in density determination, particularly in the low-density region and after the recently elevated geomagnetic activity. Indeed, the neural network-based approach does not fully remove the uncertainty in density determination. Nonetheless, it stills produces reasonable density estimates in the regions of high uncertainty that can be employed in statistical studies. Additionally, the proposed algorithm is automated, meaning that it can remove a significant part of the manual aspect of the density determination.

The analysis of the resulting electron density data has shown an agreement with the plasmasphere and trough density models by Sheeley et al. (2001). On the other hand, a large variability in the density data was observed that cannot be reproduced by empirical models based on statistical averages. Using the NURD algorithm, the electron density can be determined to a much finer resolution than using the existing empirical models.

ACKNOWLEDGMENTS

The work at GFZ was supported by Helmholtz Association Recruiting Initiative, NSF GEM AGS-1203747, NASA grant NNX12AE34G, NASA grant NNX16AF91G and project PROGRESS funded by EU Horizon 2020 No. 637302. The work at Stanford was supported by NASA award NNX15AI94G. The EMFISIS data were accessed through the official EMFISIS website hosted at the University of Iowa, and we thank the EMFISIS team, William Kurth and Principal Investigator, Craig Kletzing. The *Kp* index was provided by GFZ. John Wiley and Sons provided the license agreement to reproduce the figures from Zhelavskaya et al. (2016) (License Number 4114291495723). The electron density data set derived by the NURD algorithm is available at ftp://rbm.epss.ucla.edu/ftpdisk1/NURD.

References

Beghin, C., Rauch, J.L., Bosqued, J.M., 1989. Electrostatic plasma waves and HF auroral hiss generated at low altitude. J. Geophys. Res. Space Phys. 94 (A2), 1359–1378.

Benson, R.F., Osherovich, V.A., Fainberg, J., F-Vinas, A., Ruppert, D.R., 2001. An interpretation of banded magnetospheric radio emissions. J. Geophys. Res. Space Phys. 106 (A7), 13179–13190.

Benson, R.F., Webb, P.A., Green, J.L., Garcia, L., Reinisch, B.W., 2004. Magnetospheric electron densities inferred from upper-hybrid band emissions. Geophys. Res. Lett. 31 (20), L20803.

Breiman, L., Friedman, J.H., Olshen, R., Stone, C.J., 1984. Classification and Regression Trees. Wadsworth & Brooks/Cole Advanced Books & Software, Pacific Grove, CA.

Carpenter, D.L., 1963. Whistler evidence of a 'knee' in the magnetospheric ionization density profile. J. Geophys. Res. 68 (6), 1675–1682.

Carpenter, D.L., 1966. Whistler studies of the plasmapause in the magnetosphere: 1. Temporal variations in the position of the knee and some evidence on plasma motions near the knee. J. Geophys. Res. 71 (3), 693–709.

Carpenter, D.L., 1970. Whistler evidence of the dynamic behavior of the duskside bulge in the plasmasphere. J. Geophys. Res. 75 (19), 3837–3847.

Carpenter, D.L., Anderson, R.R., 1992. An ISEE/whistler model of equatorial electron density in the magnetosphere. J. Geophys. Res. Space Phys. 97 (A2), 1097–1108. ISSN 2156-2202. https://doi.org/10.1029/91JA01548.

Chappell, C.R., Harris, K.K., Sharp, G.W., 1970a. The morphology of the bulge region of the plasmasphere. J. Geophys. Res. 75 (19), 3848–3861.

Chappell, C.R., Harris, K.K., Sharp, G.W., 1970b. A study of the influence of magnetic activity on the location of the plasmapause as measured by OGO 5. J. Geophys. Res. 75 (1), 50–56.

Cohen, E.R., 2007. Quantities, Units and Symbols in Physical Chemistry. Royal Society of Chemistry, London.

Cybenko, G., 1989. Approximation by superpositions of a sigmoidal function. Math. Control Signals Syst. 2 (4), 303–314.

Darrouzet, F., Gallagher, D.L., André, N., Carpenter, D.L., Dandouras, I., Décréau, P.M.E., De Keyser, J., Denton, R.E., Foster, J.C., Goldstein, J., et al., 2009. Plasmaspheric density structures and dynamics: properties observed by the CLUSTER and IMAGE missions. Space Sci. Rev. 145 (1–2), 55–106.

Denton, R.E., Wang, Y., Webb, P.A., Tengdin, P.M., Goldstein, J., Redfern, J.A., Reinisch, B.W., 2012. Magnetospheric electron density long-term (> 1 day) refilling rates inferred from passive radio emissions measured by image RPI during geomagnetically quiet times. J. Geophys. Res. Space Phys. 117 (A3). https://doi.org/10.1029/2011JA017274.

Escoubet, C.P., Pedersen, A., Schmidt, R., Lindqvist, P.A., 1997. Density in the magnetosphere inferred from ISEE 1 spacecraft potential. J. Geophys. Res. Space Phys. 102 (A8), 17595–17609.

Gallagher, D.L., Craven, P.D., Comfort, R.H., 2000. Global core plasma model. J. Geophys. Res. Space Phys. 105 (A8), 18819–18833. ISSN 2156-2202. https://doi.org/10.1029/1999JA000241.

Geiger, H., Müller, W., 1928. Elektronenzählrohr zur messung schwächster aktivitäten. Naturwissenschaften 16 (31), 617–618.
Glorot, X., Bordes, A., Bengio, Y., 2011. Domain adaptation for large-scale sentiment classification: a deep learning approach. In: Proceedings of the 28th International Conference on Machine Learning (ICML-11), pp. 513–520.
Goldstein, J., Sandel, B.R., Hairston, M.R., Reiff, P.H., 2003. Control of plasmaspheric dynamics by both convection and sub-auroral polarization stream. Geophys. Res. Lett. 30 (24), 2243.
Goller, C., Kuchler, A., 1996. Learning task-dependent distributed representations by backpropagation through structure. In: IEEE International Conference on Neural Networks, 1996, vol. 1, pp. 347–352.
Grebowsky, J.M., 1970. Model study of plasmapause motion. J. Geophys. Res. 75 (22), 4329–4333.
Gringauz, K.I., 1963. The structure of the ionized gas envelope of earth from direct measurements in the USSR of local charged particle concentrations. Planet. Space Sci. 11 (3), 281–296.
Hebb, D.O., et al., 1949. The Organization of Behavior: A Neuropsychological Theory. Wiley, New York, NY.
Hochreiter, S., Schmidhuber, J., 1997. Long short-term memory. Neural Comput. 9 (8), 1735–1780.
Kletzing, C.A., Kurth, W.S., Acuna, M., MacDowall, R.J., Torbert, R.B., Averkamp, T., Bodet, D., Bounds, S.R., Chutter, M., Connerney, J., et al., 2013. The electric and magnetic field instrument suite and integrated science (EMFISIS) on RBSP. Space Sci. Rev. 179 (1–4), 127–181.
Krall, J., Huba, J.D., Fedder, J.A., 2008. Simulation of field-aligned H+ and He+ dynamics during late-stage plasmasphere refilling. Ann. Geophys. 26 (6), 1507.
Krizhevsky, A., Hinton, G.E., 2011. Using very deep autoencoders for content-based image retrieval. In: ESANN.
Kurth, W.S., De Pascuale, S., Faden, J.B., Kletzing, C.A., Hospodarsky, G.B., Thaller, S., Wygant, J.R., 2015. Electron densities inferred from plasma wave spectra obtained by the waves instrument on Van Allen Probes. J. Geophys. Res. Space Phys. 120 (2), 904–914.
LaBelle, J., Ruppert, D.R., Treumann, R.A., 1999. A statistical study of banded magnetospheric emissions. J. Geophys. Res. Space Phys. 104 (A1), 293–303.
LeDocq, M.J., Gurnett, D.A., Anderson, R.R., 1994. Electron number density fluctuations near the plasmapause observed by the CRRES spacecraft. J. Geophys. Res. Space Phys. 99 (A12), 23661–23671.
Lemaire, J.F., Gringauz, K.I., 1998. The Earth's Plasmasphere. Cambridge University Press, Cambridge, MA.
Li, H., Chen, C.L.P., Huang, H.P., 2000. Fuzzy Neural Intelligent Systems: Mathematical Foundation and the Applications in Engineering. CRC Press, Boca Raton, FL.
Marr, D., Poggio, T., et al., 1976. Cooperative computation of stereo disparity. In: From the Retina to the Neocortex. Birkhäuser, Boston, MA, pp. 239–243.
Mathworks.com, 2015. Choose a Multilayer Neural Network Training Function—MATLAB & Simulink. http://www.mathworks.com/help/nnet/ug/choose-a-multilayer-neural-network-training-function.html.
Mauk, B.H., Fox, N.J., Kanekal, S.G., Kessel, R.L., Sibeck, D.G., Ukhorskiy, A., 2013. Science objectives and rationale for the radiation belt storm probes mission. Space Sci. Rev. 179 (1–4), 3–27.
Mazzella, A.J., 2009. Plasmasphere effects for GPS TEC measurements in North America. Radio Sci. 44 (5).
McCulloch, W.S., Pitts, W., 1943. A logical calculus of the ideas immanent in nervous activity. Bull. Math. Biophys. 5 (4), 115–133.
Mohamed, A.R., Dahl, G.E., Hinton, G., 2012. Acoustic modeling using deep belief networks. IEEE Trans. Audio Speech Lang. Process. 20 (1), 14–22.
Moldwin, M.B., Thomsen, M.F., Bame, S.J., McComas, D., Reeves, G.D., 1995. The fine-scale structure of the outer plasmasphere. J. Geophys. Res. Space Phys. 100 (A5), 8021–8029.
Møller, M.F., 1993. A scaled conjugate gradient algorithm for fast supervised learning. Neural Netw. 6 (4), 525–533.
Moré, J.J., 1978. The Levenberg-Marquardt algorithm: implementation and theory. In: Numerical analysis. Springer, pp. 105–116.
Mosier, S.R., Kaiser, M.L., Brown, L.W., 1973. Observations of noise bands associated with the upper hybrid resonance by the IMP 6 radio astronomy experiment. J. Geophys. Res. 78 (10), 1673–1679.
Nishida, A., 1966. Formation of plasmapause, or magnetospheric plasma knee, by the combined action of magnetospheric convection and plasma escape from the tail. J. Geophys. Res. 71 (23), 5669–5679.
O'Brien, T.P., Moldwin, M.B., 2003. Empirical plasmapause models from magnetic indices. Geophys. Res. Lett. 30 (4), 1152.
Orlova, K., Shprits, Y., Spasojevic, M., 2016. New global loss model of energetic and relativistic electrons based on Van Allen Probes measurements. J. Geophys. Res. Space Phys. 121, 1308–1314.

Park, C.G., Carpenter, D.L., 1970. Whistler evidence of large-scale electron-density irregularities in the plasmasphere. J. Geophys. Res. 75 (19), 3825–3836.

Reeves, G.D., Spence, H.E., Henderson, M.G., Morley, S.K., Friedel, R.H.W., Funsten, H.O., Baker, D.N., Kanekal, S.G., Blake, J.B., Fennell, J.F., et al., 2013. Electron acceleration in the heart of the Van Allen radiation belts. Science 341 (6149), 991–994.

Rosenblatt, F., 1957. The Perceptron, a Perceiving and Recognizing Automaton Project Para. Cornell Aeronautical Laboratory, Buffalo, NY.

Salakhutdinov, R., Hinton, G., 2009. Semantic hashing. Int. J. Approx. Reason. 50 (7), 969–978.

Sheeley, B.W., Moldwin, M.B., Rassoul, H.K., Anderson, R.R., 2001. An empirical plasmasphere and trough density model: CRRES observations. J. Geophys. Res. Space Phys. 106 (A11), 25631–25641.

Shprits, Y.Y., Drozdov, A.Y., Spasojevic, M., Kellerman, A.C., Usanova, M.E., Engebretson, M.J., Agapitov, O.V., Zhelavskaya, I.S., Raita, T.J., Spence, H.E., et al., 2016. Wave-induced loss of ultra-relativistic electrons in the Van Allen radiation belts. Nat. Commun. 7, 12883.

Singh, N., Horwitz, J.L., 1992. Plasmasphere refilling: recent observations and modeling. J. Geophys. Res. Space Phys. 97 (A2), 1049–1079.

Singh, A.K., Singh, R.P., Siingh, D., 2011. State studies of Earth's plasmasphere: a review. Planet. Space Sci. 59 (9), 810–834.

Spasojević, M., Frey, H.U., Thomsen, M.F., Fuselier, S.A., Gary, S.P., Sandel, B.R., Inan, U.S., 2004. The link between a detached subauroral proton arc and a plasmaspheric plume. Geophys. Res. Lett. 31 (4), L04803.

Teunissen, P., 1990. Nonlinear least squares. Manuscripta Geodetica 15 (3), 137–150.

Thorne, R.M., Li, W., Ni, B., Ma, Q., Bortnik, J., Chen, L., Baker, D.N., Spence, H.E., Reeves, G.D., Henderson, M.G., et al., 2013. Rapid local acceleration of relativistic radiation-belt electrons by magnetospheric chorus. Nature 504 (7480), 411.

Trotignon, J.G., Etcheto, J., Thouvenin, J.P., 1986. Automatic determination of the electron density measured by the relaxation sounder on board ISEE 1. J. Geophys. Res. Space Phys. 91 (A4), 4302–4320.

Trotignon, J.G., Rauch, J.L., Décréu, P.M.E., Canu, P., Lemaire, J., 2003. Active and passive plasma wave investigations in the Earth's environment: the cluster/whisper experiment. Adv. Space Res. 31 (5), 1449–1454.

Trotignon, J.G., Décréau, P.M.E., Rauch, J.L., Vallières, X., Rochel, A., Kougblénou, S., Lointier, G., Facskó, G., Canu, P., Darrouzet, F., et al., 2010. The whisper relaxation sounder and the cluster active archive. In: The Cluster Active Archive. Springer, pp. 185–208.

Williams, D.R.G.H.R., Hinton, G., 1986. Learning representations by back-propagating errors. Nature 323 (6088), 533–538.

Xiong, C., Stolle, C., Lühr, H., 2016. The swarm satellite loss of GPS signal and its relation to ionospheric plasma irregularities. Space Weather 14 (8), 563–577.

Zhelavskaya, I., Spasojevic, M., Shprits, Y., Kurth, W., 2016. Automated determination of electron density from electric field measurements on the Van Allen Probes spacecraft. J. Geophys. Res. Space Phys. 121, 4611–4625. https://doi.org/10.1002/2015JA022132.

CHAPTER

13

Classification of Magnetospheric Particle Distributions Via Neural Networks

Vitor M. Souza[*], *Claudia Medeiros*[*], *Daiki Koga*[*], *Livia R. Alves*[*], *Luis E. A. Vieira*[*], *Alisson Dal Lago*[*], *Ligia A. Da Silva*[*], *Paulo R. Jauer*[*], *Daniel N. Baker*[†]

[*]National Institute for Space Research—INPE, São José dos Campos, SP, Brazil [†]University of Colorado Boulder, Boulder, CO, United States

CHAPTER OUTLINE

1 INTRODUCTION

The Earth's inner magnetosphere is a natural laboratory for exploring a vast range of physical phenomena. This region consists of multiple species of charged particles with a

Machine Learning Techniques for Space Weather
https://doi.org/10.1016/B978-0-12-811788-0.00013-5

broad range of energies, which in turn interact with plasma waves within a basically dipolar geomagnetic field configuration near the Earth, thus forming a complex scenario to be analyzed. In this regard, in situ measurements of plasma and magnetic fields by spacecraft orbiting our planet have greatly contributed to the characterization of a myriad of physical processes that ultimately helps our basic understanding of how the Earth's inner magnetosphere "works" for solar wind-terrestrial interactions. Vampola (1997) used electron data at both subrelativistic and relativistic energies acquired from instrumentation onboard the Combined Release and Radiation Effects Satellite (CRRES) spacecraft to train neural networks. The goal was the *prediction* of daily averages of energetic fluxes as a function of electron energy and radial distance in the Van Allen belts region. In fact, much of the machine learning techniques employed in the space weather context are being used for forecasting purposes (see other chapters in this book).

In this chapter, we deal with the implementation of a data clustering technique based on a special kind of neural network referred to as a self-organizing map (SOM) that helps us to identify, in an automated way, the so-called particle pitch angle distributions (PADs) in space-borne data. The algorithm can be applied to any pitch angle-resolved differential flux data from any space-borne mission. The knowledge of PAD shapes provides us with a picture of the physical processes taking place at or prior to the time of the measurement, thus enabling us to better describe the Earth's inner magnetospheric environment.

In this sense, it would be convenient to have a way of identifying an observed PAD structure, both in radial distance and in time, directly from spacecraft data. Previous attempts to identify the PADs automatically are usually based on the statistical fitting of a given PAD shape to some analytical expression that contains either a single parameter or a set of parameters that are meant to incorporate the basic features of the measured PAD shape. Thus, the observed PADs are identified by some ranges of the fitted parameters. The most common of such approaches is the fitting of an observed PAD shape by a $\sin^N(\alpha)$ form where α is the local particle pitch angle (angle of particle velocity vector with respect to the local magnetic field direction) and N the power law index obtained by a least-squares fitting of pitch angle-resolved differential particle fluxes. This method has been widely applied to characterize both ion and electron PADs data in the Van Allen belts region (see e.g., Garcia and Spjeldvik, 1985; Gannon et al., 2007; Ni et al., 2015). Another method for identifying PAD shapes that is also based on statistical fitting is the one employed by Chen et al. (2014) where Legendre polynomials are used as the fitting functions. Unlike the $\sin^N(\alpha)$ method, the Chen et al. algorithm increases the number of fitting parameters to three while providing a confident representation of a number of PAD shapes.

In this chapter we show, as an example, our algorithm's performance on the classification of electron PAD shapes at both relativistic ($\gtrsim$1 MeV) and subrelativistic (464.40 keV) energies observed in the outer Van Allen belt region. To this end, we shall use data from both the Relativistic Electron-Proton Telescope (REPT) instrument (Baker et al., 2013) and the Magnetic Electron Ion Spectrometer (MagEIS) instrument (Blake et al., 2013) onboard NASA's Van Allen Probes mission (Mauk et al., 2013), laying out the steps needed to implement our neural network-based approach on pitch angle-resolved particle flux data from any spacecraft missions. But before that, it is instructive to present a brief introduction of the near-Earth space region where we are going to focus our attention, as well as on the particle distributions present in such a region.

2 A BRIEF INTRODUCTION TO THE EARTH'S MAGNETOSPHERE

The solar wind, a constant stream of charged particles being expelled from the solar atmosphere into the interplanetary medium, travels at speeds that are far greater than the characteristic speeds of magnetized plasmas, namely, the Alfvén speed. Typically, the solar wind speed is 7–9 times greater than the Alfvén speed. Thus, when this super-Alfvénic flow of particles encounters an obstacle, whether it is magnetized or not, a shock front will develop near the obstacle, heating and slowing down the flow to sub-Alfvénic speeds, which helps to deflect the incoming particles from the obstacle. That is precisely what happens in the vicinity of our planet. A collisionless shock wave, known as the bow shock, forms between the Earth and the Sun, with typical distances from the Earth's center ranging from 13 to 15 Earth radii (1 Earth radius—R_E—is approximately 6370 km) (see e.g., Farris and Russell, 1994, and references therein). The shocked solar wind continues to move toward the Earth, where it encounters the magnetopause, which is defined as a current sheet separating the hot (~10 keV) and tenuous (~0.1 cm^{-3}) near-magnetopause magnetospheric plasma from the cold (~300 eV) and dense (~10 cm^{-3}) plasma in the magnetosheath: the region between the bow shock and the magnetopause. The magnetosphere is the near-Earth region, bounded by the magnetopause boundary, where the Earth's magnetic field plays a fundamental role in charged particle dynamics and also in a myriad of physical processes.

A 3-D view of the idealized boundaries representing both the bow shock and the magnetopause is shown in Fig. 1, along with the Geocentric Solar Magnetospheric (GSM) coordinate

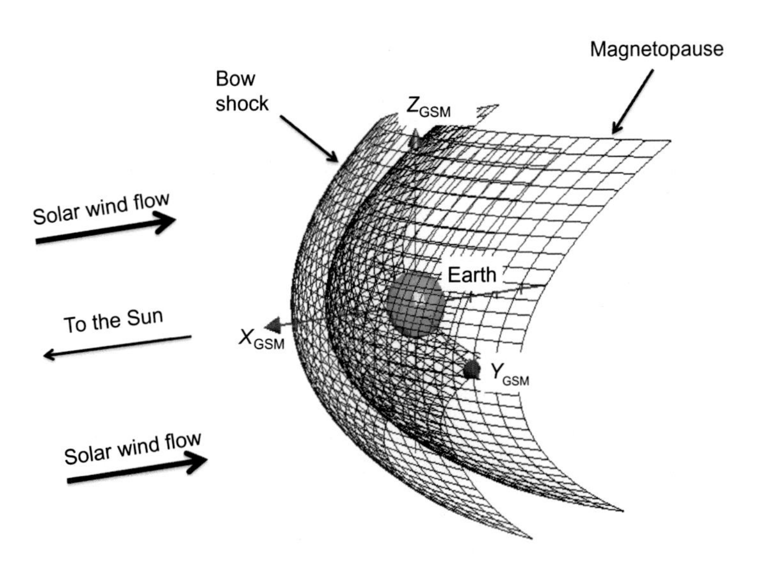

FIG. 1 Out-of-scale, 3D view of idealized bow shock (*blue grading*) and magnetopause (*black grading*) shapes. They are represented by paraboloids having the revolution axis coinciding with the X_{GSM}-axis. *Black thick arrows* indicate the direction of the bulk solar wind flow.

system used throughout this chapter. In this coordinate system, the positive *X*-direction points to the Sun, so the region where $X_{GSM} > 0\,(X_{GSM} < 0)$ is the dayside (nightside). The *Y*-direction is perpendicular to the plane containing the Earth's dipole moment and the *X*-axis, and it is positive in the direction opposite to the Earth's orbital motion around the Sun. Finally the *Z*-direction is normal to both the *X*- and *Y*-axes in a right-handed sense. In the simplified representation of Fig. 1, the magnetosheath region lies between the blue (bow shock) and black (magnetopause) boundaries. Both boundaries are shown as paraboloids having the X_{GSM}-axis as the symmetry axis. Although their actual shapes can depart from a strict paraboloid, spacecraft observations have shown that paraboloidal shapes are indeed representatives of either the bow shock (Farris and Russell, 1994) or the magnetopause (Shue et al., 1998).

The inner magnetosphere, which extends from a few thousand kilometers above the Earth's surface, up to about 7–8 R_E in geocentric distance, is an example of space environment where particles with a broad range of energies can coexist. It is populated by charged particles that can have energies as low as a few electron-volts (eV) or as large as several mega electron-volts (MeV). These particles can originate from the Earth's cold (a few eV) ionospheric population, from the more energetic (few keV) solar wind population, or even from the highly energetic (greater than multi-MeV) galactic cosmic-rays and solar protons. Alternatively, magnetospheric particles can be internally accelerated, either by adiabatic processes, that is, those that conserve the so-called adiabatic invariants (see e.g., Schulz and Lanzerotti, 1974) or by nonadiabatic processes such as the absorption of plasma-wave energy via wave-particle interactions. The Van Allen radiation belts, which are located within the Earth's magnetosphere, are an example of a region where such processes occur. The Van Allen belts typically consist of two donut-shaped regions encircling the Earth. They are mainly populated by trapped energetic electrons and protons drifting across the Earth's magnetic field lines and oscillating back and forth along magnetic field lines that connect to both northern and southern hemispheres. Protons are more frequently encountered in the region corresponding to the inner belt (between about 1 and 3 R_E in radial distance) where one can also find electrons at energies less than 1 MeV (Fennell et al., 2015). Higher energy electrons, however, are found in the outer belt ($\sim$4 to 6 R_E). A self-contained explanation of the magnetosphere's internal regions can be found elsewhere (see e.g., Cassak and Fuselier, 2016).

3 PITCH ANGLE DISTRIBUTIONS IN THE MAGNETOSPHERE

The PAD measured by an in situ spacecraft in the magnetosphere provides us with a 3-D picture of charged particles' direction of propagation relative to the local magnetic field direction at a given instant of time and for a given energy range (for more details see e.g., Fritz et al., 2003). If the instrument detects, for example, an approximately equal amount of particle flux at all pitch angle bins, the resultant shape of the PAD, that is, the geometrical figure that will emerge when plotting particle flux versus pitch angle, will be a somewhat flat line. Such a PAD is known as isotropic or flattop, because fairly equal amounts of particle flux are being detected in nearly all directions. If, on the other hand, the instrument detects more particles arriving perpendicular (90 degrees pitch angle) to the local magnetic field direction, while at the other directions the flux falls off smoothly, one will end up with a bell-shaped PAD that has a local maximum at 90 degrees pitch angle. Such a PAD is known as 90 degrees-peaked, and as with the flattop PAD, the 90 degrees-peaked PAD is associated with specific physical processes, and their identification in both space and time are essential to infer the different physical mechanisms in the magnetosphere occurring at or prior to the measurement (see e.g., Sibeck et al., 1987; Horne et al., 2003). For example, the 90 degrees-peaked PAD is characteristic of both wave-particle interactions and inward radial diffusion, a process through which charged particles further out in the magnetosphere are adiabatically accelerated as they move (drift) toward Earth (see e.g., Gannon et al., 2007, and references therein). Moreover, 90 degrees-peaked PADs can be found throughout both the dayside and nightside magnetosphere within 6 R_E from Earth (see e.g., Sibeck et al., 1987). The so-called butterfly PAD, in which there is a local flux minimum at the 90 degrees pitch angle and two local flux maxima usually located at around 45 degrees and 135 degrees, is typically associated with two mechanisms, namely magnetic field drift-shell splitting (Roederer, 1967), and magnetopause shadowing (e.g., Kim et al., 2008). In the former, particles with distinct equatorial pitch angles (i.e., pitch angle values at the magnetic equator) on the same magnetic field line at the dayside magnetosphere appear spread over a region of field lines at the nightside magnetosphere, and vice versa. In the magnetopause shadowing mechanism, however, equatorially mirroring particles (equatorial pitch angle $\approx$90 degrees) drifting toward the dayside magnetosphere may encounter the magnetopause boundary first, as compared with particles with lower equatorial pitch angles, and hence be lost from the distribution, particularly during geomagnetically disturbed periods. Additionally, quiet-time butterfly PADs can usually be found in the nightside magnetosphere around 5–9 R_E from Earth, whereas during the course of geomagnetic storms and substorms, butterfly PADs can occur at and within geosynchronous orbit, in both the dayside and nightside magnetospheres (Sibeck et al., 1987). The flattop PAD is typically attributed to pitch angle scattering due to wave-particle interactions mediated by either plasmaspheric hiss or electromagnetic ion-cyclotron (EMIC) waves (Shi et al., 2016, and references therein), or by whistler-mode chorus waves (Yang et al., 2016).

Fig. 2 shows a series of electron PADs in the outer Van Allen belt region acquired for different energy channels by the MagEIS instrument onboard the Van Allen Probe A on September 12, 2014 at 06:11:22 Universal Time (UT) at a fixed position. One can see the transition in PAD shapes as the electron energy is increased. For lower energies up to a few hundreds of kiloelectron volt, electron PADs have a pronounced peak at 90 degrees pitch angle, and for lower ($<$90 degrees) and higher ($>$90 degrees and $\leq$180 degrees) pitch angles the flux falls off smoothly. At relativistic ($>$1 MeV) energies, however, the electron PAD shapes undergo a transition from 90 degrees-peaked to nearly isotropic (PAD at 2275 keV energy channel) and then to a butterfly shape (PAD at 2651 keV energy channel). Theoretical and observational investigations have shown that magnetospheric particles with distinct energies may respond differently to the same physical process, and the measured PADs can be an indicator of such a behavior. For example, it has been shown (Shprits et al., 2017) that for the January 17, 2013 geomagnetic storm, megaelectron-volts and multi-MeV electrons in the outer Van Allen radiation belt responded differently to EMIC wave-driven pitch angle scattering. Specifically, Shprits et al. (2017) showed that for the aforementioned geomagnetic storm, only

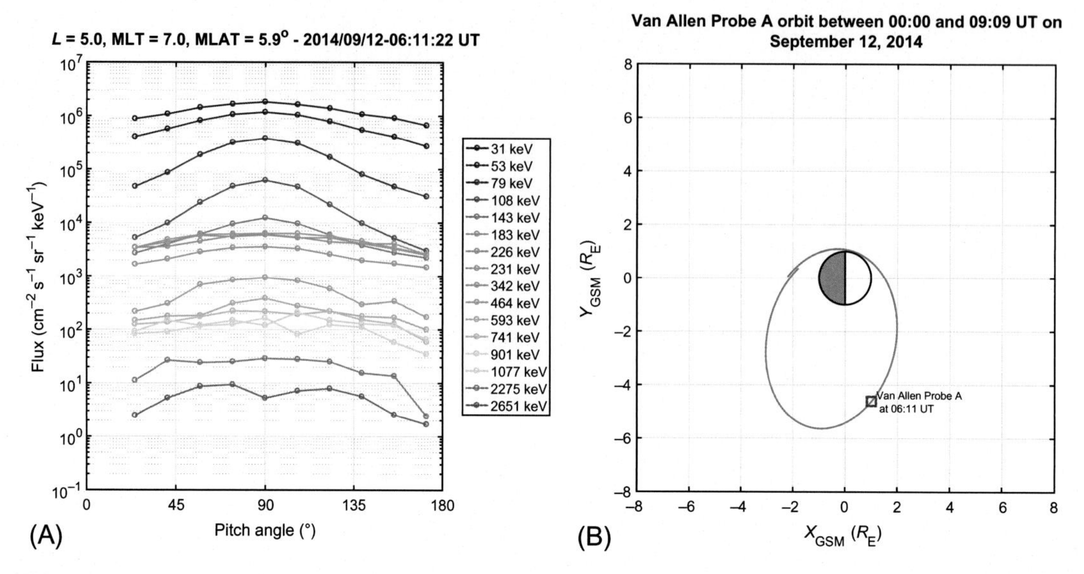

FIG. 2 (A) Electron PADs acquired for different energy channels by the Magnetic Electron Ion Spectrometer (MagEIS) instrument onboard the Van Allen Probe A on September 12, 2014. The Universal Time (UT), magnetic latitude (MLAT) in degrees, magnetic local time (MLT), and radial distance in Earth radius (L) at which these PADs were obtained are shown on the *top*. (B) One Van Allen Probe orbit track on September 12, 2014 and the spacecraft location when data shown in panel (A) was acquired.

the multi-MeV electron population was affected by EMIC waves, and the analysis of PAD shapes associated with numerical modeling was essential to reach this conclusion. Again, because each PAD shape may be associated with a signature of a given physical process, their prompt identification in the Earth's magnetosphere is important.

Nowadays, there is a whole fleet of space missions probing different regions of the magnetosphere. In principle, data from each of those missions can be used for studying particle PAD shapes. Fig. 3 shows one day-long orbital path in the equatorial (XY_{GSM}) plane of the magnetosphere for some of the near-Earth orbiting space missions; namely, one spacecraft from the European Space Agency's Cluster mission (Cluster 3, orange), the National Oceanic and Atmospheric Administration's Geostationary Operational Environmental Satellite (GOES-15, cyan), the probe A from the twin NASA's Van Allen Probes mission (formerly known as Radiation Belt Storm Probe [RBSP], in blue), one spacecraft of the NASA's Magnetospheric Multi Scale (MMS) mission (black), and finally one spacecraft of the NASA's Time History of Events and Macroscale Interactions during Substorms (THEMIS) mission, the THEMIS A spacecraft (dark blue). As an example of our neural network-based methodology we use, in this chapter, particle data from the Van Allen Probe A spacecraft, which is traversing the Van Allen radiation belts region at least four times a day.

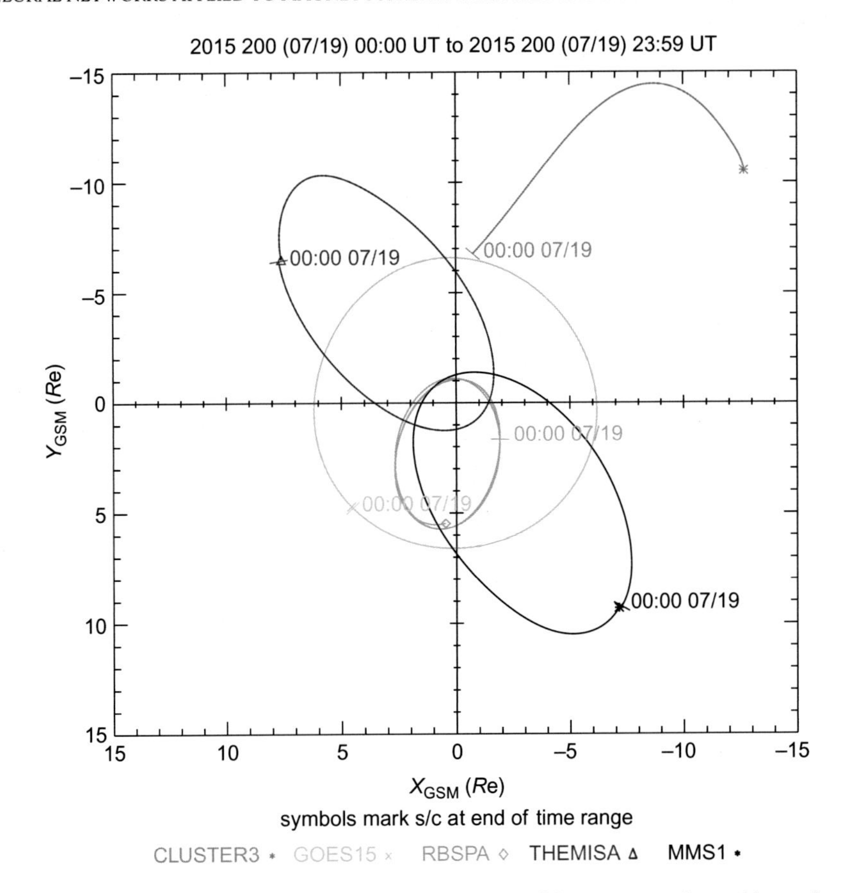

FIG. 3 One day-long orbital path in the equatorial (XY_{GSM}) plane of the magnetosphere of (*orange*) one spacecraft from the Cluster mission, the Cluster 3 spacecraft, (*cyan*) the Geostationary Operational Environmental Satellite (GOES-15), (*blue*) the probe A from the Van Allen Probes mission (formerly known as Radiation Belt Storm Probe [RBSP]), (*black*) one spacecraft of the Magnetospheric Multi Scale (MMS) mission, and finally (*dark blue*) one spacecraft of the Time History of Events and Macroscale Interactions during Substorms (THEMIS) mission, the THEMIS A spacecraft. This plot was generated at the Satellite Situation Center Web (SSCWeb) website: https://sscweb.gsfc.nasa.gov/cgi-bin/Locator_graphics.cgi, hosted at NASA/Goddard Space Flight Center. The Sun is to the *left*.

4 NEURAL NETWORKS APPLIED TO MAGNETOSPHERIC PARTICLE DISTRIBUTION CLASSIFICATION

In this section we deal with the usage of a specific kind of neural network known as Self-Organizing Map (SOM) to perform an automated classification of electron PADs at both relativistic and subrelativistic energies in the outer Van Allen radiation belt region by means of space-borne data acquired by instruments onboard the Van Allen Probes. We start out with some basic concepts and applications of neural networks. Then the SOM-based algorithm is shown to be applicable to Van Allen Probes as well as other spacecraft pitch angle-resolved particle flux data.

4.1 Basic Concepts and Applications of Neural Networks

Neural networks, or more precisely, artificial neural networks (ANNs) have the ability to process information in a similar way as the human brain does. A neural network can be understood as an adaptive machine made up by parallel-working simple processing units known as "neurons," which have the "propensity for storing experiential knowledge and making it available for use" (Haykin, 1999, p. 24). This knowledge, or learning, comes from example, the same way it happens with people in general. The experiential knowledge gained by the neural network is stored in the form of *synaptic weights* (i.e., the interneuron connection strengths). The procedure used to carry out the learning process is the learning algorithm, which is intended to modify, through a series of successive iterations, the synaptic weights of each neuron in such a way that a desired objective can be achieved. Through a learning process one can configure an ANN for a specific application, such as function fitting, clustering, data classification, time-series analysis, pattern recognition, image and data compression, market analysis, code sequence prediction, handwriting recognition, and many others.

Once the way an ANN works is borrowed from the way neurons in our brains work, the implementation of ANN into a computer language will naturally represent oversimplified idealizations of real neural networks, because the precise understanding on how the learning process actually occurs in our brains is still a matter of intense investigation, despite the remarkable breakthroughs made in the area of neurobiology. Nonetheless, research in ANNs has achieved astounding developments, particularly in the space weather context (see other chapters in this book).

4.2 Self-Organizing Map

One of the advantages of using ANNs is the so-called self-organization, whereby the ANN can create its own organization or representation of the information it receives during learning time. The SOM (Kohonen, 1995) is able to cluster together subsets of the input data that hold similar characteristics. This feature of the SOM is an attempt to mimic one of the characteristics of the human brain, which is to receive different sensory inputs, such as motor, visual, auditory, and so forth, and maps them onto corresponding areas of the cerebral cortex in a topologically ordered manner (Haykin, 1999).

The SOM is based on a competitive learning algorithm in which the neurons compete among themselves for the right to respond to a given subset of input data. The learning (training) phase of the SOM is a very important step where the SOM becomes able to *generalize*, that is, it can produce reasonable outputs for inputs not encountered during the learning (training) phase. In usual competitive learning, only a single neuron is active (i.e., responds to a given subset of input data) at any one time. This feature is what makes competitive learning highly suited to unveil statistically salient features that may be used to classify a set of input patterns (Haykin, 1999). In the following section, we provide a brief mathematical description of the SOM's learning algorithm implementation used in this chapter. We note that all neural network-related results shown in this chapter were obtained by means of the MATLAB's Neural Network Toolbox.[1]

[1] https://www.mathworks.com/products/neural-network.html.

4.2.1 Mathematical Background for the SOM's Learning Algorithm Implementation

This section provides more details on the mathematical/computational implementation of the learning algorithm used in the SOM's application presented in this chapter. We start off by describing different types of topological arrangements of the SOM's neurons. Then, we put into more quantitative terms the criteria that define the learning algorithm being used.

In an SOM layer, the neurons are arranged in physical positions according to a topology function. Fig. 4 shows examples of 2-D (A) Cartesian and (B) hexagonal grid topologies. One can also use 1-D, 3-D, or more dimensions. Such grids are used only to define how a learning neuron influences its neighbors. We note, however, that the SOM performance is not sensitive to the exact shape of the neighborhoods.

In a SOM, each neuron is fully connected to the input layer. That is to say that if we have an input vector $\mathbf{x}^t$ with dimensionality Q extracted at random from the whole set of T input vectors, that is, $\mathbf{x}^t = (x_1^t, x_2^t, \ldots, x_Q^t), t = 1, 2, \ldots, T$, each neuron k will have a weight vector $\mathbf{w}_k$[2] also with Q dimensions, $\mathbf{w}_k = (w_{k1}, w_{k2}, \ldots, w_{kQ})$. Each w_{kj} element of the weight vector can be interpreted as the "importance" attributed to the corresponding x_j^t element of the input vector to the output of the neuron k: the larger the weight, the more the influence of the corresponding input on the neuron output (see e.g., Zhelavskaya et al., 2016, and references therein for a concise and clear presentation on the background of neural networks). Here, the weight vectors are initialized by assigning them small values, between 0 and 1, picked from a random number generator.

As previously mentioned, learning in an ANN is the process by which one attempts to modify, at each iteration, the weight vectors of the neurons by applying an increment Δw_{kj} to each of their w_{kj} elements according to a learning rule. In this work, the so-called competitive learning algorithm is employed (Haykin, 1999, Chapter 9), whereby the neurons compete among themselves for the right to respond to a given subset of input data. Specifically, at any one discrete time iteration n, given an input vector $\mathbf{x}^t$, we look for the neuron i whose weight vector $\mathbf{w}_i$ is the "closest" to the input vector. The degree of "proximity" between $\mathbf{x}^t$ and a given weight vector $\mathbf{w}_k$ is conveyed by the reciprocal of the Euclidean distance $d(\mathbf{x}^t, \mathbf{w}_k)$ between them, where

$$d(\mathbf{x}^t, \mathbf{w}_k) = \left(\sum_{j=1}^{Q} (x_j^t - w_{kj})^2 \right)^{1/2} \tag{1}$$

Thus, the smaller the distance, the "nearer" the input and weight vectors are. The neuron i whose weight vector is the "closest" to the input vector $\mathbf{x}^t$ is called the *winning* (or best-matching) neuron. This neuron defines the center of a topological neighborhood within which all the neurons will have their synaptic weights updated, or alternatively where all of them will be *excited*. Consider, for the sake of argument, that the Cartesian SOM topology shown in Fig. 4A was used for a particular application, and at a given time iteration n the neuron located in the (0, 1) coordinate pair was the winning neuron. Its immediate neighbors (i.e., the neurons

[2] Here synaptic weights and weight vectors are used interchangeably. In fact, the weight vector is just a mathematical representation of what is referred to as a synaptic weight.

located at (0, 0), (0, 2), and (1, 1)) define a possible neighborhood of the winning neuron. Thus, in this example, all these neurons would have their synaptic weights updated according to a rule that will be presented below, while the remaining neurons (those at (1, 2) and (1, 0)) would remain with their weight vectors unchanged. In the very beginning of the learning phase of the SOM, however, the topological neighborhood centered on a winning neuron encompasses nearly all neurons in the whole network, and as time progresses (as n increases), the neighborhood sizes shrink considerably, and in the end of the learning process, only the nearest neighbors of the winning neuron, or only the winning neuron itself, will participate in the synaptic weight updating process. At a given iteration n, the amplitude $A_{i,j}(n)$ of the topological neighborhood centered on the winning neuron i is typically defined by a Gaussian function of the form:

$$A_{i,j}(n) = \exp\left(-\left(\frac{l_{i,j}}{\sqrt{2}\sigma(n)}\right)^2\right), \quad n = 0, 1, 2, \ldots \tag{2}$$

where $l_{i,j}$ is the distance between the winning neuron i and a neuron j, and $\sigma(n)$, which is also a function of time, provides the "effective width" of the topological neighborhood. σ measures the degree to which excited neurons in the vicinity of the winning neuron participate in the learning process. As just mentioned, the effective width of the topological neighborhood in an SOM should decrease as time goes by, thus the functional form typically attributed to $\sigma(n)$ is an exponential function:

$$\sigma(n) = \sigma_0 \exp\left(-\frac{n}{\tau_\sigma}\right), \quad n = 0, 1, 2, \ldots \tag{3}$$

where σ_0 is the effective width value at the beginning of the learning phase (i.e., at $n = 0$) and τ_σ is a time constant. For neurons arranged in a 2-D lattice, σ_0 can be set to the "radius" of the lattice, and τ_σ can be defined as $\tau_\sigma = N_{iter}/\log\sigma_0$ (Haykin, 1999), where N_{iter} is the number of iterations the learning algorithm will perform. As the learning procedure is evolved in time, the effective width decreases exponentially, and so does the topological neighborhood $A_{i,j}$.

At each time iteration n, the synaptic weight $\mathbf{w}_j(n)$ of the neuron j, which pertains to the topological neighborhood $A_{i,j}$ of winning neuron i, will have its synaptic weight updated by an amount $\Delta\mathbf{w}_j(n)$, so the new weight vector $\mathbf{w}_j(n+1)$ will be (Kohonen, 1995; Haykin, 1999):

$$\begin{aligned} \mathbf{w}_j(n+1) &= \mathbf{w}_j(n) + \Delta\mathbf{w}_j(n) \\ \Delta\mathbf{w}_j(n) &= \eta(n)A_{i,j}(n)(\mathbf{x}^t - \mathbf{w}_j(n)) \end{aligned} \tag{4}$$

where η is the learning rate that has exactly the same functional form of σ, with the only change being the η_0 parameter instead of σ_0. η_0 is the learning rate value in the beginning of the SOM's learning phase, and it is typically defined as 0.1 (Haykin, 1999). Eq. (4) simply tells us that the updated synaptic weights "move" toward the input vector $\mathbf{x}^t$. The increment $\Delta\mathbf{w}_j$ on the weight vector of the neuron j will be inversely proportional to its distance to the winning neuron i: the closer the neuron in the neighborhood is to the winning neuron, the more its weight vector will move toward the input vector $\mathbf{x}^t$. After many time iterations, the second term on the right-hand side of Eq. (4) becomes negligibly small compared with $\mathbf{w}_j(n)$, meaning that synaptic weights in the neighborhood of the winning neuron become increasingly similar to each other. In other words, neurons that are adjacent in the lattice will tend to have similar weight vectors.

Now we summarize the SOM's learning algorithm implemented here in the following steps:

1. Initialization of the weight vectors $\mathbf{w}_k(n = 0)$ by picking small values, usually between 0 and 1, from a random number generator. It is required that $\mathbf{w}_k(0)$ be different for $k = 1, 2, \ldots, N_n$, where N_n is the number of neurons in the lattice.
2. Choose, at random, a sample $\mathbf{x}^t$ of the input space that has T input vectors with dimensions equal to Q.
3. Determine the winning (best-matching) neuron i at time step n by looking for the neuron that minimizes the Euclidean distance, as given by Eq. (1), between its weight vector and the input vector $\mathbf{x}^t$.
4. Using Eq. (4), update the synaptic weights of all neurons belonging to the topological neighborhood of the winning neuron i. During the learning phase, both the learning rate $\eta(n)$ and the topological neighborhood $A_{i,j}(n)$ are dynamically changed for best results.
5. Continue with step 2 until no noticeable changes in the SOM are observed.

4.2.2 *Geometrical Interpretation of the SOM's Learning Algorithm*

In the context of electron PAD classification, once the SOM is trained, it will be able to classify (or cluster together) other sets of similar inputs. Fig. 5 shows an example that summarizes the learning phase of the SOM used here. The input data (red circles) is a set of six pairs of coordinates that may represent, for example, the locales of distinct features in a 2-D image. In the notation presented in the previous section, each of the six input vectors $\mathbf{x}^t$, $t = 1, 2, \ldots, 6$, would have dimensions Q equal to 2, and the input space would consist of these $T = 6$ input vectors. Initially, an arbitrary number of neurons (blue squares) is chosen and they are spatially arranged in a 2-D lattice according to a hexagonal topology (see Fig. 4B). The number of neurons represents the number of distinct classes/patterns one expects to distinguish in the input data. Usually, one seeks to optimize the quantity of neurons in the sense that the majority of available information in the input data can be captured and clustered into a few classes. Therefore, the number of neurons will vary for different applications. For the example at hand, we chose six neurons, each having $Q = 2$ dimensions. At each iteration of the learning process one tries to minimize the "distance," as given by Eq. (1), between the neurons and a given subset of input data. The winning neuron, that is, the one "closest" to a given subset of input data, as well as its immediate neighbors, have their synaptic weights updated according to Eq. (4), and as a result, the neurons effectively "move" in the input space toward the geometric center of the given subset of input data (see Fig. 5B).

After the learning phase is done, one can either use the trained SOM to classify other sets of similar input data, or the very same set of input data on which the SOM was trained. For the example shown in Fig. 5B, all the neurons will represent (respond to) the pair of input data (red circles) to which they are closest. In the context of pitch angle-resolved particle flux data, the pair of data points represented by neuron 1 could represent, for example, a class of data points that usually appear in electron butterfly-shaped PADs, while those represented by neuron 2 appear in flattop PADs, and so on. Thus, SOMs can be used to cluster a given subset of input data points that carry somewhat similar characteristics.

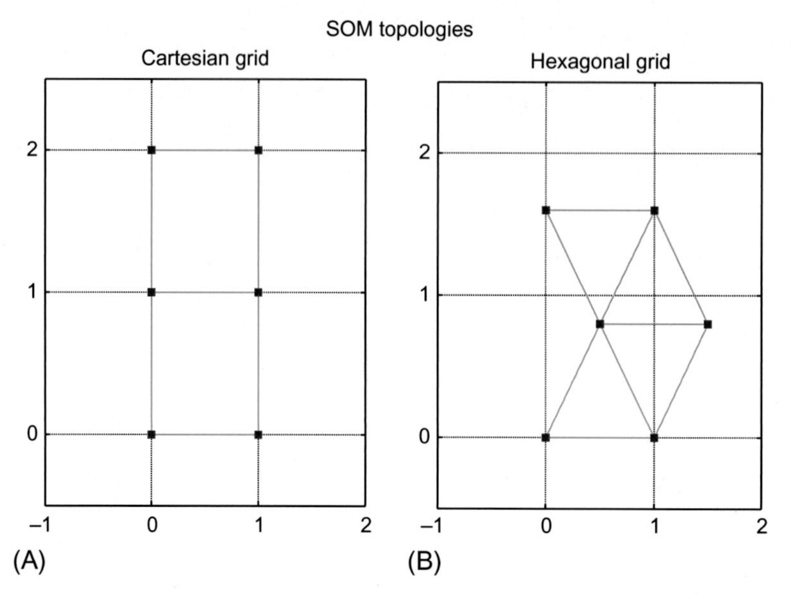

FIG. 4 Self-organized map (SOM) topologies. Example of how a two-by-three array of six neurons (*blue squares*) can be spatially arranged in (A) Cartesian and (B) hexagonal grids. It is noted that the SOM performance is not sensitive to the exact shape of the neighborhoods.

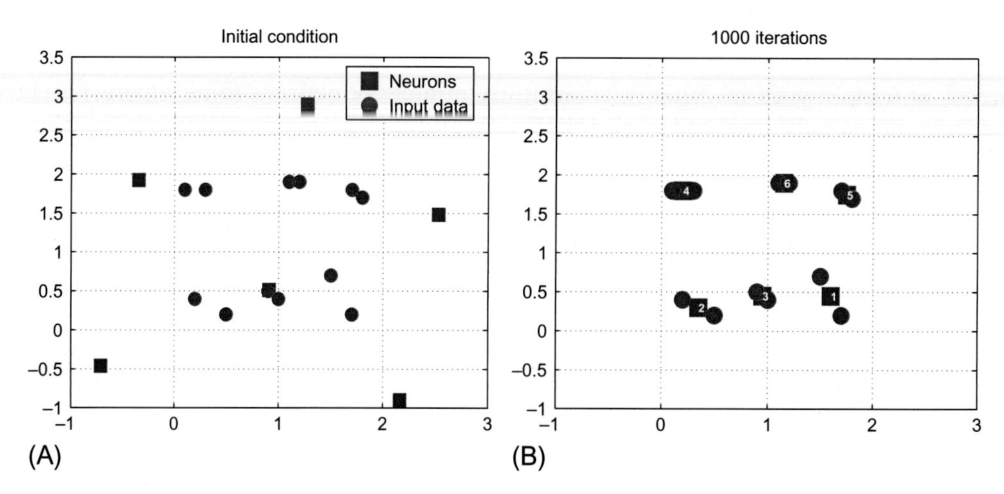

FIG. 5 Learning phase of a self-organized map (SOM) with a two-by-three array of neurons (*blue squares*) arranged in a hexagonal topology (see Fig. 4 for different network topologies). In this setup, a competitive learning algorithm is employed (see text for details). (A) Initially, the neurons are randomly placed in the input space while maintaining the chosen hexagonal topology. When the learning process starts, the neurons begin to "move" in the input space (B) toward local clusters of the input data represented in this example by pairs of *red circles*. At this stage (panel B), all neurons can faithfully represent subsets of input data to which they are closest.

4.3 PAD Classification of Relativistic and Subrelativistic Electrons in the Van Allen Radiation Belts

The preceding procedure for training the SOM and using the trained SOM to classify sets of similar input data is exactly what we do with Van Allen Probes' REPT and MagEIS data, or in fact, what we can potentially do with any PAD data from any spacecraft mission. In the case of PAD data from the Van Allen Probes, first we expose the untrained SOM to a given set of either REPT or MagEIS data that we wish to cluster or organize according to some number of classes. Here, we use electron fluxes at both relativistic and subrelativistic energies as an example, as was done by Souza et al. (2016). The SOM is trained on these particular sets of data so it can identify different classes of electron PAD shapes, for example, butterfly, 90 degrees-peaked, and flattop, within these data sets. It is desirable that the input data used to train the SOM should contain as many different PAD shapes as possible so that the SOM can faithfully classify PAD shapes in other PAD data sets. Fig. 6 addresses this point.

Fig. 6 shows electron flux data at both relativistic (panels A–E) and subrelativistic (panels F–J) energies acquired, respectively, from the REPT and MagEIS instruments onboard the Van Allen Probe A, during the month of November 2014. Panels (A) and (F) present electron flux, in units of particles/(cm^2 s sr MeV) and particles/(cm^2 s sr keV), respectively, as a function of radial distance (L-shell parameter[3]) and time along the Van Allen Probe A orbit. To generate such plots, we discretize both the L-shell range shown into 0.1 L bins, and the time domain into ~9 h bins, which is the time the Van Allen Probes take to complete one translational movement around the Earth. The electron flux is then averaged over one orbit duration (~9 h) at each time/L-shell bin, and each flux value is assigned to a color in a color map, where the bluish (reddish) colors correspond to lower (higher) values. Panels (B) and (G) in Fig. 6 show normalized electron PADs acquired at $L = 5\,R_E$ as a function of time. By "normalized" we mean that at each time stamp we find the pitch angle bin that has the highest flux and then divide all fluxes at all pitch angle bins by the highest flux found. Normalized flux values are also presented using the same color map shown in panels (A) and (F). One can see that during the course of a month, a number of PADs can be detected, particularly because there are appreciable changes in the electron fluxes (see panels A and F), that is, the presence of major decreases in the flux (such as the one on November 4) followed by increases in the flux such as the one in the midday of November 16. Panels (C–E) and (H–J) show examples of PADs obtained in the time instants marked by the vertical black lines in panels (B) and (G), respectively. The larger the amount of different PADs in the input data presented to the SOM, the better the SOM will be able to generalize.

We now elaborate more on how we apply the SOM methodology for electron PAD classification at both relativistic and subrelativistic energies in the Van Allen belts. In fact, the steps provided in the following section can be applied to any spacecraft instrument that yields pitch angle-resolved particle fluxes.

[3] The McIlwain's 1961 L is a parameter describing a particular set of planetary magnetic field lines. An L value gives the radial distance, in Earth radii, in the equatorial plane to the center of the Earth's magnetic dipole.

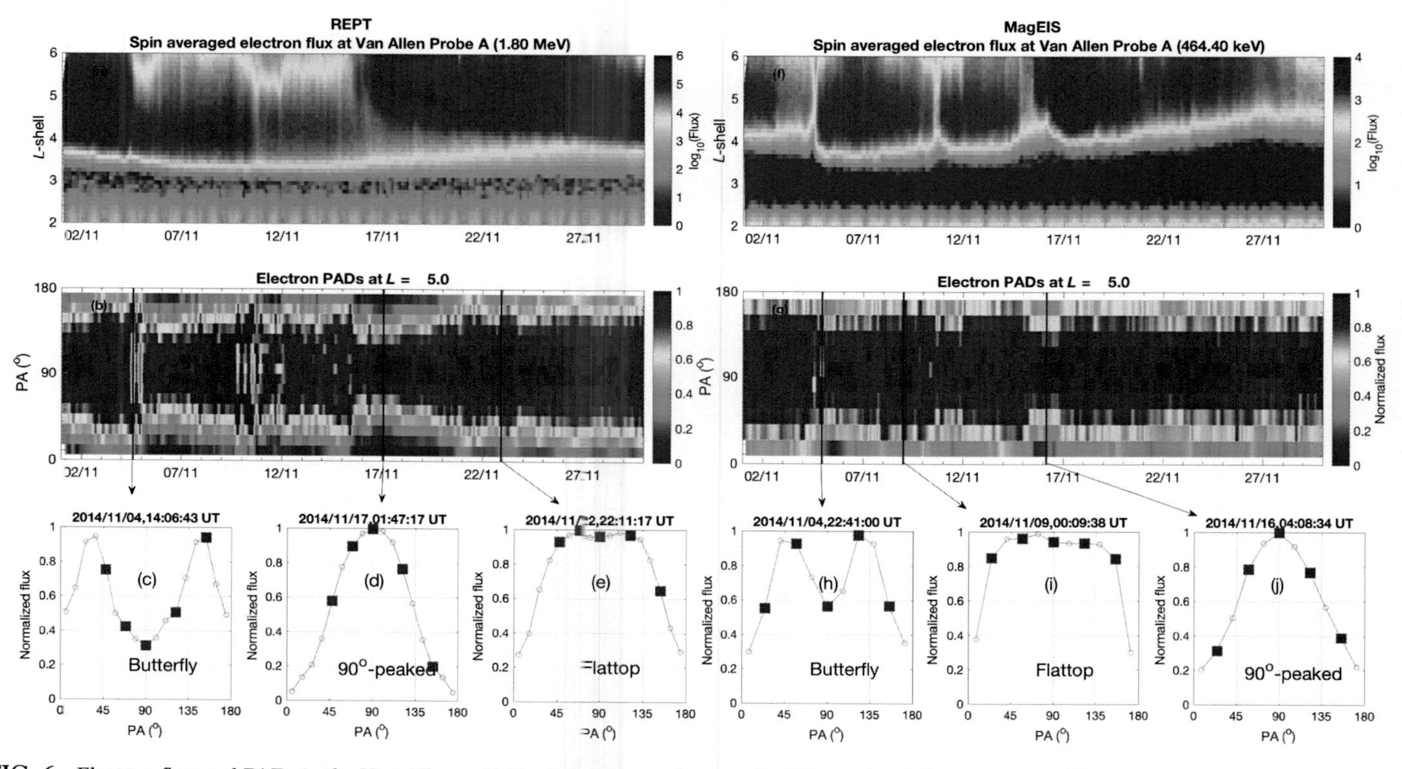

FIG. 6 Electron flux and PADs in the Van Allen radiation belts during the month of November 2014, as measured by both (A–E) Relativistic Electron-Proton Telescope (REPT) and (F–J) Magnetic Electron Ion Spectrometer (MagEIS) instruments onboard the Van Allen Probe A spacecraft. Panels (A–E) and (F–J) refer to electron flux and PADs at relativistic (1.8 MeV) and subrelativistic (464.40 keV) energies, respectively. Panels (B–E) and (G–J) show different PAD shapes that may be present during the course of a month in the Van Allen belts region. For example, one can see the three kinds of PAD shapes shown on panels (C–E) and (H–J) obtained at the moments marked by the *vertical black lines* in panels (B) and (G). The *red squares* on panels (C–E) and (H–J) indicate the pitch angles bins chosen for the learning phase of the SOM. See text for details.

4.3.1 Step 1: Data Choice for the SOM's Training Phase

The first step for applying the method is to properly choose the data set that will be used for training the SOM. By "properly" we emphasize that the input data must contain as many different PAD shapes as possible, so the SOM will be able to distinguish a larger amount of distinct PAD shapes in other data sets. The electron flux data shown in Fig. 6 provide a reasonably good example of a proper data choice for training the SOM (see also Souza et al., 2016, for another good example). In the Van Allen belts context, one would wish to choose a data interval wherein there are lots of changes in the electron flux, similar to those found in Fig. 6A and F.

There is also another criterion for choosing the dataset that has to do with its size. The dataset must not be too large, otherwise the SOM's training phase would take too much time to be completed. If we take an amount of data in the scale of months, we can achieve a reasonably short time for training the SOM ($\lesssim$3.5 min for 1 month), while exposing the SOM to a large enough dataset that may contain a reasonably large number of distinct observed particle PADs. Souza et al. (2016) also investigated the influence that the amount of input data used to train the SOM would have on the data classification. They used three similar datasets that differed mostly on their sizes. Each one of these input data was used for training a SOM with 32 neurons. At the end of each SOM's learning phase, three correspondingly different trained SOMs were obtained, and then used to classify the same dataset of 1.8 MeV energy electron PADs in the Van Allen belts region. Souza et al. noted that all trained SOMs behaved quite similarly when used to classify the same dataset, as long as the data intervals used for training the SOMs present different types of PAD shapes. For the Van Allen belts region, for example, a previous analysis (Souza et al., 2016) has shown that 1 month of either relativistic (REPT) or subrelativistic (MagEIS) electron flux data is sufficient to classify most of the electron PAD shapes found in the literature.

For pitch angle-resolved particle flux data from other spacecraft, however, the input data interval used for training the SOM will heavily depend on the existence of multiple PAD shapes on that interval. Thus, if a, say, 1-day period of data encompasses at least the three kinds of PAD shapes presented in this chapter (see Fig. 6C–E), one can use this dataset for training the SOM, and then the trained SOM to classify/cluster similar datasets.

4.3.2 Step 2: Preparing the Data to Use It as Input to the SOM

Because we are interested in analyzing the PAD shape, we normalize the flux as already mentioned in the beginning of Section 4.3. Here, we express into more mathematical terms how this normalization is done. Pitch angle-resolved particle flux data, hereafter denoted by the letter J, is, in general, a function of pitch angle (α), particle energy (E), and time (t), that is,

$$J = J(\alpha, E, t).$$

Throughout this chapter, a given PAD is classified as butterfly, 90 degrees-peaked, or flattop, according to the following quantitative criterion (Gannon et al., 2007; Ni et al., 2015; Souza et al., 2016):

$$r = \frac{J^{\alpha=90\,\text{degrees}}}{(J^{\alpha=45\,\text{degrees}} + J^{\alpha=135\,\text{degrees}})/2}, \begin{cases} r < 0.9 & \rightarrow \text{butterfly}, \\ r > 1.1 & \rightarrow 90\text{ degrees-peaked}, \\ 0.9 \leq r \leq 1.1 & \rightarrow \text{flattop} \end{cases} \tag{5}$$

where J^{α} corresponds to the in situ measured flux value at a given energy channel E_i, at a given time instant t_j, and at the pitch angle bin α (or at those nearest to the values shown in Eq. (5). The subscripts i and j, where $i = 1, 2, \ldots, M$ and $j = 1, 2, \ldots, P$, refer to the number (M) of energy channels the instrument has ($M = 12$ for REPT and $M = 25$ for MagEIS) and the number (P) of samples the dataset has, respectively.

When we normalize the flux, we do the following procedure: at a given energy channel $E = E_i$ of the instrument and at a given time instant $t = t_j$, we look for the maximum flux value among all pitch angle bins, that is, we find $J_{\max} = \max(J(*, E_i, t_j))$, where the * symbol refers to all pitch angle bins. After that we find the normalized PAD, $J_{\text{norm}}(*, E_i, t_j)$, by doing

$$J_{\text{norm}}(*, E_i, t_j) = J(*, E_i, t_j)/J_{\max}$$

From these normalized flux values, we choose data provided by only five pitch angle bins for either REPT or MagEIS instrument. The chosen pitch angle bins for each instrument are marked as red squares in Fig. 6C–E for REPT and Fig. 6H–J for MagEIS. Specifically, these pitch angles bins are, 47.65 degrees, 68.82 degrees, 90.00 degrees, 121.76 degrees, and 153.53 degrees for REPT, and 24.55 degrees, 57.27 degrees, 90.00 degrees, 122.72 degrees, and 155.45 degrees. In principle, we could have taken data from all pitch angle bins. However, the SOM training (learning) process would be cumbersome, because a larger number of neurons would be required to properly represent all datasets. By taking only the above-mentioned pitch angle bins, we have found a reasonable compromise between computational cost for training the SOM and a fair representation of the observed electron PADs.

Notice that one also needs to choose pitch angle-resolved particle flux data from one of the M possible energy channels that the instrument can provide. In the case of the REPT instrument we have chosen to work only with the first (1.8 MeV) energy channel, because it has the highest measured fluxes. More importantly, in this energy channel, one can already encounter the many types of PAD shapes expected to be found also in the higher energy channels. That is the point that matters the most for the application of the SOM methodology in the classification of particle PADs in the magnetosphere. Regarding the MagEIS instrument, we picked the 10th (464.40 keV) energy channel to train the SOM and subsequently classify the PADs at that energy level. The reason for choosing this particular value are twofold: first, because it represents electrons at subrelativistic energies and second, for the same reason just mentioned: that many types of PAD shapes are already encountered at this energy level.

A sample of the normalized pitch angle-resolved electron flux data for the month of November 2014 used in this chapter as input for training the SOM as well as to classify the corresponding PADs is shown in Table 1.

4.3.3 Step 3: Determining the Classes Outputted by the SOM

After the SOM is trained using a given input data, one can use the trained SOM to either classify/cluster subsets of the very same input data the SOM was trained on into a predefined number of classes, or use the trained SOM to perform data classification/clustering on other similar data sets. An example of the latter approach can be found in Souza et al. (2016). For illustration purposes, we implemented the former procedure using normalized pitch angle-resolved electron flux data from both REPT and MagEIS instruments (see Table 1). A SOM with 32 neurons was trained according to the competitive learning algorithm presented in

TABLE 1 Normalized Pitch Angle-Resolved Electron Flux Data From Both REPT and MagEIS Instruments for the Month of November 2014

			REPT		
Time (t)	J_{norm} **(47 degrees,** E_1**)**	J_{norm} **(68 degrees,** E_1**)**	J_{norm} **(90 degrees,** E_1**)**	J_{norm} **(127 degrees,** E_1**)**	J_{norm} **(157 degrees,** E_1**)**
1	0.699	0.917	1.000	0.791	0.394
2	0.712	0.942	1.000	0.844	0.398
⋮	⋮	⋮	⋮	⋮	⋮
P_R	0	0	0	0.999	0
			MagEIS		
Time (t)	J_{norm} **(24 degrees,** E_{10}**)**	J_{norm} **(57 degrees,** E_{10}**)**	J_{norm} **(90 degrees,** E_{10}**)**	J_{norm} **(122 degrees,** E_{10}**)**	J_{norm} **(155 degrees,** E_{10}**)**
1	0.816	1.000	0.934	0.882	0.703
2	0.810	0.970	0.996	1.000	0.713
⋮	⋮	⋮	⋮	⋮	⋮
P_M	0	0.400	0.400	0	0.150

Notes: Each dataset was used for training a SOM and each trained SOM subsequently used to classify the corresponding PADs in the dataset on which they were trained (see Fig. 9). The integer numbers $P_R = 210{,}179$ *and* $P_M = 147{,}677$ *refer to the total number of samples, per pitch angle bin, used (excluding bad data values) for the REPT and MagEIS datasets, respectively.* $E_1 = 1.80$ *MeV and* $E_{10} = 464.40$ *keV refer to the 1st and 10th energy channels of the REPT and MagEIS instruments, respectively. Whenever a zero (0) is present in the dataset, that simply means that no measurable flux was detected, that is, there was no appreciable flux measurement above the background noise, and then a zero flux value was assigned.*

Section 4.2.1. As previously mentioned, the SOM classifies, in an automated way, subsets of the input data that hold similar characteristics. Fig. 7 conveys this idea by showing the results of electron PAD classification of REPT data for the month of November 2014. Some of these 32 classes will satisfy the conditions of Eq. (5) and so we can label them as butterfly, flattop, or 90 degrees-peaked; other classes will not satisfy any of the conditions in Eq. (5) and we do not include these in the three categories of interest here. In the case at hand, the SOM clusters data characteristic of 90 degrees-peaked distributions into five classes (21, 22, 29, 30, and 31), butterfly-like distributions into seven classes (8, 10, 11, 12, 14, 16, and 24), and flattop-like distributions into one class (32). Each one of these PADs was identified according to the criterion expressed by Eq. (5). A similar analysis was made for MagEIS data as well.

Notice that the SOM method can capture nuances among the butterfly-like distributions, that is, while some of the classes present a normalized local minimum flux value at 90 degrees pitch angle above 0.5 (classes 12, 14, 16, and 24), others are below this value (classes 8 and 11). Likewise, the SOM captured subtle differences in the 90 degrees-peaked distributions (compare, for example, classes 21, 22, and 31). Because each PAD shape may be associated with specific physical processes, much of the potential of the SOM classification method can be explored to help on the characterization of the physical phenomena, giving rise to the analyzed PAD shapes. For our purposes here, however, we will not distinguish between the different types of butterflies or 90 degrees-peaked distributions, but rather gather all similar types of PAD shapes into larger clusters representing, for example, 90 degrees-peaked-like distributions or butterfly-like distributions. In what follows, classes 21, 22, 29, 30, and 31 will be considered as a single larger cluster of the input data, where 90 degrees-peaked-like distributions appear, while classes 8, 10, 11, 12, 14, 16, and 24 are considered as another single larger cluster where butterfly-like distributions are present. Again, a similar analysis was done for electron PAD classification using MagEIS data.

One can also investigate the spatial distributions of the three major classes of PADs identified by the SOM method (see Fig. 8). In this example, we used only pitch angle-resolved electron flux data from the first (1.8 MeV) energy channel of the REPT instrument onboard Van Allen Probe A. We chose the January 1, 2014 to October 1, 2015 period to analyze the occurrence rate of electron PADs throughout the outer radiation belt. Because the Van Allen Probes possess low inclination (~10 degrees) orbits, both spacecraft are often near, or at, the magnetic equator, thus the spatial distribution of electron PADs that will be shown as follows refer to those at the near equatorial magnetosphere. The aforementioned 22-month period ensures that the Van Allen Probes covered all sectors of the equatorial, inner magnetosphere, that is, both dayside and nightside regions. That is tantamount to saying that the spacecraft covered all magnetic local times (MLT). In this system of reference which is derived considering the Earth's magnetic field as a pure dipole, the magnetic equator plane is divided into 24 sectors/bins, each with a 15 degrees-width (24×15 degrees $= 360$ degrees, i.e., the whole plane). The 12 MLT (00 MLT) mark is placed along the Earth-Sun line in the dayside (nightside) region (see Fig. 8). Thus, in such a reference system, one can refers to different sectors of the magnetosphere by using the corresponding MLT.

Fig. 8 shows the occurrence rate of (A) 90 degrees-peaked, (B) butterfly, and (C) flattop electron PADs as a function of radial distance and MLT for the 1 year and 10 months period previously mentioned. The occurrence rate is obtained as follows. At a given MLT and radial distance bin, say 11–12 MLT and 4–4.5 R_E, we count the number of times N_T, during the

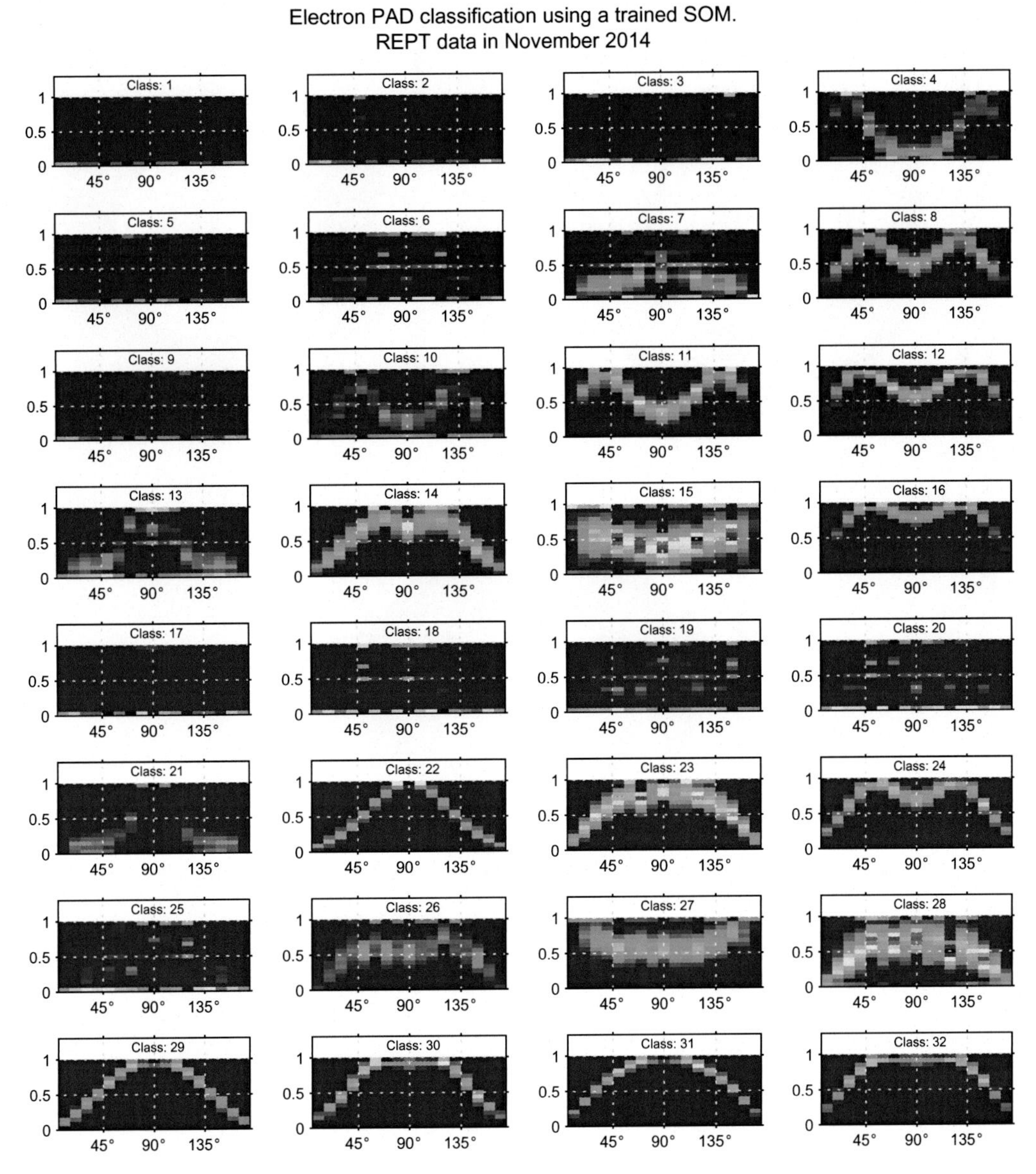

FIG. 7 Electron PAD classification for the whole month of November 2014 using the SOM method. An SOM with 32 neurons/classes was trained using normalized pitch angle-resolved electron fluxes (see Table 1) from the first (1.80 MeV) energy channel of the REPT instrument onboard the Van Allen Probe A. After the training phase, the SOM was used to classify/cluster the same input data on which the SOM was trained. The SOM classification method was able to identify three kinds of electron PADs shapes commonly found in the literature, namely, 90 degrees-peaked (classes 21, 22, 29, 30, and 31), butterfly-like distributions (8, 10, 11, 12, 14, 16, and 24), and flattop-like distributions (32). The *vertical and horizontal axes* show normalized electron flux data and pitch angles, respectively. The *colors at each plot* mean the fraction of time that the normalized flux at a given pitch angle obtains the value on the y-axis for the set of training vectors that are mapped to each class/neuron. Thus, if we add up the values at a given pitch angle for a single class, we would obtain 1.

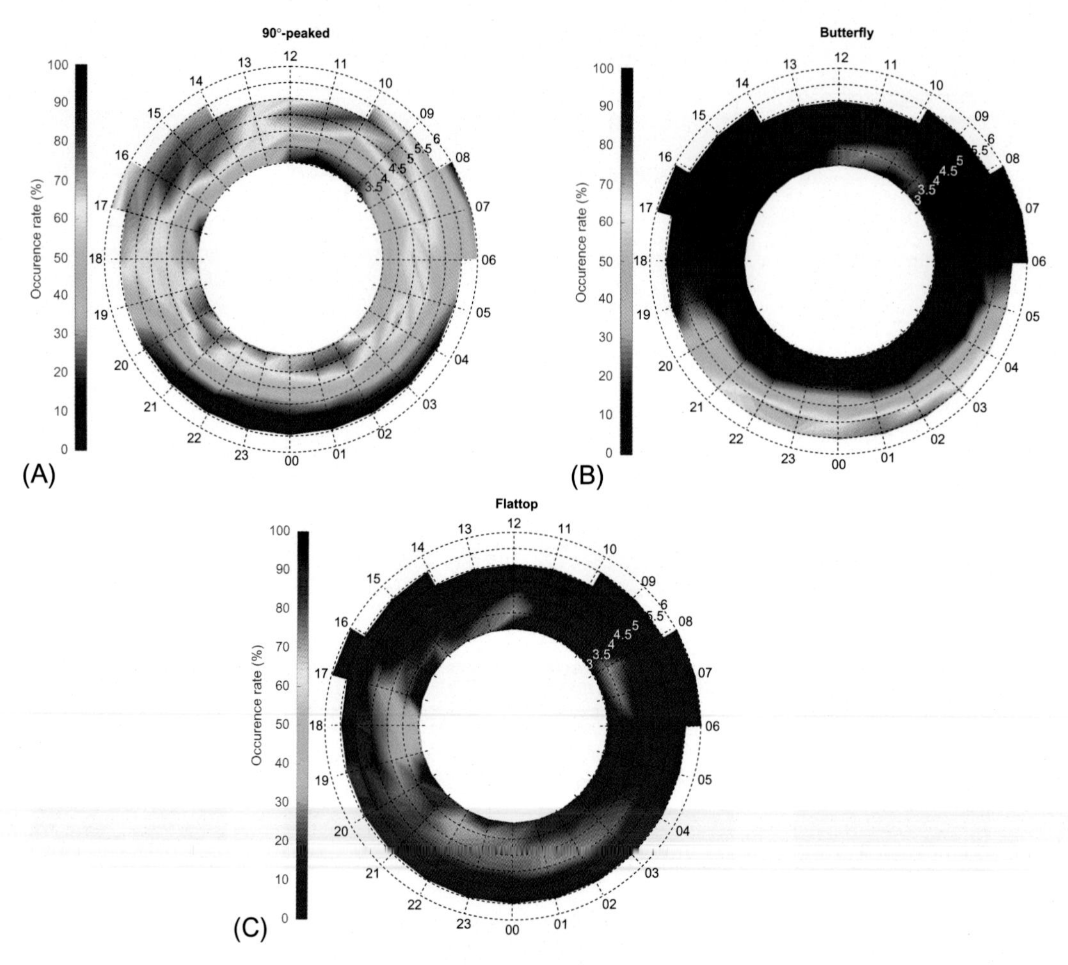

FIG. 8 Occurrence rate of (A) 90 degrees-peaked, (B) butterfly, and (C) flattop PADs in the Van Allen belts for the January 1, 2014 to October 1, 2015 period. PADs were identified using the SOM method on pitch angle-resolved 1.8 MeV energy electron flux data from the REPT instrument onboard Van Allen Probe A. In each panel the PADs are seen as a function of radial distance (from 3 to 6 R_E in 0.5 R_E steps), and magnetic local time (MLT), where the 06–12–18 MLT interval corresponds to the dayside region of the equatorial magnetosphere, whereas the 06–00–18 MLT interval to the nightside region. The Sun is located at the *top of each panel*.

22-month period just mentioned, the Van Allen Probe A was located within this bin. At each of these times, we use the SOM algorithm to determine the shapes of the measured 1.8 MeV energy electron PADs, and then to obtain the number of times a 90 degrees-peaked, butterfly, or flattop PAD is observed. The number of times that 90 degrees-peaked, butterfly, and flattop PADs are observed at a given MLT and radial distance bin are referred to as N_P, N_B, and N_F, respectively. The occurrence rate of a particular electron PAD shape at a given MLT and radial distance bin is then acquired as the ratio N_{shape}/N_T multiplied by 100 in order to provide a percentage value. The subscript "shape" can be "P" (90 degrees-peaked), "B" (butterfly), or "F" (flattop). Blank regions in Fig. 8 correspond to zones where there was an insufficient number of Van Allen Probe A passes through an MLT and radial distance bin, that is, less than 5% of the maximum N_T value for the whole analyzed interval.

Fig. 8 shows that 90 degrees-peaked distributions are often seen throughout the outer belt, except in the 05–20 MLT band at radial distances larger than about 5 R_E, where butterfly PADs are expected to have a higher occurrence rate due to drift shell-splitting (e.g., Sibeck et al., 1987, and references therein). Fig. 8 shows a good example of the capabilities of the SOM method in the characterization of particle PADs in the inner magnetosphere. The same methodology can be applied to any space platform orbiting different sectors of the vast magnetospheric environment, as long as the spacecraft is equipped with instrumentation that can provide pitch angle-resolved particle fluxes.

4.3.4 Step 4: Displaying Clustered Particle PAD Shapes as a Function of Radial Distance and Time

On one hand, the format wherein the SOM output was presented in Fig. 7 is useful to check which types of PAD shapes the SOM is able to identify within a given input data. On the other hand, one cannot localize in either space or time where a given PAD shape was detected, and that information is crucial to interpreting the underlying physical processes. Thus, the following describes how we proceed to present the SOM classification outputs as a function of the spacecraft's (radial) distance relative to the Earth's dipole center and time.

At a given instant of time the SOM outputs provide to which class a particular subset of the input data was attributed. Consider, for example, the five normalized electron flux values presented in the first row of Table 1 as being such a subset of the input data. When this particular subset is presented to the SOM, the SOM will classify this collection of data as belonging to the class number 29 of Fig. 7, which corresponds to a 90 degrees-peaked like PAD. Because we already know the time at which the particle flux measurement was made, as well as the spacecraft position relative to Earth,[4] one can assign to that space/time location a particular PAD shape, which in that case would be a 90 degrees-peaked distribution. In that way, one can build up a map that is very similar to those shown in Fig. 6A and F, but instead of a colormap representing levels of particle flux, it would represent different types of PAD shapes.

Fig. 9C and F shows how these radial distance versus time maps of particle PAD shapes look. Specifically, these maps show the results of our SOM classification method applied to (panel C) normalized pitch angle-resolved electron fluxes at the 1.8 MeV energy channel from the REPT instrument and (panel F) normalized pitch angle-resolved electron fluxes at the 464.40 keV energy channel from the MagEIS instrument. The vertical and horizontal axes represent radial distance (as given by the *L*-shell parameter) in units of R_E and time, respectively. The data set used spans the whole month of November 2014. Wherever there is a red color in Fig. 9C, electron butterfly PADs like those shown in classes 8, 10, 11, 12, 14, 16, and 24 of Fig. 7 are present. A yellow color in Fig. 9C mark electron 90 degrees-peaked PADs such as those shown in classes 21, 22, 29, 30, and 31 of Fig. 7. Light-blue colors refer to flattop distributions, while the dark-blue colors to any other PAD types that do not fall within the three major classes of PADs being analyzed here.

[4] Information regarding spatial location as well as in situ data for a large number of either past or current spacecraft missions can be readily found on the following website https://cdaweb.sci.gsfc.nasa.gov/index.html/.

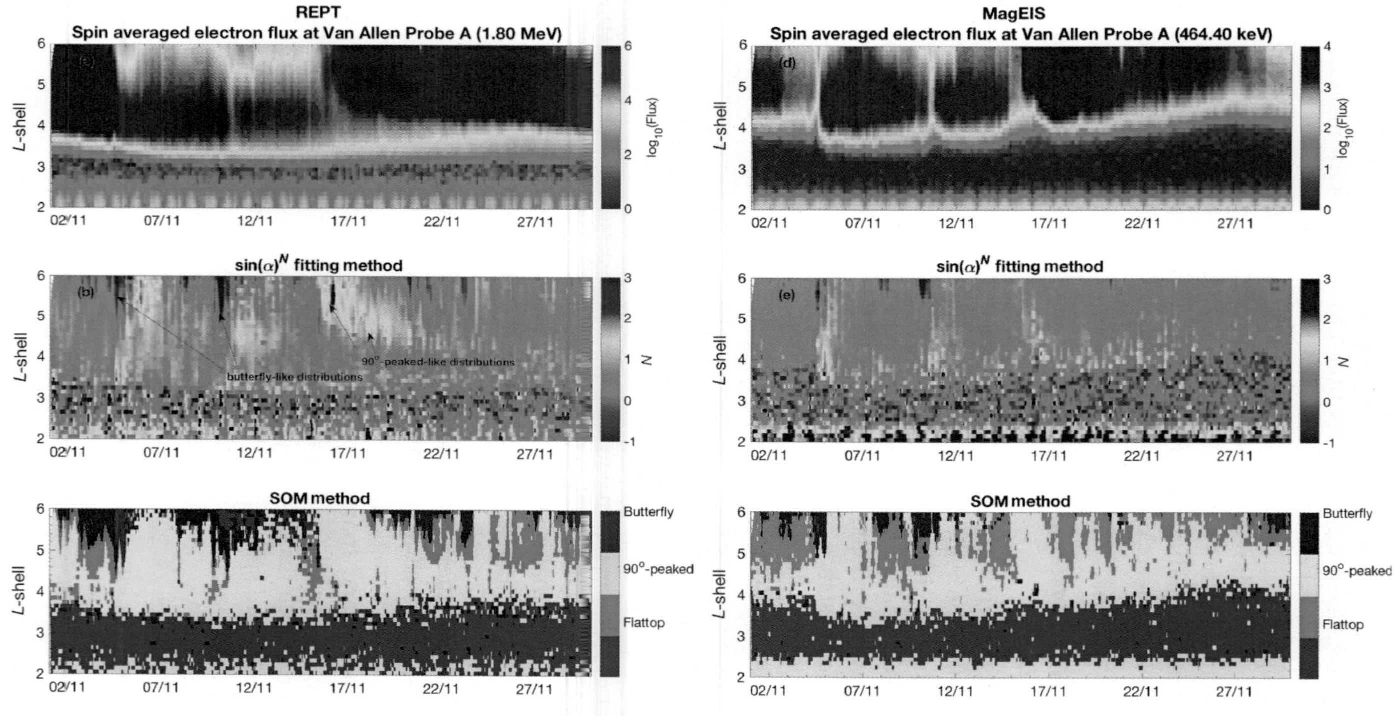

FIG. 9 Electron PAD classification in the Van Allen belts region during the month of November 2014. Two methods are used: the statistical fitting-based $\sin(\alpha)^N$ method (panels B and E), and the neural network (SOM)-based method (panels C and F). The electron PAD classification is done for PADs obtained at both relativistic (REPT, panels A–C) and subrelativistic (MagEIS, panel D–F) energies. Electron fluxes as a function of radial distance (*L*-shell) and time are also presented for reference (panels A and D).

We qualitatively compare our SOM method with a standard PAD shape classification approach that is based on the statistical fitting of PAD shapes to a $\sin^N(\alpha)$ form, where α is the particle pitch angle and N the power law index obtained by a least-squares fitting of pitch angle-resolved particle fluxes. Comparing panels (B and C) and (E and F), one can see that both methods agree reasonably well (compare, for example, the dark-blue regions in panels B and E with red regions in panels C and F). We marked with black arrows some regions in panel (B) showing different types of electron PADs present at that time and L-shell. Looking at the same regions in Fig. 9C, it is clear that the SOM method attributes the same classification for those electron PADs as the $\sin^N(\alpha)$ fitting method. The same argument can be applied when comparing the SOM and $\sin^N(\alpha)$ methods performances with MagEIS data. Also, the SOM method enables a clearer distinction between the electron PADs as compared with the $\sin^N(\alpha)$ fitting method.

With this and the previous three steps mentioned herein, one can apply the SOM method to any pitch angle-resolved particle flux data from any spacecraft mission.

5 SUMMARY

Neural networks can be applied to many areas of active scientific research such as physics, statistics, and neurobiology, among others. In this chapter, we dealt with a special kind of neural network referred to as a SOM, which is able to detect statistically salient features in a given data set provided as input to the SOM. Specifically, we showed that the SOM can be successfully applied to organize pitch angle-resolved particle flux data into at least three major classes of particle PADs commonly found in the literature, namely, 90 degrees-peaked, butterfly, and flattop distributions. The SOM can also identify subtle differences among, for example, 90 degrees-peaked-like distributions, and this is an aspect that can be further explored to investigate possible physical mechanisms occurring in the Earth's magnetosphere that may give rise to those similar, yet distinct, PAD shapes.

An important feature of the SOM method shown here is its versatility, in the sense that it can, in principle, be applied to multiple spacecraft missions, provided there is availability of pitch angle-resolved particle flux data. We showed that for electron flux data acquired by both REPT and MagEIS instruments onboard the Van Allen Probes, the SOM method was able to distinguish the different types of PAD shapes present in either data set. Moreover, our SOM particle PAD classification method was in reasonably good agreement with the results given by a standard statistical fitting-based approach, which in turn yields confidence in the SOM's results. Additionally, by using the SOM method one can promptly identify, in both radial distance and time, a given particle PAD of interest, also enabling a clearer distinction between the analyzed PADs.

ACKNOWLEDGMENTS

V.M. Souza would like to thank the São Paulo Research Foundation (FAPESP) grant 2014/21229-9 for support, and the REPT and MagEIS teams for providing high-quality science data.

References

Baker, D.N., Kanekal, S.G., Hoxie, V.C., Batiste, S., Bolton, M., Li, X., Elkington, S.R., Monk, S., Reukauf, R., Steg, S., Westfall, J., Belting, C., Bolton, B., Braun, D., Cervelli, B., Hubbell, K., Kien, M., Knappmiller, S., Wade, S., Lamprecht, B., Stevens, K., Wallace, J., Yehle, A., Spence, H.E., Friedel, R., 2013. The Relativistic Electron-Proton Telescope (REPT) instrument on board the Radiation Belt Storm Probes (RBSP) spacecraft: characterization of Earth's radiation belt high-energy particle populations. Space Sci. Rev. 179 (1–4), 337–381.

Blake, J.B., Carranza, P.A., Claudepierre, S.G., Clemmons, J.H., Crain, W.R., Dotan, Y., Fennell, J.F., Fuentes, F.H., Galvan, R.M., George, J.S., Henderson, M.G., Lalic, M., Lin, A.Y., Looper, M.D., Mabry, D.J., Mazur, J.E., McCarthy, B., Nguyen, C.Q., O'Brien, T.P., Perez, M.A., Redding, M.T., Roeder, J.L., Salvaggio, D.J., Sorensen, G.A., Spence, H.E., Yi, S., Zakrzewski, M.P., 2013. The Magnetic Electron Ion Spectrometer (MagEIS) instruments aboard the radiation belt storm probes (RBSP) spacecraft. Space Sci. Rev. 179 (1), 383–421. ISSN 1572-9672. https://doi.org/10.1007/s11214-013-9991-8.

Cassak, P.A., Fuselier, S.A., 2016. Reconnection at Earth's dayside magnetopause. In: Gonzalez, W., Parker, E. (Eds.), Magnetic Reconnection: Concepts and Applications. Springer, New York, NY.

Chen, Y., Friedel, R.H.W., Henderson, M.G., Claudepierre, S.G., Morley, S.K., Spence, H.E., 2014. REPAD: an empirical model of pitch angle distributions for energetic electrons in the Earth's outer radiation belt. J. Geophys. Res. Space Phys. 119 (3), 1693–1708. ISSN 2169-9402. https://doi.org/10.1002/2013JA019431.

Farris, M.H., Russell, C.T., 1994. Determining the standoff distance of the bow shock: Mach number dependence and use of models. J. Geophys. Res. Space Phys. 99 (A9), 17681–17689. ISSN 2156-2202. https://doi.org/10.1029/94JA01020.

Fennell, J.F., Claudepierre, S.G., Blake, J.B., O'Brien, T.P., Clemmons, J.H., Baker, D.N., Spence, H.E., Reeves, G.D., 2015. Van Allen Probes show that the inner radiation zone contains no MeV electrons: ECT/MagEIS data. Geophys. Res. Lett. 42 (5), 1283–1289. ISSN 1944-8007. https://doi.org/10.1002/2014GL062874.

Fritz, T.A., Alothman, M., Bhattacharjya, J., Matthews, D.L., Chen, J., 2003. Butterfly pitch-angle distributions observed by ISEE-1. Planet. Space Sci. 51 (3), 205–219. ISSN 0032-0633. https://doi.org/10.1016/S0032-0633(02)00202-7. Available from: http://www.sciencedirect.com/science/article/pii/S0032063302002027.

Gannon, J.L., Li, X., Heynderickx, D., 2007. Pitch angle distribution analysis of radiation belt electrons based on Combined Release and Radiation Effects Satellite Medium Electrons A data. J. Geophys. Res. Space Phys. 112 (A5), A05212. ISSN 2156-2202. https://doi.org/10.1029/2005JA011565.

Garcia, H.A., Spjeldvik, W.N., 1985. Anisotropy characteristics of geomagnetically trapped ions. J. Geophys. Res. Space Phys. 90 (A1), 047–050. ISSN 2156-2202. https://doi.org/10.1029/JA090iA01p00047.

Haykin, S., 1999. Neural Networks: A Comprehensive Foundation, second ed. Pearson Prentice Hall, Upper Saddle River, NJ.

Horne, R.B., Meredith, N.P., Thorne, R.M., Heynderickx, D., Iles, R.H.A., Anderson, R.R., 2003. Evolution of energetic electron pitch angle distributions during storm time electron acceleration to megaelectron volt energies. J. Geophys. Res. Space Phys. 108 (A1), SMP 11-1–SMP 11-13. ISSN 2156-2202. https://doi.org/10.1029/2001JA009165.

Kim, K.C., Lee, D.Y., Kim, H.J., Lyons, L.R., Lee, E.S., Öztürk, M.K., Choi, C.R., 2008. Numerical calculations of relativistic electron drift loss effect. J. Geophys. Res. Space Phys. 113 (A9), A09212. ISSN 2156-2202. https://doi.org/10.1029/2007JA013011.

Kohonen, T., 1995. Self-Organizing Maps. Springer, Berlin.

Mauk, B.H., Fox, N.J., Kanekal, S.G., Kessel, R.L., Sibeck, D.G., Ukhorskiy, A., 2013. Science objectives and rationale for the Radiation Belt Storm Probes mission. Space Sci. Rev. 179 (1–4), 3–27.

McIlwain, C.E., 1961. Coordinates for mapping the distribution of magnetically trapped particles. J. Geophys. Res. 66 (11), 3681–3691. ISSN 2156-2202. https://doi.org/10.1029/JZ066i011p03681.

Ni, B., Zou, Z., Gu, X., Zhou, C., Thorne, R.M., Bortnik, J., Shi, R., Zhao, Z., Baker, D.N., Kanekal, S.G., Spence, H.E., Reeves, G.D., Li, X., 2015. Variability of the pitch angle distribution of radiation belt ultrarelativistic electrons during and following intense geomagnetic storms: Van Allen Probes observations. J. Geophys. Res. Space Phys. 120 (6), 4863–4876. ISSN 2169-9402. https://doi.org/10.1002/2015JA021065.

Roederer, J.G., 1967. On the adiabatic motion of energetic particles in a model magnetosphere. J. Geophys. Res. 72 (3), 981–992. ISSN 2156-2202. https://doi.org/10.1029/JZ072i003p00981.

Schulz, M., Lanzerotti, L.J., 1974. Particle diffusion in the radiation belts. In: Physics and Chemistry in Space. Springer, New York, NY.

Shi, R., Summers, D., Ni, B., Fennell, J.F., Blake, J.B., Spence, H.E., Reeves, G.D., 2016. Survey of radiation belt energetic electron pitch angle distributions based on the Van Allen Probes MagEIS measurements. J. Geophys. Res. Space Phys. 121 (2), 1078–1090. ISSN 2169-9402. https://doi.org/10.1002/2015JA021724.

Shprits, Y.Y., Kellerman, A., Aseev, N., Drozdov, A.Y., Michaelis, I., 2017. Multi-MeV electron loss in the heart of the radiation belts. Geophys. Res. Lett. 44 (3), 1204–1209. ISSN 1944-8007. https://doi.org/10.1002/2016GL072258.

Shue, J.H., Song, P., Russell, C.T., Steinberg, J.T., Chao, J.K., Zastenker, G., Vaisberg, O.L., Kokubun, S., Singer, H.J., Detman, T.R., Kawano, H., 1998. Magnetopause location under extreme solar wind conditions. J. Geophys. Res. Space Phys. 103 (A8), 17691–17700. ISSN 2156-2202. https://doi.org/10.1029/98JA01103.

Sibeck, D.G., McEntire, R.W., Lui, A.T.Y., Lopez, R.E., Krimigis, S.M., 1987. Magnetic field drift shell splitting: cause of unusual dayside particle pitch angle distributions during storms and substorms. J. Geophys. Res. Space Phys. 92 (A12), 13485–13497. ISSN 2156-2202. https://doi.org/10.1029/JA092iA12p13485.

Souza, V.M., Vieira, L.E.A., Medeiros, C., Da Silva, L.A., Alves, L.R., Koga, D., Sibeck, D.G., Walsh, B.M., Kanekal, S.G., Jauer, P.R., Rockenbach, M., Dal Lago, A., Silveira, M.V.D., Marchezi, J.P., Mendes, O., Gonzalez, W.D., Baker, D.N., 2016. A neural network approach for identifying particle pitch angle distributions in Van Allen Probes data. Space Weather 14 (4), 275–284. ISSN 1542-7390. https://doi.org/10.1002/2015SW001349.

Vampola, A.L., 1997. Outer zone energetic electron environment update. In: Conference on the High Energy Radiation Background in Space. Workshop Record, pp. 128–136.

Yang, C., Su, Z., Xiao, F., Zheng, H., Wang, Y., Wang, S., Spence, H.E., Reeves, G.D., Baker, D.N., Blake, J.B., Funsten, H.O., 2016. Rapid flattening of butterfly pitch angle distributions of radiation belt electrons by whistler-mode chorus. Geophys. Res. Lett. 43 (16), 8339–8347. ISSN 1944-8007. https://doi.org/10.1002/2016GL070194.

Zhelavskaya, I.S., Spasojevic, M., Shprits, Y.Y., Kurth, W.S., 2016. Automated determination of electron density from electric field measurements on the Van Allen Probes spacecraft. J. Geophys. Res. Space Phys. 121 (5), 4611–4625. ISSN 2169-9402. https://doi.org/10.1002/2015JA022132.

CHAPTER

14

Machine Learning for Flare Forecasting

Anna M. Massone[*], *Michele Piana*[*,†], *FLARECAST Consortium*[‡]

[*]CNR—SPIN, Genova, Italy [†]Università di Genova, Genova, Italy [‡]Academy of Athens, Trinity College Dublin, Università di Genova, Consiglio Nazionale delle Ricerche, Centre National de la Recherche Scientifique, Université Paris-Sud, Fachhochschule Nordwestschweiz, Met Office, Northumbria University

CHAPTER OUTLINE

1 THE SOLAR FLARE PREDICTION PROBLEM

Solar flares are explosive phenomena in which up to 10^{33} erg of energy is released at times ranging between 10 and 1000 s (Kontar et al., 2011). According to the standard solar flare model (Priest and Forbes, 2002; Schrijver, 2009; Aulanier and et al., 2013), such energy is stored in magnetic field configurations, suddenly released when magnetic reconnection occurs, and is conveyed into mass motions, heating, particle acceleration, and radiation spanning over the entire electromagnetic spectrum. As a consequence, solar flares (eruptive or not) are one of

Machine Learning Techniques for Space Weather
https://doi.org/10.1016/B978-0-12-811788-0.00014-7

the major triggers of space weather in the heliosphere, including significant secondary effects on Earth, such as power grid malfunctions and impairments on flight navigation and satellite communications.

Prediction of solar flares relies on three fundamental ingredients: first, at an experimental level, observations of solar active regions, the undeniable hosts of major flares, must be available. These include, but are not limited to, properties of sunspot complexes, magnetic field information, and moments and proxies of the unknown (unmeasurable) coronal magnetic field. In a magnetized active region solar atmosphere, these properties are considered crucial for the ensuing fundamental properties of solar plasma (Sturrock, 1972). Second, at a computational level, these properties, encapsulated in physical and geometrical parameters, are used as input data for numerical algorithms that realize prediction by providing binary flare/no-flare or probabilistic $(0, 1)$ outcomes (Barnes and et al., 2016 and references therein). Third, at a technological level, modern prediction systems exploit software platforms realizing input/output services for big data, automatic, and user-friendly usage. This practice is followed in diverse prediction efforts at different fields.

Focusing on the first aspect, solar active regions are classified according to magnetic-complexity indicators, tracked by the National Oceanic and Atmospheric Administration (NOAA) Space Weather Prediction Center (SWPC). These are typically classified by three sets of indicators (Hale et al., 1919; Kunzel, 1960; McIntosh, 1990):

- The McIntosh scheme that uses withe-light emission to represent the sunspot structure and is composed of three independent variables.
- The Mount Wilson scheme that classifies sunspots based on the complexity of magnetic flux distribution and utilizes eight indicators.
- The total sunspot area which is computed in fractions (millionths) of a solar hemisphere.

More recently, the Solar Monitor Active Region Tracker (SMART) (Higgins et al., 2010) algorithm has been conceived in order to extract active region properties, thought to be related to flare occurrence, from line-of-sight magnetograms recorded by the Helioseismic and Magnetic Imager in the Solar Dynamics Observatory (SDO/HMI). For each active region, SMART identifies 14 properties associated with the magnetic field intensity and configuration, with a sampling time of 12 min, allowing an analysis in which input data are 14-dimensional time series.

Many contemporary flare prediction algorithms belong to the machine learning framework (Bobra and Couvidat, 2015; Colak and Qahwaji, 2009; Li et al., 2007; Yu et al., 2009; Yuan et al., 2010), where data properties utilized for prediction are named *features*. In the case of *supervised* learning, a set of historical data is used with features labeled by means of observation outcome and the forecast determines the labels associated with the new feature set. On the other hand, *unsupervised* methods do not rely on training sets, but cluster the incoming data in different groups according to similarity criteria involving data features. An important aspect of flare forecasting, characterized by notable physical implications, is to rank data features according to the intensity with which they correlate with the labels. In statistical learning theory, this practice is known as *feature selection*, although applications often refer to it as *feature importance*, which better highlights the fact that all features typically play a role in the prediction, although with different impacts. Feature selection can be realized either by using advanced implementations of standard neural network approaches (Olden et al., 2004; Garson, 1991) or by means of regularization methods (James et al., 2013). The second approach aims to optimize a function made of two terms: a discrepancy term, which measures the distance between prediction and data in the training set, and a penalty term, which constrains the number of features contributing significantly to the prediction.

A modern approach to flare prediction (or, more in general, to space weather prediction) should rely on a technology service that allows automatic computation of a *big data* amount of observations. A technological platform of this kind has been set up within the framework of the Horizon 2020 Flare Likelihood And Region Eruption foreCASTing (FLARECAST) project. The FLARECAST architecture follows the four core steps of the FLARECAST processing workflow, that is:

- The import of data from remote archives (essentially the SDO/HMI magnetograms, and the NOAA/SWPC Solar Region Summary and GOES flare occurrence).
- The extraction of features from the loaded data (this step includes loading external catalog data as well as the implementation of ad hoc algorithms for extracting features from observations).
- The interpretation of data within the framework of machine learning algorithms (in the case of supervised method, this step covers both the training and the execution phase).
- The verification of results from the previous step.

Overall, the implementation of the FLARECAST software prototype is conceived in a highly modular fashion and should therefore be considered as an extendable architecture, whereby further algorithms can be added at will to the platform and systematically validated. In the following sections, the description of the use of machine learning methods for flare prediction refers essentially to the algorithms included in the FLARECAST service. Specifically, the service includes methods that have already been used in the framework of flare prediction, more advanced algorithms that found applications in context out of solar physics, and, finally, innovative approaches whose mathematical properties are currently under investigation.

2 STANDARD MACHINE LEARNING METHODS

Most machine learning methods applied so far to the problem of flare prediction belong to the family of supervised methods. In particular, the great majority of these techniques ranges from standard multilayer perceptrons (Borda et al., 2002; Qu et al., 2003; Wang et al., 2008; Colak and Qahwaji, 2009; Mubiru, 2011; Zavvari et al., 2015) to neural networks characterized by an optimization of the computational efficiency, mainly in the learning phase (Qahwaji and Colak, 2007; Qahwaji et al., 2008; Bian et al., 2013). A few papers in more recent literature discuss the performance of support vector machines (SVMs) for flare prediction, describing rather standard implementations of this supervised approach (Qu et al., 2003; Qahwaji and Colak, 2007; Qahwaji et al., 2008; Bobra and Couvidat, 2015; Boucheron et al., 2015; Zavvari et al., 2015). Further, unsupervised clustering is used as a classification tool in combination with supervised regression algorithms (Li et al., 2007). Finally, as far as feature selection is concerned, results are obtained utilizing logistic regression, also in a hybrid combination with SVMs (Song et al., 2009; Yuan et al., 2010; Bian et al., 2013).

The current release of the FLARECAST service contains both unsupervised and supervised algorithms. Unsupervised prediction is realized by clustering methods that organize a set of unlabeled samples into meaningful clusters based on data similarity. Data partition is obtained through the minimization of a cost function involving distances between data and cluster prototypes. Optimal partitions are obtained through iterative optimization: starting from a random initial partition, samples are moved from one cluster to another until no further improvement in the cost function optimization is achieved. In a classical approach, each sample may belong to a unique cluster, while in a fuzzy clustering formulation, a different degree of membership is assigned to each sample with respect to each cluster. In particular, the standard algorithms considered in FLARECAST are the well-known Hard C-Means (or K-means) algorithm (Anderberg, 1973) and its fuzzy extension, the Fuzzy C-Means algorithm (Bezdek, 1981).

In the supervised context, multilayer perceptron is by far the most common neural network used in machine learning. The usual training algorithm is the error-back-propagation approach, which relies on a gradient descent algorithm, uses a forward and a backward pass through the feed-forward neural network, and performs weights update using the derivatives of the error function of the network with respect to the neural weights. FLARECAST also currently includes two recurrent neural networks that allow feedback loops in the feed-forward architecture. This effect is realized, specifically, by means of an Elman neural network (Elman, 1990), which permits any number of context nodes, and of a Jordan network, which constrains the number of context nodes to coincide with the number of output nodes (Jordan, 1986).

The set of currently implemented FLARECAST standard supervised methods is completed by an SVM and a LASSO algorithm that realize data regression by means of a quadratic loss function with two different penalty terms.

3 ADVANCED MACHINE LEARNING METHODS

Advanced methods in this framework refer to the developments of standard machine learning approaches aiming to fulfill more ambitious objectives, by means of more sophisticated implementations (e.g., ad hoc designed loss functions, penalty terms tailored to feature selection requirements, more automatic and more general approaches to unsupervised classification). In fact, FLARECAST perspective pointed out that (1) standard unsupervised clustering explains the data with a number of classes that must be fixed a priori; and (2) determining which physical features have the greatest predictive impact is one of the key issues in flare forecasting, mainly in the case of SDO/HMI data, which are characterized by a notable amount of parameters extractable from the vector magnetograms. Therefore, the FLARECAST pipeline presents a second group of methods that aim to account for such issues.

As already introduced in the previous section, the grouping operation in clustering aims to recognize common traits shared by the data samples of a given class and to group peculiar samples contained in an unlabeled database into different classes. Such grouping is obtained via the minimization of a given cost function on the basis of assigned criteria of similarity, anchored to the definition of a distance between samples. Among the advanced methods introduced in FLARECAST, possibilistic clustering (Massone et al., 2006) assumes that each sample could belong simultaneously to several clusters with a different degree of membership. Unlike fuzzy clustering, this approach is based on the assumption that the membership values of a sample in a cluster is absolute and does not depend on the membership values of the same sample in any other cluster, which implies that each cluster exists independently of the other ones. A typical condition encompassed by this algorithm is that a noise sample can have a very low membership value for all clusters, while representative samples could have high memberships to several clusters nonexclusively.

FLARECAST also considered two generalizations of its standard multilayer perceptron able to point out the importance with which data features contribute to the prediction task. In particular, on one hand, the Garson algorithm (Garson, 1991) assigns positive weights to each input variable of the network, the magnitude of such weights representing the significance of the prediction power of the variable. On the other hand, the Olden method (Olden et al., 2004) utilizes information concerned with both the magnitude and the sign of the connection weights in the network, with two differences with respect to the Garson approach. From a technical viewpoint, in the Olden method, the importance of every predictor can be either positive or negative, whereby the largest positive value denotes the most important predictor (differently than in the Garson algorithm, where all importances are positive and add up to 1); moreover, the Olden method can present more than a single-hidden layer, while the Garson approach is conceived to allow the presence of just a single layer.

FLARECAST can realize feature importance also by means of penalized logistic regression (Wu et al., 2009) that has been designed ad hoc for performing such task while realizing classification. This is a maximum-likelihood method, which allows the estimation of the model parameters while best-fitting the data by means of the binomial distribution. The regularization of the analysis is performed by optimizing the open parameter by means of an automatic cross-validation process.

Finally, the FLARECAST pipeline contains a more sophisticated SVM, which utilizes the hinge loss function in the fitting term of the regularization functional. This kind of loss function has proven to be optimal for classification (Rosasco et al., 2004).

4 INNOVATIVE MACHINE LEARNING METHODS

In the previous sections we have presented algorithms belonging to the state of the art in the machine learning field. More specifically, some of them were standard algorithms already employed in a number of application domains, while others were not original from a mathematical viewpoint, but presented some advanced aspects for the specific application of solar flare prediction. In this section, we briefly overview a set of algorithms included in the FLARECAST pipeline, which are innovative under both viewpoints (i.e., their core mathematical aspects and the specific application of flare forecasting). In particular, the innovative machine learning algorithms here described are based on: multitask learning approaches, strategies that are hybrid from several different perspectives, and ensemble (deep learning) methods.

Multitask learning is an approach to machine learning by means of which multiple related prediction tasks are learned jointly, sharing information across the tasks. To achieve this goal, a possible approach is based on regularized least-square regression with a penalty term designed ad hoc for selecting the best set of features explaining all prediction tasks simultaneously. The resulting multitask learning method is called Multi-Task Lasso (Yuan et al., 2006). In FLARECAST, an advanced version of Multi-Task LASSO (adaptive Multi-Task LASSO) is implemented, which guarantees two important properties when the number of data increases: (1) the method performs consistent feature selection; and (2) the estimation error is normally distributed. We point out that, in this framework, observations are assumed to be the realization of Gaussian random variables, while in several practical applications, real data follow the Poisson distribution. Therefore, FLARECAST provides two versions of Multi-Task LASSO and adaptive Multi-Task LASSO, respectively, that work in the case of Poisson measurements.

Hybrid frameworks in machine learning can be realized from several different perspectives. In particular, FLARECAST realized this approach according to two different strategies. The first strategy aimed to integrate the classification task typical of flare prediction with the ability of providing a rank of the features involved in the analysis based on their impact on the prediction effectiveness. Penalized logistic regression (L1-logit) and LASSO in all its variations are the regularization methods performing feature selection considered in FLARECAST, and both approaches present specific limitations. LASSO methods are intrinsically regression methods and therefore they are not originally conceived for applications that require a binary yes/no response. On the other hand, L1-logit, which is a classification method, predicts the flare/no-flare condition by means of a fixed threshold on the flare occurrence probability, whose choice can be critical and/or operator-dependent. Specifically, the threshold optimization can be achieved by discretizing the interval of possible thresholds in a fixed number of values, by dichotomizing the flare occurrence for each one of the possible thresholds, by computing the corresponding skill score, and by selecting the threshold value that provides the highest skill score value along the lines of Bloomfield et al. (2012). FLARECAST software prototype contains a novel, two-step, hybrid approach to flare prediction with feature selection, which aims at overtaking such limitations. The idea of this hybrid approach can be summarized as follows:

1. In the first step, LASSO is used to promote sparsity and to realize feature selection. This step provides an optimal estimate of the model parameters and corresponding predicted output.
2. In the second step, Fuzzy C-Means is applied for clustering the predicted output in two classes.

The main advantage of this approach is that step 2 utilizes fuzzy clustering to automatically classify the regression output in two classes. Indeed, Fuzzy C-Means identifies flaring/nonflaring events with a thresholding procedure, which is data adaptive, and completely independent of the skill score chosen for optimization.

The second hybrid strategy considered in FLARECAST couples Monte Carlo optimization with clustering. Specifically, the pipeline proposes a classification method based on the application of simulated annealing to the optimization of the cost functions in Hard C-means and Fuzzy C-Means. Simulated annealing (Kirkpatrick et al., 1983) implements a global search technique derived by statistical mechanics and based on the Metropolis algorithm. It iteratively simulates the behavior and small fluctuations of a system of atoms at temperature T starting from an initial configuration. Each iteration is characterized by random perturbations of the actual configuration and computation of the corresponding energy variation ΔE. If

$\Delta E < 0$, the transition is unconditionally accepted; otherwise, it is accepted with a probability given by the Boltzmann distribution. This approach can be generalized to the solution of optimization problems by using an ad hoc selected cost function, instead of the physical energy. In this perspective the temperature takes the role of a control parameter of the search area, and is gradually decreased until no further improvements of the cost function are achieved. FLARECAST applies the simulated annealing approach to the Hard C-Means and Fuzzy C-Means algorithms by replacing E with the two corresponding cost functions. The proposed algorithms, called simulated annealing C-means and simulated annealing Fuzzy C-Means, respectively, permit, in principle, avoidance of local minima, and ability to reach the global optimum of the cost functions.

Finally, FLARECAST's focus on deep learning approaches was concerned with random forest (Breiman, 2001), which belongs to the family of the so-called *ensemble methods*, that is, methods that make use of a combination of different learning models to increase the classification accuracy. In particular, the random forest ensemble works as a large collection of decorrelated decision trees, which organize a series of test questions and conditions in tree structures, recursively splitting training samples into subsets based on the value of specific attributes.

Just as an example, Fig. 1 shows a comparison of performances of six prediction algorithms currently implemented in the FLARECAST pipeline according to the values of five standard skill scores (False Alarm Ratio, Probability of Detection, Accuracy, Heidke Skill Score, and True Skill Statistics). For this example, we have considered a test set of NOAA/SWPC data.

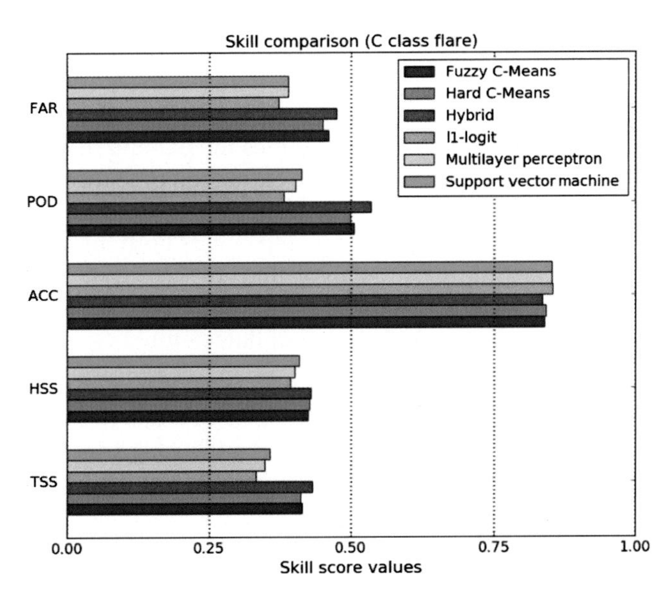

FIG. 1 Comparison of performance between some prediction algorithms in terms of the most used skill scores. The bar plots represent the skill score values obtained by applying each method to a test set of NOAA/SWPC data for the prediction of flares of class C and above.

5 THE TECHNOLOGICAL ASPECT

The main objective of FLARECAST technology is to provide an infrastructure that can be installed not only on the project's computing hardware at the Université Paris-Sud, a FLARECAST Consortium partner, but also to the local machines of developers and end users. Obviously local machines can provide limited storage space and processing power. Hence, the FLARECAST implementation relied also on the central computing hardware at Paris Sud. To install FLARECAST on different target systems, we relied on Docker containers (https://www.docker.com/whatisdocker), which can be seen as lightweight virtual machines running within the same Linux Kernel, but with an independent file system.

The design of the FLARECAST architecture started from the already existing architecture utilized in the FP7 "High Energy Solar Physics data in Europe (HESPE)" project (http://hespe.eu). And in fact some of the work done for the HESPE technological platform continues to live in FLARECAST. Specifically:

- FLARECAST reused HESPE's sophisticated concept for parallel processing of Interactive Data Language (IDL) routines in a cluster.
- FLARECAST reevaluated some of the HESPE visualization tools at a later stage of the project.

However, HESPE was designed for a closed group of developers and system administrators, while FLARECAST aimed for a plug-in system even for external users. Moreover, between the kick-off of HESPE a few years ago and the FLARECAST age, there has been a shift from cluster computing to cloud computing. Therefore, the FLARECAST technology structure is characterized by an original perspective. Such structure is shown in Fig. 2 and follows the four core steps of the FLARECAST processing workflow:

1. Import data from remote archives, such as GOES flare lists from NOAA/SWPC or magnetograms from SDO/HMI.
2. Extract features from the loaded data. This includes the parsing of external catalog data into the FLARECAST data model as well as the extraction of features from imaging observations.
3. Train and execute prediction algorithms. This step covers both the training and the execution phases of the machine learning algorithms described in the previous sections.
4. Verify the results from the previous step. Currently, this step can be seen as placeholder as we first need to accumulate sufficient amounts of data before beginning performance verification and benchmarking.

All software component boxes in Fig. 2 consist of at least one Docker container. The management infrastructure component provides several Docker containers holding different applications, such as a workflow manager, a system monitor, and a management console. All other software component boxes contain multiple Docker containers, each holding exactly one algorithm. These algorithm containers need to follow certain rules, such that the workflow manager can execute them. The algorithms may be written in any programming language. Currently tested and supported are Python 2.7 and 3.0, R, and IDL 8.3, while all other languages that can be executed on a Linux system (e.g., Java, C++, and Perl) do not cause any major problem. The management infrastructure orchestrates the execution of the FLARECAST

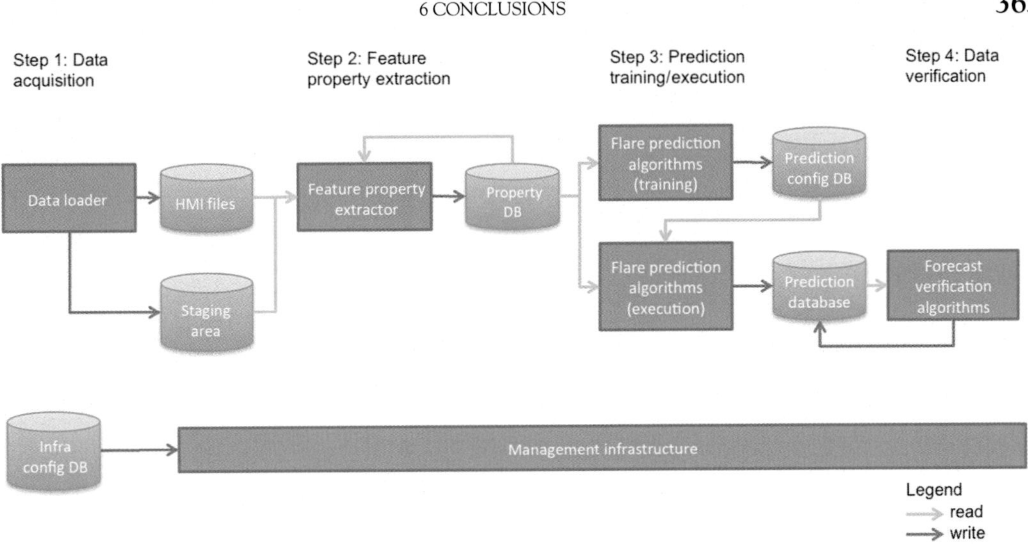

FIG. 2 The FLARECAST architecture. The *rectangular boxes* are placeholders for exchangeable software components (i.e., algorithms), while the cans are the database symbols.

workflow. It starts extraction algorithms as soon as new data arrives and triggers the flare prediction algorithms as soon as new feature properties arrive in the database. In the first release of the platform, the management infrastructure component is implemented as a simple web application, where a user can manually trigger execution of algorithms. This will later be complemented by a more sophisticated workflow engine. At some point it is intended to allow external users to plug in their algorithms (or already extracted properties) to the FLARECAST ecosystem.

Finally, the database symbols in Fig. 2 denote individual data components that define generic interfaces to read, create, update, and delete data, where the read methods include a query language for simple queries. FLARECAST is going to provide open access to all of its data by means of interfaces that are not included in the current architecture drawing. These interfaces will be implemented in a way that grants read access to the public.

6 CONCLUSIONS

This chapter illustrates some recent advances concerning the use of machine learning methods for flare prediction. Such results have been conceived and realized within the framework of the Horizon 2020 project FLARECAST and include the production of advanced and innovative algorithms that work in the context of supervised learning and unsupervised data clustering. The FLARECAST effort aimed at realizing a technological platform able to place the computational power of these algorithms at the disposal of the heliophysics community, at the same time adhering to an EU open access strategy. The high modularity of the platform design allows an effective updating of the software service for flare prediction, as well as possible future extensions to include other aspects of space weather forecasting such as, conceivably, coronal mass ejections and solar energetic particle events. In any case, the achieved fast computation enables processing of observations in line with a big data handling perspective.

The FLARECAST project has received funding from the European Union's Horizon 2020 Research and Innovation Programme under grant agreement No. 640216.

References

Anderberg, M.R., 1973. Cluster Analysis for Applications. Academic Press, London.
Aulanier, G., et al., 2013. Astron. Astrophys. 549, A66.
Barnes, G., et al., 2016. Astrophys. J. 829, 89.
Bezdek, J.C., 1981. Pattern Recognition With Fuzzy Objective Function Algorithms. Plenum Press, New York, NY.
Bian, Y., Yang, J., Li, M., Lan, R., 2013. Math. Probl. Eng. 2013, 917139.
Bloomfield, D.S., Higgins, P.A., McAteer, R.T.J., Gallagher, P.T., 2012. Astrophys. J. 747, L41.
Bobra, M.G., Couvidat, S., 2015. Astrophys. J. 798, 135.
Borda, R.A.F., Mininni, P.D., Mandrini, C.H., Gmez, D.O., Bauer, O.H., Rovira, M.G., 2002. Sol. Phys. 206, 347.
Boucheron, L.E., Al-Ghraibah, A., McAteer, R.T.J., 2015. Astrophys. J. 812, 1.
Breiman, L., 2001. Mach. Learn. 45, 5.
Colak, T., Qahwaji, R., 2009. Space Weather 7, S06001.
Elman, J.L., 1990. Finding structure in time. Cogn. Sci. 14, 179.
Garson, G.D., 1991. Interpreting neural-network connection weights. Artif. Intell. Expert 6, 47.
Hale, G.E., Ellerman, F., Nicholson, S.B., Joy, A.H., 1919. Astrophys. J. 49, 153.
Higgins, P.A., Gallagher, P.T., McAteer, R.T.J., Bloomfield, D.S., 2010. ArXiv:1006.5898.
James, G., Witten, D., Hastie, T., Tibshirani, R., 2013. An Introduction to Statistical Learning. Springer, New York, NY.
Jordan, M.I., 1986. A parallel distributed processing approach. Institute for Cognitive Science Report 8604, UC San Diego.
Kirkpatrick, S., Gelatt, C.D., Vecchi, M.P., 1983. Science 220, 661.
Kontar, E.P., Brown, J.C., Emslie, A.G., et al., 2011. Space Sci. Rev. 159, 301.
Kunzel, H., 1960. Astron. Nachr. 285, 271.
Li, R., Wang, H.N., He, H., Cui, Y.M., Du, Z.L., 2007. Chin. J. Astron. Astrophys. 7, 441.
Massone, A.M., Studer, L., Masulli F, F., 2006. Int. J. Approx. Reason. 41, 96.
McIntosh, P.S., 1990. Sol. Phys. 125, 251.
Mubiru, J., 2011. Adv. Artif. Neural Syst. 2011, 142054.
Olden, J.D., Joy, M.K., Death, R.G., 2004. Ecol. Model. 178, 389.
Priest, E.R., Forbes, T.G., 2002. Astron. Astrophys. Rev. 10, 313.
Qahwaji, R., Colak, T., 2007. Sol. Phys. 241, 195.
Qahwaji, R., Colak, T., Al-Omari, T.M., Ipson, S., 2008. Sol. Phys. 248, 471.
Qu, M., Shih, F.Y., Jing, J., Wang, H., 2003. Sol. Phys. 217, 157.
Rosasco, L., Devito, E., Caponnetto, A., Piana, M., Verri, A., 2004. Neural Comput. 16, 1063.
Schrijver, C.J., 2009. Adv. Space Res. 43, 739.
Song, H., Tan, C., Jing, J., Wang, H., Yurchyshyn, V., Abramenko, V., 2009. Sol. Phys. 254, 101.
Sturrock, P.A., 1972. Sol. Phys. 23, 438.
Wang, H.N., Cui, Y.M., Li, R., Zhang, L.Y., Han, H., 2008. Adv. Space Res. 42, 1464.
Wu, T.T., Chen, Y.F., Hastie, T., Sobel, E., Lange, K., 2009. Bioinformatics 25, 714.
Yu, D., Huang, X., Wang, H., Cui, Y., 2009. Sol. Phys. 255, 91.
Yuan, M., Lin, Y., Statist, J.R., 2006. Soc. B 68, 49.
Yuan, Y., Shih, F.Y., Jing, J., Wang, H.M., 2010. Res. Astron. Astrophys. 10, 785.
Zavvari, A., Islam, M.T., Anwar, R., Abidin, Z.Z., 2015. J. Theor. Appl. Inf. Technol. 74, 63.

CHAPTER

15

Coronal Holes Detection Using Supervised Classification

Véronique Delouille[*], *Stefan J. Hofmeister*[†], *Martin A. Reiss*[‡], *Benjamin Mampaey*[*], *Manuela Temmer*[†], *Astrid Veronig*[†]

[*]Royal Observatory of Belgium, Brussels, Belgium [†]University of Graz, Graz, Austria
[‡]Space Research Institute, Graz, Austria

CHAPTER OUTLINE

https://doi.org/10.1016/B978-0-12-811788-0.00015-9

1 INTRODUCTION

Coronal holes play an important role in geomagnetic storm activity (Tsurutani et al., 2006) and are the dominant contributors to space weather disturbances at times of quiet solar activity. The term "coronal hole" is commonly associated with low-density regions of one dominant magnetic polarity in the solar atmosphere. In extreme ultraviolet (EUV) and X-ray images of the Sun, coronal holes are visible as dark areas in the solar corona due to their lower temperature and electron density compared with the ambient coronal plasma (Munro and Withbroe, 1972). Above the bottom of coronal holes along the rapidly expanding open magnetic field lines, solar plasma is accelerated into interplanetary space to form high-speed solar wind streams (HSSs) (Gosling and Pizzo, 1999; Cranmer, 2009). The interaction between high- and slow-speed solar wind flows form stream interaction regions (SIR) that may cause geomagnetic storms when sweeping over Earth.

As SIRs are recurrent phenomena, they can cause elongated periods of increased geomagnetic activity. Therefore, predicting the characteristics of HSSs near Earth is required. Nolte et al. (1976) found that the peak velocity of HSSs correlates well with the areas of their solar source coronal holes. This leads to the possibility of predicting HSSs based on the area of coronal holes located near the solar central meridian (e.g., Vršnak et al., 2007; Verbanac et al., 2011; Rotter et al., 2012, 2015; Robbins et al., 2006; Reiss et al., 2016) and highlights the importance of a precise extraction of coronal holes for the forecasts of HSSs.

In the past, coronal holes have mostly been identified and tracked by experienced observers. Recent attempts to automate the process of identification and detection of coronal holes include (Barra et al., 2007; Kirk et al., 2009; Krista and Gallagher, 2009; Rotter et al., 2012; Henney and Harvey, 2005; Scholl and Habbal, 2008). Several of those methods are based on the lower EUV intensity observed in coronal holes as compared with the ambient corona. The detection of coronal holes purely from their low intensity in solar EUV images is, however, a challenging task because filament channels also appear as dark coronal features. Filament channels are usually interpreted in terms of the weakly twisted flux rope model, having a magnetic field that is dominated by the axial component. Those magnetic channels have high magnetic pressure, are low in density/intensity, and can be filled with cool plasma material located in the dip of the helical windings. Empty or filled channels may appear on EUV images at a similar dark intensity level as coronal holes (Mackay et al., 2010; de Toma, 2011). Hence, in order to distinguish coronal holes from filament channels, several authors use magnetogram information, such as, for example, Skewness of the magnetic field strength (Krista and Gallagher, 2009), or criteria involving several quantities computed from magnetograms (Scholl and Habbal, 2008).

In terms of space weather forecasting, a falsely extracted filament channel leads to erroneous predictions of HSSs. It is therefore necessary to improve and quantify the quality of the coronal hole detection method in order to prevent, as much as possible, filaments from being falsely identified as coronal holes in EUV images.

In this work, we use Solar Dynamics Observatory (SDO) Atmospheric Imaging Assembly (AIA) 193Å images (Lemen et al., 2012) and SDO/Helioseismic and Magnetic Imager (HMI) magnetograms (Scherrer et al., 2012) in order to analyze the properties of coronal holes and filament channels during the period August 19, 2010 until December 31, 2016.

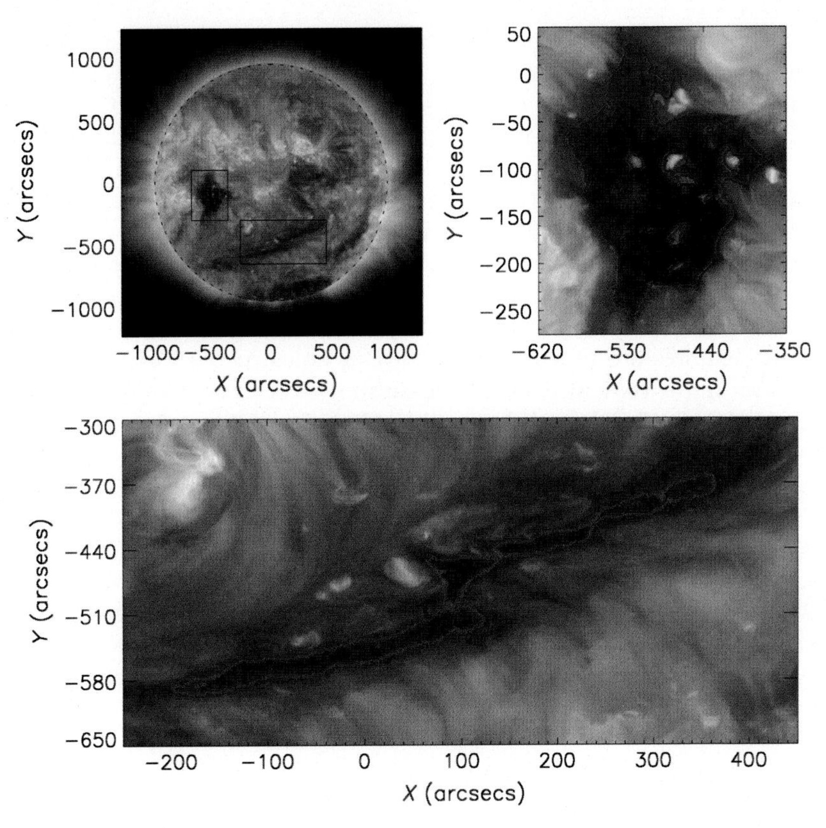

FIG. 1 (*Top left*) AIA 193Å image taken on May 13, 2015 with a coronal hole and filament outlined. (*Top right*) Zoom on the coronal hole. (*Bottom*) Zoom on the filament channel.

We first detect low-intensity regions from SDO/AIA 193Å images using a modified version of the Spatial Possibilistic Clustering Algorithm (SPoCA) (Verbeeck et al., 2014). Fig. 1 shows the detection on two low-intensity regions by SPoCA, one being a coronal hole and the other a filament. By comparing these detections with Hα filtergrams from the Kanzelhöhe Observatory (Austria), one may identify visually which regions are filaments (Pötzi et al., 2015). It is therefore possible to label these low-intensity regions detected by SPoCA as either "filament channel" (FC) or "coronal hole" (CH). These regions are then mapped onto the corresponding HMI line-of-sight (LOS) magnetograms. This allows us to calculate a set of attributes that take into account EUV intensity and binary shape information as well as the magnetic field structure for each of the regions. Finally, we insert the computed attributes into various supervised classification schemes in order to design the most suitable decision rule for a differentiation in realtime. We tested and compared through cross-validation: support vector machine (SVM), linear SVM, decision tree, random forest, *k*-nearest neighbors, as well as two ensemble learning methods based on decision trees.

The present study builds upon the work described in Reiss et al. (2015) and extends it in three directions, as follows.

First, we analyzed, by visual inspection, the performance of SPoCA-CH in detecting actual coronal holes, and identified some issues: The nonperfect extraction algorithm produced *artifacts*, defined as quiet Sun (QS) regions with a lower-than-average QS intensity that are assigned to the coronal holes class. These artifacts do not have clear boundaries, and their mean intensity is higher than that of coronal holes. To avoid the presence of artifacts, we modified the original SPoCA-CH algorithm. This new version of SPoCA is more conservative at classifying a pixel as belonging to a low-intensity region, and therefore reduces the amount of artifacts.

Second, we gathered a new dataset that spans a larger period, from August 2010 until November 2016, and where the detections were no more limited to the center of the solar of disk, as was the case in Reiss et al. (2015). With the modified SPoCA-CH algorithm, fewer filament channels are mingled with coronal holes in the low-intensity class, which already shows an improvement over the classical SPoCA-CH module. Specifically, SPoCA-CH extracted in total 5048 CH candidate maps. After visual inspection and comparison with $H\alpha$ filtergrams, we concluded that 4738 were actual CHs, whereas the remaining 310 maps featured FCs. The ratio of FCs versus CHs present in our dataset is thus 1:15 (around 6% of FCs). It is nevertheless of interest to remove the remaining FCs. As a third contribution, we propose to achieve this by applying some supervised classification schemes that take into account the imbalance between the CH majority class and the FC minority class.

Section 2 describes the preparation of the labeled datasets and the proposed attributes. The supervised classification schemes and performance metrics are introduced in Section 3. Results are given in Section 4 and their implications discussed in Section 5. This work uses the Statistics and Machine Learning Toolbox of Matlab, Version 2016b.[1]

2 DATA PREPARATION

We used a modified version of SPoCA to prepare a dataset of low intensity regions during the period August 19, 2010 until December 31, 2016. Based on these data, we created a training set to design the most suitable decision rule to differentiate coronal hole and filament channel regions.

2.1 Coronal Hole Feature Extraction

2.1.1 The SPoCA-Suite

The SPoCA-suite (Verbeeck et al., 2014) is a set of segmentation procedures that allows decomposition of an EUV image into regions of similar intensity, typically active regions, coronal holes, and QS. In the framework of the Feature Finding Team project (FFT; Martens et al., 2012), the SDO Event Detection System (SDO EDS; Hurlburt et al., 2012) runs the SPoCA-suite to extract coronal hole information from AIA images in the 193Å passband, and upload the entries every 4 h to the Heliophysics Events Knowledgebase (Hurlburt et al., 2012). These entries are searchable through the graphical interface iSolSearch, the Ontology package of IDL Solarsoft, the Python SunPy library, and the JHelioviewer visualization tool (Müller et al., 2009).

[1] Programs and datasets are available at https://bitbucket.org/vdelouille/coronal_hole_detection_ml.

The SPoCA extraction method for coronal holes relies on the Fuzzy C-Means algorithm (FCM; Bezdek, 1981). It clusters the pixel's intensity values into four classes through the minimization of the intracluster variance. Such minimization leads to an iterative algorithm, where at each iteration every pixel is assigned a membership value between 0 and 1 to each one of the classes, then the class centers are computed using these memberships. These steps are repeated until convergence of the class centers. The segmented map is obtained by attributing a pixel to the class for which it has the maximum final membership value. The coronal hole map corresponds to the class whose center has the lowest intensity value. Various preprocessing steps are performed before applying the FCM algorithm: Images are calibrated using the IDL Solarsoft `aia_prep` routine, and a correction for the limb-brightening is obtained by fitting a smooth transition function and inverting it, see Verbeeck et al. (2014). Next, on-disc pixel intensities are normalized by their median values, and finally, the square root of pixel intensities is taken in order to stabilize the variance of the noise (Anscombe, 1948).

Once the segmented maps are obtained, some postprocessing is needed to extract individual low-intensity regions. A study of low-intensity features detected by SPoCA during the month of January 2011 revealed that coronal hole candidates detected for more than 3 consecutive days exhibit the expected magnetic properties characteristic of unipolar regions, which was not the case for shorter-lived regions. Setting up a threshold on the lifetime is thus a valuable tool to eliminate spurious dark regions. Hence, only low-intensity regions that live longer than 3 days are included in the final coronal hole maps used here and in the SDO EDS pipeline (we refer to Verbeeck et al., 2014 for more details).

2.1.2 Modified SPoCA-CH Module

We visually inspected the quality of coronal hole detection obtained from June 2010 until December 2016, and in the process, identified as the main issue of the SPoCA-CH module the presence of artifacts. Those are patches of the QS with lower intensity than average QS, but higher mean intensity than CH. They appear especially when no coronal hole is on the solar disk and/or when large active regions are present on the solar disk. Indeed, the centers of class computed by the FCM clustering varies according to the intensity distribution present in the image. The center for CH class will thus be set higher when the overall intensity is higher, for example, due to the presence of numerous ARs, if the correction for limb brightness is not optimal, or if there are no CH on the solar disk. We solved this issue by taking a more conservative approach when classifying a pixel as belonging to the low-intensity class within SPoCA-CH. The modified SPoCA-CH module may be described as follows.

In the preprocessing step, all pixels above the solar limb (i.e., all pixels with a greater distance than 1 $R_\odot$ from the center of the visible solar disk) are removed and the remaining pixels are corrected for limb brightening effects as before. Intensities are normalized by their median. In addition, we want to prevent class centers from settling at the high level of active region intensities. To avoid this, intensities above the mode of the image intensity distribution (i.e., above the QS intensity) are clipped to the mode intensity itself. Such a procedure does not affect coronal holes because their intensities are always below the image mode. Finally, a square-root transformation is applied.

We visually inspected the results of the FCM classifier with three, four, and five classes on a set of preprocessed images, and concluded that four classes were sufficient while providing less artifacts than three classes. We thus chose an FCM classifier with four classes to segment the preprocessed images in the whole dataset. The final attribution of one pixel to a class is performed by using the median class center of all images within the past 27 days. This ensures having a smooth variation in the class center values over time, and prevents creation of artifacts when no coronal hole is visible on the solar disk.

In the postprocessing step, elements with a radius smaller than 6 arcsecs are removed and neighboring connected components within a 32-arcsec distance are aggregated. Coronal hole candidates having an area smaller than 3000 arcsec^2, or a lifetime of less than 3 days are completely disregarded (cf. Verbeeck et al., 2014). It takes 38 s on a standard CPU to process one $4k \times 4k$ AIA image with the modified SPoCA-CH module.

This new approach leads to fewer artifacts and more conservative coronal hole boundaries, as may be seen in Fig. 2 comparing a segmentation by the original SPoCA-CH algorithm of Verbeeck et al. (2014) and our proposed modified version. Based on intensity alone, it is not possible to completely exclude filament channels. This highlights the importance of postprocessing the results of SPoCA using a supervised classification scheme that aims at separating CH and FC.

2.2 Labeled Datasets

The extracted regions were manually labeled as either FC or CH regions by visually inspecting Hα filtergrams from the Kanzelhöhe Solar Observatory (Austria) as follows. If a filament channel was clearly observable at the position of an extracted low-intensity region in the corresponding Hα filtergram, the region was labeled as "filament." Otherwise, the region was labeled as "coronal hole."

To take into account the evolution of filament material, we checked a time span of 3 days. When an Hα filament could not be connected to an SPoCA region for the first few days,

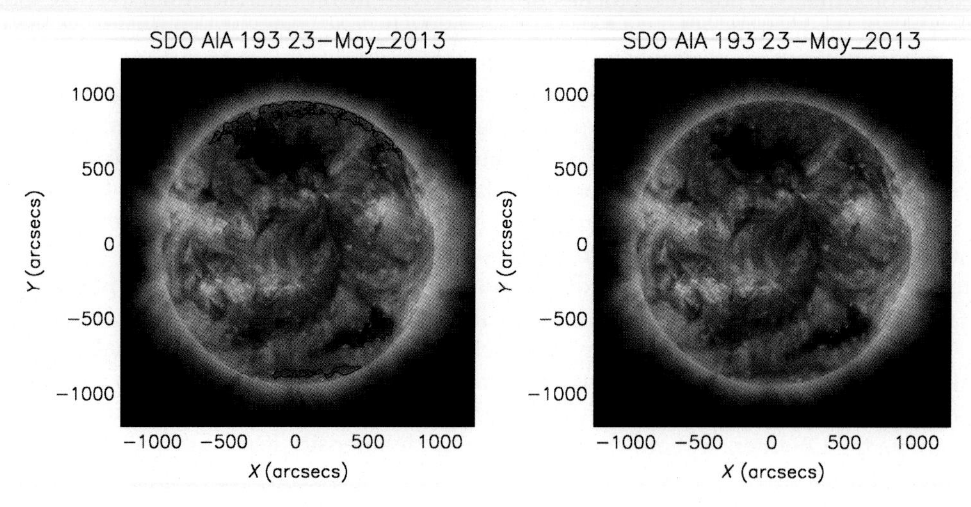

FIG. 2 (*Left*) Original SPoCA-CH maps overlaid on top of the AIA 193Å images taken on May 23, 2013. (*Right*) The same image with modified SPoCA-CH maps overlaid. Fewer pixels are classified as CH in the modified version of SPoCA-CH.

while afterward the connection was clear, we went back in time, manually checked again, and relabeled the SPoCA object seen in the first few days as FC.

This procedure was performed once per day, resulting in a training set containing 4738 coronal holes and 310 filaments. To quantify the magnetic flux characteristics in the extracted regions, we used the `hmi.M_720s` data series and mapped the extracted regions onto those HMI LOS magnetograms. The LOS magnetic field maps were further transformed to radial magnetic field maps under the assumption of a radial photospheric magnetic field (Hofmeister et al., 2017). Note that this assumption is reasonable as long as the magnetic field vector is oriented approximately along the radial direction, which is thought to be a good estimate for the magnetic field below coronal holes.

2.3 Proposed Attributes

Coronal holes are usually observed as dark regions with a dominant magnetic field polarity. Because filaments are magnetically bipolar regions orientated along a polarity inversion line, they are usually observed as elongated regions with a balanced magnetic field (Mackay et al., 2010). These properties lead us to investigate shape measures, magnetic flux properties, and first- and second-order image statistics on both EUV images and magnetograms as their characterizing attributes. We also computed attributes related to the location of the regions.

This set of attributes is briefly described as follows, and formulas are provided in Appendices A and B. For more details, we refer to Reiss et al. (2015).

2.3.1 Location

The latitude of the center of mass provides important information because coronal holes and filaments have different occurrence rates depending on the latitude; for example, objects covering the poles are almost always coronal holes. Projection effects and local image noise levels are also linked to location. We therefore added as attributes: the distance to the disk center, the minimum and maximum latitudes, the latitude of the center of mass, and the area of the extracted region.

2.3.2 Shape Measures

Shape measures were used to characterize the shape of the detected low-intensity regions. The calculated shape measures also allow us to reduce the object pixel configuration contained in a binary map of a structure to a single scalar number.

In order to investigate irregular shapes of coronal holes and filament channels in detail, two alternative shape measurements were developed: symmetry analysis and direction-dependent shape analysis. Geometrical symmetry is measured as follows: After the application of discrete geometrical transformations such as rotation, reflection, and a composition of both, the relative overlap in percentage is calculated. The direction-dependent shape analysis uses the average number of neighbors in each direction to compute the relevant shape information. Further, we used the shape attributes roundedness, roundness, and compactness. For a detailed description of the shape measures we refer to Reiss et al. (2014).

2.3.3 Magnetic Flux Imbalance

We characterize the magnetic flux imbalance in the detected features in two ways. First, we compute the ratio of the number of positive (n_+) and negative (n_-) magnetogram pixel values:

$$R_1 := \frac{n_+}{n_-} \tag{1}$$

Second, let $|\cdot|$ denote the absolute value operator, Φ_+ the total positive magnetic flux in the segmented feature, and Φ_- the total negative flux. The quantity

$$R_2 := 2\left|\frac{1}{2} - \frac{\Phi_+}{\Phi_+ + |\Phi_-|}\right| \tag{2}$$

quantifies the imbalance between the total positive and the total magnetic flux within the extracted regions. In the ideal case, this value is 1 for coronal holes, as unipolar regions ($\Phi_+ = 0$ or $\Phi_- = 0$), and 0 for filament channels ($\Phi_+ = |\Phi_-|$), as magnetically bipolar regions orientated along a polarity inversion line.

2.3.4 First- and Second-Order Statistics

First-order statistics allow characterization of the overall pixel value distribution. We consider mean, variance, energy, and entropy. Fig. 3 illustrates the difference in probability distribution function for coronal holes and filaments observed with SDO/AIA and SDO/HMI, and hence gives an intuition as to why first-order statistics could discriminate between both types of features.

On the other hand, second-order image statistics (Haralick et al., 1973; Weyn et al., 2000; Ahammer et al., 2008) focus on the spatial relation between pixel values contained in AIA EUV images and HMI LOS magnetograms. It was originally applied on gray scale images, and is adapted here to open value ranges (i.e., including negative data values) in order to calculate textural features for intensity and magnetic field configurations. This enables us to characterize the intrinsic texture information contained in the extracted structures in AIA 193Å and LOS magnetograms via the computation of a set of textural features.

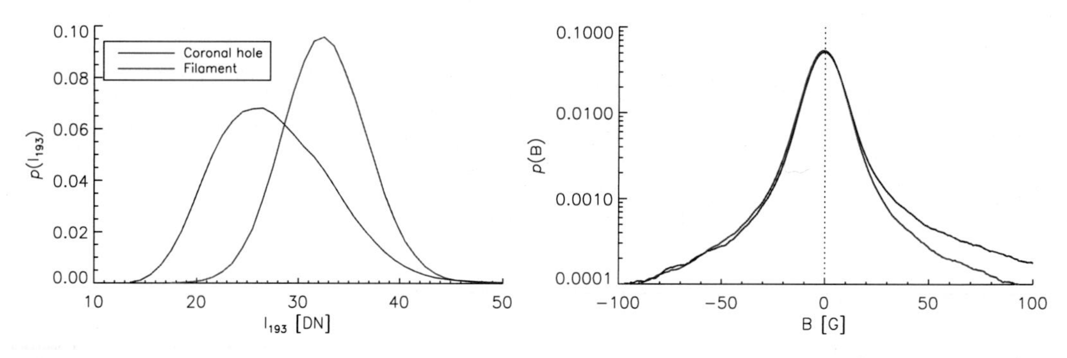

FIG. 3 Probability distribution function of a filament versus a coronal hole for (*left*) AIA 193Å intensity (*right*) HMI magnetogram.

2.3.5 *Sets of Attributes Used for Classification*

The set of attributes described in previous sections together with the labeling of the maps as FC or CH provide us with a labeled dataset of feature values. We consider four sets of attributes:

1. Set 1: Location, first- and second-order statistics on AIA images ("AIA" dataset)
2. Set 2: Location, shape measures, and first- and second-order statistics on AIA images ("AIA, shape" dataset)
3. Set 3: The "AIA" dataset augmented by the first- and second-order statistics on HMI LOS images ("AIA, HMI" dataset)
4. Set 4: The "AIA, HMI" dataset, and in addition, the shape measures ("AIA, shape, HMI" dataset)

The computation of the full set of attributes on one CH-candidate region takes less than 5 s on a standard CPU.

Our aim in considering these various sets is twofold: (i) obtain a set of rules that are easy to implement into the SDO EDS pipeline and (ii) measure the corresponding improvement in performance when using HMI information.

3 SUPERVISED CLASSIFICATION

The computed attributes were used as input for data mining in order to design the most suitable decision rule for a separation between coronal holes and filaments. Supervised classification is commonly applied when a large set of attributes is available and the interpretation of available information is arduous.

In binary supervised classification, we have access to measurements of various properties (or *features*) of an object that is further labeled as either belonging to the $\mathbf{p}$ (positive) or $\mathbf{n}$ (negative) class. Here, FC represents the positive and CH the negative class.

The goal of supervised classification is to predict the class of a new object. A classification model (or classifier) is a mapping from objects to predicted class m: $R^d \rightarrow \{\mathbf{Y}, \mathbf{N}\}$, where $m(\mathbf{x})$ indicates the decision when feature vector $\mathbf{x}$ is observed, and where we use the label $\mathbf{Y}$ ("Yes") and $\mathbf{N}$ ("No") to denote the classification produced by the model. Out of the numerous collection of possible decision rules, we aim at identifying those performing best (i.e., the those minimizing the expected cost). The expected cost depends on the error cost function, denoted by *c(classification, class)*: $c(\mathbf{Y}, \mathbf{n})$ represents the cost of classifying as a "Yes" a member from the negative class (false positive error), while $c(\mathbf{N}, \mathbf{p})$ is the cost of classifying as a "No" a member from the positive class (false negative error).

The feature dataset used in Reiss et al. (2015) contained only a mild imbalance between the two classes (1 FC for 5 CH) and could therefore be analyzed with traditional data mining algorithms. The extended dataset in the present study exhibits a more pronounced imbalance ratio of 1:15. In such cases, traditional algorithms tend to create suboptimal classification models, with a good coverage of the majority class, but where observations from the minority class are frequently misclassified (Wallace et al., 2011).

Section 3.1 lists the baseline classifiers, and Section 3.2 introduces several ways to alleviate the class imbalance issue. The resulting algorithms are tested via a common protocol

(Section 3.3) and their performance is measured via metrics that are insensitive to class imbalance (Section 3.4).

3.1 Supervised Classification Algorithms

We tested five classifiers (Duda et al., 2001): linear SVM, SVM with nonlinear kernels, decision tree, random forest, and *k*-nearest neighbor.

Linear support vector machine The key idea of linear SVM is to find hyperplanes that separate the data as much as possible, that is, with a large margin. SVM optimizes a trade-off between maximizing the margin of separability between classes and minimizing margin errors. It provides a convex approximation to the combinatorial problem of minimizing classification errors. The practical implementation is formulated as the minimization of a penalized loss function and is implemented within the `fitclinear` function in Matlab.

Support vector machine Another idea behind SVM as presented by Vapnick in his original formulation (Vapnick, 1998) is to map the feature vectors in a nonlinear way to a high (possibly infinite) dimensional space and then utilize linear classifiers in this new space. This is done through the use of a kernel function. In our case, we tried two nonlinear kernel functions: Gaussian (also called Gaussian Radial Basis Function) and polynomial. `fitcsvm` implements this classifier in Matlab.

Linear SVM is a degraded version of the Gaussian kernel, that is, the results of a linear kernel can always be found with an appropriately scaled Gaussian kernel as well. Linear SVM classifiers will have linear decision boundaries, while the nonlinear kernel models are able to accommodate more flexible nonlinear decision boundaries with shapes depending on the type of kernel and its parameter's values. Linear SVM classifiers are approximately 30 times faster to train than nonlinear SVM models.

Decision tree Decision trees produce a set of if-then-else decision rules that are selected on the basis of the expected reduction in entropy that would result from sorting a particular attribute. Various decision tree algorithms have been proposed (Breiman et al., 1984; Quinlan, 1993). In this work, CART Trees were grown to their maximum size before a pruning step is applied to reduce over-fitting. The Gini coefficient and cross-entropy were used as splitting criteria, and various depth sizes were tested. The module `fitctree` implements this classifier. Decision trees are known to overfit the data and poorly generalize to new datasets (Duda et al., 2001). To produce better results, it is most often used in an ensemble context, such as random forests.

Random forest Random forests are a bag of decision trees. Each decision tree is applied on a new training data set obtained by random sampling with replacement from the original dataset. In addition, some randomness is introduced in the construction of the decision tree: For each decision tree, a subset of attributes is picked at random for the decision splits. The prediction of the ensemble is given as the averaged prediction of the individual decision tree (Breiman, 1998). We used the module `fitcensemble` with the `bag` aggregating method to implement the random forest algorithm.

K-nearest-neighbor (k-NN) The nearest neighbor rule for classifying $\mathbf{x}$ is to assign it the label associated with its nearest neighbor. An obvious extension is the *k-nearest-neighbor* *rule*, which classifies $\mathbf{x}$ by assigning it the label most frequently represented among the k nearest samples. In other words, a decision is made by examining the labels on the k nearest neighbors and taking a vote. Such a classifier relies on a metric, or distance, between feature vectors $\mathbf{x}$. In Matlab, `fitcknn` implements 10 different metrics beside the Euclidian distance. Despite its simplicity, the k-NN classifier has shown to be efficient, especially when the decision boundary is fairly unsmooth.

3.2 Imbalanced Dataset

Class imbalance is a common issue to many application domains (López et al., 2013). Three main techniques may be used and combined to alleviate this issue: cost-sensitive learning (Elkan, 2001), resampling (He and Garcia, 2009), and ensemble learning (Galar et al., 2012). Cost-sensitive learning modifies the cost-function to take into account higher misclassification costs due to class imbalance. Sampling techniques preprocess the data in order to diminish the class imbalance. Finally, ensemble learning techniques combine the outputs of many weak classifiers to produce a more powerful aggregate classifier. We describe these approaches as follows, and demonstrate some specific implementations using the Matlab Statistics and Machine Learning Toolbox (version R2016b).

3.2.1 Cost-Sensitive Learning

Cost-sensitive learning uses a cost-matrix with different types of errors to facilitate learning from an imbalanced dataset. With adapted cost, the minority class gains importance, that is, its errors are considered more costly than those of the majority class. A theoretical analysis on optimal cost-sensitive learning for binary classification is given in Elkan (2001).

Let $p(\mathbf{p})$ denote the prior probability for class positive and $p(\mathbf{n})$ the prior for the negative class. Those can be estimated from the observed frequencies. Provost and Fawcett (2001) showed that two classifiers will have the same performance when the following ratio is kept constant

$$\frac{c(\mathbf{Y},\mathbf{n})p(\mathbf{n})}{c(\mathbf{N},\mathbf{p})p(\mathbf{p})}$$

In that case of a balanced dataset, and with equal cost $c(\mathbf{Y},\mathbf{n}) = c(\mathbf{N},\mathbf{p})$, this ratio is equal to one. Elkan (2001) showed that in order to obtain the same performance for an imbalanced dataset, $c(\mathbf{Y},\mathbf{n})p(\mathbf{n})$ must also be proportional to $c(\mathbf{N},\mathbf{p})p(\mathbf{p})$. This means that if the positive class is less observed (i.e., $p(\mathbf{p}) < p(\mathbf{n})$), then the cost of misclassifying an event from the positive class must be set higher than misclassifying an event from the negative class (i.e., $c(\mathbf{N},\mathbf{p}) > c(\mathbf{Y},\mathbf{n})$).

In Matlab, one way to achieve this proportionality is by setting the parameter `prior` to `uniform` when using a classifier.[2]

In the case of SVM, this will result in the separating hyperplane being adjusted in order to reduce the expected risk. For decision trees, modifying the prior amounts to changing the

[2] In scikit-learn (version 19), the same result is obtained by setting the *class-weight* parameter to "*balanced*."

weights used to compute the probability that an observation lies in a particular node of the tree. With the parameter `prior` set to `uniform`, observations from a majority class will have a smaller weight and therefore be less predominant in the decision of which nodes to split. In random forest and boosting methods (see Section 3.2.3), the effect will be to sample more often from the minority class when creating the various training sets via random sampling with replacements. For *k*-NN classifiers, a weighted mean and standard deviation is computed, again giving more importance to the minority class.

3.2.2 *Sampling Methods*

Sampling methods are utilized to modify the original dataset in order to provide a more balanced distribution among classes, and hence improve the performance of subsequent classifiers (Estabrooks et al., 2004).

Random undersampling (RUS) removes data from the majority class in the original dataset, while *random oversampling* augments the original data by replicating selected examples from the minority class. Both have their caveats: undersampling may throw out too many examples from the majority class (García et al., 2012), while oversampling may lead to severe overfitting (Drummond and Holte, 1993).

Other than these two basic sampling strategies, more complex synthetic data generation approaches have emerged. In this work, we use the adaptive synthetic sampling approach (ADASYN, He et al., 2008), which combines two ideas. As its core, it generates new examples from the minority class via linear interpolation between existing minority class examples, similar to what the Synthetic Minority Oversampling TEchnique (Chawla et al., 2002) does. In addition, it incorporates a criterion to decide on the number of synthetic data that needs to be generated for each observation in the minority class, with more samples being generated in the vicinity of the boundary between the two classes.

We use the Matlab implementation of the ADASYN method provided in Siedhoff (2015).[3] The ADASYN algorithm takes as an input argument a prescribed percentage indicating how much class balance should be improved in comparison with the given class balance. We set this parameter so that in the oversampled dataset, the ratio of number of instances from the majority to the minority class is equal to five. Indeed, a ratio of 1:5 is considered mild and may be reasonably well handled by traditional algorithms, while limiting the amount of change in the original dataset and hence the risk of overfitting. When partitioning a dataset into training and a test set for performance evaluation, the upsampling with ADASYN is applied on the training set only, while evaluation is done on the untouched test set.

3.2.3 *Ensemble Learning*

Ensemble-based classifiers aim at improving the performance of single classifiers by inducing several classifiers and combining them to obtain a new classifier that outperforms every one of them. *Boosting* is one such method, which proceeds by sequentially applying a weak classification algorithm to the repeatedly modified version of the data, thereby producing a sequence of weak classifiers, which are then aggregated to produce the final prediction.

[3] In scikit-learn, the contributed package *imbalanced-learn* implements a variety of synthetic sampling methods.

In the popular `AdaBoost.M1` algorithm of Freund and Schapire (1997), for every weak learner trained in a sequential manner, the algorithm computes a weighted classification. It then increases weights for observations misclassified by the learner, and reduces weights for observations correctly classified. The next learner is then trained on data with these updated weights. After training, `AdaBoostM1` computes predictions for new data by a weighted majority vote on the weak learners.

Another boosting algorithm called `RUSBoost` (Seiffert et al., 2010) combines RUS and `AdaBoost.M1`. In essence, the algorithm takes N, the number of instances in the minority class of the training data, as the basic unit for sampling during boosting. Although simple, it was shown to be effective at classifying imbalanced data. The Matlab implementation of `RUSBoost` has a `RatioToSmallest` keyword that allows specification of the desired amount of undersampling. In our case, we set `RatioToSmallest` equal to 5, which means that, with N being number of samples in the minority class, each weak learner will use $5N$ samples randomly drawn from the majority class.

We use the Matlab function `fitcensemble` with decision trees as base classifiers and compare the results of the aggregating methods `AdaBoost.M1` and `RUSBoost`.

3.3 Training and Evaluation Protocol

The supervised classification in machine learning is always performed in two steps. (1) During the training phase, a model is estimated from the data. (2) In the validation (or test-)phase, the trained model is applied on other data, and model properties such as error classification rate are estimated. This is done in practice by splitting the initial data set into a "training dataset" and a "validation dataset."

However, in the present study we want to compare the performance of several classifiers. In this case, performing the splitting only once is not enough, as the observed difference between two classifiers may depend on the chosen training and test samples. To avoid this, we need to repeat the comparison over randomly selected partitions of the data and report the average performance (see Chapter 5 in Japkowics and Shah, 2014 for a discussion on error estimation in classification problems).

3.3.1 Training

We perform 50 iterations of the following protocol. We do a stratified shuffle-split of the original dataset into a 75% development ("training") set and 25% evaluation ("test") set. This means that we shuffle the original dataset, then split it into 75%/25% subsets (shuffle-split) where each subset has approximately the same class distributions as the full dataset (stratification). The development set in each iteration is used to train and evaluate each hyperparameter combination for each algorithm. To choose the best hyperparameter combination we use stratified fivefold cross-validation. A k-fold cross-validation means that the development set is further split into k-folds. Each possible combination of $k - 1$ folds is used for training while the remaining fold serves for testing, for a total of k train/test splits. From these k train/test splits, an average performance measure is computed. The operation is repeated for each combination of hyperparameter values and the one that achieves the best performance is retained for the validation step. The use of a *stratified* k-fold cross-validation means that each fold has approximately the same class distributions as the original dataset.

In Matlab, a stratified split is obtained when giving information about the class label as an input to the `cvpartition` function.[4]

3.3.2 Hyperparameter Optimization During Training

In order to find an optimal value for the classifier hyperparameters, Matlab proposes several methods, with Bayesian optimization (`bayesopt`) as the default. For continuous functions, Bayesian optimization typically works by assuming that the unknown function was sampled from a Gaussian process and updates a posterior distribution for this function as observations are recorded. In the case of machine learning, these observations are the measure of a generalization performance under different settings of the hyperparameters to be optimized. To choose the hyperparameters of the next experiment, one may optimize the *expected improvement* or the *expected improvement per second* as proposed by Snoek et al. (2012). The latter searches for points that are not only likely to be good, but also likely to be evaluated quickly. By default, `bayesopt` optimizes with respect to this last criteria.

We refer to Appendix C for a brief description of hyperparameters related to classifiers used in this study.

3.3.3 Evaluation

Once the optimum combination of hyperparameters is found, it is used to train a classifier on the 75% development set, and is evaluated on the 25% evaluation set. This final hold-out evaluation set is necessary to accurately estimate real-world performance because the cross-validation may over-fit for the particular split. The performance is measured by computing a skill score for each of the 50 iterations, see Section 3.4. This allows us to quantify an average performance, but also to evaluate the variance in performance results across different runs.

3.4 Performance Metrics

Because FC constitutes the minority class, we consider them as the "positive" class, following a common practice in the literature.

We computed for each of the 50 runs the "confusion matrix," see Table 1, which records the results of correctly and incorrectly recognized examples of each class. It contains the elements TP (true positive, filament channel predicted and observed), FP (false positive, filament

TABLE 1 Coronal Hole and Filament Channel Classification Contingency Table (Confusion Matrix)

Predicted	Observed: Filament Channel (FC)	Coronal Hole (CH)
FC	True Positive (TP)	False Positive (FP)
CH	False Negative (FN)	True Negative (TN)

[4] In scikit-learn (version 19), the *StratifiedShuffleSplit* and *StratifiedKFold* functions perform stratified splitting.

channel predicted and coronal hole observed), FN (false negative, coronal hole predicted and filament channel observed), and TN (true negative, coronal hole predicted and coronal hole observed).

There exists a variety of skill scores (Woodcock, 1976) built by combination of the confusion matrix elements. We seek, however, a metric that is unbiased by the imbalance in class distribution. The Hanssen and Kuipers discriminant, also known as True Skill Statistics (TSS) (Hanssen and Kuipers, 1965; Bloomfield et al., 2012), has this property. It is defined as the proportion of correctly predicted filaments among all filaments (TPR) minus the proportion of coronal holes that were classified as filaments among all coronal holes (FPR):

$$\mathrm{TSS} = \mathrm{TPR} - \mathrm{FPR} = \frac{\mathrm{TP}}{\mathrm{TP+FN}} - \frac{\mathrm{FP}}{\mathrm{FP+TN}} \tag{3}$$

The TSS is defined in the range $[-1, 1]$. A TSS of 0 indicates that the algorithm is as bad at misclassifying the negative class as it is good at classifying the positive class. A perfect classifier would have the value 1 or -1 (inverse classification), respectively. The combination of TPR and FPR in the form of the TSS provides an important and intuitive indicator for the performance of the developed framework.

For the readers used to the terms "recall" and "precision," we note that recall is equivalent to the TPR, and precision denotes the fraction of correctly predicted coronal holes among all predicted coronal holes. The TSS is given by $\mathrm{TP}/(\mathrm{TP+FN}) + \mathrm{TN}/(\mathrm{TN+FP}) - 1$, the sum of recalls for coronal holes and filament channels minus a scaling factor of 1.

Another well-known approach to unify measures from the confusion matrix and to produce evaluation criteria is to use the receiver operating characteristic (ROC) graphics. They characterize how the number of correctly classified filaments (TPR) vary with the number of incorrectly classified coronal holes (FPR). To compute an ROC curve, the classifier must output a *score* proportional to its belief (or estimated posterior probability distribution) that an instance belongs to the positive class. Decreasing the "decision" threshold, above which an instance is said to belong to the positive class, will increase both TPR and FPR. By varying the decision threshold, we obtain the ROC curve.

Intuitively, the ROC illustrates the tradeoff between detecting the complete set of filaments and the contamination of the filament class with coronal holes. Our interest in this study lies in avoiding classifying a filament as a coronal hole, because this will increase errors in the expected empirical relation that links the coronal hole area to the bulk solar wind speed. This means we aim at a high TPR, at the cost of a higher FPR.

The most important statistics associated with ROC curves are the area under the curve or AUC (Huang and Ling, 2005). Because ROC curves are located in the unit square $[0, 1] \times [0, 1]$, we have $0 \leq AUC \leq 1$. $AUC = 1$ is achieved if the classifier scores every positive instance higher than every negative; $AUC = 0$ is achieved if every negative is scored higher than every positive. AUC has a useful statistical interpretation: it is the expectation that a (uniformly) randomly drawn positive instance receives a higher score than a randomly drawn negative instance.[5]

[5] This is a normalized version of the Wilcoxon Mann-Whitney sum of ranks test, which tests the null hypothesis that two samples of ordinal measurements are drawn from a single distribution.

The Matlab function `predict` takes as input the classification model estimated on the training dataset and outputs one *score* per observation. With scores and true labels as input, the function `perfcurve` then computes the AUC and ROC curve.

4 RESULTS

Section 4.1 compares the strategies of cost-sensitive learning versus upsampling, while Section 4.2 gives the results for the ensemble-based classifiers. Finally, Section 4.3 compares the performance obtained on the four sets of features and discusses their relevance.

4.1 Cost-Sensitive Learning Versus Sampling Techniques

We first consider the full set of features (Set 4) with the goal of finding the best way to handle the imbalance between FC and CH classes. For each classifier described in Section 3.1, we tested four schemes:

1. No sampling techniques of original dataset and no cost-sensitive learning (`prior` set to `empirical` in Matlab).
2. ADASYN sampling technique to obtain a ratio of 1:5 in the dataset used for training and no cost-sensitive learning.
3. No sampling techniques and cost-sensitive learning (`prior` set to `uniform`).
4. ADASYN sampling to obtain a ratio of 1:5 and cost-sensitive learning.

Fig. 4 shows the boxplot of TSS for 50 runs with these four schemes using all attributes described in Section 2.2.

As expected, results in terms of TSS are the worst when nothing is done to counteract the imbalance. Two runs of SVM even obtain null TSS in this case. The best results are obtained with cost-sensitive learning, while the synthetic sampling technique also improves the performance, but to a lesser extent. It is not advisable to perform both an upsampling with ADASYN and a cost-sensitive learning as implemented in Matlab. While in case of SVM classifiers, performance is similar to the one achieved by a cost-sensitive approach alone, it actually decreases for other classifiers. In the case of decision tree and *k*-NN, the cost-sensitive approach puts a larger weight onto the minority class, and thus also on synthetic samples, thereby increasing the possibility of overfitting. Similarly, in random forest, the cost-sensitive approach will resample more often from the minority class, thus also putting more weight on synthetically generated samples.

When using a cost-sensitive approach, the five classifiers show similar performances, with SVM classifiers producing the highest-mean TSS. A linear SVM performs as well as a nonlinear SVM in terms of TSS while being faster to train. Table 2 indicates computation time necessary to train a classifier. It is, on average, 18 times more costly to train a nonlinear SVM than a linear SVM, while decision trees and *k*-NNs are the fastest to train, almost three times quicker than a linear SVM.

Fig. 5 shows the ROC curves for the case of cost-sensitive learning, and ADASYN sampling (without cost-sensitive learning). Although the cost-sensitive approach produced a slightly

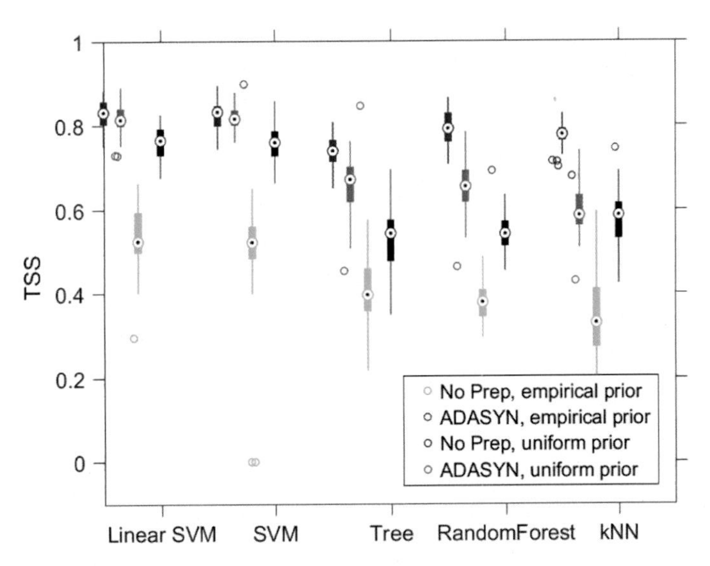

FIG. 4 Boxplot of True Skill Statistics for 50 runs on the full set of features. Median is represented as a *black dot inside a white circle* and the interquartile interval $q_{0.25}, q_{0.75}$ as a *thick line*, where $q_{0.25}$ (respectively, $q_{0.75}$) denotes the 25th (respectively, 75th) quantiles. The *thin line* represents the whiskers, which are set to 2.7σ, with σ the standard deviation of the TSS. Outliers are shown with *circles*. The four strategies to handle imbalance are: (1) no resampling, empirical prior (no cost-sensitive learning) (shown in *green*), (2) ADASYN resampling, empirical prior (in *black*), (3) no resampling, uniform prior (cost-sensitive learning) (in *blue*), (4) ADASYN resampling, uniform prior (in *magenta*).

TABLE 2 Mean and Standard Deviation Computation Time (in Seconds) Necessary to Train a Classifier With the Cost-Sensitive Approach, and Mean Ratio With Respect to Linear SVM Computing Time

	Linear SVM	SVM	Tree	Random Forest	*k*-NN
Mean (in s)	106.8	1904.4	33.3	284.3	43.1
SD (in s)	32.3479	654.463	8.669	162.8966	9.7101
Ratio w.r.t. linear SVM	1	18.8013	0.3354	2.8544	0.4288

Notes: The mean and standard deviations are computed over 50 runs. Calculations are done on an Intel(R) Xeon(R) processor with 10 cores, a CPU at 2.80 GHz, and 256 GB RAM.

higher average TSS, the estimated ROC curves show that, in case of SVM classifiers, a TPR of 0.9 yields an FPR of 0.082 for the cost-sensitive approach, and a slightly lower value of 0.076 for the sampling approach.

4.2 Ensemble Learning

We compare the ADABoost and RUSBoost methods described in Section 3.2.3 for the full set of features when using both types of prior. Fig. 6 illustrates via TSS and the ROC curve the better performance of ADABoost over RUSBoost. The performance is, however, inferior to the one achieved by an SVM classifier, as illustrated by the higher value of FPR incurred for a TPR of 0.9.

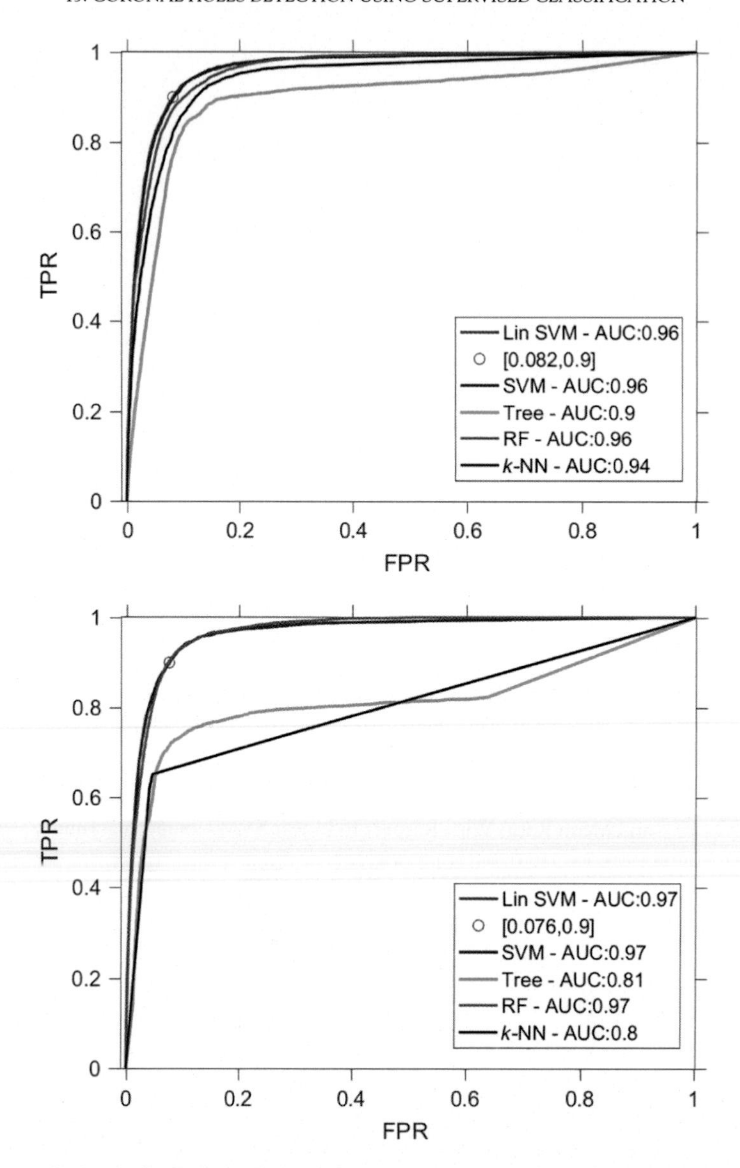

FIG. 5 ROC curves for linear SVM, SVM, decision trees, random forests, and *k*-nearest neighbors, applied on the full set of features, with corresponding AUC in the legend (*top*) cost-sensitive learning. Linear SVM achieves an FPR of 0.082 for a TPR of 0.9 (*bottom*) ADASYN applied on the training set (no cost-sensitive learning). Linear SVM achieves an FPR of 0.076 for a TPR of 0.9.

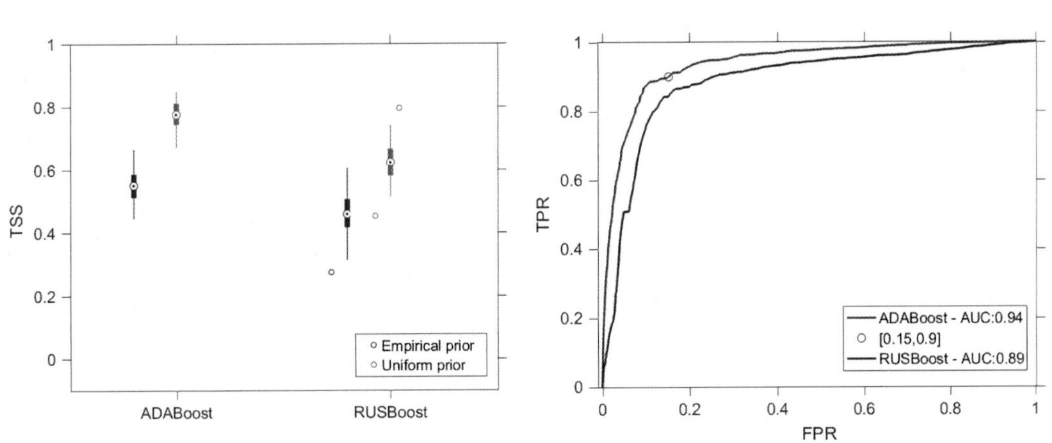

FIG. 6 Comparison of ADABoost and RUSBoost ensemble algorithms on the full set of features (*left*) True Skill Statistics, with and without cost-learning (*right*) ROC curve for cost-sensitive learning. With ADABoost, a TPR of 0.9 yields an FPR of 0.15.

4.3 Importance of Attributes

We analyze the loss incurred when fewer attributes are employed. Our goal is to obtain a set of rules that are easy to implement within the SDO EDS pipeline, where only AIA data series are inputted to the SPoCA-CH module.

Fig. 7(top) shows the TSS for the four sets of attributes described in Section 2.3.5 handled with the five base classifiers in case of cost-sensitive learning.

The use of HMI attributes seems to improve significantly the TSS, while adding shape attributes does not seem to be critical. There is still some benefit in performing supervised classification with AIA information though. In that case, all five classifiers but the decision tree exhibit similar performance, as further illustrated by the ROC curves in Fig. 7(bottom).

The linear SVM model estimates a vector of coefficient β, which is used to compute the classification score f for classifying a feature vector $\mathbf{x}$ into the positive class:

$$f_{\mathrm{LinSVM}}(\mathbf{x}) = \beta\mathbf{x} + \alpha$$

where α is the intercept of the model. A null value of a coefficient β thus indicates that the corresponding attributes are irrelevant for discrimination purposes.

Appendix D lists the first 15 attributes with the largest linear SVM median coefficient values, for sets 2 and 4. The median is computed over the 50 runs of simulations. Variability in first-order statistics and textural information regarding HMI and AIA are given a high importance, while location information also shows some discriminating power. Note that different classifiers (e.g., *k*-NN) might favor different attributes, and hence a comprehensive feature selection analysis is left for future work.

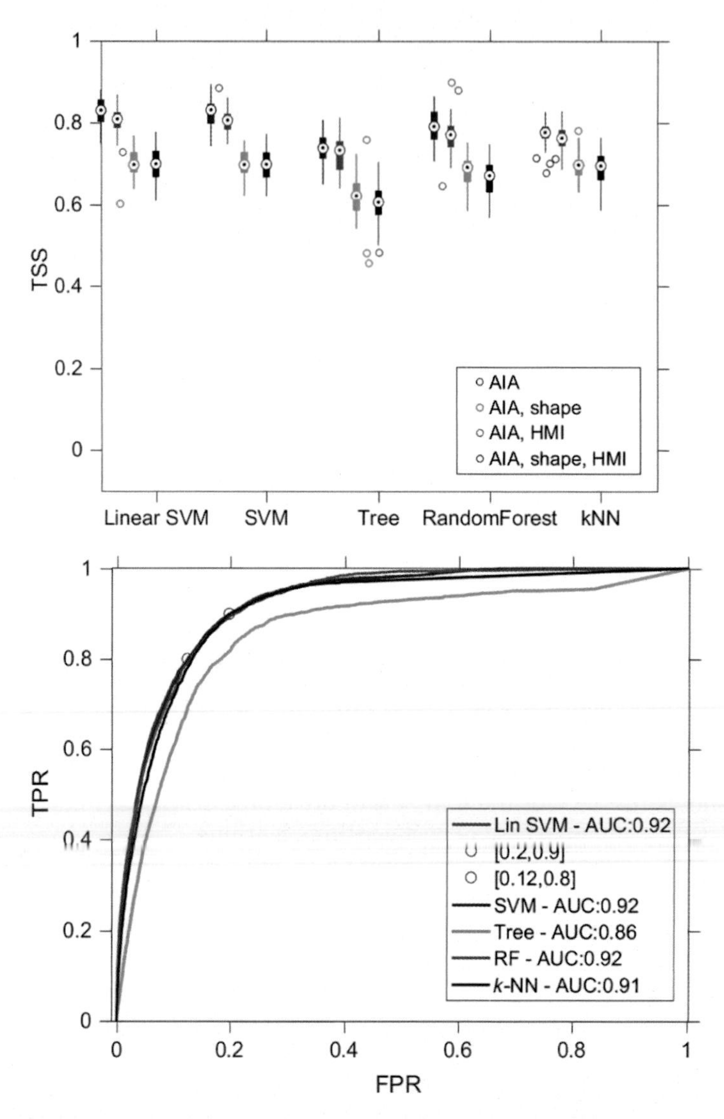

FIG. 7 (*Top*) TSS for linear SVM, SVM, decision trees, random forests, and *k*-NN for the four sets of attributes and with costs-sensitive learning. (*Bottom*) ROC curves for the same five classifiers, the set of attributes containing AIA and shape information, and with cost-sensitive learning. Linear SVM induces an FPR of 0.2 for a TPR of 0.9, and an FPR of 0.12 for a TPR of 0.8.

5 DISCUSSION AND CONCLUSION

In this paper, we provided an improved way of detecting coronal holes with a modified version of the SPoCA algorithm. Next, we used a labeled dataset of coronal holes and filament channels, computed a set of attributes related to these detected regions, and trained a set of classifiers in order to separate the coronal holes from the filament channels. Our focus was on finding the classifier that performs the best. We also compare ways to handle the imbalance in the number of CH versus FC.

Our study indicates that the cost-sensitive learning approach is the best at handling class imbalance, while (in the case when no-cost sensitive approach is available) upsampling the minority class also increases the performance in terms of TSS.

We find that adding magnetogram information allows for a better distinction between both types of solar features. When using HMI information and cost-sensitive learning, linear SVM and SVM classifiers provide the highest TSS values. When only AIA data are available, linear SVM, SVM, random forest, and *k*-NN achieved similar TSS.

As computation time is often an issue, linear SVM has the advantage over SVM that it is faster to train and requires only one hyperparameter to be optimized. When only AIA data are available, *k*-NN could be considered as the best classifier, as it performs as well as linear SVM while being three times faster to train. A comprehensive feature selection analysis could lead to improved computation time, and is left for future work.

In Reiss et al. (2016) we found that the empirical model over-forecasts the number of high-speed streams (bias, BS = 1.20) during the years 2011–14 and concluded that the high number of false alarms (false alarm ratio, FAR = 0.43) might partially be attributed to the false detection of coronal holes. Our interest in this study thus lies in avoiding the false detection of coronal holes, that is, to classify a filament as a coronal hole. On the contrary, misclassifying a coronal hole as a filament might be less costly. Indeed, such "false" coronal holes are likely to have attributes resembling filaments. For example, relatively bright CHs with low magnetic field density and small areas are supposed to result in low-velocity high-speed streams (Nolte et al., 1976; Abramenko et al., 2009; Karachik and Pevtsov, 2011).

According to the ROC curve on Fig. 5, a linear SVM classifier applied on a dataset preprocessed with ADASYN may achieve a TPR of 90% for an FPR of 7.6% for an appropriate threshold on the scores provided by the classifier. Given the high ratio between coronal holes and filaments, such values of TPR and FPR might represent a reasonable balance between potential false alarms and misses in terms of forecasting.

This study paves the way for establishing a decision rule to distinguish filament channels from coronal holes. As we made 50 runs, we further need to analyze how hyperparameter values change over these runs. We also need to perform a comprehensive feature selection and ranking in order to interpret the results, and to look into characteristics of misclassified coronal holes. We expect those to have a small area, which does not affect space weather predictions.

The final step will then be to implement this decision rule in near real time. HMI data and AIA data exist in near real time at JSOC as quicklook data, whereas we used science-grade data in the present study (level 1.5 images). We would thus need to check the quality of SPoCA-CH detection, computed attributes, and precision of the classifier on these quicklook data. When tracking one region over time, we may want to add information about previous classifications, in order to avoid alternating classifications between coronal holes and filaments. The optimal decision rule may also vary with the solar cycle, and will need to be reevaluated regularly. We note that our study already covers more than half of a solar cycle, and therefore should be relatively stable over time.

Further improvements in the SPoCA-CH could be achieved by an improved EUV limb brightening correction. A better knowledge of the height of coronal holes and filament channels as seen by EUV images may also help in separating them because such height plays an important role in the projection of the coronal hole boundaries to the photospheric magnetic field data.

Finally, the coronal hole detection and machine learning modules could be combined with the Hα automatized filament detection currently running at Kanzelhöhe Solar Observatory.

ACKNOWLEDGMENTS

Reiss and Temmer gratefully acknowledge the NAWI Graz funding *Förderung von JungforscherInnengruppen 2013–15*. The research leading to these results has received funding from the European Commission's Seventh Framework Programme (FP7/2007-2013) under the grant agreement No. 284461 (eHEROES). Delouille and Mampaey acknowledge support from the Belgian Federal Science Policy Office through the BRAIN.be and the ESA-PRODEX programs. Hofmeister acknowledges support from the JungforscherInnenfonds der Steiermärkischen Sparkassen, as well as from the association *Dynamics of the Solar System* to finance a 1-month stay at the Royal Observatory of Belgium. We thank P. Dupont, R. D'Ambrosio, and the two anonymous reviewers for their comments that helped improve the manuscript.

APPENDIX A FIRST-ORDER IMAGE STATISTICS

Pixel values contained in the extracted regions were rounded to the nearest integer value. The probability $P(i)$ for the occurrence of pixels with integer value i is defined as

$$P(i) = \frac{n(i)}{N} \tag{A.1}$$

where $n(i)$ is defined as the number of pixels with the actual pixel value i and N is the total number of image pixels. The following standard first-order attributes were used:

$$\text{Mean:}\ \ \mu = \sum_i i\, P(i) \tag{A.2}$$

$$\text{Variance:}\ \ \sigma^2 = \sum_i (i-\mu)^2 P(i) \tag{A.3}$$

$$\text{Standard deviation (contrast):}\ \ C_1 = \sqrt{\sigma^2} \tag{A.4}$$

$$\text{Energy:}\ \ E_1 = \sum_i P(i)^2 \tag{A.5}$$

$$\text{Entropy:}\ \ S_1 = \sum_i -P(i)\log(P(i)) \tag{A.6}$$

Variance σ^2 and contrast C_1 are associated with the width of the pixel value distribution. High variance or contrast indicate large pixel value differences. Energy E_1 is high for unevenly distributed pixel values independent of the designated pixel value itself. Entropy S_1 values reflect the degree of information content within the pixel value distribution. Low entropy values correspond to coherent pixel values with low information content. High values reflect highly disordered image values with a high amount of intrinsic information content.

APPENDIX B SECOND-ORDER IMAGE STATISTICS

B.1 Cooccurrence Matrix

Analogous to the probability $P(i)$ for the occurrence of pixels with value i, we define $P_{\phi,d}(i,j)$ indicating the probability for the cooccurrence of pixel value i and pixel value j in a given distance d and direction ϕ. $n_{\phi,d}(i,j)$ is the total number of cooccurrences of pixel value i with pixel value j within the distance d and direction ϕ. The matrix $P_{\phi,d}(i,j)$ is defined as

$$P_{\phi,d}(i,j) = \frac{n_{\phi,d}(i,j)}{N} \tag{B.1}$$

where N is the total number of object pixels. Because there is no consensus about the choice of angles ϕ and distance d, the calculations have been done with a distance of one pixel and 8 directions considering only the nearest neighborhood of each pixel. The used set of angles $\Omega = \{\phi_0, \ldots, \phi_7\}$ is given by

$$\phi_l = l \cdot 45 \,\text{degrees}, \quad l = 0, \ldots, 7, \quad \phi_l \in \Omega \tag{B.2}$$

This set of angles provides eight matrices $P_{\phi,d}(i,j)$, one matrix for each of the directions. The cooccurrence matrix $C(i,j)$ is finally defined as

$$C(i,j) = \text{avg}\left(P_{\phi,d}(i,j)\right) \tag{B.3}$$

The entry (i,j) in the cooccurrence matrix $C(i,j)$ therefore indicates the probability for the cooccurrence of pixel value i with pixel value j within the nearest neighborhood.

B.2 Notations

We suggest to use a set of textural features which can be calculated from the cooccurrence matrix. We need the following notations.

$C(i,j)$	(i,j)th entry in the normalized cooccurrence matrix
$p_x(i)$	ith entry is obtained by summing the columns of $C(i,j)$
$p_y(j)$	jth entry is obtained by summing the rows of $C(i,j)$
$p_{x-y}(k)$	kth entry of $p_{x-y}(k)$ corresponding to the sum over all entries of $C(i,j)$ with absolute pixel value difference of i and j equal to k
$p_{x+y}(k)$	kth entry of $p_{x+y}(k)$ corresponding to the sum over all entries of $C(i,j)$ where the addition of i and j is equal to k
μ_x, μ_y	means of p_x and p_y
σ_x, σ_y	standard deviations of p_x and p_y
N_g	Number of distinct pixel values
$\sum_i, \sum_j$	Convention indicating $\sum_{i=1}^{N_g}, \sum_{j=1}^{N_g}$
HX, HY	Entropies of p_x and p_y

$$p_x(i) = \sum_j C(i,j), \quad p_y(j) = \sum_i C(i,j) \tag{B.4}$$

$$p_{x+y}(k) = \sum_i \sum_j C(i,j),\ i+j=k,\ k=2,3,\ldots,2N_g \tag{B.5}$$

$$p_{x-y}(k) = \sum_i \sum_j C(i,j),\ |i-j|=k,\ k=0,1,\ldots,N_g-1 \tag{B.6}$$

$$\mathrm{HX} = -\sum_i p_x(i)\ \log(p_x(i)), \quad \mathrm{HY} = -\sum_i p_y(i)\ \log(p_y(i)) \tag{B.7}$$

$$\mathrm{HXY1} = -\sum_i \sum_j C(i,j) \log(p_x(i) p_y(j)) \tag{B.8}$$

$$\mathrm{HXY2} = -\sum_i \sum_j p_x(i)\ p_y(j) \log(p_x(i)\ p_y(j)) \tag{B.9}$$

B.3 Textural Features

The following textural features can be calculated:
Energy:

$$H_1 = \sum_i \sum_j C^2(i,j) \tag{B.10}$$

Contrast:

$$H_2 = \sum_i \sum_j (i-j)^2 C(i,j) \tag{B.11}$$

Correlation:

$$H_3 = \sum_i \sum_j \frac{(i-\mu_x)(j-\mu_y)}{\sigma_x \sigma_y} \tag{B.12}$$

Sum of squares – Variance:

$$H_4 = \sum_i \sum_j (i-\mu)^2 C(i,j) \tag{B.13}$$

Homogeneity:

$$H_5 = \sum_i \sum_j \frac{C(i,j)}{1+(i-j)^2} \tag{B.14}$$

Sum average:

$$H_6 = \sum_{i=2}^{2N_g} i\, p_{x+y}(i) \tag{B.15}$$

Sum entropy:

$$H_8 = -\sum_{i=2}^{2N_g} p_{x+y}(i)\log\left(p_{x+y}(i)\right) \tag{B.16}$$

Entropy:

$$H_9 = -\sum_i \sum_j C(i,j)\log\left(C(i,j)\right) \tag{B.17}$$

Sum variance:

$$H_7 = \sum_{i=2}^{2N_g} (i - H_8)^2 p_{x+y}(i) \tag{B.18}$$

Difference variance:

$$H_{10} = \mathrm{var}\left(p_{x-y}\right) \tag{B.19}$$

Difference entropy:

$$H_{11} = -\sum_{i=0}^{N_g-1} p_{x-y}(i)\log(p_{x-y}(i)) \tag{B.20}$$

Information measures of correlation (I):

$$H_{12} = \frac{H_9 - \mathrm{HXY1}}{\max(\mathrm{HX},\mathrm{HY})} \tag{B.21}$$

Information measures of correlation (II):

$$H_{13} = \sqrt{1 - \exp(-2\,(\mathrm{HXY2} - H_9)\,)} \tag{B.22}$$

APPENDIX C CLASSIFIER HYPERPARAMETER RANGE

C.1 Base Classifiers

Here we give for each base classifier a description of their hyperparameters and on which range of values they are searched. In the following, we let n be the total number of observations and d be the number of predictors.

Linear SVM Linear SVM has only one free parameter, the penalization factor (called λ in Matlab) that controls how much to add as a penalty to misclassified training samples. The optimization searches for optimal λ over the range 10^{-8} to 10^{2}.
SVM In case of (possibly nonlinear) SVM, the penalization parameter is called "box constraint" within Matlab. Its optimization searches the range [10^{-3} to 10^{3}].
Because a linear kernel will give similar results to the linear SVM above, we consider here only two (nonlinear) kernels: Gaussian and Polynomial. The "kernel scale" is optimized within the range [10^{-3} to 10^{3}], and for Polynomial kernels the function tests polynomials of order 2 to 4.

Decision tree Two splitting criteria are tested: Gini coefficient and cross-entropy. The tree depth size is further controlled by two parameters:

- `MinLeafSize` is minimum number of observations within a leaf node. `MinLeafSize` values between 1 and $\lfloor n/2 \rfloor$ were tested.
- `MaxNumSplits` : If the number of branch nodes within a layer exceeds `MaxNumSplits`, the algorithms does some merging of leaves, or return the grown tree. A layer is the set of nodes that are equidistant from the root node. The algorithm search for the best number of maximum splits within the range $[1, n-1]$

***K*-nearest neighbor** The following parameters are searched during hyperparameter optimization:

- Number of nearest neighbors: `fitcknn` searches for the optimal number of neighbors to consider within the range $[1, n/2]$.
- Distance: `fitcknn` searches among "cityblock," "chebychev," "correlation," "cosine," "Euclidean," "hamming," "jaccard," "mahalanobis," "minkowski," "seuclidean," and "spearman."
- Distance-weight: the mean and variance used to calculate distances may be computed with "equal," "inverse," or "squared inverse" weights.
- The Minkowski distance exponent is searched within the range [0.5, 3].

C.2 Ensemble Methods

Random forest We use `fitcensemble` with `Method` set to *bag* and *Tree* as `Learner`. As for decision tree, the Gini coefficient and cross-entropy criteria are tested. The tree depth size is controlled by three parameters as follows:

- Minimum leaf size (`MinLeafSize`) takes values between 1 and $\lfloor (n/2) \rfloor$.
- `MaxNumSplits`: The algorithm search for the best number of maximum splits within the range $[1, n-1]$.
- `NumVariablesToSample`: This is the number of attributes to select at random for each decision tree of the random forest. It is optimized over the range 1 to d.

Boosting methods We use `fitcensemble` with `Method` set to `ADABoost.M1` and `RUSBoost`, respectively. In both cases the same set of hyperparameters is used. Three of such parameters are similar to the random forest, namely: the split criteria, minimum leaf size, and maximum number of splits. In addition, the number of learning cycle (`NumLearningCycles`) is to be chosen within $[10, 500]$ and the learning rate (`LearnRate`) within $[10^{-3}, 1]$.

APPENDIX D RELEVANCE OF ATTRIBUTES

Tables D.1 and D.2 lists the attributes, which contribute the most to the determination of the linear SVM decision boundaries.

TABLE D.1 Attributes With the 15 Largest Median Coefficient Values for the Linear SVM Model, Ranked From Highest to Smallest Value

	AIA, HMI, Shape Attributes (Set 4)	
Rank	**Name in Dataset**	**Definition**
1	Energy1HMI	HMI first-order energy E_1, Eq. (A.5)
2	Entropy1HMI	HMI first-order entropy S_1, Eq. (A.6)
3	Entropy1AIA	AIA first-order entropy S_1, Eq. (A.6)
4	Contrast1AIA	AIA first-order contrast, Eq. (A.4)
5	EntropyAIA	AIA Haralick entropy H_9, Eq. (B.17)
6	Distancetodiskcenterarcsec	Distance to disk center
7	minlat	Minimum latitude
8	DiffVarianceHMI	HMI Difference Variance, Eq. (B.19)
9	Energy2HMI	HMI Energy H_1, Eq. (B.10)
10	SumEntropyHMI	HMI Sum Entropy, Eq. (B.16)
11	MAGNETICFLUXHMI	R_2 value defined in Eq. (2)
12	Contrast2AIA	AIA Haralick contrast H_2, Eq. (B.11)
13	SumVarianceAIA	AIA Haralick sum of variance, Eq. (B.18)
14	StdDevXAIA	σ_x defined in Appendix B.2 computed on AIA
15	Contrast2HMI	HMI Haralick contrast H_2, Eq. (B.11)

Notes: The median of coefficient values is computed over 50 runs of simulation. Dataset contains 56 attributes in total.

TABLE D.2 Attributes With the 15 Largest Median Coefficient Values for the Linear SVM Model, Ranked From Highest to Smallest Value

	AIA, Shape Attributes (Set 2)	
Rank	**Name in Dataset**	**Definition**
1	absMeanAIA	AIA absolute mean value, Eq. (A.2)
2	Contrast1AIA	AIA first-order contrast, Eq. (A.4)
3	StdDevXAIA	σ_x defined in Appendix B.2 computed on AIA
4	meanXAIA	μ_x defined in Appendix B.2 computed on AIA
5	SumAverageAIA	AIA Haralick sum average, Eq. (B.15)
6	Contrast2AIA	AIA Haralick contrast H_2, Eq (B.11)
7	DiffVarianceAIA	AIA Haralick difference variance, Eq. (B.19)
8	SqVarianceAIA	AIA Haralick sum of square variance, Eq. (B.13)
9	Entropy1AIA	AIA first-order entropy S_1, Eq. (A.6)

Continued

TABLE D.2 Attributes With the 15 Largest Median Coefficient Values for the Linear SVM Model, Ranked From Highest to Smallest Value—cont'd

	AIA, Shape Attributes (Set 2)	
Rank	**Name in Dataset**	**Definition**
10	HomogeneityAIA	AIA Haralick homogeneity H_5, Eq. (B.14)
11	minlat	Minimum latitude
12	VarianceAIA	AIA variance, Eq. (A.3)
13	EntropyAIA	AIA Haralick entropy H_9, Eq. (B.17)
14	Areakm2	Area in km^2
15	HXY1AIA	AIA entropy, Eq. (B.8)

Notes: The median of coefficient values is computed over 50 runs of simulation. Dataset contains 32 attributes in total.

References

Abramenko, V., Yurchyshyn, V., Watanabe, H., 2009. Parameters of the magnetic flux inside coronal holes. Sol. Phys. 260, 43–57. https://doi.org/10.1007/s11207-009-9433-7.

Ahammer, H., Kröpfl, J.M., Hackl, C., Sedivy, R., 2008. Image statistics and data mining of anal intraepithelial neoplasia. Pattern Recogn. Lett. 29 (16), 2189–2196. ISSN 0167-8655.

Anscombe, F.J., 1948. The transformation of Poisson, Binomial and Negative-Binomial data. Biometrika 35 (3–4), 246. https://doi.org/10.1093/biomet/35.3-4.246.

Barra, V., Delouille, V., Hochedez, J.F., 2007. Segmentation of extreme ultraviolet solar images using a multispectral data fusion process. In: 2007 IEEE International Fuzzy Systems Conference, pp. 1–6.

Bezdek, J.C., 1981. Pattern Recognition With Fuzzy Objective Function Algorithms. Kluwer Academic Publishers, Norwell, MA.

Bloomfield, D.S., Higgins, P.A., McAteer, R.T.J., Gallagher, P.T., 2012. Toward reliable benchmarking of solar flare forecasting methods. Astrophys. J. Lett. 747, L41. https://doi.org/10.1088/2041-8205/747/2/L41.

Breiman, L., 1998. Arcing classifier (with discussion and a rejoinder by the author). Ann. Statist. 26 (3), 801–849. https://doi.org/10.1214/aos/1024691079.

Breiman, L., Friedman, J., Olshen, R., Stone, C., 1984. Classification and Regression Trees. Chapman & Hall, New York, NY.

Chawla, N., Bowyer, K.W., Hall, L.O., Kegelmeyer, W.P., 2002. SMOTE: synthetic minority over-sampling technique. J. Artif. Intell. Res. 16, 321–357.

Cranmer, S.R., 2009. Coronal holes. Living Rev. Sol. Phys. 6 (3). https://doi.org/10.12942/lrsp-2009-3.

de Toma, G., 2011. Evolution of coronal holes and implications for high-speed solar wind during the minimum between cycles 23 and 24. Sol. Phys. 274, 195–217. https://doi.org/10.1007/s11207-010-9677-2.

Drummond, C., Holte, R.C., 1993. C4.5, Class Imbalance, and Cost Sensitivity: Why Under-Sampling Beats Over-Sampling. Morgan Kaufmann Publishers Inc., San Francisco, CA, pp. 1–8.

Duda, R.O., Hart, P.E., Stork, D.G., 2001. Pattern Classification, second ed. Wiley, Hoboken, NJ.

Elkan, C., 2001. The foundations of cost-sensitive learning. In: Proceedings of the 17th International Joint Conference on Artificial Intelligence. IJCAI'01, Seattle, WA, vol. 2. Morgan Kaufmann Publishers Inc., San Francisco, CA, pp. 973–978, Available from: http://dl.acm.org/citation.cfm?id=1642194.1642224.

Estabrooks, A., Jo, T., Japkowicz, N., 2004. A multiple resampling method for learning from imbalanced data sets. Comput. Intell. 20 (1), 18–36. ISSN 1467-8640. https://doi.org/10.1111/j.0824-7935.2004.t01-1-00228.x.

Freund, Y., Schapire, R.E., 1997. A decision-theoretic generalization of on-line learning and an application to boosting. J. Comput. Syst. Sci. 55, 119–139.

Galar, M., Fernandez, A., Barrenechea, E., Bustince, H., Herrera, F., 2012. A review on ensembles for the class imbalance problem: bagging-, boosting-, and hybrid-based approaches. IEEE Trans. Syst. Man Cybern. Part C Appl. Rev. 42 (4), 463–484. ISSN 1094-6977. https://doi.org/10.1109/TSMCC.2011.2161285.

García, V., Sánchez, J.S., Mollineda, R.A., 2012. On the effectiveness of preprocessing methods when dealing with different levels of class imbalance. Knowl. Based Syst. 25 (1), 13–21. ISSN 0950-7051. https://doi.org/10.1016/j.knosys.2011.06.013. Special Issue on New Trends in Data Mining, Available from: http://www.sciencedirect.com/science/article/pii/S0950705111001286.

Gosling, J.T., Pizzo, V.J., 1999. Formation and evolution of corotating interaction regions and their three dimensional structure. Space Sci. Rev. 89, 21–52. https://doi.org/10.1023/A:1005291711900.

Hanssen, A.W., Kuipers, W.J.A., 1965. On the relationship between the frequency of rain and various meteorological parameters: (with reference to the problem of objective forecasting), Available from: https://books.google.be/books?id=nTZ8OgAACAAJ.

Haralick, R.M., Shanmugam, K., Dinstein, I., 1973. Textural features for image classification. IEEE Trans. Syst. Man Cybern. SMC-3 (6), 610–621. ISSN 0018-9472. https://doi.org/10.1109/TSMC.1973.4309314.

He, H., Garcia, E.A., 2009. Learning from imbalanced data. IEEE Trans. Knowl. Data Eng. 21 (9), 1263–1284. ISSN 1041-4347. https://doi.org/10.1109/TKDE.2008.239.

He, H., Bai, Y., Garcia, E.A., Li, S., 2008. ADASYN: adaptive synthetic sampling approach for imbalanced learning. In: International Joint Conference on Neural Networks, IJCNN Hong Kong, China, June 1–6, 2008. IEEE, pp. 1322–1328.

Henney, C.J., Harvey, J.W., 2005. Automated coronal hole detection using He 1083 nm spectroheliograms and photospheric magnetograms. In: Sankarasubramanian, K., Penn, M., Pevtsov, A. (Eds.), Large-scale structures and their role in solar activity, Astronomical Society of the Pacific Conference Series, vol. 346, 261 pp.

Hofmeister, S.J., Veronig, A., Reiss, M.A., Temmer, M., Vennerstrom, S., Vršnak, B., Heber, B., 2017. Characteristics of low-latitude coronal holes near the maximum of Solar Cycle 24. Astrophys. J. 835, 268. https://doi.org/10.3847/1538-4357/835/2/268.

Huang, J., Ling, C.X., 2005. Using AUC and accuracy in evaluating learning algorithms. IEEE Trans. Knowl. Data Eng. 17, 299–310.

Hurlburt, N., Cheung, M., Schrijver, C., Chang, L., Freeland, S., Green, S., Heck, C., Jaffey, A., Kobashi, A., Schiff, D., Serafin, J., Seguin, R., Slater, G., Somani, A., Timmons, R., 2012. Heliophysics event knowledgebase for the solar dynamics observatory (SDO) and beyond. Sol. Phys. 275, 67–78. https://doi.org/10.1007/s11207-010-9624-2.

Japkowics, N., Shah, M., 2014. Evaluating Learning Algorithms: A Classification Perspective. Cambridge University Press, Cambridge.

Karachik, N.V., Pevtsov, A.A., 2011. Solar wind and coronal bright points inside coronal holes. Astrophys. J. 735, 47. https://doi.org/10.1088/0004-637X/735/1/47.

Kirk, M.S., Pesnell, W.D., Young, C.A., Hessï¿½Webber, S.A., 2009. Automated detection of EUV Polar Coronal Holes duringï¿½Solar Cycle 23. Sol. Phys. 257 (1), 99–112. ISSN 1573-093X. https://doi.org/10.1007/s11207-009-9369-y.

Krista, L.D., Gallagher, P.T., 2009. Automated coronal hole detection using local intensity thresholding techniques. Sol. Phys. 256 (1), 87–100. ISSN 1573-093X. https://doi.org/10.1007/s11207-009-9357-2.

Lemen, J.R., Title, A.M., Akin, D.J., Boerner, P.F., Chou, C., Drake, J.F., Duncan, D.W., Edwards, C.G., Friedlaender, F.M., Heyman, G.F., Hurlburt, N.E., Katz, N.L., Kushner, G.D., Levay, M., Lindgren, R.W., Mathur, D.P., McFeaters, E.L., Mitchell, S., Rehse, R.A., Schrijver, C.J., Springer, L.A., Stern, R.A., Tarbell, T.D., Wuelser, J.P., Wolfson, C.J., Yanari, C., Bookbinder, J.A., Cheimets, P.N., Caldwell, D., Deluca, E.E., Gates, R., Golub, L., Park, S., Podgorski, W.A., Bush, R.I., Scherrer, P.H., Gummin, M.A., Smith, P., Auker, G., Jerram, P., Pool, P., Soufli, R., Windt, D.L., Beardsley, S., Clapp, M., Lang, J., Waltham, N., 2012. The Atmospheric Imaging Assembly (AIA) on the Solar Dynamics Observatory (SDO). Sol. Phys. 275, 17–40. https://doi.org/10.1007/s11207-011-9776-8.

López, V., Fernández, A., García, S., Palade, V., Herrera, F., 2013. An insight into classification with imbalanced data: empirical results and current trends on using data intrinsic characteristics. Inf. Sci. https://doi.org/10.1016/j.ins.2013.07.007.

Mackay, D.H., Karpen, J.T., Ballester, J.L., Schmieder, B., Aulanier, G., 2010. Physics of solar prominences: II—magnetic structure and dynamics. Space Sci. Rev. 151 (4), 333–399. ISSN 1572-9672. https://doi.org/10.1007/s11214-010-9628-0.

Mackay, D.H., Karpen, J.T., Ballester, J.L., Schmieder, B., Aulanier, G., 2010. Physics of solar prominences: II—magnetic structure and dynamics. Space Sci. Rev. 151, 333–399. https://doi.org/10.1007/s11214-010-9628-0.

Martens, P.C.H., Attrill, G.D.R., Davey, A.R., Engell, A., Farid, S., Grigis, P.C., Kasper, J., Korreck, K., Saar, S.H., Savcheva, A., Su, Y., Testa, P., Wills-Davey, M., Bernasconi, P.N., Raouafi, N.E., Delouille, V.A., Hochedez, J.F., Cirtain, J.W., Deforest, C.E., Angryk, R.A., de Moortel, I., Wiegelmann, T., Georgoulis, M.K., McAteer, R.T.J., Timmons, R.P., 2012. Computer vision for the Solar Dynamics Observatory (SDO). Sol. Phys. 275, 79–113. https://doi.org/10.1007/s11207-010-9697-y.

Müller, D., Dimitoglou, G., Caplins, B., Garcia Ortiz, J.P., Wamsler, B., Hughitt, K., Alexanderian, A., Ireland, J., Amadigwe, D., Fleck, B., 2009. JHelioviewer—visualizing large sets of solar images using JPEG 2000. Comput. Sci. Eng. 11 (5), 38–47. https://doi.org/10.1109/MCSE.2009.142.

Munro, R.H., Withbroe, G.L., 1972. Properties of a coronal "hole" derived from extreme-ultraviolet observations. Astrophys. J. 176, 511. https://doi.org/10.1086/151653.

Nolte, J.T., Krieger, A.S., Timothy, A.F., Gold, R.E., Roelof, E.C., Vaiana, G., Lazarus, A.J., Sullivan, J.D., McIntosh, P.S., 1976. Coronal holes as sources of solar wind. Sol. Phys. 46, 303–322. https://doi.org/10.1007/BF00149859.

Pötzi, W., Veronig, A.M., Riegler, G., Amerstorfer, U., Pock, T., Temmer, M., Polanec, W., Baumgartner, D.J., 2015. Real-time flare detection in ground-based Hα imaging at Kanzelhöhe observatory. Sol. Phys. 290, 951–977. https://doi.org/10.1007/s11207-014-0640-5.

Provost, F., Fawcett, T., 2001. Robust classification for imprecise environments. Mach. Learn. 42 (3), 203–231. ISSN 1573-0565. https://doi.org/10.1023/A:1007601015854.

Quinlan, J.R., 1993. C4.5: Programs for Machine Learning. Morgan Kaufmann Publishers Inc., San Francisco, CA. ISBN 1-55860-238-0.

Reiss, M.A., Temmer, M., Rotter, T., Hofmeister, S.J., Veronig, A.M., 2014. Identification of coronal holes and filament channels in SDO/AIA 193Å images via geometrical classification methods. Cent. Eur. Astrophys. Bull. 38, 95–104.

Reiss, M.A., Hofmeister, S.J., De Visscher, R., Temmer, M., Veronig, A.M., Delouille, V., Mampaey, B., Ahammer, H., 2015. Improvements on coronal hole detection in SDO/AIA images using supervised classification. J. Space Weather Space Climate 5 (27), A23. https://doi.org/10.1051/swsc/2015025.

Reiss, M.A., Temmer, M., Veronig, A.M., Nikolic, L., Vennerstrom, S., Schöngassner, F., Hofmeister, S.J., 2016. Verification of high-speed solar wind stream forecasts using operational solar wind models. Space Weather 14 (7), 495–510. ISSN 1542-7390. https://doi.org/10.1002/2016SW001390.

Robbins, S., Henney, C.J., Harvey, J.W., 2006. Solar wind forecasting with coronal holes. Sol. Phys. 233, 265–276. https://doi.org/10.1007/s11207-006-0064-y.

Rotter, T., Veronig, A.M., Temmer, M., Vršnak, B., 2012. Relation between coronal hole areas on the Sun and the solar wind parameters at 1 AU. Sol. Phys. 281, 793–813. https://doi.org/10.1007/s11207-012-0101-y.

Rotter, T., Veronig, A.M., Temmer, M., Vršnak, B., 2015. Real-time solar wind prediction based on SDO/AIA coronal hole data. Sol. Phys. 290, 1355–1370. https://doi.org/10.1007/s11207-015-0680-5.

Scherrer, P.H., Schou, J., Bush, R.I., Kosovichev, A.G., Bogart, R.S., Hoeksema, J.T., Liu, Y., Duvall, T.L., Zhao, J., Title, A.M., Schrijver, C.J., Tarbell, T.D., Tomczyk, S., 2012. The Helioseismic and Magnetic Imager (HMI) investigation for the Solar Dynamics Observatory (SDO). Sol. Phys. 275, 207–227. https://doi.org/10.1007/s11207-011-9834-2.

Scholl, I.F., Habbal, S.R., 2008. Automatic detection and classification of coronal holes and filaments based on EUV and magnetogram observations of the solar disk. Sol. Phys. 248, 425–439. https://doi.org/10.1007/s11207-007-9075-6.

Seiffert, C., Khoshgoftaar, T.M., Hulse, J.V., Napolitano, A., 2010. RUSBoost: a hybrid approach to alleviating class imbalance. IEEE Trans. Syst. Man Cybern. Part A 40 (1), 185–197.

Siedhoff, D., 2015. ADASYN algorithm to reduce class imbalance by synthesizing minority class examples. Available from: https://nl.mathworks.com/matlabcentral/fileexchange/50541-adasyn-improves-class-balance-extension-of-smote-.

Snoek, J., Larochelle, H., Adams, R.P., 2012. Practical Bayesian optimization of machine learning algorithms. In: Pereira, F., Burges, C.J.C., Bottou, L., Weinberger, K.Q. (Eds.), Advances in Neural Information Processing Systems 25. Curran Associates, Inc., pp. 2951–2959.

Tsurutani, B.T., Gonzalez, W.D., Gonzalez, A.L.C., Guarnieri, F.L., Gopalswamy, N., Grande, M., Kamide, Y., Kasahara, Y., Lu, G., Mann, I., McPherron, R., Soraas, F., Vasyliunas, V., 2006. Corotating solar wind streams and recurrent geomagnetic activity: a review. J. Geophys. Res. Space Phys. 111 (A7), A07S01. ISSN 2156-2202. https://doi.org/10.1029/2005JA011273.

Vapnick, V., 1998. Statistical Learning Theory. Wiley, New York, NY.

Verbanac, G., Vršnak, B., Veronig, A., Temmer, M., 2011. Equatorial coronal holes, solar wind high-speed streams, and their geoeffectiveness. Astron. Astrophys. 526, A20. https://doi.org/10.1051/0004-6361/201014617.

Verbeeck, C., Delouille, V., Mampaey, B., De Visscher, R., 2014. The SPoCA-suite: software for extraction, characterization, and tracking of active regions and coronal holes on EUV images. Astron. Astrophys. 561, A29. https://doi.org/10.1051/0004-6361/201321243.

Vršnak, B., Temmer, M., Veronig, A.M., 2007. Coronal holes and solar wind high-speed streams: I. Forecasting the solar wind parameters. Sol. Phys. 240, 315–330. https://doi.org/10.1007/s11207-007-0285-8.

Wallace, B.C., Small, K., Brodley, C.E., Trikalinos, T.A., 2011. Class imbalance, redux. In: Proceedings of the 2011 IEEE 11th International Conference on Data Mining. ICDM '11. IEEE Computer Society, Washington, D.C., pp. 754–763.

Weyn, B., Tjalam, W., de Wouwer, G.V., Daele, A.V., Scheunders, P., Jacob, W., Marck, E.V., Dyck, D.V., 2000. Validation of nuclear texture density, morphometry and tissue syntactic structure analysis as prognosticators of cervical carcinoma. Anal. Quant. Cytol. Histol. 22 (5), 373–382.

Woodcock, F., 1976. The evaluation of yes/no forecasts for scientific and administrative purposes. Mon. Weather Rev. 104, 1209. https://doi.org/10.1175/1520-0493(1976)104<1209:TEOYFF>2.0.CO;2.

CHAPTER

16

Solar Wind Classification Via k-Means Clustering Algorithm

Verena Heidrich-Meisner, Robert F. Wimmer-Schweingruber

Institute of Experimental and Applied Physics, Kiel, Germany

CHAPTER OUTLINE

1 INTRODUCTION

The most prominent space weather events that have the highest potential to affect the Earth's magnetosphere are solar energetic particle events and interplanetary coronal mass ejections (ICMEs). These eruptive phenomena are embedded in the ubiquitous solar wind. During their travel time from the Sun to the Earth they interact with the preceding and surrounding solar wind. Therefore, the quiet background solar wind influences space weather events and, in particular, space weather prediction.

Although ICMEs represent the more prominent space weather phenomena visible in the solar wind, here we only consider the background solar wind. The main reason for this lies in

Machine Learning Techniques for Space Weather
https://doi.org/10.1016/B978-0-12-811788-0.00016-0

the variable and complex structure of ICME plasma. Depending on the propagation speed an ICME might or might not contain multiple compression regions. Some, but not all, ICMEs are associated with magnetic clouds, and typically, the charge states and collisional age are high in ICMEs, while the proton temperature is lower than in solar wind with a comparable solar wind speed. But usually these criteria do not hold for the whole duration of the ICME and they can contain pockets of hotter and less highly charged plasma. All these ICME signatures share the property that they require context information to correctly identify complete ICME periods. Without additional context-dependent features, point-wise classification (as is used, e.g., in Xu and Borovsky (2015) and Zhao and Fisk (2010)) is simply not sufficient for the identification of ICME plasma. This can be resolved by incorporating appropriate context-dependent features into the feature space. However, this might also bias the classification of non-ICME plasma. Therefore, as a first approach, and to illustrate the benefit of the machine learning method itself for solar wind categorization, we remove ICMEs from the dataset (based on the Jian ICME list) and focus on the background solar wind first Schwadron et al., 2005; Hefti et al., 2000; Sakao et al., 2007; Antiochos et al., 2011.

The solar wind that is continuously emitted from the Sun is historically divided into at least two categories, namely, slow and fast solar wind. This is motivated by, for example, Ulysses observations (McComas et al., 2000) that revealed two distinct modes of the solar wind that differ in solar wind speed. However, further observations have exposed that it is not the solar wind speed, but the elemental and charge state composition of the solar wind that distinguishes between these two categories (von Steiger et al., 2000; Zhao and Fisk, 2010). Consequently, charge-state composition-based solar wind categorization schemes have emerged (e.g., von Steiger et al., 2000; Zhao and Fisk, 2010). But like the solar wind speed, the charge state composition is not the only characteristic signature of solar wind. Another approach broadens the selection of solar wind parameters relevant for solar wind categorization to specific proton entropy, Alfvén speed, and the ratio of velocity-dependent expected proton temperature and observed proton temperature (Xu and Borovsky, 2015). This categorization scheme defines decision boundaries in a 3D feature space spanned by these parameters.

From a machine learning point of view, all of the preceding solar wind categorization schemes are heuristic and they are driven by expert knowledge alone. These categorization schemes agree (1) that fast solar wind represents a distinct solar wind type, (2) on the existence of at least one slow solar wind type, and (3) that ICMEs/ejecta plasma form a distinct third category. However, they differ considerably in the position of the decision boundary between, in particular, slow solar wind and fast solar wind. So far, no uniformly accepted definition of the boundaries between the slow solar wind and the fast solar wind has emerged. To reflect the origin of fast solar wind, we follow the emerging terminology and interchangeably also call this solar wind type coronal hole wind in the following. Further, it is still under debate how many solar wind types can reasonably be distinguished. For example, the interface regions between slow solar wind and coronal hole wind are sometimes considered as distinct solar wind plasma categories as well, and different candidates for subtypes of slow solar wind are considered (Stakhiv et al., 2015; Sanchez-Diaz et al., 2016; Xu and Borovsky, 2015). Unlike for fast solar wind, or coronal hole wind, where the source regions have been uniquely identified as coronal holes (Tu et al., 2005), the origin and release mechanism of slow solar wind is still under debate. The assumption that several different release mechanisms might play a role for slow solar wind motivate us to look for different subtypes of slow solar wind (Stakhiv et al., 2015; Sanchez-Diaz et al., 2016).

The simple division of the solar wind into two categories has frequently been challenged as insufficient, in particular for slow solar wind. For example, very slow solar wind sometimes is considered a separate solar wind type (Sanchez-Diaz et al., 2016), as well as highly Alfvénic slow solar wind (D'Amicis and Bruno, 2015). In Xu and Borovsky (2015), slow solar wind is divided into two categories, depending on whether it is associated with helmet streamers or pseudo-streamers. But also for fast, or rather, coronal hole wind, the assumption of a single type of coronal hole wind is overly simplifying. Zhao and Landi (2014) find different properties in polar and equatorial coronal hole wind and Heidrich-Meisner et al. (2016) discuss two types of coronal hole wind that differ in their respective average Fe charge state. So far, the different existing solar wind categorization schemes do not reflect this complexity. The simplified perspective of two solar wind types also ignores the effect of extended stream interaction regions that show not only a mixture of the properties of coronal hole and slow solar wind, but are also changed by the interaction. Compression regions are heated, and show a stronger magnetic field and higher densities than the surrounding coronal hole and slow solar wind. The properties of a solar wind package also determine how it interacts with the Earth's magnetosphere and with ICMEs. Thus, identifying subtypes of solar wind with well-defined and distinct properties allows a more precise description and prediction of how space weather phenomena that are embedded in the solar wind affect Earth. None of the existing categorization schemes takes these considerations fully into account. The different solar wind categorization schemes are, in particular, very inconsistent when applied to solar wind from stream interaction regions. From a solar wind science perspective, a reliable data-driven identification of subtypes of slow solar wind can play a crucial role in answering the unresolved question of the origin of slow solar wind.

From a machine learning perspective, this has the following consequence. No ground truth is available. Thus, labels are, at best, available for a subset of the data, and this subset is biased in the sense that it contains only data points far away from the decision boundaries. Using this as a training data set for a supervised learning method would inherit the same bias. This rules out the application of any supervised learning methods or reinforcement learning to solar wind classification. Thus, unsupervised learning methods are here the most appropriate approach. A classification task in an unsupervised learning scenario is called clustering. In the following, we use both terms, as well as solar wind categorization, interchangeably. However, lack of an available ground truth also complicates the evaluation of the applied unsupervised machine learning methods. As always, the performance of different methods or algorithms should be compared on a test data set that has not been used previously in any way. In addition, the training and test data should be generated by the same hidden distribution. In a real-world application such as solar wind classification, these distributions are not known a priori, but it is still necessary to ensure that this assumption is plausible. If these principles were violated, the resulting comparison would be biased.

To establish a baseline for future research, it is usually beneficial to first apply the simplest possible approach to a given new problem. Therefore, we here employ linear k-means clustering to solar wind classification as an illustrative and conceptually very simple approach to unsupervised classification.

2 BASIC ASSUMPTIONS AND METHODOLOGY

Although other approaches exist (see, e.g., Neugebauer et al., 2016), the three aforementioned solar wind categorization schemes, the solar wind speed-based scheme, the composition-based scheme from Zhao and Fisk (2010) and the Xu and Borovsky (2015) scheme, share the property that they are applied point-wise. Thus, they share the underlying assumption that the available observation of solar wind parameters at any given point in time is sufficient to determine the solar wind type. Here, we are adopting this assumption as well. However, in a space weather context, this has an inexpedient consequence. Correct and comprehensive identification of ICMEs requires at least some context information. Therefore, the approach discussed here is only directly applicable to the background solar wind and not to eruptive events such as ICMEs. There are standard methods to tackle this problem such as, for example, the inclusion of context-sensitive features. But the selection of such context-sensitive features can be expected to considerably influence the results. Thus, for the purpose of illustrating the capabilities of our chosen unsupervised machine learning method, it is more instructive to solely focus on point-wise categorization. Consequently, because we cannot expect ICMEs to be classified correctly in this approach, we exclude known ICMEs (based on the Jian ICME list; Jian et al., 2006, 2011) from the data set. The Jian ICME list is a manually selected list of ICME candidates that takes the following criteria into account: high total perpendicular pressure, low proton temperature, stronger ambient magnetic field, relatively quiet and smooth rotation in the magnetic field, helium abundance enhancement, and bi-directional solar wind electron strahls (BDE). However, not all ICMEs need to exhibit all of these criteria and some criteria are also associated with other phenomena. Any decision about which subset of the preceding criteria is used as a necessary criterion to identify ICMEs introduces a bias. In particular, the start and end times of an ICME need to be defined in a different way depending on which of the criteria are fulfilled for a specific ICME candidate. Neither the Jian ICME list nor alternatives (see, e.g., Cane and Richardson, 2003; Richardson and Cane, 2010) can guarantee a complete list of all ICMEs. These lists typically focus on avoiding false positive instances and as a result, false negatives (events that might be ICMEs but that are excluded from the list) are more likely. For example, ICMEs that have a similar speed than the surrounding solar wind plasma can be difficult to detect (Iju et al., 2014).

The ability to derive an appropriate clustering result and thus, in our case, a reliable and physically meaningful solar wind categorization scheme, depends sensitively on the solar wind parameters provided to the clustering method. Therefore, the so-called feature selection plays an important role in machine learning and can have a stronger influence on the results than the chosen machine learning method. In the following, we use a seven-dimensional feature space as the baseline scenario that we refer to as the *basic* feature space F_{basic}. We consider typical solar wind parameters that can be determined by the instrumentation on the Advanced Composition Explorer (ACE), namely the Solar Wind Electron, Proton and Alpha Monitor (ACE/SWEPAM, McComas et al., 1998), the Solar Wind Ion Composition Spectrometer (ACE/SWICS, Gloeckler et al., 1998), and the magnetometer (ACE/MAG, Smith et al., 1998) observed from 2001 to 2010. The time resolution of 12 min of ACE/SWICS determines the time resolution of the dataset. Our *basic* feature space is comprised of: (1) solar wind proton speed v_{p} (in km s^{-1}), (2) solar wind proton density n_{p} (in cm^{-3}), (3) solar wind proton temperature T_p (in K), (4) magnetic field strength B (in nT), (5) O charge state ratio $n_{O^{7+}}/n_{O^{6+}}$ (where $n_{X_{y+}}$ denotes the density of the $y+$ charged ion of species X), (6) C charge state ratio $n_{C^{6+}}/n_{C^{5+}}$, as well as (7) (motivated by considerations in Kasper et al., 2008) the proton-proton collisional age $a_c = \frac{r}{v_p \tau_{col}}$ (where r is the distance from the Sun to ACE and $\tau_{col} \sim n_p T_p^{-3/2}$ is the proton-proton collisional time scale). In Section 7, we additionally consider different feature spaces, including the following parameters: the average Fe charge state $\tilde{q}_{Fe} = \sum_{c=7}^{13} c n_{Fe^{c+}} / \sum_{c=7}^{13} n_{Fe^{c+}}$ (in units of the elementary charge e), the average Si ion charge state $\tilde{q}_{Si} = \sum_{c=7}^{12} c n_{Si^{c+}} / \sum_{c=7}^{12} n_{Si^{c+}}$ (in units of e), the magnetic field components in geocentric solar ecliptic coordinates B_x, B_y, B_z (in nT), the variability of each magnetic field component measured as $\frac{\Delta B_x}{B}, \frac{\Delta B_y}{B}, \frac{\Delta B_z}{B}$, and the temporal gradients of solar wind proton speed $\frac{dv_p}{dt}$ (in km s^{-2}) and solar wind proton density $\frac{dn_p}{dt}$ (in cm^{-3} s^{-1}). Table 1 summarizes the solar wind parameters and feature selection sets that are used to train k-means clustering as a solar wind classifier in the following.

Typically machine learning algorithms assume that the training and test data sets are independently identically distributed. But because the ground truth for solar wind categorization is unknown, this requirement is difficult to ensure. Instead of randomly dividing the available data into test and training data, we set aside the equivalent of 20 Carrington rotations (two from each year) as the test data set X_{test} and use the remaining data as the training data set X_{train}. One Carrington rotation typically contains several coronal hole and slow solar wind streams as well as the stream interaction regions between them. Taking two Carrington rotations per year ensures that solar-cycle-dependent changes in the underlying distributions are captured as well.

TABLE 1 Summary of Feature Selection Scenarios

Feature Space	Dim.	Solar Wind Parameters
F_{basic}	7	$v_p, n_p, T_p, B, a_c, n_{O^{7+}}/n_{O^{6+}}, n_{C^{6+}}/n_{C^{5+}}$
$F_{basic \setminus a_c}$	6	$v_p, n_p, T_p, B, n_{O^{7+}}/n_{O^{6+}}, n_{C^{6+}}/n_{C^{5+}}$
$F_{basic\ only\ a_c}$	4	$B, a_c, n_{O^{7+}}/n_{O^{6+}}, n_{C^{6+}}/n_{C^{5+}}$
$F_{no\ v_p}$	5	$n_p, T_p, B, n_{O^{7+}}/n_{O^{6+}}, n_{C^{6+}}/n_{C^{5+}}$
$F_{no\ comp.}$	5	v_p, n_p, T_p, B, a_c
$F_{only\ comp.}$	4	$n_{O^{7+}}/n_{O^{6+}}, n_{C^{6+}}/n_{C^{5+}}, \tilde{q}_{Fe}, \tilde{q}_{Si}$
F_{SOHO}	3	v_p, n_p, T_p
$F_{linear\ Xu}$	4	v_p, n_p, T_p, B
$F_{speed\ +\ comp.}$	3	$v_p, n_{O^{7+}}/n_{O^{6+}}, n_{C^{6+}}/n_{C^{5+}}$
$F_{only\ B}$	7	$B, B_x, B_y, B_z, \frac{\Delta B_x}{B}, \frac{\Delta B_y}{B}, \frac{\Delta B_z}{B}$
$F_{basic\ +\ grad.}$	9	$v_p, n_p, T_p, B, a_c, n_{O^{7+}}/n_{O^{6+}}, n_{C^{6+}}/n_{C^{5+}}, \frac{dv_p}{dt}, \frac{dn_p}{dt}$
$F_{only\ context}$	5	$\frac{\Delta B_x}{B}, \frac{\Delta B_y}{B}, \frac{\Delta B_z}{B}, \frac{dv_p}{dt}, \frac{dn_p}{dt}$
$F_{only\ derived}$	3	$a_c, \frac{dv_p}{dt}, \frac{dn_p}{dt}$

Whenever at least one of the considered solar wind parameters is invalid or missing, the respective data point is excluded. This leads to a training data set X_{train} of size $N_{\text{train}} = 305{,}034$ and a test data set X_{test} of size $N_{\text{test}} = 51{,}022$. We use the *k*-means C++ implementation of the Shark machine learning library (Igel et al., 2008). Because the initial cluster assignment is randomized, but can influence the final results, we perform 100 independent clustering trials and consider the median clusters as the result. The only source of randomness in *k*-means is the initial clustering.

If a single dimension contains higher absolute values than all other dimensions, the distance between two data points is dominated by the difference in this single dimension. This can bias the results of distance-based clustering approaches such as *k*-means. To exclude this effect, the training data are normalized to unit variance and zero mean before they are presented to the machine learning clustering method. For the *basic* feature set, each unnormalized vector $\tilde{\mathbf{x}} \in X_{\text{test}} \cap X_{\text{train}}$ in the training and test data set is normalized to a vector $\mathbf{x}$ according to

$$\mathbf{x} = \begin{pmatrix} 104.2 & 0 & 0 & 0 & 0 & 0 & 0 \\ 0 & 5.437 & 0 & 0 & 0 & 0 & 0 \\ 0 & 0 & 72{,}230 & 0 & 0 & 0 & 0 \\ 0 & 0 & 0 & 2.616 & 0 & 0 & 0 \\ 0 & 0 & 0 & 0 & 5.370 & 0 & 0 \\ 0 & 0 & 0 & 0 & 0 & 0.9221 & 0 \\ 0 & 0 & 0 & 0 & 0 & 0 & 10.45 \end{pmatrix} \tilde{\mathbf{x}} + \begin{pmatrix} 433.9 \\ 6.252 \\ 84{,}800 \\ 5.066 \\ 2.263 \\ 0.1771 \\ 1.310 \end{pmatrix}$$

3 K-MEANS

Typically the application of a machine learning method discerns between a training phase (which exploits a priori available test data) and a testing phase (in which a method operates on previously unseen data). Machine learning techniques are applicable to different types of learning tasks depending on the availability of feedback signals. If full feedback information is available, that is, if the correct answer is known for all samples available during the training phase, the learning task is called supervised. In contrast, in the unsupervised scenario, the correct answer is unknown during the training phase. In the third type of machine learning methods, reinforcement learning, only a (potentially vague) qualitative feedback signal is available. Whereas reinforcement learning problems are often formulated as control problems with frequent agent-environment interactions, all three cases can be applied to classification tasks, where each data point is associated with a label indicating a class, or to regression tasks, where the goal is to estimate a continuous target function. Here, we regard solar wind categorization as an unsupervised classification task.

Before we provide a short description of the *k*-means algorithm, we briefly motivate the choice of this particular clustering method. A variety of clustering methods exists, and several of these methods would have been applicable in this case (see Jain et al., 1999 for a review on clustering methods). *k*-means makes the implicit assumption that the clusters are convex and isotropic. Both assumptions are also underlying the heuristic solar wind categorization scheme described in Xu and Borovsky (2015) (although the decision boundaries are derived in a different space). Therefore, k-means is well suited for a fair comparison to this existing method. k-means has only one adjustable hyperparameter, the number of classes k. As discussed previously, the true number of solar wind types is unknown. Therefore, in this case, standard methods to determine an appropriate value for k in k-means have the additional advantage of inferring a data-driven estimation of the relevant number of types in the solar wind. In an ideal use-case for k-means, all clusters are represented with the same number of samples in the training data set. However, because the clusters are not known a priori in our case, this cannot be assured and is likely not the case. k-means is best suited for a small-to-medium number of clusters, as is the case for solar wind categorization.

Mean-shift (Comaniciu and Meer, 2002) is conceptually similar to k-means, but operates on neighborhoods rather than on individual samples. However, unlike k-means, it is not scalable with the number of samples. Gaussian mixtures (Spall and Maryak, 1992; Lindsay, 1995) are particularly beneficial for density estimation (which we are not interested in here). Spectral clustering (Shi and Malik, 2000; Von Luxburg, 2007) first conducts a low-dimensional embedding of the affinity matrix between samples before a k-means algorithm is applied in this low-dimensional space. Spectral clustering is best suited for image processing, and, because the first step is computationally expensive, for few clusters and at most a medium sample size. DBSCAN (Ester et al., 1996) aims at clustering the data into regions with high density with regions of low density in between. This is likely not achievable for solar wind categorization. Frequent stream interaction regions between different solar wind streams assure that observations from transitions between different solar wind types are always included in the data set. If we do not want to exclude the option that these transition regimes can be identified as distinct solar wind types as well, DBSCAN is not well-suited for this task. Hierarchical clustering methods, such as, for example, Ward hierarchical clustering (Ward, 1963), are best suited for many clusters (which are unlikely to be contained in the solar wind). Self-organizing maps (SOMs) are a neural network approach to clustering. With neighborhood relations between the cluster centers in such a map, an additional objective is introduced. As a result, SOMs do not necessarily converge to the best clustering result and they are better suited for visualization purposes.

The isotropy assumption of k-means can be removed by applying the kernel trick to k-means. However, this also leads to more hyperparameters that need to be determined. Therefore, this step is left for a later study. Because in this study we are mainly interested in the clustering result and not in, for example, a density estimation of each cluster, the inherent but not explicit approximation of each cluster as done by k-means is sufficient. We expect few clusters and would like to apply a method that can be transferred easily to other and larger data sets. The aim of this study is to provide an example of how machine learning can be used in the context of solar wind categorization. For this, a method that is as simple and easy to apply as k-means is well suited. The methodology described here is transferable to the application of other classification methods as well.

k-means clustering (MacQueen et al., 1967; Steinhaus, 1956; Arthur and Vassilvitskii, 2007; Jain et al., 1999) is a simple unsupervised clustering approach that is based on the following principles.

In k-means each of k clusters is represented by a cluster center. In the training phase, k clusters are extracted from the training data. For a later application of the derived clustering, only the cluster centers are required. Both during training and testing, each data point is assigned to the nearest cluster center or cluster center candidate. Thus, the objective of the algorithm is to find a partition $S = (S_1, \ldots, S_k)$ of the training data set X_{train} with N_{train} elements and corresponding cluster centers $\mathbf{c}_1, \ldots, \mathbf{c}_k$ that minimize the innercluster distances for all clusters (see, e.g., Arthur and Vassilvitskii (2007), indices indicating the iteration are omitted in the following):

$$\underset{S}{\text{argmin}} \sum_{i=1}^{k} \sum_{\mathbf{x} \in S_i} \|\mathbf{x} - \mathbf{c}_i\|^2 \tag{1}$$

A pseudocode is given in Algorithm 1. Initially, the cluster centers are chosen randomly (or possibly based on available expert knowledge). Then, the distance of each data point to each cluster center is computed, each data point is associated with a cluster center, and is assigned a respective label $l_i \in \{1, \ldots, k\}$. In this way, partitions as cluster candidates are formed. Typically, the squared norm is used as the distance measure. In the next step, new cluster centers are defined as the center of mass of each cluster candidate. Unless the following termination criterion is met, this process is repeated. The algorithm terminates if the last iteration did not lead to changes in the assignment of each data point to the current cluster centers.

We restrict ourselves to the case of linear k-means. This is equivalent to defining the decision boundaries directly in the space of the observables. Although this most likely does not constitute the most appropriate feature space, this scenario represents a necessary baseline that allows comparison with existing solar wind categorization schemes in an unbiased way.

Algorithm 1 K-MEANS ALGORITHM

Data: number of clusters k, data set X
Result: cluster centers $\mathbf{C} = \{\mathbf{c_1}, \ldots, \mathbf{c_k}\}$
begin
  Randomly select k data points as initial cluster centers;
  repeat
    Reinitialize all partition subsets as empty:
    $S_1 = S_2 = \cdots = S_k = \{\}$;
    Assign each data point to the closest cluster center:
    for $i \in \{1, \ldots, N\}$ **do**
      $l = \text{argmin}_{j \in \{1,\ldots,k\}} \|\mathbf{x}_i - c_j\|^2$;
      $S_l = S_l \cup \{x_i\}$;
    end
    Define new cluster centers based on current partition:
    for $j \in \{1, \ldots, k\}$ **do**
      $c_j = \sum_{i \in \{1,\ldots,N\} x_i \in S_j} \mathbf{x_i} / |S_j|$
    end
  until *the cluster assignment converges*;
end

In each iteration, k-means visits all N_{train} d-dimensional training data points once per cluster. If l is an upper bound on the number of iterations until convergence and d the dimension of the search space, linear k-means scales with $O(kN_{\text{train}}dl)$ (see, e.g., Arthur and Vassilvitskii, 2007). k-means clustering is NP-hard. This limits the practical applicability of k-means. But with currently available computational power, applying k-means to solar wind data sets is well within reasonable computational costs.

The number of clusters k is not determined by the k-means algorithm itself and has to be defined a priori to be able to apply k-means clustering.

4 COMPARING 2-MEANS CLUSTERING TO EXISTING SOLAR WIND CATEGORIZATION SCHEMES

We applied k-means clustering to our dataset comprised of 10 years of solar wind observations from ACE. In this section, we compare the obtained clustering results of k-means clustering with the composition-based categorization scheme based on Zhao and Fisk (2010), the four-type categorization scheme following Xu and Borovsky (2015), and a basic solar wind speed-dependent solar wind categorization. Table 2 summarizes the criteria used in each of the considered solar wind categorization schemes.

To facilitate such a comparison, we can only consider the solar wind types that are contained in all four considered solar wind categorization schemes. For the Xu and Borovsky (2015) case,

TABLE 2 Overview of Considered Solar Wind Categorization Schemes

Scheme/Type	Parameters/Criteria
Speed-Only	Proton solar wind speed: v_{p}
Slow	$v_{\text{p}} \leq 450\,\text{km s}^{-1}$
Coronal hole	$v_{\text{p}} > 450\,\text{km s}^{-1}$
Zhao and Fisk (2010)	O and C charge state ratios: $\frac{n_{\text{O7+}}}{n_{\text{O6+}}}, \frac{n_{\text{C6+}}}{n_{\text{C5+}}}$
Slow	$\frac{n_{\text{O7+}}}{n_{\text{O6+}}} \frac{n_{\text{C6+}}}{n_{\text{C5+}}} > 0.01$
Coronal hole	$\frac{n_{\text{O7+}}}{n_{\text{O6+}}} \frac{n_{\text{C6+}}}{n_{\text{C5+}}} < 0.01$
Xu and Borovsky (2015)	Alfvén speed: v_{A}, proton-specific entropy: S_{p}, ratio of expected proton temperature T_{exp}, and observed proton temperature T_{p}
Coronal hole	$\log 10(v_{\text{A}}) \leq 0.277 \log 10(S_{\text{p}}) + 0.055 \log 10(T_{\text{exp}}/T_{\text{p}}) + 1.83$ and $\log 10(S_{\text{p}}) > -0.525 \log 10(T_{\text{exp}}/T_{\text{p}}) - 0.676 \log 10(v_{\text{A}}) + 1.74$
Sector-reversal plasma	$\log 10(v_{\text{A}}) \leq 0.277 \log 10(S_{\text{p}}) + 0.055 \log 10(T_{\text{exp}}/T_{\text{p}}) + 1.83$ and $\log 10(S_{\text{p}}) < -0.658 \log 10(v_{\text{A}}) - 0.125 \log 10(T_{\text{exp}}/T_{\text{p}}) + 1.04$
Streamer-belt plasma	$\log 10(v_{\text{A}}) \leq 0.277 \log 10(S_{\text{p}}) + 0.055 \log 10(T_{\text{exp}}/T_{\text{p}}) + 1.83$ and $\log 10(S_{\text{p}}) \leq -0.525 \log 10(T_{\text{exp}}/T_{\text{p}}) - 0.676 \log 10(v_{\text{A}}) + 1.74$ and $\log 10(S_{\text{p}}) \geq -0.658 \log 10(v_{\text{A}}) - 0.125 \log 10(T_{\text{exp}}/T_{\text{p}}) + 1.04$
Slow	Sector-reversal plasma or streamer-belt plasma

Note: The criteria for ICME/ejecta plasma are omitted.

we combine the two slow solar wind types, streamer-belt plasma and sector-reversal plasma, into one slow solar wind plasma type. Because the point-wise approach is not suitable for identifying ICMEs reliably, we omit ejecta/CME types from the comparison. Therefore, we can only use slow solar wind and coronal hole wind for this comparison. Thus, we do not use the optimal number of k here (see next section), but for the sake of a fair comparison also restrict k-means to the $k = 2$ case. On average, 2-means clustering required only 29 iterations until the clustering converged.

As comparison metric, we use the pairwise agreement $A(t)$ of the predictions $p_{\text{test}}\colon X_{\text{test}} \longrightarrow L$ of a test categorization scheme with the predictions $p_{\text{ref}}\colon X_{\text{test}} \longrightarrow L$ of a reference categorization scheme on the test data set X_{test} per solar wind type $t \in L$ which is (similar to Neugebauer et al., 2016) defined as

$$A(t, p_{\text{test}}, p_{\text{ref}}) = \frac{\sum\limits_{\mathbf{x} \in X_{\text{test}}, p_{\text{test}}(\mathbf{x}) = p_{\text{ref}}(\mathbf{x}) = t} 1}{\sum\limits_{\mathbf{x} \in X_{\text{test}}, p_{\text{ref}}(\mathbf{x}) = t} 1} \tag{2}$$

Here, L denotes the set of class labels.

Table 3 summarizes the results. Because the agreement of a test method p_{test} is always computed based on the prediction of a reference method p_{ref} and a solar wind type, the agreement is not symmetric. Therefore, for two methods a and b $A(t, p_a, p_b) \neq A(t, p_b, p_a)$. The three heuristic categorization schemes are deterministic. Because the results of k-means can depend on the initial cluster assignment, for 2-means we give the mean of 100 independent trials as well as the 15.9th and 83.1st percentiles. For most cases the variability is small. This indicates that 2-means shows a stable performance in this application and is here not very sensitive to the initial cluster assignment. The fact that in some cases the mean lies outside the confidence intervals given by the 15.9th and 83.1st percentiles is an indication of outliers.

The mean relative agreement ranges between 0.501 and 0.98, depending on the respective reference and test methods. With the speed-only solar wind categorization as the reference method, the agreement is high (at least 0.854) for all other solar wind categorization schemes for slow solar wind but it is lower for coronal hole wind. The lowest mean agreement (0.713–0.720) occurs for slow solar wind if the composition-based scheme from Zhao and Fisk (2010) is the reference method and this holds for all three test schemes. This means that the Zhao and Fisk (2010) scheme classifies more test data points as slow solar wind than all the other methods. Accordingly, if the Zhao and Fisk (2010) scheme is the test scheme for coronal hole wind, the agreement with both the Xu and Borovsky (2015) scheme and 2-means is low as well. All four solar wind categorization schemes operate on different feature sets of different dimensions. Because the 2-means classifier is based on a seven-dimensional feature space, it has, in principle, more information available to derive its decision boundary than the other three methods. That, nevertheless, most methods show high agreements with all others is evidence that they all represent the same basic structure of the data set and thus the solar wind.

Fig. 1 shows a projection of the feature space to the $v_{\text{p}} - n_{\text{p}}$ plane for slow solar wind clusters from the 2-means and the three other categorization schemes. The middle panels show the difference between the clusters shown in the left and right panels. The clusters differ most for

TABLE 3 Relative Agreement per Solar Wind Type

	Speed-Only			Zhao and Fisk (2010)			Xu and Borovsky (2015)			2-Means		
	Mean	15.9th Perc.	83.1st Perc.	Mean	15.9th Perc.	83.1st Perc.	Mean	15.9th Perc.	83.1st Perc.	Mean	15.9th Perc.	83.1st Perc.
Slow												
Speed- only	–	–	–	0.914	–	–	0.854	–	–	0.945	0.944	0.944
Zhao and Fisk (2010)	0.713	–	–	–	–	–	0.719	–	–	0.720	0.717	0.717
Xu and Borovsky (2015)	0.826	–	–	0.892	–	–	–	–	–	0.848	0.846	0.846
2-Means	0.923	0.908	0.936	0.767	0.601	0.911	0.807	0.737	0.867	–	–	–
Coronal hole												
Speed-only	–	–	–	0.600	–	–	0.658	–	–	0.900	0.895	0.895
Zhao and Fisk (2010)	0.741	–	–	–	–	–	0.624	–	–	0.732	0.729	0.729
Xu and Borovsky (2015)	0.976	–	–	0.501	–	–	–	–	–	0.980	0.980	0.980
2-Means	0.918	0.908	0.936	0.749	0.601	0.911	0.822	0.670	0.992	–	–	–

Notes: Each row shares the same reference method and each column the same test method. For each comparison involving 2-means, the mean and the 15.9th and 83.1st percentiles are listed.

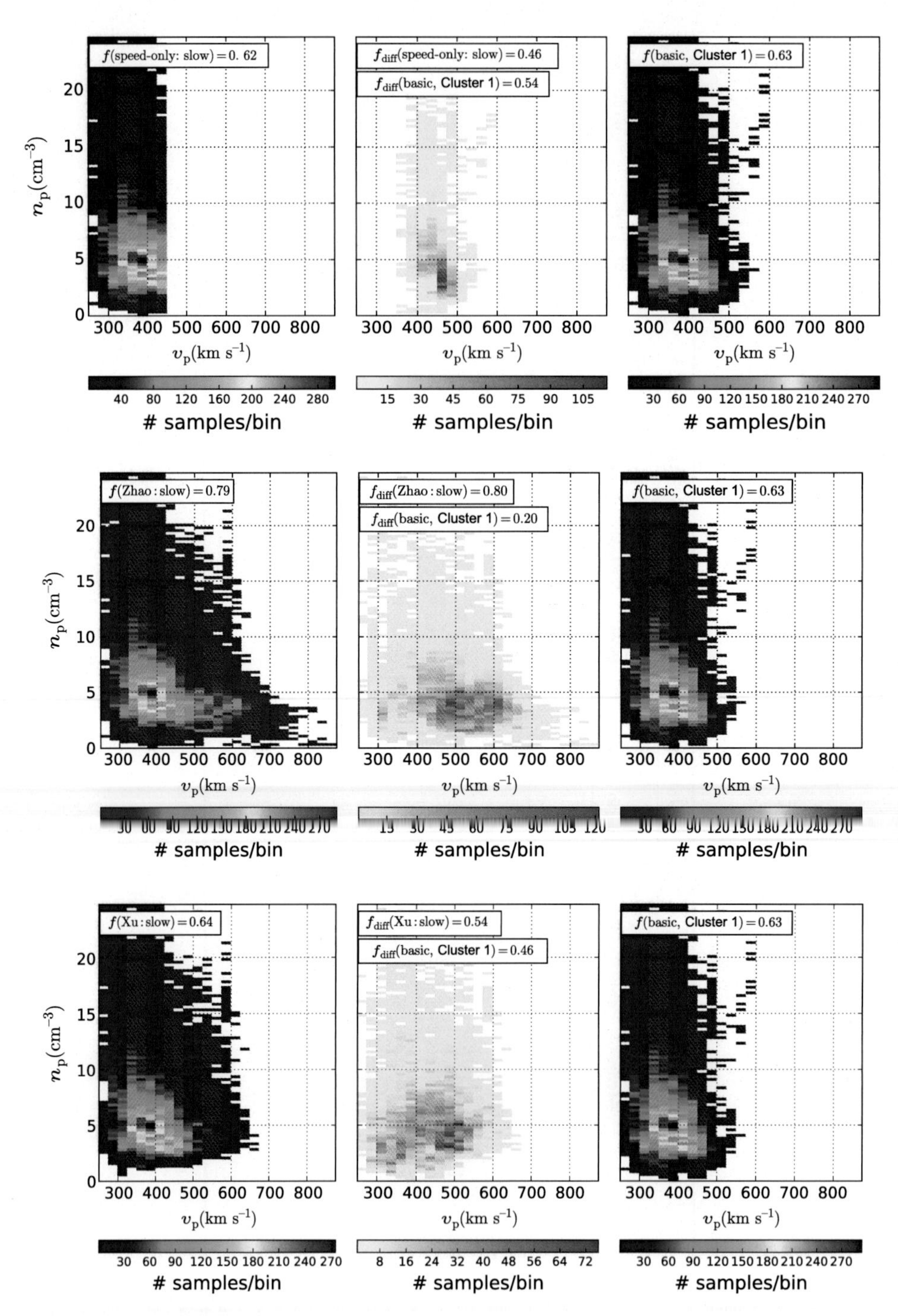

FIG. 1 Two-dimensional histograms of 12-min resolution solar wind observations from the test data set in the feature space projected to the $v_p - n_p$ plane for slow solar wind. *Left column*: Slow solar wind categorized by solar wind speed only (*top*), slow solar wind based on the Zhao and Fisk (2010) scheme (*middle*), and slow solar wind derived from the Xu and Borovsky (2015) scheme (*bottom*). *Right column*: Slow solar wind derived from 2-means clustering. In the *middle column*, the difference between the two slow solar wind clusters is shown. In each panel, the fraction f of the test data set shown in the panel is given as an *inset*. In the *middle panels*, the fraction of data points from the first and second scheme relative to the number of data points on which the two schemes differ is indicated.

intermediate solar wind proton speeds ($450 \text{ km s}^{-1} \leq v_p \leq 630 \text{ km s}^{-1}$). This is not surprising (and has been shown for the Zhao and Fisk (2010) and Xu and Borovsky (2015) schemes in Neugebauer et al. (2016)) because this is the regime in which the simplified view of only two solar wind types is most debatable and, in particular, their explicit decision boundary is not well defined from expert knowledge. But this is not the only solar wind regime where differences are found. The difference maps also contain solar wind packages with higher solar wind speeds that are considered as slow solar wind by both solar wind categorization schemes. In particular, the slow solar wind from the composition-based Zhao and Fisk (2010) scheme includes a fraction of the solar wind with solar wind proton speeds $v_p > 650 \text{ km s}^{-1}$.

5 MODEL SELECTION, OR HOW TO CHOOSE K

Machine learning techniques can typically be described on an abstract level as extracting a model from the available data set. This process is based on the assumptions that (1) the resulting model is representative not only for the training data set it was based on but for all data generated by the same (unknown) probability distribution (training and test data set are assumed to be independent identically distributed), and (2) the optimal model is a model that generalizes well, which means that the model describes only the essential structure of the training data. A model that generalizes well is expected to perform well not only on the training data set used to generate the model but also on unseen test data. This represents the central objective of model selection. Finding the right degree of model complexity that also generalizes well for a given problem can be a nontrivial task. A too-simple model fails to represent relevant structure in the data. A simple example of this so-called underfitting would be to fit a linear model to a third-order polynomial. A first-order model might represent an underlying trend of the polynomial reasonably well, but its complexity is too low to capture the true graph of the polynomial. But an overly complex model is undesirable as well. In real-world applications that usually show signatures of noise and uncertainty, an overly complex model tends to model the noise as well or even instead of the underlying structure of the data. This effect is called overfitting. Both overfitting and underfitting manifest in the case of k-means clustering with respect to inappropriate choices for the number of clusters k. If the data was generated from the optimal number of clusters k^*, any k-means clustering result for $k < k^*$ or $k > k^*$ cannot be expected to represent the data as well. Thus, one should always take care to choose the best value for k.

However, in real-world applications such as solar wind categorization, the true number of clusters contained in the data is typically unknown. Therefore, model selection tools are necessary to choose an appropriate approximation for k^*. For this, first, a measure of the generalization error is required. The easiest way to estimate the generalization error is to apply the method of interest to previously unseen data points. We cannot simply use the already defined test data set X_{test} because then our model selection procedure would no longer be independent from this data set and we would not be able to use this data set as a test bed to compare different approaches. Thus, instead, we need to set aside a subset of the available training data set X_{train} that is no longer used to train the model, but that is only used for testing its generalizing ability. However, which part of the training data set is used to estimate the generalization error can affect the results and we would like to remove or at least reduce the influence of the particular partition into training and test data set. A commonly used concept to achieve this is n-fold cross-validation (Geisser, 1993; Kohavi et al., 1995; Devijver and Kittler, 1982). In n-fold cross-validation, the data set is randomly partitioned into n subsets. Then, from these n subsets, instead of only one, n pairs of training and test data sets are constructed in the following way: One of the n subsets is chosen as test data set $X^{CV}_{i,\text{test}}$ and the remaining $n-1$ subsets form the training data set $X^{CV}_{i,\text{train}}$. Because there are n different possible choices for the test data set, this leads to n pairs of training and test data set $(X^{CV}_{1,\text{train}}, X^{CV}_{1,\text{test}}), \ldots, (X^{CV}_{n,\text{train}}, X^{CV}_{n,\text{test}})$. For each of these, an error measure $\tau_i, i \in \{1, \ldots, n\}$, is computed, and the cross-validation error τ^{CV} is then given as $\tau^{CV}(k) = \frac{1}{n}\sum_{i=1}^{n} \tau_i(k)$.

Before we can apply n-fold cross-validation to our solar wind clustering data, we need to define an appropriate error measure $\tau_i(k), i \in \{1, \ldots, n\}$. A simple choice that is consistent with the optimization criterion for k-means clustering is

$$\tau_i(k) = \frac{1}{N_{i,\text{test}}} \sum_{j=1}^{k} \sum_{x \in S_j} \|\mathbf{x} - c_j\|^2 \tag{3}$$

where $N_{i,\text{test}}$ is the number of data points in the ith test data set $X_{i,\text{test}}$.

Thus, we define (similar to, e.g., Arlot et al., 2010) the error of a clustering as the average squared innerclass distance per data point. Then (again as in Arlot et al., 2010), the n-fold cross-validation error τ^{CV} is given as

$$\tau^{CV}(k) = \frac{1}{n} \sum_{i=1}^{n} \tau_i(k) \tag{4}$$

Choosing the best number of folds n for cross-validation represents a trade-off between low variance and low computational cost (Kohavi, 1996; Fushiki, 2011; Zhang, 1993). The variance can be reduced by increasing the number of folds or repeating the cross-validation process for different partitions of the training data set. In the following, we use as a common choice $n = 10$ (Kohavi, 1996), and we, therefore, partition our training data X_{train} into 10 subsets, sort these into 10 pairs

$$(X^{CV}_{1,\text{train}}, X^{CV}_{1,\text{test}}), \ldots, (X^{CV}_{10,\text{train}}, X^{CV}_{10,\text{test}})$$

apply for all $k \in \{2, \ldots, 20\}$ k-means clustering to each training data set $X^{CV}_{i,\text{train}}$, determine a clustering error $\tau_i(k)$ for each corresponding test data set $X^{CV}_{i,\text{test}}$, and average over these to obtain the cross-validation error $\tau^{CV}(k)$. This process is then repeated for 50 random partitions of the training data set.

Fig. 2 shows the cross-validation error for k-means with $k \in \{2, \ldots, 20\}$. The cross-validation error decreases with an increasing number of clusters. This is expected because more cluster centers distributed in the same hypervolume lead to shorter average distances to the closest cluster center. However, this does not mean that larger values of k are always more appropriate because overfitting also becomes more likely the larger k is. The extreme overfitting case of one cluster center per data point would trivially lead to a cross-validation error of 0.

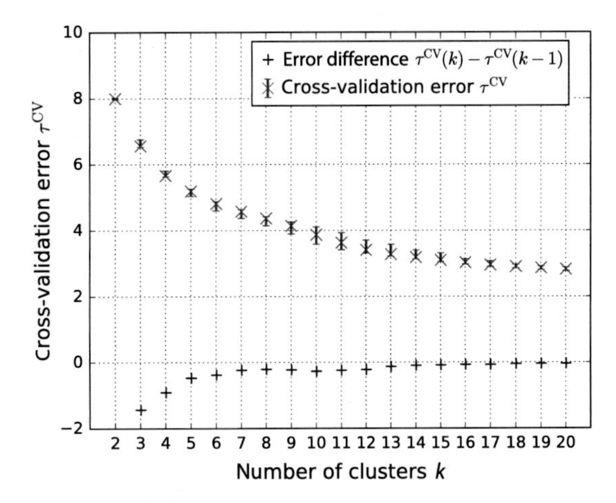

FIG. 2 Tenfold cross-validation error τ^{CV} depending on the number of clusters k are shown as ×-shaped symbols. The median of 50 random initial partitions is shown as well as the 1st and 99th percentiles. In addition, the difference between the median values for two consecutive values of k is shown with +-shaped symbols.

This is not a desirable solution. But where to stop? This is a nontrivial question without an easy answer (see Pham et al., 2005 for a short overview). A simple approach is the so-called *elbow method* (Thorndike, 1953). This heuristic approach assumes that overfitting becomes the dominant effect as soon as adding more clusters has a smaller and smaller influence on the result. Consequently, the elbow method suggests selecting the largest number of clusters k that still leads to a noticeable improvement. Therefore, Fig. 2 also shows the error difference $\tau^{CV}(k) - \tau^{CV}(k-1)$ of two consecutive values of k. Based on this, we consider $k = 7$ as the best compromise between small generalization error and avoiding overfitting. However, rare solar wind types could be missed by this approach because they are, compared with the other solar wind types, underrepresented in the data set.

6 INTERPRETING CLUSTERING RESULTS

From the point of view of solar science, a solar wind categorization produced by a machine learning method is only reasonable if the resulting solar wind types are physically meaningful. For the case of two solar wind types derived from 2-means clustering, Section 4 already showed that the 2-means predictions agree reasonably well with established solar wind categorization methods. Thus, the simple machine learning technique indeed recovered the familiar types of slow solar wind and coronal hole wind. Because the model selection procedure suggests ~7 distinct solar wind types based on the available ACE solar wind data, the question arises whether these solar wind types are physically meaningful as well. Within the scope of this study, we cannot answer this question comprehensively. Instead, we restrict ourselves to low-dimensional visualizations of the emerging clusters. On average, 7-means clustering required 141 iterations until the clustering converged.

Visualizing the complete seven-dimensional feature space is not feasible. Instead, we can only look at low-dimensional projections directly. In the following, we mainly focus on projections to the $v_p - n_p$ plane because this provides a familiar framework to illustrate solar wind types and is thus readily interpretable. However, this does not provide a complete, representative picture of the relevant clusters. Fig. 3 shows projections of the seven solar wind types identified by 7-means clustering and Table 4 gives the coordinates of the cluster centers. Although some of the solar wind types in Fig. 3 are easily recognizable in this projection, this is not the case for all of them. Clusters 4 and 7 both represent coronal hole wind. Whereas Cluster 4 represents the bulk of coronal hole wind, Cluster 7 contains even faster and less dense solar wind plasma. The difference between them becomes more apparent in a different projection. Fig. 4 shows the projection of all seven clusters to the $T_p - B$ plane. Here, the difference between Clusters 4 and 7 is visible in the different proton temperatures. The plasma assigned to Cluster 7 is hotter and the magnetic field strength is higher than the coronal hole wind plasma in Cluster 4.

Cluster 3 represents the bulk of slow solar wind; 42% of all training data points are assigned to this cluster. Clusters 2 and 6 both represent interstream solar wind, wherein Cluster 6 is denser and thus probably represents compression regions, while the less dense Cluster 2 represents rarefaction regions. This is supported by the high magnetic field strength and temperatures in Cluster 2 as shown in Fig. 4.

The Clusters 1 and 5 are more difficult to interpret. Both represent a small fraction of the training data points (2% for Cluster 1 and 3% for Cluster 7) and both are comprised of dense and slow, and in the case of Cluster 1, very slow solar wind. Fig. 4 illustrates that Cluster 1 also represents very cool solar wind. This can be interpreted as the very slow solar wind discussed in Sanchez-Diaz et al. (2016). Because both clusters are also characterized by high O and C charge state ratios these could also represent or contain undetected slow ICMEs that are not contained in the Jian et al. ICME list. Although *k*-means is best suited for clusters with equal sample sizes, in this case, *k*-means nevertheless robustly inferred clusters that are underrepresented in the training and test data sets.

7 USING K-MEANS FOR FEATURE SELECTION

A machine learning approach to solar wind categorization also allows posing a different kind of question: Which parameters are relevant for solar wind categorization? In machine learning, this is called feature selection and is a complex research topic in itself. Here, we follow a simple approach and evaluate how the resulting clustering changes if certain selected solar wind parameters are included or omitted as features in the training data set. In the following, we always use the 2- and 7-means clustering trained on the F_{basic} feature space as the reference scheme and determine the relative agreement with all other feature selection scenarios as test schemes for all $k \in \{2,7\}$ clusters. All considered feature selection scenarios are summarized in Table 1. To generalize the agreement given in Eq. (2) from the binary classification case to k clusters, we define the agreement of reference and test scheme per reference and test solar wind type (t_{ref} and t_{test}, respectively) as

$$A(t_{test}, t_{ref}, p_{test}, p_{ref}) = \frac{\sum\limits_{\mathbf{x} \in X_{test}, p_{test}(\mathbf{x}) = t_{test} \wedge p_{ref}(\mathbf{x}) = t_{ref}} 1}{\sum\limits_{\mathbf{x}\ in X_{test}, p_{ref}(\mathbf{x}) = t_{ref}} 1} \quad (5)$$

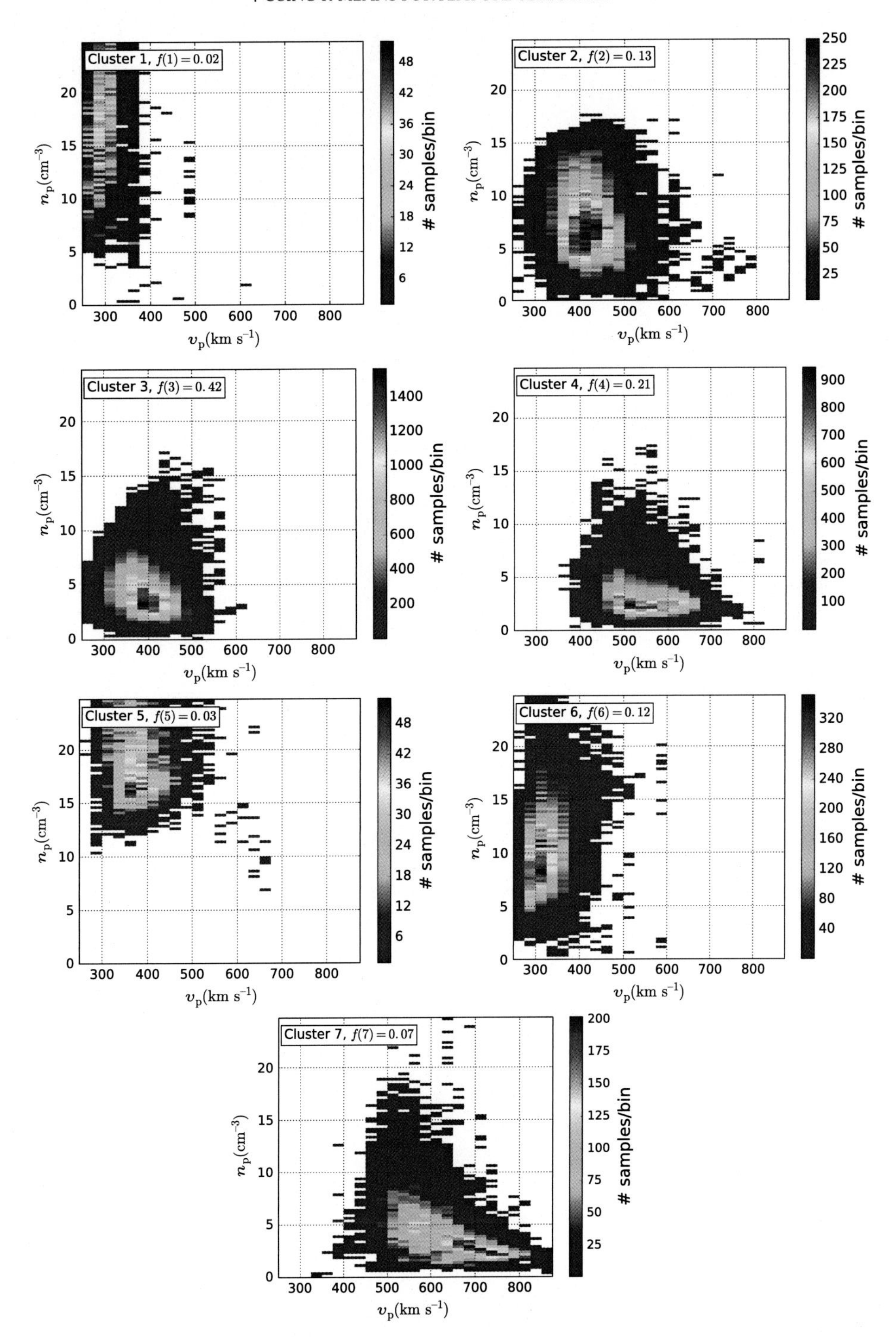

FIG. 3 Projections to the $v_p - n_p$ plane of training data cluster assignments for all seven clusters derived by 7-means clustering on the basic feature set. For each panel, the number of counts per bin is color-coded. The fraction f of the training data set that is represented in each panel is given as an *inset*.

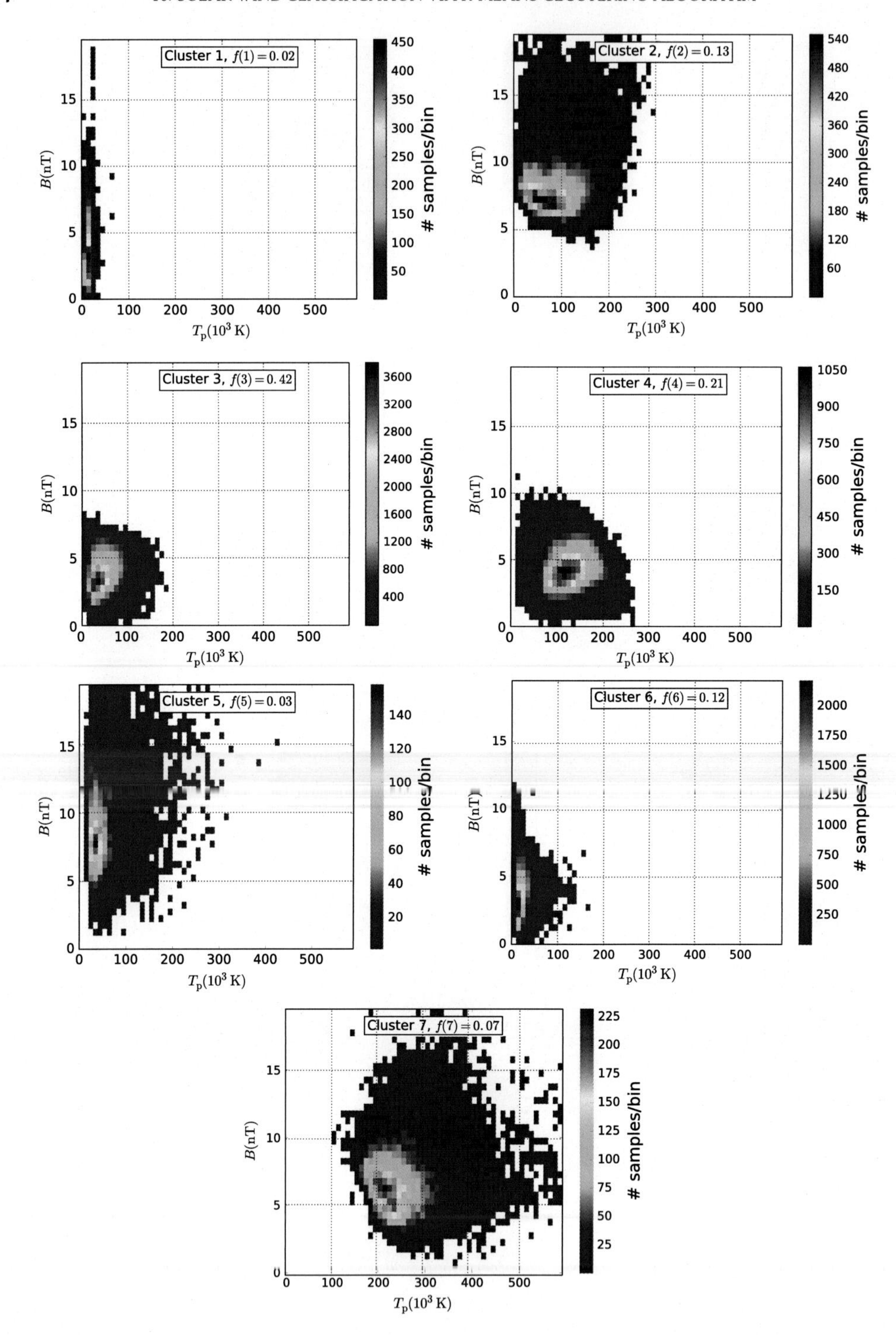

FIG. 4 Projections to the $T_P - B$ plane of training data cluster assignments for all seven clusters derived by 7-means clustering on the basic feature set. For each panel, the number of counts per bin is color-coded. The fraction f of the training data set that is represented in each panel is given as an *inset*.

TABLE 4 Median Cluster Centers Derived With 7-Means Clustering on 10 Years of ACE Data and the *Basic* Feature Selection

Cluster	v_p (km s^{-1})	n_p (cm^{-3})	T_p (10^3 K)	B (nT)	a_c	$\frac{n_{O^{7+}}}{n_{O^{6+}}}$	$\frac{n_{C^{6+}}}{n_{C^{5+}}}$
1	304.5	21.6	13.9	3.8	29.03	0.59	1.73
2	417.0	8.2	92.9	9.0	1.01	0.23	1.27
3	388.3	4.5	49.1	3.8	1.12	0.17	1.24
4	550.2	3.2	130.3	4.4	0.10	0.07	0.56
5	388.9	22.3	77.1	9.7	4.33	0.30	1.09
6	324.7	10.8	23.	4.0	7.86	0.29	3.20
7	616.2	4.6	260.5	7.3	0.04	0.06	0.50

with p_{test}, p_{ref}: $X_{test} \longrightarrow L$ both mapping to the set of class labels L.

For different feature selection scenarios, the resulting clusters are typically not ordered in the same way and in some cases altogether different clusters emerge. The mean relative agreement of 100 independent trials is given in Table 5. Before averaging, the test clusters are sorted by their agreement to the reference cluster. Consequently, the full results in Table 5 can be difficult to interpret.

As a more accessible comparison metric, we define an average agreement of a reference scheme with a test scheme in the following way: First, we determine for each cluster in the test scheme to which cluster of the reference scheme the highest agreement $\max_{t_{test}} A(t_{test}, t_{ref}, p_{test}, p_{ref})$ is observed. Second, we provide the mean $\tilde{A}(p_{test}, p_{ref})$ of these as the mean over all reference types t_{ref}. These averaged agreements are summarized in Table 6 for all considered feature selection scenarios and $k \in \{2, 7\}$. If all clusters had the same size (which is clearly not the case here) and if cluster assignment was purely random, this would lead to an expected average agreement with any model of 0.5 for the binary case and to an expected average agreement of 0.14 for the seven cluster case.

Both for 2- and 7-means clustering, the confidence intervals (as indicated by the 15.9th and 83.1st percentiles in Table 6) are small. Thus, the clustering performance is robust and the results do not sensitively depend on the initial clustering.

The comparison with the results of the $F_{no\ comp.}$ feature set reveals that omitting all charge state composition information from the feature set has only a small effect on 2- and 7-means clustering. Still, the same respective two and seven clusters are determined and the average agreement of the respective clusters is 0.998 for the $k = 2$ case and 0.981 for the $k = 7$ case. This can also be seen from the detailed results in Table 5. For each cluster in the F_{basic} scenario, exactly one cluster derived from the feature set $F_{no\ comp.}$ has an agreement with it of at least 0.956. Thus, the other solar wind parameters still contain the same information with respect to solar wind classification. Although charge state composition information is very useful for solar wind classification it is not necessary—as long as solar wind proton speed, solar wind proton density, solar wind proton temperature, magnetic field strength, and the collisional age are available. The opposite, however, does not hold. Providing only charge

TABLE 5 Relative Agreements Between 7-Means Solar Wind Type Predictions Based on Different Feature Selections

Cluster	Ref. Scheme	Test Scheme	Agreement A (Resorted)	Cluster	Ref. Scheme	Test Scheme	Agreement A (Resorted)
1	F_{basic}	$F_{no\ comp.}$	0.000 0.000 0.000 0.000 0.00[illegible] 0.028 0.965	1	F_{basic}	$F_{speed\ +\ comp.}$	0.000 0.000 0.002 0.022 0.115 0.272 0.589
2	F_{basic}	$F_{no\ comp.}$	0.000 0.000 0.000 0.000 0.00[illegible] 0.034 0.958	2	F_{basic}	$F_{speed\ +\ comp.}$	0.000 0.000 0.002 0.019 0.112 0.290 0.577
3	F_{basic}	$F_{no\ comp.}$	0.000 0.000 0.000 0.001 0.00[illegible] 0.034 0.958	3	F_{basic}	$F_{speed\ +\ comp.}$	0.000 0.000 0.001 0.016 0.109 0.290 0.584
4	F_{basic}	$F_{no\ comp.}$	0.000 0.000 0.000 0.000 0.00[illegible] 0.037 0.956	4	F_{basic}	$F_{speed\ +\ comp.}$	0.000 0.000 0.002 0.020 0.125 0.290 0.563
5	F_{basic}	$F_{no\ comp.}$	0.000 0.000 0.000 0.000 0.00[illegible] 0.036 0.958	5	F_{basic}	$F_{speed\ +\ comp.}$	0.000 0.000 0.003 0.025 0.117 0.291 0.565
6	F_{basic}	$F_{no\ comp.}$	0.000 0.000 0.000 0.001 0.00[illegible] 0.037 0.956	6	F_{basic}	$F_{speed\ +\ comp.}$	0.000 0.000 0.002 0.017 0.117 0.276 0.588
7	F_{basic}	$F_{no\ comp.}$	0.000 0.000 0.000 0.001 0.00[illegible] 0.035 0.959	7	F_{basic}	$F_{speed\ +\ comp.}$	0.000 0.000 0.002 0.021 0.109 0.272 0.596
1	F_{basic}	$F_{only\ comp.}$	0.002 0.040 0.080 0.118 0.18[illegible] 0.241 0.340	1	F_{basic}	$F_{basic\ \backslash a_c}$	0.000 0.000 0.000 0.004 0.057 0.200 0.739
2	F_{basic}	$F_{only\ comp.}$	0.002 0.037 0.071 0.112 0.17[illegible] 0.244 0.357	2	F_{basic}	$F_{basic\ \backslash a_c}$	0.000 0.000 0.000 0.003 0.046 0.201 0.749
3	F_{basic}	$F_{only\ comp.}$	0.001 0.036 0.066 0.113 0.18[illegible] 0.254 0.345	3	F_{basic}	$F_{basic\ \backslash a_c}$	0.000 0.000 0.000 0.002 0.052 0.203 0.743
4	F_{basic}	$F_{only\ comp.}$	0.002 0.039 0.071 0.119 0.18[illegible] 0.249 0.336	4	F_{basic}	$F_{basic\ \backslash a_c}$	0.000 0.000 0.000 0.004 0.053 0.194 0.749
5	F_{basic}	$F_{only\ comp.}$	0.003 0.042 0.081 0.127 0.18[illegible] 0.237 0.327	5	F_{basic}	$F_{basic\ \backslash a_c}$	0.000 0.000 0.000 0.003 0.055 0.211 0.729
6	F_{basic}	$F_{only\ comp.}$	0.002 0.039 0.067 0.117 0.18[illegible] 0.249 0.346	6	F_{basic}	$F_{basic\ \backslash a_c}$	0.000 0.000 0.000 0.004 0.050 0.198 0.748
7	F_{basic}	$F_{only\ comp.}$	0.001 0.039 0.080 0.118 0.18[illegible] 0.247 0.334	7	F_{basic}	$F_{basic\ \backslash a_c}$	0.000 0.000 0.000 0.004 0.062 0.198 0.736
1	F_{basic}	$F_{basic\ only\ a_c}$	0.000 0.001 0.004 0.025 0.1[illegible] 0.315 0.549	1	F_{basic}	$F_{no\ v_p}$	0.000 0.000 0.004 0.018 0.082 0.195 0.701
2	F_{basic}	$F_{basic\ only\ a_c}$	0.000 0.001 0.005 0.028 0.09[illegible] 0.320 0.548	2	F_{basic}	$F_{no\ v_p}$	0.000 0.000 0.003 0.016 0.073 0.184 0.724
3	F_{basic}	$F_{basic\ only\ a_c}$	0.001 0.003 0.010 0.042 0.1[illegible] 0.331 0.506	3	F_{basic}	$F_{no\ v_p}$	0.000 0.000 0.002 0.014 0.094 0.199 0.691
4	F_{basic}	$F_{basic\ only\ a_c}$	0.001 0.003 0.007 0.028 0.09[illegible] 0.341 0.528	4	F_{basic}	$F_{no\ v_p}$	0.000 0.000 0.003 0.017 0.087 0.189 0.705
5	F_{basic}	$F_{basic\ only\ a_c}$	0.001 0.004 0.008 0.032 0.09[illegible] 0.329 0.529	5	F_{basic}	$F_{no\ v_p}$	0.000 0.000 0.005 0.018 0.090 0.208 0.679
6	F_{basic}	$F_{basic\ only\ a_c}$	0.001 0.003 0.006 0.036 0.11[illegible] 0.331 0.507	6	F_{basic}	$F_{no\ v_p}$	0.000 0.000 0.002 0.014 0.095 0.206 0.682
7	F_{basic}	$F_{basic\ only\ a_c}$	0.000 0.001 0.005 0.030 0.1[illegible] 0.338 0.525	7	F_{basic}	$F_{no\ v_p}$	0.000 0.000 0.004 0.017 0.079 0.209 0.690

1	F_{basic}	F_{SOHO}	0.000 0.001 0.004 0.031 0.093 0.297 0.574	1	F_{basic}	$F_{linear\ Xu}$	0.000 0.000 0.001 0.004 0.063 0.209 0.723
2	F_{basic}	F_{SOHO}	0.000 0.000 0.003 0.025 0.094 0.279 0.598	2	F_{basic}	$F_{linear\ Xu}$	0.000 0.000 0.001 0.004 0.055 0.207 0.733
3	F_{basic}	F_{SOHO}	0.000 0.000 0.002 0.011 0.079 0.291 0.617	3	F_{basic}	$F_{linear\ Xu}$	0.000 0.000 0.000 0.003 0.058 0.204 0.735
4	F_{basic}	F_{SOHO}	0.000 0.001 0.003 0.021 0.096 0.277 0.602	4	F_{basic}	$F_{linear\ Xu}$	0.000 0.000 0.000 0.003 0.057 0.195 0.744
5	F_{basic}	F_{SOHO}	0.000 0.001 0.005 0.039 0.104 0.306 0.544	5	F_{basic}	$F_{linear\ Xu}$	0.000 0.000 0.000 0.004 0.058 0.212 0.726
6	F_{basic}	F_{SOHO}	0.000 0.001 0.003 0.019 0.078 0.302 0.598	6	F_{basic}	$F_{linear\ Xu}$	0.000 0.000 0.000 0.004 0.056 0.202 0.738
7	F_{basic}	F_{SOHO}	0.000 0.000 0.004 0.032 0.099 0.300 0.565	7	F_{basic}	$F_{linear\ Xu}$	0.000 0.000 0.000 0.004 0.065 0.206 0.724
1	F_{basic}	$F_{basic + grad.}$	0.000 0.002 0.004 0.014 0.055 0.174 0.751	1	F_{basic}	$F_{only\ context}$	0.001 0.007 0.016 0.045 0.115 0.289 0.528
2	F_{basic}	$F_{basic + grad.}$	0.000 0.002 0.004 0.011 0.061 0.216 0.706	2	F_{basic}	$F_{only\ context}$	0.001 0.007 0.015 0.048 0.139 0.290 0.501
3	F_{basic}	$F_{basic + grad.}$	0.000 0.001 0.003 0.013 0.082 0.201 0.701	3	F_{basic}	$F_{only\ context}$	0.001 0.007 0.018 0.060 0.133 0.277 0.504
4	F_{basic}	$F_{basic + grad.}$	0.000 0.001 0.003 0.013 0.083 0.197 0.703	4	F_{basic}	$F_{only\ context}$	0.001 0.007 0.017 0.056 0.125 0.287 0.509
5	F_{basic}	$F_{basic + grad.}$	0.000 0.002 0.006 0.019 0.069 0.204 0.699	5	F_{basic}	$F_{only\ context}$	0.001 0.009 0.020 0.060 0.126 0.280 0.504
6	F_{basic}	$F_{basic + grad.}$	0.000 0.001 0.003 0.016 0.071 0.186 0.722	6	F_{basic}	$F_{only\ context}$	0.001 0.008 0.018 0.052 0.126 0.288 0.507
7	F_{basic}	$F_{basic + grad.}$	0.000 0.002 0.005 0.015 0.058 0.198 0.722	7	F_{basic}	$F_{only\ context}$	0.001 0.008 0.018 0.048 0.122 0.285 0.518
1	F_{basic}	$F_{only\ B}$	0.016 0.036 0.072 0.122 0.151 0.252 0.351	1	F_{basic}	$F_{only\ derived}$	0.001 0.009 0.019 0.042 0.145 0.268 0.517
2	F_{basic}	$F_{only\ B}$	0.016 0.041 0.074 0.119 0.150 0.256 0.344	2	F_{basic}	$F_{only\ derived}$	0.001 0.007 0.020 0.046 0.140 0.270 0.517
3	F_{basic}	$F_{only\ B}$	0.018 0.041 0.076 0.123 0.154 0.244 0.344	3	F_{basic}	$F_{only\ derived}$	0.002 0.010 0.029 0.058 0.150 0.260 0.491
4	F_{basic}	$F_{only\ B}$	0.018 0.036 0.074 0.120 0.154 0.247 0.351	4	F_{basic}	$F_{only\ derived}$	0.001 0.011 0.022 0.046 0.144 0.280 0.496
5	F_{basic}	$F_{only\ B}$	0.016 0.036 0.068 0.123 0.159 0.253 0.344	5	F_{basic}	$F_{only\ derived}$	0.001 0.014 0.031 0.054 0.142 0.263 0.494
6	F_{basic}	$F_{only\ B}$	0.017 0.038 0.070 0.126 0.156 0.255 0.339	6	F_{basic}	$F_{only\ derived}$	0.002 0.014 0.029 0.056 0.141 0.255 0.504
7	F_{basic}	$F_{only\ B}$	0.016 0.039 0.073 0.121 0.152 0.254 0.345	7	F_{basic}	$F_{only\ derived}$	0.001 0.010 0.020 0.048 0.142 0.265 0.514

Notes: To increase the readability, the agreements from each test class are always ordered from smallest to largest agreement.

TABLE 6 Mean Relative Agreement With the F_{basic} Feature Set as Reference Scheme Averaged Over All Clusters for 2- and 7-Means

		2-Means: $\tilde{A}$			7-Means: $\tilde{A}$		
Reference Feature Set	**Test Feature Set**	**Mean**	**15.9th Perc.**	**83.1st Perc.**	**Mean**	**15.9th Perc.**	**83.1st Perc.**
F_{basic}	$F_{no\ comp.}$	0.998	0.998	0.998	0.981	0.914	0.993
F_{basic}	$F_{speed\ +\ comp.}$	0.911	0.911	0.911	0.585	0.544	0.602
F_{basic}	$F_{only\ comp.}$	0.630	0.630	0.630	0.335	0.323	0.345
F_{basic}	$F_{basic \setminus a_c}$	0.958	0.958	0.958	0.749	0.721	0.767
F_{basic}	$F_{basic\ only\ a_c}$	0.691	0.691	0.969	0.517	0.504	0.535
F_{basic}	$F_{no\ v_p}$	0.728	0.728	0.728	0.707	0.671	0.710
F_{basic}	F_{SOHO}	0.941	0.941	0.941	0.576	0.573	0.578
F_{basic}	$F_{linear\ Xu}$	0.958	0.958	0.958	0.734	0.724	0.764
F_{basic}	$F_{basic\ +\ grad.}$	0.966	0.966	0.966	0.703	0.697	0.730
F_{basic}	$F_{only\ context}$	0.971	0.971	0.971	0.508	0.490	0.515
F_{basic}	$F_{only\ B}$	0.597	0.597	0.597	0.347	0.307	0.357
F_{basic}	$F_{only\ derived}$	0.796	0.791	0.967	0.511	0.479	0.511

Note: For each comparison, the mean of 100 independent trials as well as the 15.9th and 84.1st percentiles are given.

state composition information (feature set $F_{only\ comp.}$) leads to different clustering results and the agreement with the F_{basic} scenario drops to 0.630 in the two-cluster case and 0.335 in the seven-cluster case. Although the average agreement of the $F_{only\ comp.}$ scenario with the F_{basic} scenario is still above the random guessing threshold values, this indicates that the clustering results are qualitatively different. This is supported by the full results in Table 5. All clusters from the F_{basic} results overlap with all seven clusters derived from the $F_{only\ comp.}$ scenario. This is different in the $F_{speed\ +\ comp.}$ case. Although the maximum agreement of any cluster from the F_{basic} scenario with any cluster based on $F_{speed\ +\ comp.}$ is only 0.596, for all clusters from the F_{basic} scenario at least two clusters from the $F_{speed\ +\ comp.}$ scenario do not overlap with the reference cluster at all. If charge state composition and solar wind speed (feature set $F_{speed\ +\ comp.}$) are provided, this increases the agreement to the reference model compared with the $F_{only\ comp.}$ case. But although they agree on the most frequent observations of slow solar wind, the clustering is nevertheless still qualitatively different. This is illustrated in Fig. 5 for projections of the feature space to the $v_p - n_p$ plane and the main slow solar wind type from the F_{basic} scenario, Cluster 3, compared with two slow solar wind types derived from the $F_{speed\ +\ comp.}$ feature selection scenario. Although the feature set $F_{speed\ +\ comp.}$ also led to several slow solar wind types (only the two representing the highest number of data points are shown in Fig. 5), they differ considerably from those in the F_{basic} scenario. This illustrates that the solar wind speed v_p plays an important role in determining solar wind type. The clustering in the $F_{speed\ +\ comp.}$ scenario appears to be dominated by the solar wind speed and not by

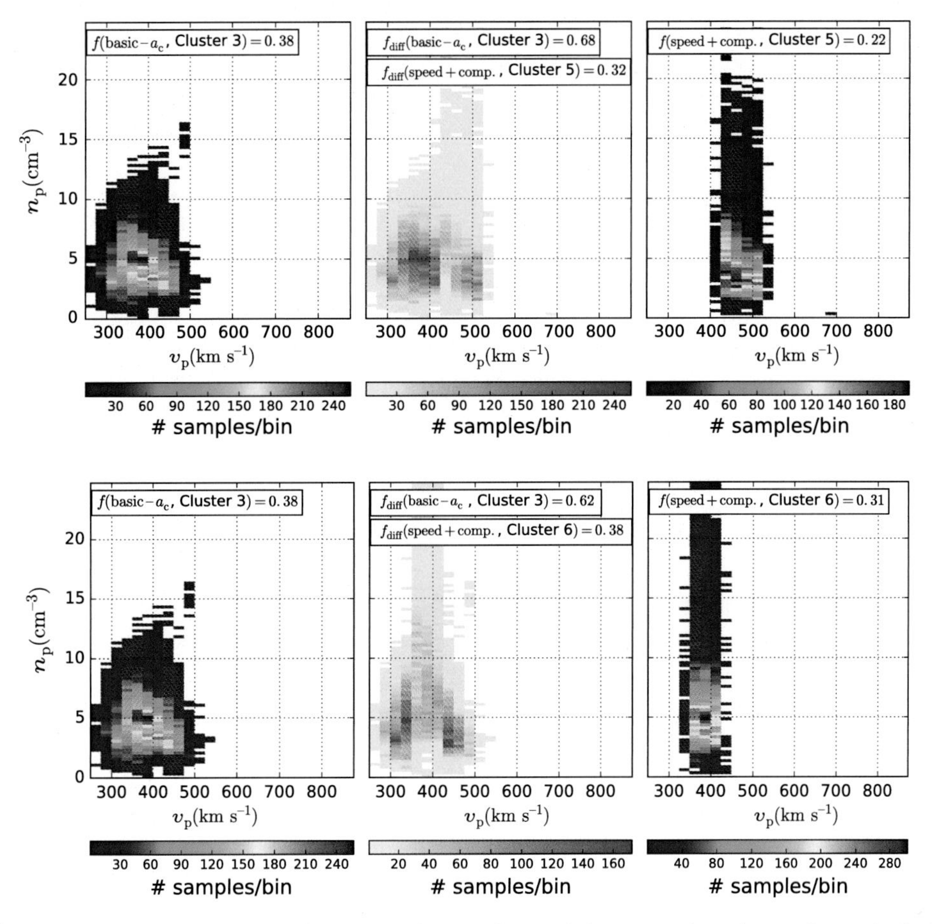

FIG. 5 Two-dimensional histograms of 12-min resolution solar wind observation from the test data set in the feature space projected to the $v_p - n_p$ plane for slow solar wind. *Left*: Cluster 3 taken from the F_{basic} feature selection scenario. *Top right*: Cluster 6 taken from the $F_{speed + comp.}$ feature selection scenario. *Bottom right*: Cluster 7 taken from the $F_{speed + comp.}$ feature selection scenario. In the *middle panels*, the difference between the two slow solar wind clusters is shown.

the charge state composition. However, this does not mean that composition information is irrelevant for solar wind categorization. In combination with other parameters, such as, for example, the solar wind proton density, composition information could be better suited for solar wind categorization. This needs to be investigated further.

The solar wind speed is represented by two parameters in the F_{basic} scenario, with v_p and the collisional age a_c. If only the collisional age is omitted from the data set, this has a small influence on the two-cluster case but in the seven-cluster case, the agreement is reduced to 0.749. If no solar wind speed information is used (neither v_p nor a_c, feature set $F_{no\ v_p}$) the two-cluster agreement is reduced to 0.728 while the seven-cluster agreement is less affected (compared to only omitting the collisional age a_c) and changes to 0.707. This is understandable because, in the two-cluster case, a simplistic perspective on the data is enforced. Because solar wind speed is particularly useful for a rough separation of solar wind data into two classes, omitting it has a larger effect for the two-cluster case. In the $F_{\text{basic only } a_c}$ scenario the collisional age replaces all three parameters it depends on: solar wind proton speed, proton density, and proton temperature. As this leads to large changes in the resulting clustering this indicates that the linear parameters indeed play a relevant role for solar wind classification.

The instrumentation of SOHO does not include a magnetometer, which is unfortunate for solar wind research. Here, we can pose the question: how much is lost with respect to solar wind classification if only the solar wind parameters that are directly available from SOHO instruments, v_{p}, n_{p}, and T_{p}, are used as a feature set? The resulting average agreement with the F_{basic} scenario is similar to the $F_{\text{speed + comp.}}$ case that has the same number of parameters but replaces solar wind proton density and proton temperature with the O and C charge state ratios. However, the resulting clusters in these two scenarios are qualitatively different as is illustrated in Fig. 6. Interestingly, in the $F_{\text{speed + comp.}}$ scenario the solar wind speed has a particularly strong impact on the resulting solar wind clusters. Additionally, including the magnetic field (as in the $F_{\text{linear Xu}}$ case) leads to a similar agreement with the F_{basic} scenario as the F_{SOHO} feature set for 2-means clustering but has a stronger impact on the 7-means clustering. This indicates that the mean magnetic field is—even without a measure of wave activity—an important parameter for a refined solar wind categorization.

The $F_{\text{basic + grad.}}$ feature set incorporates some context information by adding the two gradients $\frac{dv_{\text{p}}}{dt}$, $\frac{dn_{\text{p}}}{dt}$ to the F_{basic} feature set. This again has a stronger impact on 7-means clustering than on 2-means clustering. The remaining three scenarios $F_{\text{only B}}$, $F_{\text{only context}}$, and $F_{\text{only derived}}$ lead for 7-means clustering to solar wind clusters that are very different from the F_{basic} scenario and the agreement to the F_{basic} scenario is low for all three cases. For 2-means clustering, the context information is sufficient to recover the same two clusters as in the $F_{\text{basic + grad.}}$ case.

Although this procedure allows estimating the influence of different solar wind parameters on identifying the solar wind type, this does not provide an indication of which of the considered feature selection scenarios is most appropriate for solar wind categorization. This is out of the scope of this study and an additional model selection step would be required to determine this.

8 SUMMARY AND CONCLUSION

It is well known that the solar wind comes in at least two varieties and this distinction has been applied frequently in the solar wind community as well as in adjacent research fields such as, for example, space weather. Nevertheless, it is still under debate how many solar wind types can and should be distinguished and which solar wind parameters best describe their respective different properties (McComas et al., 2000; Zhao and Fisk, 2010; Xu and Borovsky, 2015; Stakhiv et al., 2015; D'Amicis and Bruno, 2015; Sanchez-Diaz et al., 2016). Arguably, the stream interaction regions around the decision boundaries can be regarded as a mixture of the two major solar wind types, which might be of less interest than either slow solar wind or coronal hole wind themselves. In practice, interstream solar wind is observed frequently and makes up a significant portion of all solar wind data and should, from our

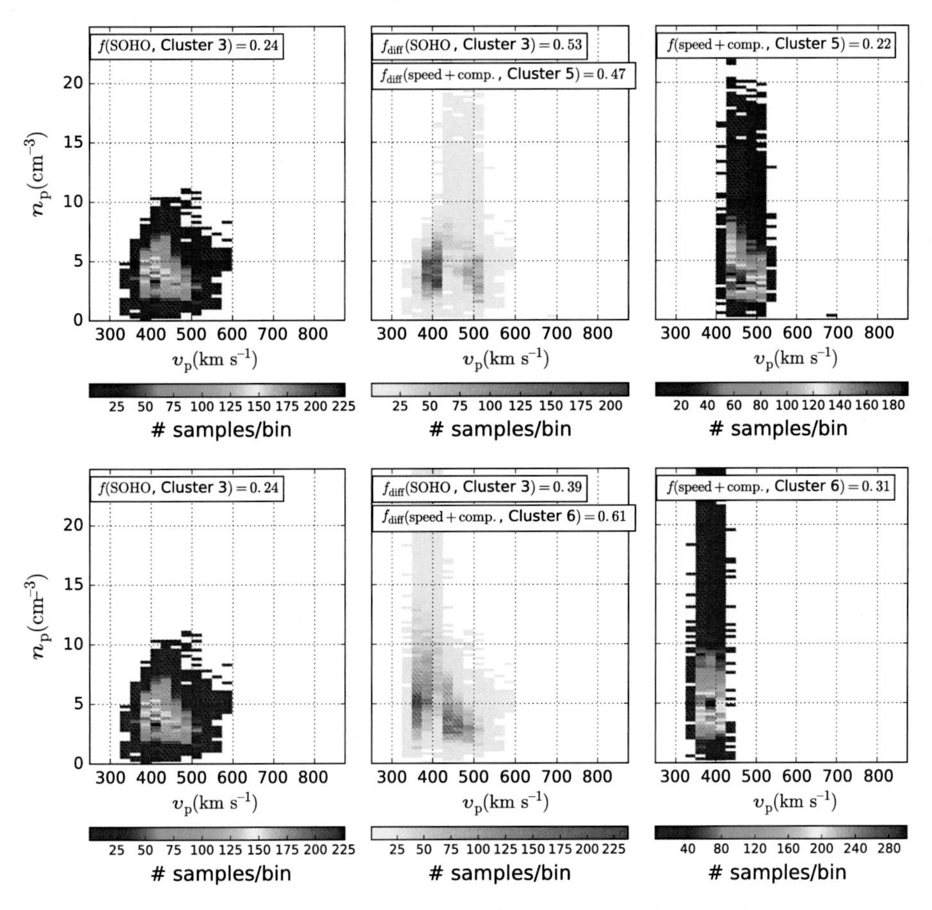

FIG. 6 Two-dimensional histograms of 12-min resolution solar wind observation from the test data set in the feature space projected to the $v_p - n_p$ plane for slow solar wind. *Left*: Cluster 3 taken from the F_{SOHO} feature selection scenario. *Top right*: Cluster 5 taken from the $F_{speed\,+\,comp.}$ feature selection scenario. *Bottom right*: Cluster 6 taken from the $F_{speed\,+\,comp.}$ feature selection scenario. In the *middle panels*, the difference between the two slow solar wind clusters is shown.

point of view, be at least reliably separated from slow solar wind and coronal hole wind. The recent comparison of solar wind categorization methods in Neugebauer et al. (2016) illustrated that the exact decision boundary between slow solar wind and coronal hole wind is frequently defined differently and no widely accepted definition is currently available. In this situation, machine learning tools are well suited to provide an additional purely data-driven perspective. Here, we regard solar wind categorization as an interesting and relevant application to showcase the capabilities and limitations of an unsupervised machine learning method. We applied a simple unsupervised clustering method, *k*-means clustering, to the problem of solar wind categorization. Linear *k*-means was chosen because it is conceptually similar to an existing heuristic solar wind categorization scheme (Xu and Borovsky, 2015), its computational cost is easily affordable for the sample size in this application, it is easy to implement; in fact, implementations of *k*-means are freely available in several popular programming languages, it is easy to apply, and is in this case, sufficient to produce insightful results.

As a first step, we compared the solar wind types derived from well-known solar wind categorization schemes, a simple solar wind speed-based scheme, the composition-based scheme from Zhao and Fisk (2010) and the recent Xu and Borovsky (2015) solar wind categorization scheme, with the results of *k*-means clustering. The comparison was restricted to the slow solar wind and coronal hole wind solar wind types from all four considered categorization schemes. As expected, the largest differences were found for intermediate solar wind speeds.

In a second step, we applied a model-selection tool to determine the most appropriate number of solar wind types based on the available solar wind data from ACE. The best compromise between low cross-validation error and the need to avoid overfitting was procured for seven solar wind types.

A drawback of applying machine learning methods can be that the results are often difficult to interpret. The interpretation of the results is not part of the task of machine learning, instead, this can only be supplied by expert knowledge. Therefore, in the next step, we analyzed the physical properties of the seven derived solar wind types. We found two coronal hole wind types, one that represents the majority of coronal hole wind and a second cluster that contained the fastest and least dense solar wind observations. This second type could be interpreted as an unconsidered type of coronal hole wind, but it could also be comprised of ICME plasma with unusually low O and C charge states. The *k*-means clustering assigned the bulk of slow solar wind to one cluster, but in addition, four special cases are also represented. Two of them can be understood as interstream solar wind with a distinction between solar wind from compression regions and from rarefaction regions. The third subtype of slow solar wind that *k*-means clustering extracted from the solar wind data is comprised of very slow, very dense, and very cool solar wind. This matches the very slow solar wind discussed in Sanchez-Diaz et al. (2016). The last solar wind type contains even denser plasma with high charge states and is also cooler than the bulk of the slow solar wind plasma. These could be undetected slow ICMEs. These are very difficult to distinguish from the solar wind they are embedded in and can, therefore, be expected to be frequently missing from available ICME lists.

Finally, we briefly used the *k*-means clustering as a tool to investigate the effect of different sets of solar wind parameters as input to a solar wind categorization method. In machine learning, the feature selection process is well known to have a large impact on the results of an algorithm. Here, we have shown that the choice of solar wind parameters on which a solar wind categorization scheme bases its decision can indeed have a significant impact on the results. A surprising and unexpected result is that while composition information is valuable for solar wind categorization, it is not necessary if all other basic solar wind parameters are supplied to the clustering method. The solar wind speed alone is insufficient to capture the real structure of the feature space spanned by solar wind parameters, but is very valuable if used in combination with magnetic field strength, solar wind proton temperature, solar wind proton density, and the charge state composition.

Unsupervised clustering has been demonstrated to be not only applicable to solar wind categorization, but the derived clusters have interesting properties that can augment and change our understanding of the physics of solar wind.

References

Antiochos, S.K., Mikić, Z., Titov, V.S., Lionello, R., Linker, J.A., 2011. A model for the sources of the slow solar wind. Astrophys. J. 731 (2), 112.

Arlot, S., Celisse, A., et al., 2010. A survey of cross-validation procedures for model selection. Stat. Surv. 4, 40–79.

Arthur, D., Vassilvitskii, S., 2007. *k*-Means++: the advantages of careful seeding. In: Proceedings of the Eighteenth Annual ACM-SIAM Symposium on Discrete Algorithms, pp. 1027–1035.

Cane, H.V., Richardson, I.G., 2003. Interplanetary coronal mass ejections in the near-Earth solar wind during 1996–2002. J. Geophys. Res. Space Phys. 108 (A4), 1156.

Comaniciu, D., Meer, P., 2002. Mean shift: a robust approach toward feature space analysis. IEEE Trans. Pattern Anal. Mach. Intell. 24 (5), 603–619.

D'Amicis, R., Bruno, R., 2015. On the origin of highly Alfvénic slow solar wind. Astrophys. J. 805 (1), 84.

Devijver, P.A., Kittler, J., 1982. Pattern Recognition: A Statistical Approach, vol. 761. Prentice-Hall, London.

Ester, M., Kriegel, H.P., Sander, J., Xu, X., et al., 1996. A density-based algorithm for discovering clusters in large spatial databases with noise. KDD 96, 226–231.

Fushiki, T., 2011. Estimation of prediction error by using K-fold cross-validation. Stat. Comput. 21 (2), 137–146.

Geisser, S., 1993. Predictive Inference, vol. 55. CRC Press, Boca Raton, FL.

Gloeckler, G., Cain, J., Ipavich, F.M., Tums, E.O., Bedini, P., Fisk, L.A., Zurbuchen, T.H., Bochsler, P., Fischer, J., Wimmer-Schweingruber, R.F., et al., 1998. Investigation of the composition of solar and interstellar matter using solar wind and pickup ion measurements with SWICS and SWIMS on the ACE spacecraft. In: The Advanced Composition Explorer Mission. Springer, pp. 497–539.

Hefti, S., Grünwaldt, H., Bochsler, P., Aellig, M.R., 2000. Oxygen freeze-in temperatures measured with SOHO/CELIAS/CTOF. J. Geophys. Res. Space Phys. 105 (A5), 10527–10536.

Heidrich-Meisner, V., Peleikis, T., Kruse, M., Berger, L., Wimmer-Schweingruber, R., 2016. Observations of high and low Fe charge states in individual solar wind streams with coronal-hole origin. Astron. Astrophys. 593, A70.

Igel, C., Heidrich-Meisner, V., Glasmachers, T., 2008. Shark. J. Mach. Learn. Res. 9, 993–996.

Iju, T., Tokumaru, M., Fujiki, K., 2014. Kinematic properties of slow ICMEs and an interpretation of a modified drag equation for fast and moderate ICMEs. Sol. Phys. 289 (6), 2157–2175.

Jain, A.K., Murty, M.N., Flynn, P.J., 1999. Data clustering: a review. ACM Comput. Surv. 31 (3), 264–323.

Jian, L., Russell, C.T., Luhmann, J.G., Skoug, R.M., 2006. Properties of interplanetary coronal mass ejections at one AU during 1995–2004. Sol. Phys. 239 (1–2), 393–436.

Jian, L.K., Russell, C.T., Luhmann, J.G., 2011. Comparing solar minimum 23/24 with historical solar wind records at 1 AU. Sol. Phys. 274 (1–2), 321–344.

Kasper, J.C., Lazarus, A.J., Gary, S.P., 2008. Hot solar-wind helium: direct evidence for local heating by Alfvén-cyclotron dissipation. Phys. Rev. Lett. 101 (26), 261103.

Kohavi, R., 1996. Scaling up the accuracy of naive-Bayes classifiers: a decision-tree hybrid. KDD 96, 202–207.

Kohavi, R., et al., 1995. A study of cross-validation and bootstrap for accuracy estimation and model selection. In: IJCAI, vol. 14, Stanford, CA, pp. 1137–1145.

Lindsay, B.G., 1995. Mixture models: theory, geometry and applications. In: NSF-CBMS Regional Conference Series in Probability and Statistics, pp. 1–163.

MacQueen, J., et al., 1967. Some methods for classification and analysis of multivariate observations. In: Proceedings of the Fifth Berkeley Symposium on Mathematical Statistics and Probability, vols. 1–14, Oakland, CA, pp. 281–297.

McComas, D.J., Bame, S.J., Barker, P., Feldman, W.C., Phillips, J.L., Riley, P., Griffee, J.W., 1998. Solar wind electron proton alpha monitor (SWEPAM) for the advanced composition explorer. In: The Advanced Composition Explorer Mission. Springer, pp. 563–612.

McComas, D.J., Barraclough, B.L., Funsten, H.O., Gosling, J.T., Santiago-Muñoz, E., Skoug, R.M., Goldstein, B.E., Neugebauer, M., Riley, P., Balogh, A., 2000. Solar wind observations over Ulysses' first full polar orbit. J. Geophys. Res. Space Phys. 105 (A5), 10419–10433.

Neugebauer, M., Reisenfeld, D., Richardson, I.G., 2016. Comparison of algorithms for determination of solar wind regimes. J. Geophys. Res. Space Phys. 121 (9), 8215–8227.

Pham, D.T., Dimov, S.S., Nguyen, C.D., 2005. Selection of *k* in *k*-means clustering. Proc. IME C. J. Mech. Eng. Sci. 219 (1), 103–119.

Richardson, I.G., Cane, H.V., 2010. Near-Earth interplanetary coronal mass ejections during solar cycle 23 (1996–2009): catalog and summary of properties. Sol. Phys. 264 (1), 189–237.

Sakao, T., Kano, R., Narukage, N., Kotoku, J., Bando, T., DeLuca, E.E., Lundquist, L.L., Tsuneta, S., Harra, L.K., Katsukawa, Y., et al., 2007. Continuous plasma outflows from the edge of a solar active region as a possible source of solar wind. Science 318 (5856), 1585–1588.

Sanchez-Diaz, E., Rouillard, A.P., Lavraud, B., Segura, K., Tao, C., Pinto, R., Sheeley, N.R., Plotnikov, I., 2016. The very slow solar wind: properties, origin and variability. J. Geophys. Res. Space Phys. 121 (4), 2830–2841.

Schwadron, N.A., McComas, D.J., Elliott, H.A., Gloeckler, G., Geiss, J., von Steiger, R., 2005. Solar wind from the coronal hole boundaries. J. Geophys. Res. 110 (A4), A04104.

Shi, J., Malik, J., 2000. Normalized cuts and image segmentation. IEEE Trans. Pattern Anal. Mach. Intell. 22 (8), 888–905.

Smith, C.W., L'Heureux, J., Ness, N.F., Acuña, M.H., Burlaga, L.F., Scheifele, J., 1998. The ACE magnetic fields experiment. In: The Advanced Composition Explorer Mission. Springer, pp. 613–632.

Spall, J.C., Maryak, J.L., 1992. A feasible Bayesian estimator of quantiles for projectile accuracy from non-IID data. J. Am. Stat. Assoc. 87 (419), 676–681.

Stakhiv, M., Landi, E., Lepri, S.T., Oran, R., Zurbuchen, T.H., 2015. On the origin of mid-latitude fast wind: challenging the two-state solar wind paradigm. Astrophys. J. 801 (2), 100.

Steinhaus, H., 1956. Sur la division des corp materiels en parties. Bull. Acad. Pol. Sci. 1 (804), 801.

Thorndike, R.L., 1953. Who belongs in the family? Psychometrika 18 (4), 267–276.

Tu, C.Y., Zhou, C., Marsch, E., Xia, L.D., Zhao, L., Wang, J.X., Wilhelm, K., 2005. Solar wind origin in coronal funnels. Science 308 (5721), 519–523.

Von Luxburg, U., 2007. A tutorial on spectral clustering. Stat. Comput. 17 (4), 395–416.

von Steiger, R., Schwadron, N., Fisk, L.A., Geiss, J., Gloeckler, G., Hefti, S., Wilken, B., Wimmer-Schweingruber, R.F., Zurbuchen, T.H., 2000. Composition of quasi-stationary solar wind flows from Ulysses/Solar Wind Ion Composition Spectrometer. J. Geophys. Res. 105, 27.

Ward Jr., J.H., 1963. Hierarchical grouping to optimize an objective function. J. Am. Stat. Assoc. 58 (301), 236–244.

Xu, F., Borovsky, J.E., 2015. A new four-plasma categorization scheme for the solar wind. J. Geophys. Res. Space Phys. 120 (1), 70–100.

Zhang, P., 1993. Model selection via multifold cross validation. Ann. Stat. 21, 299–313.

Zhao, L., Fisk, L., 2010. Comparison of two solar minima: narrower streamer stalk region and conserved open magnetic flux in the region outside of streamer stalks. In: SOHO-23: Understanding a Peculiar Solar Minimum, vol. 428, pp. 229.

Zhao, L., Landi, E., 2014. Polar and equatorial coronal hole winds at solar minima: from the heliosphere to the inner corona. Astron. J. 781 (2), 110.

Index

Note: Page numbers followed by *f* indicate figures, *t* indicate tables and *b* indicate boxes.

D

E

H

I

J

K

O

P

T

U

V